SUSTAINABLE AGRICULTURE

Advances in Technological Interventions

SUSTAINABLE AGRICULTURE

Advances in Technological Interventions

Edited by

Ajoy Kumar Singh

Vishwa Bandhu Patel

Apple Academic Press Inc.
4164 Lakeshore Road
Burlington ON L7L 1A4, Canada

Apple Academic Press Inc.
1265 Goldenrod Circle NE
Palm Bay, Florida 32905, USA

Exclusive worldwide distribution by CRC Press, a member of Taylor & Francis Group

International Standard Book Number-13: 978-1-77188-853-0 (Hardcover)
International Standard Book Number-13: 978-0-42932-583-0 (eBook)

Library and Archives Canada Cataloguing in Publication

Title: Sustainable agriculture : advances in technological interventions / edited by Ajoy Kumar Singh, Vishwa Bandhu Patel.

Other titles: Sustainable agriculture (Burlington, Ont.)

Names: Singh, Ajoy Kumar (Agronomist), editor. | Patel, Vishwa Bandhu, editor.

Description: Includes bibliographical references and index.

Identifiers: Canadiana (print) 20200177184 | Canadiana (ebook) 20200177214 | ISBN 9781771888530 (hardcover) | ISBN 9780429325830 (ebook)

Subjects: LCSH: Sustainable agriculture.

Classification: LCC S494.5.S86 S87 2020 | 630.2/086—dc23

Library of Congress Cataloging-in-Publication Data

Names: Singh, Ajoy Kumar (Agronomist), editor. | Patel, Vishwa Bandhu, editor.

Title: Sustainable agriculture : advances in technological interventions / edited by Ajoy Kumar Singh, Vishwa Bandhu Patel.

Description: Palm Bay, Florida, USA : Apple Academic Press, [2020] | Includes bibliographical references and index. | Summary: "This new volume looks at the evolution and challenges of sustainable agriculture, a field that is growing in use and popularity. Sustainable Agriculture: Advances in Technological Interventions discusses some of the important ideas, practices, and policies that are essential to an effective sustainable agriculture strategy. The book features 25 chapters written by experts in related fields, including crop improvement, natural resource management, crop protection, social sciences, and product development. The volume aims to provide a good understanding of the use of sustainable agriculture and the sustainable management of agri-horticultural crops, focusing on eco-friendly approaches, such as the utilization of waste materials. Topics include ecofriendly plant protection measures, climate change and natural resource management, tools to mitigate the effect of extreme weather events, agrochemical research and regulation, soil carbon sequestration, water and nutrient management in agricultural systems, and more. Key features: Discusses sustainable agriculture within the framework of recent challenges in agriculture Looks at the development and diversification of crops and cultural practices to enhance biological and economic stability Features the application of innovative nanotechnologies to agricultural research and production technologies Highlights the development of new varieties in agri-horticultural crops in relation to biotic and abiotic management Discusses the application of recent technologies for soil-plant-microbe-environment interactions"-- Provided by publisher.

Identifiers: LCCN 2020003856 (print) | LCCN 2020003857 (ebook) | ISBN 9781771888530 (hardcover) | ISBN 9780429325830 (ebook)

Subjects: LCSH: Sustainable agriculture.

Classification: LCC S494.5.S86 S835 2020 (print) | LCC S494.5.S86 (ebook) | DDC 338.1--dc23

LC record available at https://lccn.loc.gov/2020003856

LC ebook record available at https://lccn.loc.gov/2020003857

Apple Academic Press also publishes its books in a variety of electronic formats. Some content that appears in print may not be available in electronic format. For information about Apple Academic Press products, visit our website at **www.appleacademicpress.com** and the CRC Press website at **www.crcpress.com**

About the Editors

Ajoy Kumar Singh, PhD
Vice-Chancellor, Bihar Agricultural University, Sabour, India

Ajoy Kumar Singh, PhD, is presently a Vice-Chancellor of Bihar Agricultural University, Sabour, India. Dr. Singh earned his master's and PhD degrees in Agronomy from the Indian Agricultural Research Institute, New Delhi, India. He also served as Principal Scientist and Director at the Indian Council of Agricultural Research institutions. Dr. Singh worked extensively on diara land development and cropping systems, farming systems, and On-Farm Research under varied agro-climatic zones. As Zonal Co-ordinator (Zone II), Zonal Project Director, and Director, ATARI, Kolkata, Dr. Singh has monitored, evaluated, and guided 83 Krishi Vigyan Kendra (Farm Science Centres) spread across A&N Island and Eastern India. He is a Fellow of the Indian Society of Agronomy, Indian Society of Extension Education, and Bioved Research Society, and a recipient of several national awards, including lifetime achievement awards from five different organizations.

Vishwa Bandhu Patel, PhD
Principal Scientist, Division of Fruits and Horticultural Technology, Indian Council of Agricultural Research–Indian Agricultural Research Institute, New Delhi, India

Vishwa Bandhu Patel, PhD, is presently working as an Associate Professor at the Division of Fruits and Horticultural Technology, Indian Agricultural Research Institute, New Delhi, India. He has worked on developing leaf nutrient guides, nutrient management strategies, and the use of arbuscular mycorrhiza fungi (AMF) for biohardening and stress tolerance, and surveyed indigenous germplasm of fruit crops. He developed leaf sampling techniques and standards as well as made fertilizer recommendations for several fruits. He has worked out nutrient management through organic means for high-density planted mango. He has guided several MSc and PhD students. Dr. Patel organized several national and international seminars and workshops as a convener, associate convener, and core team member to identify the constraints and researchable areas for prioritizing and planning the agenda for the improvement of horticultural crops. These seminars include four Indian Horticulture Congresses (held during 2004, 2006, 2008, and 2010), the International Seminar on Precision Farming and Plasticulture (2005), the National Seminar on Organic Farming (2004), the Seminar on Hi-Tech Horticulture (2012), and the National Seminar on Climate Change and Indian Horticulture (2013). Dr Patel has received recognition and a number of national awards for his work in the field of horticulture research and development, such as Fellow, The Horticultural Society of India, New Delhi; Associate, National Academy of Agricultural Sciences (NAAS), New Delhi, India; Agricultural Leadership Award (2012) by Agricultural Today, Centre for Agricultural and Rural Development, New Delhi; Hari Om Ashram Trust Award (2007); Lal Bahadur Shastri Young Scientist Award, ICAR, New Delhi (2009); Young Scientist Award, NAAS (2005-06); Yuva Vigyanic Samman from Council of Science and Technology, Govt. Uttar Pradesh (2005); AAAS Junior Award (2005) from the Indian Society for Plant Physiology, New Delhi; and five best research paper/poster paper presentation awards by different organizations.

Contents

Contributors .... *xi*

Abbreviations .... *xv*

Preface .... *xix*

Part I: Crop Improvement **1**

1. **Genetic Improvement of Major Cereals in Prospect to Bihar State** **3**
   P. K. Singh, Anand Kumar, and Ravi Ranjan Kumar

2. **Status, Opportunities, and Challenges of Spice Research and Development** **39**
   Sangeeta Shree and Paramveer Singh

3. **Perspective of Plantation Crops in Bihar with Special Reference to Coconut and Palmyra** **63**
   Ruby Rani, Vishwa Bandhu Patel, and H. P. Maheshwarappa

4. **Overview of the Floriculture Sector: Performance, Problems, and Prospects** **89**
   Paramveer Singh, Sangeeta Shree, and Ajay Bhardwaj

Part II: Natural Resource Management **119**

5. **Climate Change and Natural Resource Management** **121**
   Suborna Roy Choudhury and Syed Sheraz Mahdi

6. **Soil Carbon Sequestration: With a Particular Reference to Bihar** **135**
   Rajeev Padbhushan, Anupam Das, and Swaraj Kumar Dutta

7. **Agromet Advisory Services: Tool to Mitigate the Effect of Extreme Weather Events** **149**
   Sunil Kumar

8. **Soil Pollution: Studies with a Specific Reference to Bihar** **173**
   M. K. Dwivedi, S. Kumar, A. Kohli, Y. K. Singh, Shweta Shambhavi, and R. Kumar

9. **Nanotechnology in Agricultural Science** **189**
   Nintu Mandal, Kasturikasen Beura, and Abhijeet Ghatak

10. **Endomycorrhizal Fungi: Phosphorous Nutrition in Crops....................203**
Mahendra Singh, Rajiv Rakshit, and Kasturikasen Beura

11. **Water Management in Horticultural Crops..211**
K. Karuna and Abhay Mankar

12. **Orchard Floor Management..227**
Manoj Kundu

13. **High-Density Planting in Fruit Crops for Enhancing Fruit Productivity..253**
V. K. Tripathi, Sanjeev Kumar, Vishal Dubey, and Md. Abu Nayyer

14. **Enriching Vermicompost Using P-Solubilizing and N-Fixing Biofertilizers and Eco-Friendly Low-Cost Minerals..............................269**
Arun Kumar Jha

15. **Protected Horticulture in India: An Overview 295**
Paramveer Singh, Ajay Bhardwaj, and Randhir Kumar

16. **Use of Fly Ash in Agriculture..319**
Sankar Ch. Paul

17. **Natural Resource Management and Land Use Planning.......................335**
C. D. Choudhary, B. K. Vimal, Shweta Shambhavi, and Rajkishore Kumar

Part III: Crop Protection 365

18. **Plant Disease Management Through Application of Nanoscience.........367**
Abhijeet Ghatak

19. **Status of Insect Pests of Cereals in India and Their Management 379**
Anil, Tarak Nath Goswami, and Sanjay Kumar Sharma

20. **Major Insect Pests of Vegetable Crops in Bihar and Their Management 395**
Tamoghna Saha, Nithya Chandran, and B. C. Anu

21. **Emerging Viral Diseases of Vegetable Crops: An Outline and Sustainable Management..431**
Mohammad Ansar, A. Srinivasaraghavan, and Aniruddha Kumar Agnihotri

22. **Advancement of Agrochemical Research and Regulation 457**
Pritam Ganguly

Part IV: Product Development and Extension Education....................479

23. Institutional Outreach Through Innovative Approaches.........................481
Aditya Sinha and R. K. Sohane

24. An Appraisal on Quality Honey Production..501
Ramanuj Vishwakarma and Ranjeet Kumar

25. Empowering Rural Youth Through Agripreneurship..............................517
Ram Datt

Index...539

Contributors

Aditya Sinha
Department of Extension Education, Bihar Agricultural University, Sabour , Bhagalpur, Bihar, India

Aniruddha Kumar Agnihotri
Department of Plant Pathology, Bihar Agricultural University, Sabour, Bhagalpur, Bihar, India

Anil
Department of Entomology, Bihar Agricultural University, Sabour, Bhagalpur, Bihar, India

Mohammad Ansar
Department of Plant Pathology, Bihar Agricultural University, Sabour, Bhagalpur, Bihar, India

B. C. Anu
Department of Entomology, Bihar Agricultural University, Sabour, Bhagalpur, Bihar, India

Kasturikasen Beura
Department of Soil Science and Agricultural Chemistry, Bihar Agricultural University, Sabour, Bhagalpur, Bihar, India

Ajay Bhardwaj
Department of Horticulture (Olericulture, Vegetable and Floriculture), Bihar Agricultural University, Sabour, Bhagalpur, Bihar, India

Nithya Chandran
Division of Entomology, Indian Agricultural Research Institute, New Delhi, India

C. D. Choudhary
Department of Soil Science and Agricultural Chemistry, Bihar Agricultural University, Sabour, Bhagalpur, Bihar, India

Suborna Roy Choudhury
Department of Agronomy, Bihar Agricultural University, Sabour, Bhagalpur, Bihar, India

Anupam Das
Department of Soil Science and Agricultural Chemistry, Bihar Agricultural University, Sabour, Bhagalpur, Bihar, India

Ram Datt
Department of Extension Education, Bihar Agricultural University, Sabour, Bhagalpur, Bihar, India

Vishal Dubey
Department of Applied Plant Sciences (Horticulture), B.B.A. University, Lucknow, Uttar Preadesh, India

Swaraj Kumar Dutta
Department of Agronomy, Bihar Bihar Agricultural University, Sabour, Bhagalpur, Bihar, India

M. K. Dwivedi
Department of Soil Science and Agricultural Chemistry, Bihar Agricultural University, Sabour, Bhagalpur, Bihar, India

Pritam Ganguly
Department of Soil Science and Agricultural Chemistry, Bihar Agricultural University, Sabour, Bhagalpur, Bihar, India

Abhijeet Ghatak
Department of Plant Pathology, Bihar Agricultural University, Sabour, Bhagalpur, Bihar, India

Tarak Nath Goswami
Department of Entomology, Bihar Agricultural University, Sabour, Bhagalpur, Bihar, India

Arun Kumar Jha
Department of Soil Science and Agricultural Chemistry, Bihar Agricultural University, Sabour, Bhagalpur, Bihar, India

Monika Karn
Department of Plant Pathology, Bihar Agricultural University, Sabour–813210, Bhagalpur, Bihar, India

K. Karuna
Department of Horticulture, Bihar Agricultural University, Sabour, Bhagalpur, Bihar, India

A. Kohli
Department of Soil Science and Agricultural Chemistry, Bihar Agricultural University, Sabour, Bhagalpur, Bihar, India,

Anand Kumar
Department of Plant Breeding and Genetics, Bihar Agricultural University, Sabour, Bhagalpur, Bihar, India

R. Kumar
Department of Soil Science and Agricultural Chemistry, Bihar Agricultural University, Sabour, Bhagalpur, Bihar, India

Rajkishore Kumar
Department of Soil Science and Agricultural Chemistry, Bihar Agricultural University, Sabour, Bhagalpur, Bihar, India

Randhir Kumar
Department of Horticulture (Olericulture and Floriculture), Bihar Agricultural University, Sabour, Bhagalpur, Bihar, India

Ranjeet Kumar
Department of Entomology, Bihar Agricultural University, Sabour, Bhagalpur, Bihar, India

Ravi Ranjan Kumar
Department of Molecular Biology and Genetic Engineering, Bihar Agricultural University, Sabour, Bhagalpur, Bihar, India

S. Kumar
Department of Soil Science and Agricultural Chemistry, Bihar Agricultural University, Sabour, Bhagalpur, Bihar, India

Sanjeev Kumar
U.P. Council of Agricultural Research, Gomti Nagar, Lucknow, Uttar Pradesh, India

Sunil Kumar
Department of Agronomy, Bihar Agricultural University, Sabour, Bhagalpur, Bihar, India

Manoj Kundu
Department of Horticulture (Fruit and Fruit Technology), Bihar Agricultural University, Sabour, Bhagalpur, Bihar, India

Syed Sheraz Mahdi
Department of Agronomy, Bihar Agricultural University, Sabour, Bhagalpur, Bihar, India

H. P. Maheshwarappa
Project coordinator, AICRP, Palms, CPCRI, Kasaragod, Kerala, India

Nintu Mandal
Department of Soil Science and Agricultural Chemistry, Bihar Agricultural University, Sabour, Bhagalpur, Bihar, India

Abhay Mankar
Department of Horticulture, Bihar Agricultural University, Sabour, Bhagalpur, Bihar, India

Md. Abu Nayyer
Department of Agriculture, Integral University, Lucknow, Uttar Pradesh, India

Rajeev Padbhushan
Department of Soil Science and Agricultural Chemistry, Bihar Agricultural University, Sabour, Bhagalpur, Bihar, India

Paramveer Singh
Department of Horticulture (Vegetable and Floriculture), Bihar Agricultural University, Sabour, Bhagalpur, Bihar, India

Vishwa Bandhu Patel
Division of Fruits and Horticultural Technology, Indian Council of Agricultural Research–Indian Agricultural Research Institute, New Delhi, India

Sankar Ch. Paul
Department of Soil Science and Agricultural Chemistry, Bihar Agricultural University, Sabour, Bhagalpur, Bihar, India

Rajiv Rakshit
Department of Soil Science and Agricultural Chemistry, Bihar Agricultural University, Sabour, Bhagalpur, Bihar, India

Ruby Rani
Department of Horticulture, Bihar Agricultural University, Sabour, Bhagalpur, Bihar, India

Tamoghna Saha
Department of Entomology, Bihar Agricultural University, Sabour, Bhagalpur, Bihar, India

Shweta Shambhavi
Department of Soil Science and Agricultural Chemistry, Bihar Agricultural University, Sabour, Bhagalpur, Bihar, India

Sanjay Kumar Sharma
Department of Entomology, Bihar Bihar Agricultural University, Sabour, Bhagalpur, Bihar, India

Sangeeta Shree
Department of Horticulture (Vegetable and Floriculture), Bihar Agricultural University, Sabour, Bhagalpur, Bihar, India

Mahendra Singh
Department of Soil Science and Agricultural Chemistry, Bihar Agricultural University, Sabour, Bhagalpur, Bihar, India

P. K. Singh
Department of Plant Breeding and Genetics, Bihar Agricultural University, Sabour, Bhagalpur, Bihar, India

Paramveer Singh
Department of Horticulture (Olericulture, Vegetable and Floriculture), Bihar Agricultural University, Sabour, Bhagalpur, Bihar, India

Y. K. Singh
Department of Soil Science and Agricultural Chemistry, Bihar Agricultural University, Sabour, Bhagalpur, Bihar, India

R. K. Sohane
Department of Extension Education, Bihar Agricultural University, Sabour, Bhagalpur, Bihar, India

A. Srinivasaraghavan
Department of Plant Pathology, Bihar Agricultural University, Sabour, Bhagalpur, Bihar, India

V. K. Tripathi
Department of Horticulture, C.S. Azad University of Agriculture and Technology, Kanpur, Uttar Pradesh, India

B. K. Vimal
Department of Soil Science and Agricultural Chemistry, Bihar Agricultural University, Sabour, Bhagalpur, Bihar, India

Ramanuj Vishwakarma
Department of Entomology, Bihar Agricultural University, Sabour, Bhagalpur, Bihar, India

Abbreviations

AA	atomic absorption
AAS	agromet advisory services
AFLPs	amplified fragment length polymorphisms
AFM	atomic force microscopy
AGT	Assam green tall
AICRP	All India Co-ordinated Research Project
ALR 1	Aliyar Nagar 1
ALR 2	Aliyar Nagar 2
AM	*Arbuscular mycorrhiza*
AMFUs	agro meteorological field units
BAU	Bihar Agricultural University
BEC	base exchange capacity
BIPM	bio-intensive pest management
BLB	bacterial leaf blight
BOD	biochemical oxygen demand
BPH	brown planthopper
Bt	*Bacillus thuringiensis*
CaCV	capsicum chlorosis virus
CaO	calcium hydroxide
CCC	chlormequat/cycocel
CGD	Chowghat green dwarf
CH_4	methane
CIB	central insecticides board
CIPET	Central Institute of Plastic Engineering and Technology
CMV	*cucumber mosaic virus*
CO_2	carbon dioxide
COD	Chowghat orange dwarf
CP	coat protein
CPCB	Central Pollution Control Board
CPCT	Center for Protected Cultivation Technology
CSI	chitin synthesis inhibitors
CSISA	cereal system initiatives for South Asia
DBM	diamond back moths
DCD	dicyandiamide

DH	double haploid
DIP	digital image processing
DSR	direct seeded rice
EAS	extension advisory services
ECM	*ectomycorrhizae*
ECT	east coast tall
EDP	Entrepreneurship Development Program
EMT	Entrepreneurial Motivation Training
EPA	Environmental Protection Agency
ESE	entrepreneurial self-efficacy
EU	Europe Union
EXAFS	extended x-ray absorption fine structure
FCO	fertilizer control order
FIFRA	Federal Insecticide, Fungicide, and Rodenticide Act
FQPA	Food Quality Protection Act
FSSAI	Food Safety and Standards Authority of India
GA	gibberellic acid
GAP	good agriculture practice
GBNV	groundnut bud necrosis virus
GDP	gross domestic product
GHGs	greenhouse gases
GKMS	Gramin Krishi Mausam Sewa
GWP	global warming potential
HACCP	hazard analysis critical control point
HDP	high-density planting
HTMA	heat-tolerant maize for Asia
ICP	inductively coupled plasma
ICTs	Information and Communication Technologies
IFS	integrated farming system
IGP	Indo-Gangetic plains
IGR	insect growth regulators
IISR	Indian Institute of Spices Research
IMD	meteorological department
IPM	integrated pest management
ISO	International Standard Organization
ISSRs	inter SSRs
IYSV	iris yellow spot virus
JH	juvenile hormones
KVK	*Krishi Vigyan Kendras*
LCT	Lakshadweep ordinary

LS-IR	late sown irrigated
LUP	land use planning
MAI	moisture adequacy index
MAS	marker-assisted selection
MIDH	mission for integrated development of horticulture
MNCPC	multinutreint nanoclay polymer composite formulation
MPL	maximum permissible limit
MSW	municipal solid waste
MYD	Malayan yellow dwarf
NAA	naphthalene acetic acid
NBMW	need-based manual weeding
NCMRWF	National Center for Medium Range Weather Forecast
NCPCs	nanoclay polymer composites
NH_3	ammonia
NHM	National Horticulture Mission
NHRDF	National Horticultural Research and Development Foundation
NNI	National Nanotechnology Initiative
NSKE	neem seed kernel seed extract
NSs	non-structural protein
NTPC	National Thermal Power Corporation
OECD	Organization for Economic Cooperation and Development
OPs	organophosphates
ORFs	open reading frames
PAM/clay	polyacrylamide/clay
PBNV	peanut bud necrosis virus
PCPA	para-chlorophenoxy acetic acid
PCR	polymerase chain reaction
PE	pendimethalin
PE	pre-emergence
PGRs	plant growth regulators
PHI	pre-harvest interval
PoE	post-emergence
PP	pre-planting
PPP	public-private partnership
PRD	partial root drying
PVA	polyvinyl alcohol
PVMV	pepper veinal mottle virus
QPM	quality protein maize
RAPDs	random amplified polymorphic DNAs

RC	registration committee
RD	recommended dose
RdRp	RNA-dependent RNA polymerase
RKVY	Rashtriya Krishi Vikas Yojana
RMSE	root mean square error
RNPs	ribonucleoproteins
RWC	relative water content
SAPSRPF	slow-release phosphate fertilizer
SAU	state agricultural university
SD	secure digital
SEM	scanning electron microscopy
SHM-1	Sabour hybrid maize-1
SHM-2	Sabour hybrid maize-2
SNPs	single nucleotide polymorphism
SOC	soil organic carbon
SPAC	soil-plant-atmosphere continuum
SPI	standardized precipitation index
SSRs	simple sequence repeats
STRASA	Stress Tolerant Rice for Asia & South Africa
TEM	transmission electron microscopy
TGA	total geographical area
TiO_2	titanium dioxide
TLCV	tomato leaf curl virus
TOT	transfer of technology
TPT	Tiptur tall
TQM	total quality management
TSWV	tomato spotted wilt virus
UV	ultraviolet
VAM	vesicular-arbuscular mycorrhizae
WA	water absorbency
WBNV	watermelon bud necrosis virus
WCE	weed control efficiency
WCT	west coast tall
WI	weed index
WTCER	Water Technology Center for Eastern Region
WUE	water use efficiency
XANES	x-ray absorption near edge structure
YVM	yellow vein mosaic
ZNCPC	zincated nanoclay polymer composites
ZYMV	zucchini yellow mosaic virus

Preface

Agriculture has changed spectacularly over the years, especially in the last five decades. Agricultural productivity has increased due to innovative technologies, mechanization, increased application of fertilizers and pesticides, and modified government policies that resulted in maximized production. This increasing trend allowed farmers to reduce their need for labor.

A rising association has appeared during the past few decades to question the function of the agricultural establishment in promoting practices that contribute to different social problems. Now sustainable agriculture is gathering increasing support and acceptance within mainstream agriculture. Sustainable agriculture not only addresses many environmental and social apprehensions, but it offers innovative and economically viable opportunities for growers, laborers, consumers, policymakers, and many others in the entire food system. Sustainable agriculture includes three major goals, namely environmental health, economic profitability, and social and economic equity.

Because the concept of sustainable agriculture is still evolving, this book is an attempt to discover and discuss the ideas, practices, and policies that are essential to a sustainable agriculture strategy. The book is comprised of 29 chapters written by experts of related fields, including crop improvement, natural resource management, crop protection, social sciences, and product development.

Highlights of the Book

- Discusses sustainable agriculture within the framework of recent challenges in agriculture.
- Looks at the development and diversification of crops and cultural practices to enhance the biological and economic stability.
- Features the application of innovative nanotechnologies to agricultural research and production technologies.
- Highlights the development of new varieties in agri-horticultural crops in relation to several biotic and abiotic management.
- Discusses the application of recent technologies to soil-plant-microbe-environment interactions.
- Describes the efficient and gentle use of inputs as well as consideration of farmers' goals and lifestyle choices.

PART I

Crop Improvement

CHAPTER 1

Genetic Improvement of Major Cereals in Prospect to Bihar State

P. K. SINGH,[1] ANAND KUMAR,[1] and RAVI RANJAN KUMAR[2]

[1]*Department of Plant Breeding and Genetics, Bihar Agricultural University, Sabour, Bhagalpur, Bihar, India*

[2]*Department of Molecular Biology and Genetic Engineering, Bihar Agricultural University, Sabour, Bhagalpur, Bihar, India*

ABSTRACT

Agriculture is major sector in Bihar state since it contributes about 16% to State Gross Domestic Product and provides employment to about 70% of working force in rural area. More than 90% of farm households belong to marginal farm category (less than 1 hectare land) but own about 44% of cultivated land in Bihar. Genetic improvement of major cereals in Bihar through plant breeding has accounted for more than 50 percent of increases in yield of major cereal crops and has the ability to solutions of several challenges such as food security, hunger alleviation, increasing nutritional values, and higher input costs. Genetic improvement of crops consists of analytical frameworks that allow researchers to create and select plants that are consistently outstanding in desired traits. Centuries of selection for preferred traits, accelerated dramatically with the development of scientific plant breeding, have had tremendous positive impacts on food security and an improved quality of life. Today's scientific advances in genomics and genetics are exploring new frontiers in crop breeding, including rapid and targeted advances in specific traits. But there is still more work to do and many goals to achieve. Constantly evolving plant pests and pathogens, global climate change, and changing social needs make plant breeding, genetics, and genomics ever more vital. We hope genetic improvement in crops can contribute to sustainable agriculture and to the improvement of food in quantity and quality as well as security and safety.

1.1 INTRODUCTION

Before the 20th century, farmers used the selection of seeds from superior plants for the genetic improvement of crops. As human populations expanded into new regions, farmer selection produced increased genetic diversity in the form of distinct 'landraces,' or traditional varieties, with different genetic characteristics within the cultivated crop species. In the first half of the 20th century, specialized crop breeding programs were developed to exploit this farmer-created diversity to produce 'modern' crop varieties through systematic crossing and selection. These crop breeding programs had by mid-century produced several generations of modern crop varieties in most cultivated species, suitable primarily for richer and more developed—and incidentally or not, temperate zone-countries (Mba et al., 2012).

With the advent second half of the 20th-century, economic development policies and investments came into prominence. This period saw the creation of international (multilateral) institutions for meeting the objectives of economic development (e.g., The World Bank, Regional Banks, and other agencies of UN). Individual countries also established bilateral aid agencies (e.g., USAID, IDRC, and GTZ). As a result of the end of colonial regimes, the expectations were high. Many observers expected the subsequent decades to be characterized by per capita income 'convergence,' in which the highest growth rates in per capita income would be achieved by countries with the lowest initial levels of per capita income. By the 1960s, the prevailing conditions clearly depicted that the maintenance of food production per capita was a challenging task. It was impressive to see the improvements in health in the 1950s and 1960s. There was a decline in infant mortality rate and an increase in life expectancy. Even though in most of the countries, birth rates experienced a decline shortly after the decline in death rates, still there was an unprecedented increase in population in all developing countries. Agricultural policymakers in developing countries had experimented in the 1950s and 1960s with extension-led programs predicated on the concept of the 'inefficient farmer.' The central idea was that technology was available to farmers, but that farmers' 'ignorance'-combined with the lack of community education and information programs, as well as credit constraints and a high degree of risk refusal were preventing the effective and full use of modern technology. The agricultural aid programs of the 1950s and 1960s also recognized the need for capacity building in universities, both for purposes of training extension and education specialists and for developing agricultural experimental station capabilities in national agricultural research systems (Miflin, 2000).

The popularized view of the green revolution was based on the patchy data which showed rapid adoption of HYV rice and wheat in Asia and Latin America from 1968 to the early 1980s. Till the day today, only a few data have been available publicly on other crops or regions, or on more recent time periods. For example, until recent years, little has been known about the development or diffusion of new rice and wheat varieties in the 1980s and 1990s. In the same manner, few data have been available on varietal adoption of rice and wheat in sub-Saharan Africa or in West Asia and North Africa. Finally, the data on the improved varieties in other crops have been very few. As a result, there has been little systematic work attempting to evaluate crop improvement in developing countries until this volume (Miflin, 2000).

In a developing country like India, the agriculture sector continues to be the backbone of the Indian economy contributing approximately 27.4% to the gross domestic product (GDP), and accounts for about 18% share, of the total value of the country's export. The agricultural production growth rate is 21% per annum. Today we are the largest exporter of spices and cashew as well as the second-largest producer of wheat, rice, fruits, vegetables, and freshwater aquaculture. Per capita availability of food grains went up to 528.77 g per day in 1996–1997 when compared to 395 g in the early fifties. Fertilizer consumption has also increased and India has become fourth in the world after the USA, USSR, and China (Mahadevan, 2003). In the world, the Indian pulse area harbors the largest area for its production. India is the first to develop a cotton hybrid. The domestic demands and requirements of export have lead to a change in cropping pattern with an increase in the importance of commercial and nontraditional crops (moong, summer groundnut, soybean, sunflower) production. The introduction of short-duration varieties allows the use of residual moisture available from post-Kharif and post-rabi cultivation.

Bihar state, endowed with appropriate climatic conditions, with its bountiful natural resources of fertile soil, abundant water, varied climate, and rich cultural and historical heritage is one of the most fascinating states of India. Agriculture is the vital source of employment in the state with about 79% of its population engaged in agricultural activities. Bihar's productive contribution to food grain, fruit, vegetables, spices, and flowers can increase manifold with improved methods and system management (Bansil, 2011). In this chapter, we will discuss the various strategies used in the crop improvement and development in major field crops in Bihar state.

1.2 MAJOR CEREALS IN BIHAR STATE

Bihar has a geographical area of approximately 94.2 thousand square km which is divided by river Ganges into the north Bihar (53.3 thousand square km) and the south Bihar (40.9 thousand square km). In Bihar, on the basis of rainfall, soil characters, temperature and terrain, four main agro-climatic zones have been identified *viz.* Zone-1, North Alluvial Plain; Zone II, north-east Alluvial Plain; Zone III, A South East Alluvial Plain Zone, and Zone III B, South West Alluvial Plain. Zones I and II are situated on the north of the river Ganges whereas Zone III is situated on its south. Zone I is located in the northwestern part and Zone II on the northeastern part of the state with Zone I and Zone II being flood-prone and Zone III drought-prone. Bihar is mostly of subtropical climate as it is located between 25 to 27 north latitude. Across the state soil texture is varies from sandy loam to heavy clay, the majority type being loam category. The crop seasons are Kharif, Rabi, and Zaid. Cropping pattern in Bihar is dominated by cereals. The major cropping system is rice-wheat which occupies 70% of the gross cropped area. Maize occupies around 7% of the gross cropped area in the state. Rice, wheat and maize are grown in all the districts however the choice of the crop and crop rotation varies across the agro-climatic zone (Table 1.1).

1.3 BREEDING METHODS IN CROP PLANTS

Plant breeding has integrated the latest innovations in biology and genetics for the enhancement of crop improvement (Moose and Mumm, 2008). Crop improvement methods have also changed dramatically. Mass and pure line selections in landraces, consisting of genotype mixtures, were the popular breeding techniques until the 1930s for most crops. There has been a splendid progress in crop productivity; still greater progress is needed to meet the food requirements of an additional 2 billion people by the early part of the 21st century. Approximately 800 million people are facing hunger and another 185 million pre-school children are malnourished due to lack of food and water, or disease. Hence as suggested by the Nobel Peace Laureate, Norman Borlaug (1997), new bio-techniques, in addition to conventional plant breeding, are needed to boost yields of the crops that feed the world. In this section, we discussed about the different conventional and molecular methods involved in crop improvement.

TABLE 1.1 Area Production and Productivity of Major Cereals in Bihar State (Directorate of Statistical and Evaluation, Government of Bihar), *4th Advance Estimate

Year	Rice			Wheat			Maize		
	Area (Lakh ha)	Production (Lakh Metric tons)	Productivity (kg/ha)	Area (Lakh ha)	Production (Lakh Metric tons)	Productivity (kg/ha)	Area (Lakh ha)	Production (Lakh Metric tons)	Productivity (kg/ha)
2011–12	33.24	71.63	2155	21.42	47.25	2206	7.12	26.67	3745
2012–13	32.99	83.22	2523	22.08	61.743	2797	6.93	27.56	3975
2013–14	31.51	66.50	2110	21.49	61.347	2855	7.32	29.04	3966
2014–15	32.63	82.42	2525	21.54	35.702	1657	7.07	24.79	3508
2015–16	32.32	68.02	2104	21.10	47.36	2244	7.04	25.17	3571
2016–17	33.40	82.38	2467	21.05	59.85	2843	7.20	38.45	5335
2017–18*	32.84	79.11	2409	20.38	57.41	2816	6.68	26.10	3904

1.3.1 CONVENTIONAL PLANT BREEDING METHODS

Conventional breeding is the oldest method of changing genetic constituent of species through successive generations. Whereas nature changes a species through natural selection, human apply *artificial* selection; we identify the traits we want in individuals, and then breed those individuals to pass on their genetics. In due course of time, the desired characteristics become the usual character with complete alteration of species genetic makeup. Conventional plant breeding has been in practice for over thousands of years, since the advent of human civilization, recoded evidence being as early as 9000–11,000 years ago. Domestication in ancient times has lead to the development of the present-day crops. The domesticated varieties have given a way to the development of all major food crops varieties. The various methods used in classical/conventional plant breeding are Mass selection, pure line selection, hybridization, pedigree breeding, etc.

The above-mentioned breeding methods have been going on for hundreds of years, and are still commonly used today. Early farmers found that artificial mating or cross-pollination of crop plants leads to an increase in yield. Further development of plant breeding in the 20th century allowed plant breeders to create new and improved varieties. The productivity and quality of the plants were dramatically increased. In plant breeding, the most important aspect is the art of recognizing desirable traits and incorporating them into future generations. Breeders travel long distances to search the individual plants exhibiting desirable traits. The selection for features such as faster growth, higher yields, pest and disease resistance, larger seeds, or sweeter fruits has dramatically changed domesticated plant species compared to their wild relatives.

1.3.2 MOLECULAR METHOD OF CROP IMPROVEMENT

The advances in biotechnology hold great promise for crop improvement. For instance, molecular breeding, the integration of molecular biology techniques in plant breeding, through enhanced efficiencies, has great potentials for changing permanently the science and art of plant breeding. Molecular breeding uses molecular profiles for a selection of breeding materials and application of rDNA technology, i.e., genetic transformation. The efficiency of plant breeding can be enhanced using various molecular biology-based techniques.

The Polymerase Chain Reaction (PCR) development (PCR, Saiki, et al., 1988) has made DNA marker-techniques easy, convenient and cheaper.

Several PCR-based markers such as random amplified polymorphic DNAs (RAPDs), amplified fragment length polymorphisms (AFLPs), simple sequence repeats (SSRs or microsatellites), inter SSRs (ISSRs) and most lately single nucleotide polymorphism (SNPs) have been developed and applied to a wide range of crop species including cereals. Under the past decades, the molecular marker technology has rapidly evolved into a valuable tool able to dramatically enhance the efficiency of conventional plant breeding (Peleman and van der Voort, 2003).

Modern plant breeding is an amalgamation of genotype building and manipulating variation within the gene-pools. DNA-fingerprinting of breeding lines using molecular markers, as well as detailed genome analysis of plants, provides in this aspect a very powerful and efficient tool to characterize, monitor and protect germplasm (Lombard et al., 2000). The availability of molecular markers segregating with candidate genes, marker-assisted selection (MAS) can improve the efficiency of simple traits selection in conventional plant breeding programs (Knapp, 1998; Podlich et al., 2004). The MAS approach is not only a tool of speeding up the process of gene transfer, but also allows the pyramiding of desirable genes and QTLs from different genetic backgrounds. MAS strategies facilitate the development of lines with stacked resistance genes, giving the cultivar more durable protection than that afforded by a single resistance gene. Also, genes controlling resistance to different races or biotypes of a pest or pathogen, or genes contributing to agronomic or seed quality traits can be pyramided together to maximize the benefit of MAS through simultaneous introgression (Dwivedi et al., 2007).

Genetic transformation is another approach that offers direct access to a vast pool of useful genes not previously accessible to plant breeders. Current genetic engineering such as Agrobacterium-mediated transformation (Vergunst and Hooykaas, 1998) and biolistic/gene gun method (Srivastava et al., 2004) allow the simultaneous use of several desirable genes in a single event, thus allowing coordinated approaches to the introduction of novel genes/traits into the elite background. The priorities for applied transgenic research are similar to those of conventional plant breeding, aiming to selectively alter, add or remove a specific character in order to address regional constraints to productivity. Genetic engineering broadens the possibility for the introduction of a desirable character from related plants without associated deleterious genes. In many species, the development of rapid, highly efficient, and routine transformation systems is still in progress and thus represents a bottleneck in the development of stable high yielding transgenic plants. The sustainable and economic use of biotechnology can be achieved through the development and deployment of transgenic plants.

The advancements in the field of genetic transformation and gene expression have led to rapid progress in genetic engineering and crop improvement in terms of herbicide tolerance, pest resistance, and male-sterility systems. The potential of this technology has now been widely recognized and extensively adopted in the plant breeding of temperate crops (Albert et al., 1995; Vergunst and Hooykaas, 1998; Vergunst et al., 1998; Srivastava and Ow, 2002; Srivastava et al., 2004; Chawla et al., 2006; Louwerse et al., 2007).

The GM techniques used for the development of good quality food supply to the world is meeting public worries about the security of the derived food and their resulting products. The controversy spotlights the probable hazards due to the agglomeration of new substances in crop plants conferring allergy toxicity and genetic threats in human nutrition. Cisgenic (Schouten et al., 2006) plants are presumably considered safer than those produced through conventionally bred plants because of the lack of linkage drag. In cisgenesis, there is an introduction of desired genes without the inference of undesirable genes. It furnishes no hazard when compared to induced translocation or mutation breeding. Thus, it excludes linkage drag and, therefore, prevents hazards arising from unidentified genes. Therefore, cisgenesis is safer than traditional breeding programs. To provide long-lasting and wider forms of resistance, the various biotic and abiotic stress resistance genes can be pyramided (Holme et al., 2013).

1.4 DEVELOPMENT OF DIFFERENT CROPS FOR DIFFERENT AGRO CLIMATIC CONDITION OF BIHAR

Agriculture is a major sector in Bihar state since it contributes about 16% to State GDP and provides employment to about 70% of the working force in a rural area. More than 90% of farm households belong to the marginal farm category (less than 1-hectare land) but own about 44% of cultivated land in Bihar. Bihar is having a great history for developing different new and improved varieties in rice, wheat, and maize. The varietal improvement program got accelerated after the establishment of the first agricultural university, i.e., Rajendra Agricultural University, Pusa, in this state. In this section, we would discuss the historical perspective and development of various cereal crops of this state.

1.4.1 RICE BREEDING

With the separation of Bihar from Bengal in 1911 and the establishment of the Department of Agriculture in Bihar in 1914 the work of rice improvement of this state was undertaken by the Deputy Directors of Agriculture in different ranges viz., (1) South Bihar with headquarters at Patna, (2) South East Bihar with headquarters at Sabour, (3) Chotanagpur at Kanko (Ranchi), and (4) Cuttack with headquarters at Cuttack, which is now the Central Rice Research Institute, of course, with a much larger area. These Deputy Directors, with some improved varieties selected from the local varieties. Mr. D.R. Sethi evolved Dahia and an early Katika variety and Latisal, a late Aghani variety. Mr. C.B. Machean evolved from Kanke, which was released and distributed to Bihar farmers up to 1932.

Organized research work on a rice was undertaken in 1932. Late Mr. Madam, who was then working as an economic botanist for improvement of all cultivated crops, was appointed as the Rice Specialist Bihar with headquarters at Sabour, supported by two assistant botanists (rice) and a contingent of senior and junior research assistant, fieldsmen, and plant collectors.

The main objective of rice improvement work was undertaken on the following lines:

1. Complete botanical and agricultural survey of the paddy varieties of the province.
2. Isolation of pure line and their classification and maintenance besides the selection of improved strains.
3. Conduct of cultural and manurial improvements.
4. Genetical studies.

As a result of research work from 1932–38 a few improved varieties were recommended after trials of these varieties with local varieties in different departmental farms. They are (1) (early Aman (Katika) 115 BK and 141BK (2) medium Aman 88BK and 16 BK and (3) late Aghani 36 BK 498–2A. These new varieties rapidly replaced earlier released varieties and became very popular with the farmers all over the State. Two more fine-grained varieties viz., Tulsi-Manjari and Badshabhog were also released.

Along with varietal improvement suitable manurial and cultural recommendations were also released for higher production of rice in different rice-growing areas of the state. Apart from the work of research on rice undertake at the main station at Sabour, a few substations were also started to test the results evolved at Sabour. They were (1) Patna (South Bihar), (2) Kanke (Chotanagpur), (3) Purnea (South-East Bihar), (4) Sipaya (North Bihar), and Dumka (Santhal Parganas) The sub-station at Cuttack and Dumka discontinued after the separation of Orissa from Bihar in 1938 and Jharkhand in 2000.

The entire rice research work at Sabour was cut down drastically in 1941 when I.C.A.R. reduced their grant significantly and the work was somehow maintained at Sabour with one assistant botanist, one junior research assistant, and a few field overseers. They could only maintain the pure lines; somehow research work came to an almost standstill in 1944 when the I.C.A.R. stopped the grant.

In 1951–52 the agricultural Development and Research work got another big boost with the launching of the first five-year plan. The post of Rice Specialist was again revived with headquarters at Sabour with a contingent of assistant botanists, senior research assistants, junior research assistants, and fieldmen. Rice substations were also sanctioned at Patna, Sipaya, Purnea, Kanke, Dumka with one assistant botanist, one senior research assistant and two field overseers for each sub-station.

It may be mentioned here that during 1942 and 1951 when the rice work was again being supervised by the Economic Botanist, Bihar with headquarters at Sabour some exotic varieties were released for farmers *viz*., CH10 (BR24) and CH1039, etc. These were very early maturing 'Aus' varieties and replaced mostly some of the low yielding 'Gora' varieties of Chotanagpur uplands. These became very popular with early 'Áus' growing farmers of Bihar. Other early Aus varieties released were Sona and Sathika (Big 19 and BR20, respectively) but they did not become very popular. The variety, namely, 2206 B was released in 1953–54 which was later named BR34. This was a selection from the local variety of Munger district. The variety BR34 became very popular as early Katika variety replacing 115-BK and 141-BK later named as BR-4 and BR-5, respectively. A number of rice varieties were evolved by selection and hybridization as Aus paddy, Aman paddy, early Aman paddy, medium Aman paddy, late Aman paddy and christened as "BR" varieties. All evolved were fine-grained and scented varieties, purple varieties and deep-water varieties. Some of these varieties like BR-9 (fine-grained and scented), BR-13 (flood resistant) and BR-14 (Deepwater) served the farming community for a long period. During sixty decades of 20th century high yielding varieties like Taichung native-1, Padma, BCS, IR-8, Jaya, etc., were selected for commercial cultivation. These varieties on an average yielded 50 q/ha in the research farmer although IR-8 yielded up to 80 q/ha under good management.

During 1952–1970 pure line selections continued to occupy an important place in the breeding program. The number of exotic cultures began to swell the genetic stock, besides germplasm from a number of rice-growing countries were introduced. Large numbers of Japonica X Indica hybrids received from C.R.R.S. Cuttack were tried against the local standard varieties. Most of the earlier selections continued to occupy important placed in the rice varieties of Bihar. No success, however, was obtained from Japonica X Indicia crosses.

In the year 1964 two Indicas namely, Taiching Native-1 and Deo-geo-woo-gen from Taiwan were introduced. A very short duration, very high yielding and with coarse grain and summer paddy due to its excessive high yielding as high as 8 tons per hectare of rice. This variety, however very susceptible to bacterial leaf blight (BLB) particularly when grown in Kharif. But this variety was later used as parents. Evolution of IR8 and later Jaya completely replaced TN-1. Varieties IR-8 and Jaya are equally high yielding with better grain quality and definitely very much less susceptibility to BLB and much better fertilizer response. These two varieties at that time occupied very large rice areas of almost all the rice-growing states of India.

The year 1963–64 saw the introduction of summer paddy (sown in February-March and harvested in June-July) in Bihar in the irrigated areas. Summer paddy in those years occupies about 1–2 lakh hectares yielding about 2–3 tons of rice per hectare thereby adding about 6–10 lakh tons of rice in the total food grain production of the state previously the variety of rice used in the three crop sequence (Paddy- paddy-wheat) was N-136 but it was soon replaced by TN-1. The rice research center, Sabour has a great contribution to the evolution of recommended varieties such as CR-44–35, Archana, Sita, Deepa, Panidhan-1, Panidhan-11, and IET-3257. Two very early maturing cultures of rice namely, ESI-2-3 and Es-29-3-2 evolved through hybridization had been found to mature in 65–70 days. They were suitable for normal as well as late sowing up to 10 September, their yield potential being 25–30 quintals per hectare. They were particularly suitable for flood-prone areas where they may be sown in the field direct when flood recedes. Radha variety of rice (BR-51-46-5) had been identified tolerant to rice tungro virus at Sabour.

Rice improvement work was given greater part after the opening of the All India Coordinated Rice Improvement project at Hyderabad under the Joint auspices of I.C.A.R., Govt. of India, USAID, Rockefeller Foundation and Ford Foundation of U.S.A.

Thereafter work was being pursued with the main objective of evolving suitable varieties of different agro-climatic regions of Bihar by combining the high yielding potential of the dwarf indicas with the agronomical base tolerance to diseases and pests and superior grain quality of the traditional local varieties. In April 1979 a group of scientists from ICAR and biochemists from CRRI, Cuttack have laid down the criteria for quality grain rice for the export purpose (Table 1.2).

TABLE 1.2 Different Parameters Involved in Quality Grain Rice Selection

Parameters	Range
Length	More than 6.8 mm
L/B ratio	More than three times
Classification	LS
Scent	Strongly scented
Volume expansion	More than 3.8 times
Elongation ratio	More than 1.8 times of preferably twice
Amylose content (%)	20–25
Alkali value	4–5

Several scented rice like Tulsi-Manjari (BR-9), Katarni, kamod Khirsapati, Badshahbhog (BR-10), Cuttack Basmati have been growing by farmers, however, these varieties have not satisfies the export quality parameters. With the concerted efforts of the rice research group, several varieties (non-scented / scented) for different rice ecosystems have been identified for commercial cultivation to enhance the production and productivity of rice in the state.

In general early to medium early group varieties like Turanta, Prabhat, R. Bhagwati, etc., are being grown in upland, Mid early and Medium maturing varieties (120–135 days maturity) like, MTU 1010, MTU 1001, Sita R. Sweta, etc. under non-scented group and Rajendra Bhagwati, R. Suwasini, R. Kasturi, under scented group are being grown in medium low land irrigated condition. In low land irrigated and rainfed low land condition, late-maturing varieties (more than 135 days maturity) like MTU 7029, R. Mahasuri-1, MTU 1001, Kanak, traditional aromatic varieties, etc., are being grown (Table 1.3).

TABLE 1.3 List of Rice Varieties Released in Bihar (Up to 2010)

S. No.	Name of Variety	Days to Maturity	Grain Type	Yield (q/ha)
For Up Land				
1.	Turanta	75–80	Long-bold	25–30
2.	Prabhat	90–95	Long-slender	35–40
3.	Richharia	90–95	Long-slender	30–35
4.	Dhan Lakshmi	95–100	Long-slender	30–35
5.	Saroj	115–120	Long-slender	40–50
6.	Rajendra Bhagwati	115–120	Long-slender (scented)	40–45
For Medium Land				
8.	Sita	130–135	Long-slender	45–50
9.	Santosh	130–135	Long-slender	45–50
10.	Rajendra Suwasini	120–125	Long-slender (scented)	40–45

TABLE 1.3 *(Continued)*

S. No.	Name of Variety	Days to Maturity	Grain Type	Yield (q/ha)
11.	Rajendra Kasturi	125–130	Medium-slender (scented)	35–40
12.	Rajendra Sweta	130–135	Medium-slender	40–45
For Low Land				
14.	Rajshree	140–145	Medium-slender	40–45
15.	Rajendra Mahsuri-1	150–155	Medium-slender	55–60
16.	Satyam	140–145	Long-bold	40–45
17.	Shakuntala	140–145	Long-slender	40–45
18.	Swarna Sub-1	145–150	Medium-slender	40–45
For Deep and Chaur Land				
19.	Sudha		Long-bold	30–35
20.	Vaidehi		Long-bold	30–35
21.	TCA – 177		Long-bold	25–30
22.	Janaki		Long-bold	20–25

1.4.1.1 NEW ACCOMPLISHMENT IN RICE RESEARCH SINCE INCEPTION OF BIHAR AGRICULTURAL UNIVERSITY (BAU) (AFTER 2010)

The Bihar Agricultural University (BAU) was established in the year 2010 with the objective of improving the quality of life of people of this state especially farmers constituting more than two-thirds of the population. This university is dedicated to field crop improvement through various researches that could improve overall agriculture production in Agro-climatic zone II, IIIA and IIIB comprising larger parts of the state. The university has collaborated with various national and international centers in order to improve the present cultivar as well for the development of new varieties.

1.4.1.1.1 Research Priorities for Rice Research

The university identified several priorities area of research on the basis of the problem faced by local farmers and burning issues in rice development. After several brainstorming sessions with farmers and experts, the university identified the following research priorities area:

- Collection, characterization, documentation of rice germplasm.
- Development of aerobic rice.
- Hybrid rice development for mid and early maturity group.
- Breeding hybrid, parents and varieties of medium and late maturity duration.
- Development of aromatic rice.

- Herbicide resistance rice.
- Pure line varieties tolerant to drought and waterlogging to address the needs of the rainfed ecosystem.
- Development of resistance varieties against sheath blight, false smut, and BLB.
- Tagging of the gene linked to specific traits through MAS.

1.4.1.1.2 Scheme in Operation

After identification of research priorities, the researchers of this university started several projects funded by the state, national and international funding agencies. The prime aim of these projects was to carry out the basic and applied research on specific problems as well as the development of improved/new rice varieties suitable for different ecological conditions of this state (Table 1.4).

TABLE 1.4 List of Research Projects Under Implementation in Rice (2012–2016)

Sl. No.	Project	Funding Agency
1.	Stress Tolerant Rice for Asia and South Africa (STRASA)	BMGF (IRRI)
2.	Cereal System Initiatives for South Asia (CSISA, Breeding component)	IRRI
3.	Development of high yielding aerobic rice through marker-assisted breeding	Institutional
4.	Development of Early and Medium Maturity High Yielding Rice Hybrids Suitable for the Ecosystem of Bihar	Institutional
5.	Pvt. Paddy hybrid testing	Pvt. Co.
6.	Mutational analysis of calmodulin-binding protein gene family involved in abiotic stress responses in rice	DAE-BRNS
7.	Engineered resistance in rice against fungal pathogens (rice blast and sheath blight)	Institutional
8.	Marker-assisted introgression of *Sub-1* locus to transfer submergence tolerance in rice	Institutional
9.	Breeding for the architectural modification for the Katarnirice through marker-assisted selection	DST-SERB
10.	Identification and characterization of root physiological and biochemical trials in significance of water uptake in rice grown under drought	Institutional
11.	AICRP on Rice	ICAR
12.	Molecular and morphological characterization of local germplasm of the rice	Institutional
13.	Development of transgenic rice resistant to glyphosate herbicide	Institutional

TABLE 1.4 (Continued)

Sl. No.	Project	Funding Agency
14.	Identification of high yielding wheat genotypes suitable for limited irrigation	Institutional
15.	Maintenance breeding of crops	Institutional

1.4.1.1.3 Varietal Development in Rice

Rice research gained momentum after the inception of this university and several rice varieties were developed which were suitable for different ecology. All credit of the rice research goes to the team who worked for it in the past as well as working present on rice varietal improvement and development. The development rice varieties by BAU, Sabour are:

1. **Sabour Surbhit:** Under the eastern region Bihar is the rice bowl of India particularly for the aromatic rice. This aromatic rice (basmati or non-basmati) fetches premium prices in the export market due to its quality. Farmers of the many districts of Bihar are still growing their local scented cultivars which are low yielding, short grain and photo-period sensitive. Hence, to compete with the rice export market there was a need of high yielding, fine grain varieties for this community of farmers. By keeping farmers demand in the view, Sabour Surbhit variety was developed which was the semi-dwarf with long and superfine slender grain and strong aroma. This variety was developed through the mutation of Rajendra Suwasini. The yield of this variety is 30–35 q/ha. This variety can be grown in irrigated medium upland and medium land situation of Bihar and can perform well under limited irrigated conditions also (Figure 1.1).
2. **Sabour Ardhjal:** This variety was developed in the year 2013 as aerobic rice suitable for both rainfed and irrigated conditions. This aerobic rice variety was tailored for high yield potential (50–55 q/ha), input responsive, drought-tolerant (save 50% water) and weed competitive to attain high yields under aerobic soil conditions. This variety was developed through the selection of Dhagaddeshi/IR78584-98-2-2 cross. This variety is a promising approach for dealing with the emerging water shortage due to climatic change and sustained rice production. It can be direct seeded in upland as well as unpuddled, non-flooded medium irrigated condition with intermittent irrigation. However, under transplanted condition, this variety is also performing well (Figure 1.2).

FIGURE 1.1 Sabour Surbhit (field view).

FIGURE 1.2 Sabour Ardhjal.

3. **Sabour Deep:** Most of the rice varieties released for commercial cultivation in the state of Bihar are either late or medium in maturity and are poor yielder with coarse grain. Therefore, there was a need to develop a high yield potential variety with long slender grain and good cooking quality in the early group. Sabour Deep is fit for this group of rice varieties. The parentage of this variety is VG-56 and Type-3 and method used was the pedigree method of breeding. This variety has extra-long slender grain, good cooking quality, and high yield potential and early maturity. Under the late onset of monsoon and low precipitation during the crop season, the variety may be a good option. Therefore, this variety is suitable for contingent plan in case of failure of monsoon in Bihar. The variety may fetch a high price in the market due to high milling and head rice recovery and good cooking quality. It can be grown in irrigated medium upland and medium land situation of Bihar. This variety is suitable for contingent plan in case of late-onset of monsoon (Figure 1.3).

FIGURE 1.3 Sabour Deep.

4. **Sabour Shree (RAU 724-48-33):** It was developed by crossing Haryana Basmati/Mahsuri. This variety recorded an average yield of 6 tonnes/ha under the irrigated medium favorable land agro-climatic situation of Bihar. It has a semi-dwarf plant type (100–105 cm) with the brown color of husk and white kernel. The spikelet is awnless. The variety is of medium duration (135–140 days) and is moderately resistant to Sheath Blight and BLB and highly resistant to BPH, GLH, and Stem borer. The variety has non-aromatic medium slender grain with very good cooking and a high degree of milling (70.6%) and head rice recovery (63.0%). The variety can sustain drought at the early vegetative stage and waterlogging at the latter stage. The variety is suitable for the irrigated medium favorable land of Bihar (Figure 1.4).

FIGURE 1.4 Sabour Shree.

5. **Bhagalpur Katarni:** Katarni rice was famous for its unique taste and aroma. It was Grown natively in the Bhagalpur and Banka

districts of Bihar, Katarni rice is not only in demand in Bihar, but throughout the country. Despite its uniqueness, Katarni rice was facing the threat of extinction due to its genetic impurities over time. The main aim of the development of Bhagalpur Katarina rice was mainly to popularize this aromatic rice in this region as well as to generate quick income for the farmers. This variety was developed through pure line selection method of local landraces collected from Jagdishpur Block of Bhagalpur district. This variety is suitable for medium irrigated ecology of Agro-climatic zone IIIA of this state. This variety is tall, erect culm, well exerted long panicle with unique aroma having a maturity period of 150–155 days. This variety has the medium erected slender kernel and excellent cooking quality due to its strong aroma, soft texture. Besides rice, its Chura is highly aromatic. It has an excellent export potential as the producers get a premium price by selling the product (Figure 1.5).

1.4.2 WHEAT BREEDING

Wheat research in Bihar dates back to the first decade of the twentieth century when the Imperial Agricultural Research Institute was established at Pusa (Bihar) in 1905. Sri Albert Howard and his wife Mrs. G.L.C. Howard started their pioneering work in wheat improvement in 1906 at I.A.R.I., Bihar. They made a comprehensive collection of local types. Pure lines were isolated from the local types and that led to the release of Pusa 4, Pusa 6 and Pusa 12. Pusa 4 (NP4), a pure line selected from the local type "Mundia" was excellent with regard to maturity duration, adaptability, yield potential, and grain quality. The variety won the first prize for grain quality in several international exhibitions.

After the shifting of I.A.R.I., from Bihar to Delhi in 1935, the work on the improvement of wheat was taken up at Sabour under the aegis of State Department of Agriculture. Several varieties viz. BR 147 (medium maturing), BR 152–2 (late maturing), BR 164 and BR 166 (very late maturing), and BR 319 (very early maturing) were evolved to suit different agro-climatic situations. BR 319 was one of the most predominant wheat varieties of the state for its yield and quality under late sown conditions. It was a selection from the local wheat of Munger district.

FIGURE 1.5 Bhagalpur Katarni.

Subsequently, the wheat research was reorganized in 1957, and systematic work on developing rust-resistant and high yielding varieties were started. Three rust-resistant strains viz. HBR 2, HBR 3 and HBR 5 were developed through hybridization. They had excellent grain qualities with yield ranging from 25 to 30 q/ha. Some drought-resistant lines like HBR 13 and HBR 42 were also evolved and incorporated in the All India Coordinated wheat trials. Wheat Research in India as well as in Bihar had taken a great leap with the advent of Mexican dwarf wheat in 1966–67 which heralded green revolution.

By 1970 the following improved varieties were under cultivation in Bihar. Mid tall and tall varieties BR 147, BR 152–2, BR 164, BR 166, BR 319, NP 720, NP 761, NP 798, NP 799, NP 835, NP 846, NP 852, NP 884 and C 306. BR varieties were developed by scientists of this state. Semi-dwarf varieties: Sonora 64, Kalyan Sona, Sonalika, Chhoti Lerma, Safed Lerma, and Sharbati Sonora.

With the establishment of R.A.U., Bihar in 1970, the wheat improvement was reoriented with the main objective of developing early maturing drought and disease-resistant wheat varieties for rainfed conditions and fertilizer responsive semi-dwarf wheat varieties for irrigated conditions of the state. Systematic wheat breeding work in 1970 led to the development of "Desharma" (BR 104) which was released in 1974. Desharma (S503/NP835) was suited to late sown, high fertility irrigated condition. Grains were amber, hard and lustrous. Multi-location field trials in different centers of the university had resulted in the acceptance of varieties viz; C 306, NP 852 for rainfed situation: Sonalika, Kalyansona, Sharbati Sonora, Janak (HD 1982), UP 262, HP 1102, UP 115, HP 1209, HUW 206 HUW 234 and K 7410 for irrigated situation of this state. Scientists of this university had developed materials like BR 326, BR 380, BR 1012, BR 1015 and BR 1019 which performed well in All India Coordinated trials. The list of wheat varieties released up to the year 2010 is provided in Table 1.5.

1.4.2.1 WHEAT RESEARCH AT BIHAR AGRICULTURAL UNIVERSITY (BAU), SABOUR

Similar to rice research, the university also identified various research priorities area after several discussions with the farmers and experts. These research areas were formulated to solve the burning issues and farmers problems. The following research priorities were identifies:

- Collection, characterization, documentation of germplasm.
- Evaluation of germplasm against biotic and abiotic stresses to identify resistant genes for utilizing in variety developmental program.
- Development of varieties against leaf rust, leaf blight and black point diseases with high yield potential.
- Development of varieties for heat and drought tolerance including *duram* wheat.
- Tagging of the gene linked to specific traits through MAS.

TABLE 1.5 List of Wheat Varieties Released in Bihar (Up to Year 2010)

Sl. No.	Variety	Pedigree	Yield (q/ha)	Condition	Maturity (in Days)	Characteristics
1.	RW 346	JANAK/SA42	40–42	Irrigated, Timely sown	120–125	Semi-dwarf variety with slightly nonsynchronous, brown ear color at maturity. Grains bold and amber color; Resistant to brown rust
2.	RW 3016	NP852/S308	28–30	Marginal rainfed	125–130	Tall non-lodging variety. Glume surface glabrous, awn white color, medium panicle length, Resistant to brown rust
3.	RW 3413	–	40–42	Irrigated, Timely sown	120–125	Semi-dwarf variety with green foliage and white fusi form ear. Grains amber color and semi-hard, Resistant to brown rust and tolerant to kernel bunt

1.4.2.1.1 *Scheme in Operation*

After identification of research priorities, the researchers of this university started several projects funded by the state, national and international funding agencies. The prime aim of these projects was to carry out the basic and applied research on specific problems as well as the development of improved/new wheat varieties with high yield and disease resistance (Table 1.6).

TABLE 1.6 List of Projects Under Implementation in Wheat (2012–2016)

Sl. No.	Project Title	Funding Agency
1.	AICRP on wheat	ICAR
2.	Identification of high yielding wheat genotypes suitable for limited irrigation	Institutional
3.	Development of spot blotch resistance genotypes of spring wheat for eastern Gangetic plain of India using double haploid technology	DST-SERB
4.	Development and identification of suitable double haploid (DH) line for terminal heat and drought tolerance in spring wheat for eastern Gangetic plains of India.	Institutional
5.	Identification of high yielding wheat genotypes suitable for limited irrigation	Institutional
6.	Maintenance breeding of crops	Institutional
7.	Development of Heat tolerant Wheat through marker-assisted backcrossing	Institutional
8.	Wheat bio-fortification with reference to Fe and Zn	Institutional

1.4.2.1.2 *Wheat Variety Developed by Bihar Agricultural University (BAU), Sabour*

Similar to rice research, wheat research and development was also a dedicated and scientific effort by the wheat research team. The research team also developed high yielding, timely/late sown variety suitable for irrigated and rainfed conditions.

1. **Sabour Nirjal:** This variety is a selection from an international nursery, EIGN-I 2009–10. Parentage: ACHYUT/BL 1887. It is a medium-tall (105–110 cm) variety suitable for timely sown (Last

week of October to 1st fortnight of November) rainfed condition. The plant is semi-erect, pale green foliage, flag leaf long and drooping attitude. Strong waxiness on leaf sheath. Ear shape is tapering, medium dense and very long. Grains are bold (42–46 g/1000), amber-colored and semi-hard. Protein content: 10–12%. It is matured in 125 to 135 days and tolerant to major diseases and insect pests. It is also tolerant of lodging. The yield potential of this variety is 25–30 q/ha. This variety has an advantage over other rainfed varieties in the sense that other rainfed varieties do lodge if there is winter rain whereas this variety does not lodge and giving higher yield. It is recommended for timely sown rainfed ecological conditions and adaptable to the wider agro-climatic situations of Bihar (Figure 1.6).

FIGURE 1.6 Sabour Nirjal.

2. **Sabour Samriddhi:** This variety is a selection from an international nursery, SAWYT 2008–09, Parentage: PASTOR/ MILAN// MILAN/SHA 7. It is a semi-dwarf (90–100 cm) variety suitable for timely sown (15th to 30th November) irrigated condition. The plant is semi-erect, green foliage, flag leaf long and

drooping attitude. Ear shape is tapering, dense and medium in length. Grains are medium bold (37–41 g/1000), amber-colored and semi-hard. Protein content: 10–12%. It is matured in 120 to 125 days and tolerant to major diseases and insect pests. It is also tolerant of lodging. The yield potential of this variety is 45–50 q/ha. It is recommended for timely sown irrigated ecological conditions and adaptable to the diverse agro-climatic situations of Bihar (Figure 1.7).

FIGURE 1.7 Sabour Samriddhi.

3. **Sabour Shreshtha (BRW 934):** Wheat variety Sabour Shreshtha (BRW 934) is developed by BAU, Sabour. The variety was evolved by crossing HUW 234/CBW12-Sel. The variety is suitable for general cultivation under the irrigated late sown situation of Bihar. It recorded an average yield of 4.3 tonnes/ha. The variety has semi-dwarf (72–75 cm) and erect plant type with green and semi-erect flag leaf, white

tapering and awned ear, bent peduncle, elliptical, amber and semi-hard grain with test grain weight of 34.7 g. It matures in 105–110 days. It has resistance to lodging, brown rust and loose smut along with tolerance to leaf blight. It is non-shattering and easily threshable and fertilizer responsive. The variety has high protein content (11.03%) and Zinc (36.8 ppm) contents along with good chapatti making quality (7.60/10.0 score). The variety is suitable for the irrigated late sown condition (15–31 December) of Bihar (Figure 1.8).

FIGURE 1.8 Sabour Shreshtha.

1.4.3 MAIZE

A scheme on Maize and Millet's research was initiated in the year 1951, as one of the postwar research scheme with a view to augment research on Maize and Millet crops. It was financed by the State Government. The headquarters of the scheme was located at Sabour. As a result of the reorientation of Agriculture Research in 1955 so as to serve five ecological zones of the State, Maize and Millet Specialist was transferred to Pusa. Maize is mainly cultivated during the Kharif season as a rainfed crop on the onset of monsoon. Open-pollinated varieties of maize viz; Tulbulia, Mungetis, Herbaria, Mukwa, Sona tikkar, etc. were cultivated by the farmers.

Tinpakhia and Jaunpur were found to be quite promising ones. Tinpakhia is an early manuring variety, maturing in 75 days only. It is yield is 10 to 15 quintals/ha. It is suitable for flood-prone areas and can escape the flood. Jaunpur is a late-maturing variety. It matures in 90–95 days. Its yield is about 30 to 35 quintals/ ha. ICAR sanctioned Maize Improvement Scheme for diara area at Sabour in 1976. A major break-through in maize cultivation could be possible with the release of a series of hybrids and composites.

1.4.3.1 HYBRID MAIZE

As a result of co-operative researches, the first set of 4 maize hybrids having yellow grain, i.e., Ganga Hybrid Makka No.1, Ganga Hybrid Makka No.101, Ranjit Hybrid Makka and Deccan Hybrid Makka were release in the year 1961. These four hybrids were tested at Sabour from 1961 to 1963. Ganga Hybrid Makka No.1, early maturing hybrid having 85 days maturity was found suitable for flood-prone areas. Ganga Hybrid Makka No.101 having about 105 days maturity was found suitable for upland areas where flood is not a common feature.

In 1963 two hybrids having white grain viz; Ganga Safed-2 and Histarch were released by I.C.A.R. Their suitability was tested at Sabour. Ganga Safed-2 has been found most suitable for cultivation in the Kharif season. It matures in 95 to 100 days yield is 45 to 50 q/ha.

Ganga Hybrid No.5 having yellow grain and Ganga Hybrid Makka No.4 having white grain were released by ICAR in 1963. These were also tested at Sabour. Ganga-5 is suitable for both Kharif and Rabi. It matures in 95 to 100 days. Ganga 4 has a yield potential of 55 to 60 q/ha. It matures in 105 to 110 days.

One of the biggest successes in the development of maize hybrids was achieved by developing a Quality Protein Maize (QPM) hybrid. Shaktiman-1, Shaktiman-2, Shaktiman-3 and Shaktiman-4 QPM hybrids were developed between the years 2001 to 2007.

1.4.3.2 COMPOSITE MAIZE

The hybrid maize seed has to be replaced every year with new seed to maintain the same hybrid vigor. To overcome this difficulty, a number of composite varieties have been developed in which the same seed can be utilized for three terms by following proper isolation. Six composites viz;

Vijay, Vikram, Jawahar, Sona, Kisanand Amber were tested at Sabour and released by ICAR in 1967 for commercial cultivation. Vijay was found suitable for both Kharif and Rabi seasons; it matures in 100–150 days.

Diara composites have been released in the year 1978 as a result of experiments at Sabour. It was released for Diara areas for summer seasons. It matures in 75 to 80 days. Yields about 25 to 30 q/ha. Suwan composite has been released in 1981 for cultivation in the Kharif season. Its yield is about 40–45 q/ha. The farmers were desirous of having white composites with the potentials of Ganga Safed-2.

Two white composites viz; M-9 and M-13 have been developed at Dholi. White composite M-9 has been found suitable for the Kharif season. Composite M-8 and M-7 (pool 17) are of early maturing types. They have given a higher yield than Diara composite but are about 4 to 7 days late. Pool-17 may be suitable for summer maize cultivation in the Saharsa district.

Composite maize like Lakshmi, Heamant and Dewaki were developed for the late sown condition in Rabi season. These maize varieties were much popular among maize growers and breeders.

1.4.3.3 RABI MAIZE

Experiments on Rabi maize was started from Rabi 1961–62. The farmers saw the performance of maize in the winter season and were quite astonished to see the vigorous and healthy plants.

Hybrids Ganga-1, Ganga-101, Deccan, Ranjit, Ganga, Safed-2, and Hi-starch were tested. Hi-starch was found to be the best for Rabi cultivation. Composite maize varieties namely Amber, Kisan, Sona, Jawahar, Vikram, and Vijay were also tested. Vijay was found suitable for Rabi season. White composite Laxmi has been found at par with Hybrid Hi-starch and released for Rabi cultivation. Several hybrids suitable for the Rabi season were developed in which Rajendra Hybrid Makka-1, Rajendra Hybrid Makka-2, Shaktiman-2, Shaktiman-3, and Shaktiman-4 was high yielder.

A brief detail of the maize varieties developed up to the year 2010 is provided in Table 1.7.

1.5 WHEAT RESEARCH AT BIHAR AGRICULTURAL UNIVERSITY (BAU), SABOUR

Similar to rice and wheat research, the university also identified various research priorities area after several discussions with the farmers and

TABLE 1.7 List of Maize Varieties Released in Bihar (Up to the year 2010)

Sl.No.	Season	Variety	Parentage	Maturity (in Days)	Yield (q/ha)	Characteristics
Kharif Composite						
1.		Suwan	—	85–90	40–45	Yellow, semi dent
2.		Dewaki	American early dent × Tuxpeno	100–110	35–40	White, semi-dent, Early
3.		Hybrid				
4.		Rajendra Hybrid Makka-3	Dholi inbred 32 × Dholi inbred 40	110–115	50–55	Bold grain, Orange-yellow color
5.		Shaktiman-1	(CML 142 × CML 150) × CML 186	110–115	50–55	White, flint, QPM hybrid
6.		Shaktiman-2	CML 176 × CML 186	110–115	60–65	White, flint, QPM hybrid
Rabi Composite						
7.		Lakshmi	American Dent × Tuxpeno Yellow	150–155	60–65	White, semi dent, Late
8.		Hemant	Dholi 7744 (AE) × Tuxpeno	165	60–65	White, bold grain, semi dent
9.		Dewaki	American early dent × Tuxpeno	155	50–55	White, semi-dent
		Hybrid				
10.		Rajendra Hybrid Makka-1	(CM 400 × CM 300) × P 7421	155–160	65–70	Yellow, semi-dent
11.		Rajendra Hybrid Makka-2	EVM 13 × Jogia Local	155–160	65–70	White, semi-dent
12.		Shaktiman-2	CML 176 × CML 186	145–150	80–85	White, flint, QPM hybrid
13.		Shaktiman-3	CML 161 × CML 163	150–155	85–90	Yellow-orange, semi-flint
14.		Shaktiman-4	CML 161 × CML 169	150–155	85–90	Yellow-orange, semi-flint
Summer Composite						
15.		Suwan	—	85–90	35–40	
16.		Dewaki	American early dent × Tuxpeno	100–105	40–45	White, semi-dent, Early

experts. These research areas were formulated to solve the burning issues and farmers problems. The following research priorities were identified:

- Collection, characterization, documentation of locally adopted short-duration cultivars.
- Development of high yielding short duration hybrid and open-pollinated varieties for Kharif and full-season hybrids for the winter season.
- Inbred line development for abiotic stresses.
- Maintenance of inbreds.

1.5.1 SCHEMES IN OPERATION

The maize research of this university is supported by the state, national as well as international center like CIMMYT. In the short span of time, the maize research has achieved remarkable progress. The prime aim of these projects is to development of high yielding hybrids and specialty corn (Table 1.8).

TABLE 1.8 List of Projects Under Implementation in Maize (2012–2016)

Sl.No.	Project	Funding Agency
1.	Heat Tolerant Maize for Asia (HTMA)	USAID and CIMMYT
2.	Breeding and seed production of Maize	Institutional
3.	Development of hybrids for specialty corn	Institutional
4.	AICRP on Maize	ICAR
5.	Short duration single cross hybrid maize for Kharif season in Bihar	Institutional
6.	Maintenance breeding of crops	Institutional

1.5.2 MAIZE HYBRIDS DEVELOPED/IDENTIFIED BY BIHAR AGRICULTURAL UNIVERSITY (BAU), SABOUR

Similar to rice and wheat research, the development/identified maize research and development was also a dedicated and scientific effort by the maize research team. The research team also developed high yielding suitable for Kharif/Rabi/Summer season.

1. **Sabour Hybrid Maize-1 (SHM-1):** BAU covers 65% arable land of Bihar that includes Agro-climatic Zone-II (northern-eastern part)

and Zone III A and Zone III B (southern part). The northern part is mostly affected by floods during the Kharif season whereas the southern part of Bihar faces periodic drought. Production of Kharif maize in the northern part of the state is uncertain due to high rainfall and flood. In zone III-A in case of low rainfall or delayed monsoon farmers require short duration maize hybrids. In the state, more than 30% of the total acreage (2.75 lakh hectares) of the maize is being grown during the Kharif season where crop faced drought stress during the early growth stage in one hand and waterlogging in another hand. Presently, available varieties are well suited only for good environment and suffered badly under aforesaid stress conditions. SHM-1 is developed for the Kharif season. This variety is developed by the crossing of SML-1 × VQL-1 inbred. It is an early duration (85–90 days) hybrid with the Yield potential of 55–60 q/ha. SHM-1 is of yellow grain color, tolerant to major diseases and insect pests. Being a short duration hybrid SHM-1 escapes both stresses (Figure 1.9).

2. **Sabour Hybrid Maize-2 (SHM-2):** Many farmers grow potato, mustard, and tobacco in Bihar. Mostly, the northern part is a flood-prone area in Kharif and water recession is late in some of the areas. There is the demand for spring maize after harvesting of potato, mustard, tobacco and late recession of floodwater. Therefore, the need was felt for high yielding spring maize such as SHM-2 for the spring season. This hybrid is developed by the crossing of CML-451 × CLO-2450 inbred. It is heat tolerant, medium duration (100–105 days) hybrid with the Yield potential of 65–70 q/ha. SHM-1 is of orange grain color, tolerant to major diseases and insect pests. This spring hybrid can sustain at high temperatures. It can be adopted in all Agro-Climatic Zones of Bihar after harvest of potato, mustard, tobacco and in a flood-prone area (Figure 1.10).
3. **DHM-117:** There is the demand of rabi maize especially in those areas where a single crop is grown that covers around 5.0 lakh hectares of the state. Simultaneously, the northern part is mostly affected by floods during the Kharif season whereas the southern part of Bihar faces a periodic drought. Production of Kharif maize in the northern part of the state is uncertain due to high rainfall and flood. In zone III A in case of low rainfall or delayed monsoon, farmers require short duration maize hybrids. Therefore, there was a need of high yielding rabi and Kharif maize such as DHM-117 for both the seasons. This

variety was developed by ANGRAU, Hyderabad by the crossing of BML-6 × BML-7 inbred. The maize research team of BAU, Sabour identified the potentiality of this hybrid for both Kharif and Rabi season for this state. It is a medium duration hybrid, sturdy plant type with a yield potential of 55–60 q/ha in Kharif and 85–90 q/ha in Rabi season. It is lodging resistant and stays green at brown husk stage or harvest. Its grain is a very attractive orange-yellow seed color, bold and flint grain type. It has field resistance to major foliar disease viz., MLB and TLB under Bihar conditions. It is a rabi and Kharif hybrid that can sustain in waterlog conditions for few days during Kharif. It can be adopted in all Agro-climatic Zones of Bihar (Zone II, Zone IIIA, Zone IIIB) (Figure 1.11).

FIGURE 1.9 SHM-1.

FIGURE 1.10 SHM-2.

FIGURE 1.11 DHM-117.

KEYWORDS

- **Bihar**
- **cereals**
- **genetic improvement**
- **genomics**

REFERENCES

Albert, H., Dale, E. C., Lee, E., & Ow, D. W., (1995). Site-specific integration of DNA into wild-type and mutant lox sites placed in the plant genome. *The Plant Journal*, *7*(4), 649–659.

Bansil, P. C., (2011). *Bihar Agriculture: A Perspective*. Concept Publishing Company.

Chawla, R., Ariza-Nieto, M., Wilson, A. J., Moore, S. K., & Srivastava, V., (2006). Transgene expression produced by biolistic-mediated, site-specific gene integration is consistently inherited by the subsequent generations. *Plant Biotechnology Journal*, *4*(2), 209–218.

Dwivedi, S. L., Crouch, J. H., Mackill, D. J., Xu, Y., Blair, M. W., Ragot, M., & Ortiz, R., (2007). The molecularization of public sector crop breeding: Progress, problems, and prospects. *Advances in Agronomy*, *95*, 163–318.

Handbook, (2008). *Bihar's Agriculture Development: Opportunities and Challenges—A Report of the Special Task Force on Bihar*. Government of India, New Delhi.

Handbook, B. S., (2014). *Directorate of Statistics and Evaluation*. Government of Bihar.

Holme, I. B., Dionisio, G., Pedersen, H. B., Wendt, T., Madsen, C. K., Vincze, E., & Holm, P. B., (2012). Cisgenic barley with improved phytase activity. *Plant Biotechnology*, *10*, 237–247.

Knapp, S. J., (1998). Marker-assisted selection as a strategy for increasing the probability of selecting superior genotypes. *Crop Science*, *38*(5), 1164–1174.

Lombard, V., Baril, C. P., Dubreuil, P., Blouet, F., & Zhang, D., (2000). Genetic relationships and fingerprinting of rapeseed cultivars by AFLP: Consequences for varietal registration. *Crop Science*, *40*(5), 1417–1425.

Louwerse, J. D., Van Lier, M. C., Van Der Steen, D. M., De Vlaam, C. M., Hooykaas, P. J., & Vergunst, A. C., (2007). Stable recombinase-mediated cassette exchange in arabidopsis using *Agrobacterium tumefaciens*. *Plant Physiology*, *145*(4), 1282–1293.

Mahadevan, R., (2003). Productivity growth in Indian agriculture: The role of globalization and economic reform. *Asia Pacific Development Journal*, *10*(2), 57–72.

Mba, C., Guimaraes, E. P., & Ghosh, K., (2012). Re-orienting crop improvement for the changing climatic conditions of the 21st century. *Agriculture and Food Security*, *1*(1), 1.

Miflin, B., (2000). Crop improvement in the 21st century. *Journal of Experimental Botany*, *51*(342), 1–8.

Moose, S. P., & Mumm, R. H., (2008). Molecular plant breeding as the foundation for 21st-century crop improvement. *Plant Physiology*, *147*(3), 969–977.

Peleman, J. D., & Van Der Voort, J. R., (2003). In: Van Hintum, T. H. J. L., Lebeda, A., Pink, D., & Schut, J. W., (eds.), *The Challenges in Marker Assisted Breeding* (pp. 125–130). Eucarpia leafy vegetables. CGN.

Podlich, D. W., Winkler, C. R., & Cooper, M., (2004). Mapping as you go. *Crop Science*, *44*(5), 1560–1571.

Saiki, R. K., Gelfand, D. H., Stoffel, S., Scharf, S. T., Higuchi, R., Horn, G. T., & Ehrlich, H. A., (1988). Primer-directed enzymatic amplification of DNA. *Science*, *239*, 487–491.

Schouten, H. J., Krens, F. A., & Jacobsen, E., (2006). Cisgenic plants are similar to traditionally bred plants. *Science and Society*, *7*, 750–753.

Srivastava, V., & Ow, D. W., (2002). Biolistic mediated site-specific integration in rice. *Molecular Breeding*, *8*(4), 345–349.

Srivastava, V., & Ow, D. W., (2004). Marker-free site-specific gene integration in plants. *TRENDS in Biotechnology*, *22*(12), 627–629.

Vergunst, A. C., Jansen, L. E., & Hooykaas, P. J., (1998). Site-specific integration of Agrobacterium T-DNA in Arabidopsis thaliana mediated by Cre recombinase. *Nucleic Acids Research*, *26*(11), 2729–2734.

CHAPTER 2

Status, Opportunities, and Challenges of Spice Research and Development

SANGEETA SHREE and PARAMVEER SINGH

Department of Horticulture (Vegetable and Floriculture), Bihar Agricultural University, Sabour, Bhagalpur, Bihar, India

ABSTRACT

Spices can be said to be a strongly flavored or aromatic substance of vegetable origin, obtained from tropical plants, commonly used as a condiment. Spices constitute any part of plants which may be root, leaves, seeds, stems, barks and even flower or fruit. The main function of spices is that they add flavor to the food. Besides, they also add some color to the food and also help in preserving them.

2.1 HISTORY OF SPICES

The history of the world is never complete without the information and knowledge of the importance and role spices occupied in the olden times. The discovery of countries and continents, economic rivalries, trade and commerce, important war and ventures and many more such events and advents were linked in one way or the other to search and hunt for spices. Simultaneously study of spices will remain incomplete if its historical background is not diagnosed and unearthed. The history of spices dates back to the pyramid age of Egypt, roughly 2600 to 2100 B.C. The existence of a high quality of spices in South India is reflected from the Babyloman and Assyrian era (Balraman et al., 1989). The sources of the spices were not known to most of Europe till the Christian era began. Cinnamon was supposed to be the aromatics that were utilized by the queen of Egypt, Hatshepsut, (Parry, 1969). Spices and aromatics were among important commodities in the second-third millennium used by Arabian traders. Likewise, remnants of Indus Valley also validate the truth that spices used to be the prized possession of the Harrappans. Besides, Kautilya's Arthasasthra also mentions plentiful usage of spices including coriander, ginger, pepper cardamom fenugreek, and mustard in the third century B.C. Hippocrates (460–377 B.C.), Theophrastus (372–287 B.C.), and Dioscorides, the father of Botany (A.D. too had cited about the worth of spices in their literature during their respective periods. Thus even in the early era of the history of mankind spices were very important (Rosengarten, 1969). Romans used spices lavishly both in kitchen as well as for cosmetics reasons. It is also mentioned that in the olden days peppercorns were considered to be the ready money and were used to pay taxes, excise, rental fee, even dowries and for various social causes (Purseglove, 1981). Vedas and epics too have cited the practice of using spices in various ways and for plenty of purposes. Rigvedas, Yajurvedas, and Atharvedas have recognized the worth and value of all types of spices and their medicinal properties (Rosengarten, 1969). The epics like Ramayana and Mahabharatha (Mahindru, 1982) both old and new testaments of the Bible have the mention of spices in several ways. Even a great prophet of Islam, himself was a spice merchant (Rosengarten, 1969). In the post-Vedic period, Charak and Susruta mention the role of spices play in surgery in the classic book, Susruth Samhita. There are references in the books of Vatsyayana on spices too. Theophrastus, the Greek Philosopher in the fourth century B.C. pointed out that the sizzling flavor of spices came out from the hotter parts of the Asian world. Pliny also has mentioned about the value of pepper in some of his writings, as early as the first century B.C. There was intensive trade and transaction of spices between the Western world and India. The spice trade and commerce between Hindustan and Rome was well organized. Petite in volume, soaring in cost and stable in demand, spices were the most attractive object of business. India was the most sought after place in olden, as navigators were lured by spices possessed by India. In fact, rummage around for spices showed the way to India via the Cape of Good Hope by Vasco da Gama in 1498. Subsequently, a monopoly on the spice trade was captured by Portuguese which continued onto the 16th century. The domination in this area was passed on to the Dutch, until the British finally took over. In the modern-day world, the acquiring spices are not as complicated or death-defying as it was in the past but the charm of Hindustani spices has still not faded as it was in those days. Gradually spices in a way became started affecting lifestyle, culture, business financial and diplomatic dealings among countries (shodhganga.inflibnet.ac.in/bitstream/10603/156/16/09_chapter2.pdf)

2.2 IMPORTANCE OF SPICES

Spices have ever been closely associated with cultural traditions, social customs, rites and rituals, magic and charms, preservation, medicines cosmetics and what not since early human history. Spices are associated with human values and culture in a variety of ways. They are part and parcel of kitchen. They are exploited as various medicinal and cosmetic reasons. They are used as relaxants as well stimulants. They occupy special privileges in various rituals of ancient people of not only Indians but also the Egyptians, Arabians and the Romans irrespective of caste, creed, religion, and culture. Ancient peoples such as the Egyptians, the Arab and the Roman made extensive uses of spices, not only to add essence to foods and drinks, but also as medicines, disinfectants, incenses, stimulants and cosmetics. Spices are loaded with nutraceuticals, photochemical and several secondary metabolites. They possess chemopreventive, antioxidant, antimicrobial, anti-inflammatory properties and facilitate indigestion. They serve as a raw material in the treatment of stomach problems, diarrhea, dysentery, and nausea. Thus they act like functional food. They also enhance shelf life, advance storage and help preservation of foodstuff, etc. No wonder in the bygone days they were craved as gold like other precious metals and stones. India is thought to be the exquisite land of spices. Peninsular India is endowed with the great diversity of herbs and spices grown naturally. Ginger, turmeric, cinnamon, black pepper, capsicum, clove, nutmeg, cardamom, tamarind, and vanilla are the important spices. Seed spices like coriander, caraway, fennel, fenugreek, dill, cumin, anise and herbal spices like saffron, lavender, thyme, oregano, celery, anise, sage, and basil are also important. The Indian peninsula is the native home of many key spices, viz., tamarind, curry leaf, black pepper, cardamom, and to a certain amount, ginger, cinnamon turmeric and where a good diversity exists.

In fact, there is no state in India that does not grow spices. This plays a crucial role in providing livelihood security and economic sustainability to the people in and around that area. From the Indian subcontinent, these spices spread over other parts of the tropical world. Each country and each region has its own traditional races/cultivars or spices types. As many as 109 spices have been listed by IS out of which 75 types of spices are produced in India.

2.3 DEVELOPMENT OF SPICE CROPS

Directorate of Areca nut and Spices Development, Kozhikode, Kerala. The importance and worth of spice crop was being felt and recognized by officials at various levels since a very long time but the first worthwhile initiative regarding spice development was taken in 1951 when a high-level Spices Enquiry Committee was set up by the Planning Commission in view of the significant role spices play among the agricultural commodities produced in the country. The committee considered that the spice crops were high-value crops and could have enormous contributions in the construction of the national wealth and economy. However, organized efforts were needed to boost the production and strengthen the marketing structure as in other cash crops. The committee emphasized the need of better planning, investigations, explorations and joint efforts are required for the appropriate development spice crops. Henceforth, the Government of India funded ICAR was for implementing various schemes on Research as well as Development and Marketing throughout the country. At present Directorate of Areca nut and Spices Development at Calicut in Kerala which was established in 1966 acts as a secondary office under the Ministry of Agriculture for paying enough interest in different aspects of crop development. Simultaneously, the Council for Development of Indian Spices was created to deal with various issues related with the development programs on these crops. These institutions laid stress upon improvement in production, the fulfillment of the export demand, and enhancement of quality of spices and related products. The scientific post-harvest procedure was encouraged to guarantee quality improvement. The State Horticulture or Agriculture Departments act as line departments for the execution of different programs. National Horticulture Mission (NHM) was launched during 2005–06 by of Indian Government as a holistic approach for the development of different horticultural crops together with spices, aromatic and, medicinal crops. However, at present, for the holistic growth of the horticulture sector including spices this scheme functions as a sub-scheme under one Centrally Sponsored Plan called Mission for Integrated Development of Horticulture (MIDH). The demand for spices, grown organically is on the increase since spices are consumed invariably as a food additive and also as a medicine or even as aromatics. Some farmer participatory program has also been initiated by DASD to encourage most recent technologies in spices during 2008–09 which are practiced effectively by different organizations. During 2015–16, it was planned to establish 78 frontline demonstration plots at various University centers/ICAR Institutes/Farmers fields for organic cultivation of various spice crops. ICAR Institutes, some reputed NGOs and several SAUs, jointly implemented the programs. Calicut Team observed that seed spice still remains one segment, where output was not very significant. This was primarily due to the non-implementation of the latest technologies and current know-how evolved for various crops. Drip as well as sprinkler irrigation technologies save water enhance water use efficiency (WUE). Because of over-exploitation of the agricultural land fertility status has gone down and developing water wealth or, implementation of water-saving methods, etc., would make possible availability of water to the crop and help in the growth and development of the plants. Growth and yield parameters of spices like Black pepper, Chili, etc., showed a significant increase in response to irrigation. A significant increase has been observed yield of rhizomes of turmeric and ginger and growth and yield of Black and pepper, with proper irrigation. The DASD has set up program namely, on management of farm water which aims at increasing efficiency of water use by encouraging efficient technologies and equipments of water management in spices crops Spices Board (Ministry of commerce and industry, Government of India) Spices Board, Ministry of Commerce and Industry is the apex body responsible for the progress and promotion of spices in India on global basis. The one-time Cardamom Board (1968) and Spices Export Promotion Council (1960) merged into Spices Board in 1987] The Board serves to connect the exporters in India and the importers overseas. The Board along with industrial sector caters to different activities for maintaining quality and appeal of spices of India. It autonomously controls the export promotion of all the listed spices and development of Cardamom (Small and Large). Main functions include research, growth, expansion and regulation of domestic marketing of small and large cardamom, post-harvest enhancement of all spices, promotion of organic spices, value addition, curing and certification of spices, development of spices in North East, provision of quality assessment services, export encouragement of all spices by means of technology up-gradation, quality up-gradation, brand promotion, research and product development. Other responsibilities include collection and keeping records of trade information and provide guiding principles to the central government on policy matters related to import and export of spices. It also promotes exports of spice and their products, encourages growing of spices organically, facilitates basic facilities for value addition, helps in registration and licensing of all spice exporters, provides assistance for investigations and research for enhancing quality Indian Society for Spices, Kozhikode, Kerala, the responsibility of any scientific society in promoting indigenous research and exploring scientific developments has ever since been immense. The Western countries have greatly contributed towards development of science and technologies through such societies. In India, too scientific societies have played a crucial task in bringing forth novel research outputs and the latest technologies in diverse fields. The scientific societies have played an important role in bringing together multifarious and diverse scientific structures. Therefore, for the progress and expansion of science and research in spices, aromatic and related crops, the Indian Society for Spices was formed in 1991. The society aims at providing a forum for research workers on these crops for sharing ideas and exists as a link between research and related organizations and the industry. The society organizes the symposium, convention, seminar and other scientific meets in association with other organizations on aspects pertinent to spices, medicinal and aromatic crops. Aims and objectives of this Society are to provide common platform for all those linked with spices, aromatic plants and allied crops for fertile interactions and update of the latest technologies, to publish an International Journal called Journal of Spices and Aromatic Crops, as a means for speedy publication of research results, to encourage communication among scientists of different institutes and groups in India, to organize symposia, conferences, workshops, etc., on spice, aromatic and related crops, to connect the research organizations with the industry and trade, and to keep brotherhood among research workers, farmers, industrialists, traders and the consumers of spices, aromatic plants and allied crops. The ISS is a tough pillar for attaining scientific literacy and this helps in understanding science, its nature, and ethics that control scientists in their work, the interrelationships of science and society and also the interrelationships of science and the humanities (Pella et al., 1966). Production of Spices India is the home of a large number of spices. The present status of different spices production in India is around 3.2 million tonnes which are approximately 4 billion US $ worth, and holds a highflying position in world spice production. Because of the changing climate from tropical to subtropical and temperate, almost all spices grow superbly well in India. As many as 109 spices are reported in the ISO list. However, only 52 spices are under the jurisdiction of the Spices Board. The area and production of spices in the country has shown increase up to 3% and 4% respectively in the year 2013–14 to 2014–15 as compared to their respective previous years (Source: Department of Agriculture and Cooperation). According to the Economics and Statistics Department, the Ministry of Agriculture, the current scenario of the production of spices in the country is approximately over 5 million tons out of which about 10% is exported. Because of the social awareness, enhanced economic status and boost in the living standard of the people and in the country and their shifting food patterns, the domestic utilization of spices has augmented significantly in recent years.

India holds a worldwide reputation of being the only nation where all types of spices are produced. Spices serve as an important foreign exchange earner and thus play a key role in the trade and commerce of India. With the modernization in transport systems and the growing global market, spices of

India are savored by the people throughout the world and this definitely has a role in improving the mutual trade and business relations. Spices Board, Ministry of Commerce and Industry, is the highest authority for encouraging export and international trade regarding Hindustani spices and their different products. It is accountable for post-harvest enhancement and valorizations as an export promoting measure for the spices under list. The last five years registered significant growth in exports of spices leading to 14% increase with respect to Indian currency and 5% in the dollar. India has a respectable place in the world trade of spices. During the year 2015–16, 843,255 tons of spices and products of spices worth Rs.16,238.23 crores was exported by India while the export was 893,920 tons worth Rs.14,899.68 crores in 2014–15, with a record of 9% and 2% increase in terms of rupees and dollar respectively. During 2015–16 the total export surpassed the target both with respect to volume and price.

2.4 IMPORT OF SPICES

Foreign Trade Policy of India under recent scenario does not impose any substantial boundary on spice import except for 'seed quality' of seed spices, fresh ginger, and poppy seed. Gradually the duty for import has also been lowered. There is no tariff for imports from Sri Lanka. There exists Free Toll Trade Pact with Sri Lanka on spices. Under the Advance Authorization System tax allow free imports for valorization and then re-export of the product after value addition. The work on spice research is primarily looked after by The Indian Institute of Spices Research (IISR) Kozikhkode; (Calicut) is a constituent unit of ICAR. It is a vital wing dedicated to spice research since 1976. IISR works for mankind and owes all its achievements to the farmers and industries, involved in one way or the other with spices production, handling, post-harvest procedural and marketing. The research programs are carried out under various projects, which are time-bound, and with specific objectives like collection, conservation, screening, assessment and categorization of germplasm, breeding varieties having high yield and quality employing all traditional and biotechnological methods, production of nucleus seeds and planting materials of superior cultivars, identification, characterization and standardization of techniques for the detection the diseases and losses caused by infestation of insects, fungi, bacteria, viruses and pathogens specific to spice crops, standardizing production and breeding methods for bulk production and circulation of improved genotypes, development of agro techniques for augmenting production and for optimum exploitation of integrated management of pest as well as disease. Post-harvest know-how, socio-economic part of cultivation, distribution, and sale also some extension activities and investigation on nutrition and medicinal features of spices are also researched upon by IISR.

2.5 CROP-WISE IMPROVEMENT IN SPICES

2.5.1 CHILI

Chili is a versatile and significant spice cultivated all over India. Chili is a resourceful crop and has varied usages. It is chiefly utilized as spice, condiment, culinary complement, medicines, vegetable, and ornamental plant. It is one of the commercial crops and cultivated throughout the country. Chili is most commonly cultivated in, Orissa, Andhra Pradesh, Maharashtra, West Bengal, Karnataka, Rajasthan, and Tamil Nadu. Andhra Pradesh and about 46% of the Chili are produced in this area.

2.5.1.1 PRODUCTION IN INDIA

Various research works to study genetic variability and genetic diversity in chili have been worked upon by various universities in India for the enhancement in yield, quality, nutraceuticals and pharmaceutical properties in Chili. Kumar in 1992 carried out heterosis study in F1 and F2 population in Chili at BAC, Sabour and found that variety Sabour Angar, Pusa Jwala and 85–2 were the good combiners for yield as well as contributing characters. Sabour Angar and Sabour Anal were the premiere varieties released from Sabour. In the past studies were done on combining ability, heterosis, development of CMS based hybrids and evaluation, path analysis, association of characters and diversity study for genetic enhancement in yield in addition to quality as reported by Gaddagimath (1992); Shirshat (1994); Giritammannavar (1995); Hiremath (1997); Lankesh Kumar and Sridevi (2005), and Srilakshmi (2006). Savita (2013) conducted a study to evaluate the extent of heterosis as well as to know good general combiners for yield in addition to quality. Later molecular l approaches were used to attain improvement in Chili. Improved varieties of Chili include N.P. 46, Pant Jwala, G-3, CA-960, Pant C-t X-235, AKC-79-18, Parbhani Tejas K 1, K 2, CO_2, CO_4 (vegetable type), PKM 1, PMK 1, PLR1, and KKM (Ch) 1.

2.5.2 GARLIC

Garlic, a member of the Amaryllidaceae family is a bulb crop and is invariably used as spices due to its typical garlicky flavor. The intense aroma and strong taste in garlic are due to the sulfur-containing compounds in garlic. These compounds form the active principles in garlic which are responsible for their characteristic odor and pungency (Robinowitch and Currah, 2002). Garlic is propagated exclusively by the vegetative method. Improvement in garlic is possible primarily through clonal selection of existing types in various regions. Conventional research on garlic-breeding has been restricted to screening and characterization and evaluation for yield as well as morphological traits (Figliuolo et al., 2001; Khar et al., 2005a, 2005b). A positive correlation between plant height, bulb weight, bulb diameter and mean clove weight has been reported by Zhila (1981). Baghalian et al., (2005) reported significant positive correlation between mean weight of cloves and mean weight of bulbs and negative correlation between mean weight cloves and number of cloves. In their work on garlic Baghalian et al., (2006) reported that these traits could be helpful in improvement program in garlic for yield and quality. Significant amount of variability with respect to morphological character of garlic has been observed by Shashidhar and Dharmatti (2005) as well as Khar et al., (2006). Bulb weight, diameter, yield, number of cloves per bulb, maturity, plant height, green leaves number and bulbing period are considered as main traits responsible for to genetic diversity in garlic. Morphological characterization of garlic intended diversity study was done by Baghalian et al., in 2005 and Panthee et al. in 2006. The physico-chemical characters including allicin content and molecular characteristics was reported by Baghalian et al. (2006). Resende et al. (2003) reported quality traits and chemotaxonomic categorization was specified by Storsberg et al., (2003). Allicin is a key chemical component of garlic as well is invariably exploited by pharmaceutical industries for making several formulations of medicinal importance. Huchette et al. (2005) reported that allicin content in garlic is varied by several factors, like light, temperature, genotype, location and also S fertilization. Similarly, Yang et al. (2005) similarly reported that relative humidity, soil type as well as the date of harvest affects the allicin content in garlic. Contrary to these reports, Baghalian (2005) found no significant interrelationship between the environment and allicin content of genotypes of garlic. The true seed production in garlic (*A. sativum* and *A. longicuspis*) is not recent phenomenon. It was with the discovery of fertile clones by

Etoh (1986) first of all discovered fertile clones of garlic. He utilized the same to bring flowering and seeds in garlic. With the knowledge of flowering garlic, Jenderek and Hannan (2004) first time produced S1 bulbs some fertile clones of flowering garlic. This served to be precious stuff for research towards garlic genetics (Jenderek, 2004). Jenderek and Zewdie (2005) in their study on genetic variability of garlic concluded that garlic over the years due to clonal propagation, have become greatly heterozygous in nature. Attempt was also made by Koul et al. (1979) to study genetics as well as breeding systems for garlic improvement. Koul et al. (1979) studied prospects for garlic improvement for genetic studies and complete review about origin, flowering in addition to seed production was carried out by Simon and Jenderek in 2004. Reports on of flowering in garlic, helped further in the study of linkage as documented by Ipek et al. (2005) and Zewdie et al. (2005). This helps tagging of significant genes in future. Improvement of garlic was taken in BAC, Sabour under varietal trials of AICRP. Recently project on varietal improvement in garlic for yield along with storability is in headway. Over 142 genotypes are under study with the objectives to find out variety having high yield and longer storability. The biochemical basis of storability is also being studied at BAC, Sabour. Morphological, biochemical plus molecular characterization of genotypes was also done. MTA was signed with DOGR, Rajgurunagar, Pune. Few genotypes are in advance stage of study and expected to be released in very recent future. Garlic is primarily a vegetatively propagated crop and therefore improvement in garlic is limited to either clonal selection or by mutation breeding among conventional methods and by somaclonal variations created through biotechnology. In India, most varieties are bred by conventional methods. Most of the varieties of garlic have been released under the umbrella of National Horticultural Research and Development Foundation (NHRDF). Various agricultural universities like Gujarat Agricultural University, Punjab Agricultural University, MPKV, Rahuri, etc., also have greatly have developed superior varieties of garlic which are normally short day types, suitable for tropical and sub-tropical climates. Some temperate varieties released at the national level are Agrifound Parvati, VLG-1 (VPKAS, Almora), SKUAG 1 (SKUAST, Srinagar), DARL 52 and Solan Local (YSPUHF, Solan). Besides, varieties, G41, G1, G50, G282, are the classical soft neck (short-day) released by NHRDF, Pune. Jamnagar Local, Ooty Local, Jeur Local are the varieties selected by farmers, showing high performance over the years.

2.5.2.1 MOLECULAR MARKERS IN GARLIC

Garlic is being cultivated since ages, but the taxonomic origins of this domestication process are not well-known. According to modern taxonomy, the world's garlic germplasm has been subdivided into five distinct groups: Sativum, Ophioscordon, Longicuspis, Subtropical, and Pekinense as per Fritsch and Friesen, 2002. The Longicuspis group from central Asia is documented to be most primitive, and all other groups were known to be derived from this group (MaaB and Klaas, 1995; Etoh and Simon, 2002; Fritsch and Friesen, 2002). Central Asia was thought as primary center for evolution and garlic diversity (Fritsch and Friesen, 2002). This proposition is very strongly supported by modern studies on native garlic types in Tien-Shan Mountains (Etoh, 1986; Kamenetsky et al., 2003). Simon and Jenderek, 2003 have reported a broad array of morphological diversity in garlic for flowering capacity, leaf character, bulb characters, plant development, influence of temperature and photoperiod on bulb formation, bulbil and flowering pattern. According to MaaB and Klaas (1995), the subtropical clones were undoubtedly different from all other types and Pekinense subgroup was rather alike flowering type of type. Researchers, across the world have reported RAPD techniques for characterization of garlic germplasm. Bradley et al., in 1996 have described the use of RAPD techniques for Australian garlic germplasm characterization, Hsu et al. (2006) of Taiwanese garlic germplasm, Buso et al. (2008) of Brazilian garlic germplasm, Xu et al. (2005) of Chinese garlic germplasm, Paredes et al. (2008) of Chilean garlic germplasm, Rosales et al. (2007) of Guatemalan garlic germplasm and Khar et al. (2008) for Indian garlic. Ipek et al. (2003, 2005); Lampasona et al. (2003); Volk et al. (2004); have reported AFLP (Amplified Fragment Length Polymorphism) technique to characterize garlic. A comparative study of different markers like AFLPs, RAPD, and isozymes for diversity estimation of garlic and for polymorphism study was done by Ipek et al. (2003). He found good correlation between the markers and established that there existed genetic diversity amongst closely-related clones, which could not be distinguish with RAPD markers and isozymes but was identified by AFLPs. Lampasona et al., 2003 have maintained that there exists correlation between geographical locations and the diversity. Duplicity in commercial collections maintained at various centers has been reported by Volk et al. (2004). Therefore speedy characterization of garlic accession is essential for keeping away from duplicity. Therefore a number of locus-specific polymerase chain reaction (PCR) based DNA markers were developed by Ipek et al. (2008) for this purpose. Besides, markers have been exploited to explain the taxonomic status of some local cultivars of garlics as observed by Ipek et al. (2008) and Figliuolo and Stefano, (2007). Al Zahim et al., (1997, 1999) reported genetic purity of micro propagated crops. A character like pollen fertility in garlic was reported by Etoh et al. (2001) while Nabulski et al. (2001) reported marker associated to white rot.

2.5.3 BLACK PEPPER

Piper nigrum belongs to the family Piperaceae. Most of the species of the genus Piper are woody climbers which are perennial in nature, or maybe herbs or shrubs and are pan tropically distributed. Work on Chili improvement is mostly confined to Southern India. Well-organized and well-thought-out research work in the last three decades led to the development of superior breed of black pepper. Several breeding techniques like hybridization, open pollination or clonal selection from the popular cultivars have been adopted for the purpose. PRS (KAU) under AICRP has released seven varieties of spices viz., Panniyur-1, 2, 3, 4, 5, 6 and 7 yielding between 12.7 and 25.7 quintal/ha. IISR have released four varieties viz., Subhakara, Panchami, Sreekara, and Pournami with a yield ranging from 23 to 28 quintal/ha. Palode, a Regional station of CPCRI, released a variety PLD-2 with a yield capacity of 24 quintal/ha as reported by Ravindran et al., 2000. Black pepper vines should be kept under shade (7% incident light) to keep it green and vigorous whereas when kept in sunlight they turned pale and turned sickly and developed necrotic lesions during the summer months (Vijayakumar and Mammen, 1990). Black pepper flourishes on soils, clayey to light sandy clays which are rich and friable, well-drained and must have the capacity to hold enough water. Sadanandan, 1993 observed that soils having neutral pH and high organic matter were conducive high yield. Multiplication technique using bamboo method to fulfill the demand of planting material was developed (Ravindran et al., 2000). Spray of planofix 40 ppm decreased the shedding of spike up to 20% (Ravindran et al., 2000). Salvi et al., 2000 have opined that use of plant growth regulator like NAA and 2,4-D of 10 ppm strength might be beneficial in promoting spike initiation and berry setting. Application of biofertilizers and vermicompost improved growth, biomass, absorption of nutrition, yield as well as quality of black pepper (Kandiannan et al., 1998; Kandasamy et al., 1998; Kannan and Thangaselvabai, 2006). Thangaselvabai et al., 2006 have concluded that multi-storeyed cropping system of cinnamon, pepper, and pineapple as constituent crops under forest-agro ecology was the most remunerative among all the cropping systems.

2.5.4 GINGER

Ginger (*Zingiber officinale*) belongs to the Zingiberaceae family. It is a huge family consisting of 47 genera and 1400 species. Ginger is a monocot underground rhizome. It is an important spice and has medicinal importance too. Muralidharan and Velayudhan (1983) and also Sasikumar et al. (1995) have informed about the diversity in wild Western Ghats, India. Nonetheless, utmost variation among ginger under cultivation is prevalent in North East India. The huge diversity may be accounted to the widespread from its center of origin in Southeast Asia followed by local adaptation due to mutations or other factors as suggested by Ravindran et al. (1994). Besides yield quality is an important determinant for the ginger to be desirable. The important quality traits include essential oil oleoresin, and fiber along with volatile as well as non-volatile constituents according to Jaleel and Sasikumar, 2011. Wohlmuth et al. (2006), studied oil content of fresh rhizomes of 17 Australian ginger clones and established that the cultivar Jamaican contained the oil which had a considerably different composition. Datta et al. (2003) in his study on quality estimation of ginger found that the cultivars 'Suravi,' Uttar Pradesh, Suprabha, 'Tura,' and Gorubathan suitable for producing dry ginger, cultivars Suravi, Suprabha, Taffingiva and Jughijan for oleoresins content and Suravi, Suprabha, Mazulay, Jughijan and Tura for essential oil extraction. Goyal and Korla (2001) observed a negative correlation existed between essential oil and oleoresins with the dry matter. Tiwari (2003) in his study on ginger production under the mid-hill conditions of Himachal Pradesh in India observed that genotypes, BDJR 1054, SG 687 SG 61 and SG 62, were fit for ginger oil as well as oleoresin removal, while SG 692 was as a rule apt for dry ginger. Furthermore, SG 646 was accepted as a replacement variety for Himgiri for fresh ginger production is concerned.

2.5.5 TURMERIC

Turmeric (*Curcuma longa*) belongs to the family Zingiberaceae. This is grown for its underground rhizome. It is perennial in nature and is native to India and Indonesia. It is popularly grown all over the world and is called 'Golden Spice of life.' It is an essential spices and used in a number of dishes world over. It adds a rich golden color to the curry. It is a commercial crop in India. Indian. Over 100 species of Curcuma are there out of which 41 are known to be there in India. However, only 6% of the total area under spices and condiments is occupied by turmeric. India accounts for 78% world's total turmeric production and holds the topmost position in the production and export of turmeric in the world. Moreover, turmeric in the second major foreign exchange earner among the Indian spices. The world's greatest consumption of turmeric is done by India, which amounts to nearly 80% of total turmeric consumed throughout the world. The main turmeric producing states in India are Maharashtra, Karnataka, Kerala; Andhra Pradesh occupies the maximum area under turmeric cultivation and accounts for very high production. Sustainable and dedicated research and development activities have boosted the production potential of the country. The prospects of Turmeric production in India along with the state wise production of turmeric in India has been analyzed by several workers. Turmeric exhibits very high heritability with significant genetic advance for rhizome yield, number of leaves, number of primary fingers, the yield of secondary fingers, the height of the pseudostem and crop duration. Singh et al., (2003) recommended promising genotypes must be identified through selection based on the number and weight of mother, primary and secondary rhizomes. According to Reddy (1987) and Panja et al. (2002) positive correlation of number of leaves, number of primary fingers and crop duration was observed with yield of rhizomes. Singh (1995) studied the production of turmeric in Bihar and observed that area, production, productivity of turmeric in the Bihar is likely to increase. The most important yield contributing character in turmeric is the number of rhizomes and their size (Chadha, 2001). Studies on varietal performance of turmeric was made by Chaudhary and Singh (2006) and it was found that More number of rhizomes per plant were produced by Krishna (11.48) followed by Rajendra Sonia (10.22). As an outcome of the large number of quality research performed by some of the most qualified scientists and research personnel, IISR Kozhikode – Kerala has given some proven and suitable technologies for quality turmeric production. For raising the seedling it recommends selection of most healthy turmeric rhizomes for seed purpose and that selected rhizomes must be treated with mancozeb (0.3%), and quinolphos (0.075%) for 30 minutes and stored in well-ventilated place. Pramila et al. carried out an experiment in the lab to study the influence of growth regulators, strength of growth media used and photoperiod on induction of micro rhizome in turmeric. BAP 1 mg l^{-1} with NAA 0.2 mg l^{-1} (31–33 days) gave the best response for early induction of microrhizomes. High heritability with appreciable genetic advance was reported for rhizome yield, number of leaves, height of pseudostem number of primary fingers and yield of secondary fingers. Singh et al. (2003) suggested superior lines may be obtained through selection based on the number and weight of mother, primary and secondary rhizomes.

2.5.6 SEED SPICES

Seed spices are the largest group among spices consisting of coriander, fenugreek, cumin, fennel, celery, ajowan, nigella, dill, anise, etc. India stands first in production, consumption, and export of seed spices and occupies key position in our economy because of its large household consumption and ever increasing demand for export. Seed spices mostly are annual and can easily be grown in rotation with food crops and also as intercrops or mixed crops under rained or irrigated conditions. Rajasthan and Gujarat in India are major states which produce seed spices and no wonder states of these two states and parts of Madhya Pradesh are said to be the 'bowl of seed spices' and accounts for 80% of the annual production of the country (Balraj and Solanki, 2015). Crop breeding, production and plant protection technologies and recommendations were the major thematic areas which were researched upon and as an outcome it had a profound effect in changing the national seed spice picture. Multidisciplinary research activities are performed at the Center. The major research outputs include weed management by the application of herbicides like pendimethalin and oxadiargyl, which had shown significant impact in reducing crop-weed competition and cost of cultivation (Meena et al., 2013; Sundaria et al., 2014). Irrigation management using sprinklers particularly for cumin in parts of western Rajasthan and drip methods in fennel had played a noteworthy job in boosting yield and quality (Ravindran et al., 2006; Sundaria et al., 2014). Disease as well as pest management by integrated approach using both bioagents and pesticides were effective in the management of downy mildew wilt, blight, aphids, powdery mildew, in major seed spices (Israel and Lodha, 2004; Khare et al., 2014a, b; Lodha and Mawar, 2014). Rathore et al. (2013) reported some unique features and industrially important metabolites of these crops. Promising cultivars of many seed spices besides some outstanding varieties such as extra early line of ajwain, AA-93 of about 110 days have been found to be satisfactory for conserved moisture farming (Meena et al., 2014) Meena et al. (2013) advocated some biopesticides for control of aphid pest in many seed spices.

Postharvest management and latest packaging systems to preserve quality and flavor of the crop has been recommended by Lal et al. (2013). Varieties rich in essential oil content and showing resistance to stem gall should be encouraged for boosting production and for meeting the criteria for the export purposes (Meena et al., 2013). The state of Bihar is naturally endowed with large stretch of fertile land and congenial subtropical climate apt for raising seed-spices like fennel, coriander, fenugreek, nigella, and onum. Fennel, coriander, and onum are commercially grown in Champaran, Muzzaffarpur, Samastipur, Begusarai, Munger and Bhagalpur but nigella is restricted to Rohtas districts (Kumar, 2005). However, lack of high yielding varieties, standard agro-technology and plant protection measures, improved processing technology, organized marketing system and organized spice extension programs for farmers are major limitations that impede spice production in Bihar. Spice improvement and production programs in the two agricultural universities in Bihar have been taken up under various schemes like Macromode, NHM and very recently under MIDH. The Spice improvements project in the agricultural university in Bihar has developed coriander variety, like Rajendra Swati, Rajendra Dhania-1, fenugreek variety Rajendra Kanti, Rajendra Abha and fennel variety Rajendra Saurabh, nigella variety, Rajendra Shyama and ajwain variety, Rajendra Mani and has also standardized dates of sowing, seed rate, spacing, fertilizer dose and plant protection measures. The seed production of coriander, fenugreek, fennel, dill has been in practice at the agricultural university, Sabour since along back. Of late nigella and ajwain has been added to the list of spice. In spite of many such remarkable achievements, much more has to be done to meet the constraints of the spice growers of Bihar (Kumar, 2005).

2.6 OPPORTUNITIES OF SPICE RESEARCH

Time has seen the continually rising demand for spices and its value-added products throughout the world which implicates a huge possibility for crop improvement like the development of genotypes resistant to biotic and abiotic stresses and also responsive to low input management through conventional breeding and biotechnological approaches. There is the immense scope of large-scale multiplication of quality planting materials of released varieties with strict quality regulation and certification. Also, there is room for the establishment of Advanced Centers for Research on Biotechnology, Phytophthora Research, Biocontrol and Biosystematics, high-value compounds, etc. The production and productivity of spices can be sustained by popularizing the soil conservation/water management technologies and encouraging organic farming and IPM approaches at the community level. There is also potential for establishment of cooperative movement to regulate production and marketing to increase competitiveness of Indian products in the international market. Vast employment opportunities for trained manpower in spice industry and spice farming cannot be overlooked. There is substantial scope for value addition and diversification in spices. The envisaged increase in share of value-added products in the export basket of spices needs strengthening of processing facilities. Spices are increasingly being noticed for their pharmacological activities and therefore their potential as functional food has magnified scope. The scientific validation of the medicinal properties of spices using state of the art technology like drug modeling, molecular biology and nanotechnology holds great promise and will provide greater avenues for medicinal uses of spices.

2.7 CHALLENGES IN SPICE RESEARCH

The major challenges involved in spices production are, Emergence and epidemics of pests and diseases, vagaries of monsoon resulting in drought, emergence of other chief spice producing countries which compete with India in the International market, shifting of interests of growers to more profitable/less risky crops, adulteration of spices, cyclic market fluctuations at international and national level, lack of awareness about pesticide residues and mycotoxin contaminants in the products and lack of MRL and ADI standards in some of the pesticides used in spices. The research programs for the future has to be geared to meet the challenges that can arise from these risk factors.

2.8 FUTURE STRATEGIES FOR RESEARCH

The work on genetic enhancement will play a major role in the near future. Collection, conservation and characterization (molecular markers, barcoding) of germplasms and establishment of a global gene bank of spices genetic resources should be taken up on a priority basis in the future. The breeder will have to own up the responsibility of locating the source for resistance to all types of stresses, be it biotic or abiotic with the help of a traditional and modern biotechnological tool for and evolving new varieties which would ultimately increase production. Work of developing molecular bar codes for all germplasm accessions, pests and pathogens their natural enemies and molecular profiles of all released varieties and molecular farming to identify desired genes must be prioritized. Convergent improvement of black pepper for multiple resistance genes (Pollu beetle, Phytophthora, and nematodes) and developing crop ideotypes suited for uniform ripening (black pepper), synchronous flowering (cardamom) should be propelled. Breeding of varieties of ginger for a specific purpose such as ginger candy, ginger shreds, ginger wine and also curcuminoids rich turmeric has to be taken up. Accessibility of an excellent superior grade of planting materials is important for the success of spice cultivation. Hence the development of novel techniques for accelerating yield of quality planting material and certification systems for planting materials has to taken up on priority basis in future. Horticultural interventions (high tech horticulture) to maximize productivity of spice based cropping systems and expansion of precision farming for spices to boost productivity especially for ginger and turmeric must be worked upon as the matter of foremost concern. Vertical farming, hydroponics and aeroponic cultivation for ginger and turmeric, protected cultivation, alternate cropping systems with spices/farming system approach, high-density planting (HDP) in black pepper, cardamom and tree spices including dwarf plant types must be exploited to sustain high production of spices in future. Development of varieties suitable for organic farming (through organic plant breeding) and extreme climate situations must also be taken up. Development of agro-ecosystem based insect pest control strategies using novel selfish gene drive systems; gene silencing must be taken up in the times to come. Study of crop weather soil relation based simulation models for extreme climatic conditions and impact of climate change on productivity, emergence of new diseases and pests and adaptation and mitigation studies will be the need of the future. Futuristic research for dryland/rain fed cropping system for spice crops; study on increasing the water up-take ability of plants, drought management studies of spice crops, designing smart fertilizer and pesticide delivery systems shall have to be taken up in the days to come. Molecular profiling, marker aided selection for desired traits, allele mining and identifying genes controlling superior quality traits, pest and disease resistance, genome sequencing of *P. capsici* and comparative transcriptomics/genomics for identification of species-specific markers will play an important role in the future spice improvement program. Application of nanotechnology-enabled devices for real-time monitoring of soil conditions and crop growth, development of super sensors for detecting a specific molecule for detection of a pathogen in the rhizosphere, rhizoplane and in plant parts and biosensors for detecting spoilage of post-harvest spice products by sensing metabolic products of spoilage bacteria or direct detection of spoilage bacteria should be explored. The development and popularization of cost-effective agricultural practices (INM/IPM) for increasing productivity and carbon sequestration potential in spice-based cropping systems should be prioritized. Value addition through microencapsulation, extrusion and other techniques, bioprospecting using bioinformatics tools, chemical modification, synthesis and appropriate packaging of spice(s) derived phytochemicals at defined dosages in conditions supporting optimum bioavailability, minimum toxicity, encapsulation of spice extracts, nutraceutical, drug, agricultural and other novel applications, with improved physicochemical properties should be taken up in future. Developing electronic devices for monitoring quality and adulteration of spices should be taken up on a priority basis. Development of implements/tools for harvesting and processing for value-added spice products, developing precision farming models for the management of nutrient and water to get optimum production from unit of water and nutrient used, development of solar dryers and solar cookers for post-harvest processing to enhance and economize the production. Policy issues to be exerted upon in the future are commercialization of techniques/technologies, genetic fingerprinting of germplasm and its registration, registration of released varieties, patenting technologies related to spices, documentation of ITKs and conversion of agriculture into a business venture with maximization of profit. Giving a prime attention to market promotion for these low-volume high-value crop species has also been essential. There is a serious gap in the demonstration and popularization of available technologies of these invaluable crops of economic significance. Transfer of technology (TOT) systems would have to be strengthened by way of, establishing 24 × 7 knowledge centers on spices, large scale demonstration of proven technologies through KVK's as FLDs, establishing technology incubation center, participatory seed production of major spices and use of next-generation ICT for knowledge updating of farmers to develop the technopreneurs.

KEYWORDS

- **high-density planting**
- **irrigation management**
- **polymerase chain reaction**
- **transfer of technology**

REFERENCES

Al Zahim, M., Newbury, H. J., & Ford, L. B. V., (1997). Classification of genetic variation in garlic (*Allium sativum* L.) revealed by RAPD. *Hort. Sci., 32*, 1102–1104.

Al Zahim, M., Newbury, H. J., & Ford, L. B. V., (1999). Detection of somaclonal variation in garlic (*Allium sativum* L.) using RAPD and cytological analysis. *Plant Cell Rep., 18*, 473–477.

Baghalian, K., Sanei, M. R., Naghavi, M. R., Khalighi, A., & Badi, H. A. N., (2005). Post-culture evaluation of morphologicaldivergence in Iranian garlic ecotypes. *Acta Hort., 688*, 123–128.

Balraman, N., Nair, M., & Madhusoodhanakurup, P., (1989). *Sugandhakeralam, Cochin: Spices Board* (p. 41).

Bradley, K. F., Rieger, M. A., & Collins, G. G., (1996). Classification of Australian garlic cultivars by DNA fingerprinting. *Aust. J. Exptl. Agric., 36*, 613–618.

Buso GSC, Paiva, M. R., Torres, A. C., Resende, F. V., Ferreira, M. A., Buso, J. A., & Dusi, A. N., (2008). Genetic diversity studies of Brazilian garlic cultivars and quality control of garlic clover production. *Genet. and Mole. Res., 7*, 534–541.

Chadha, K. L., (2001). *Turmeric, Handbook of Horticulture*. ICAR, New Delhi.

Chaudhary, A. S., Sachanl, S. K., Singh, R. L., (2006). *Studies on Varietal Performance of Turmeric (Curcuma longa L.),* Krishi Gyan Kendra of SVBPUAT, Krishi Utpadan Mandi Samiti Parisar, Pilibhit Road, Izatnagar, Bareilly-243122 (U.P.), India, Department of Entomology, Sardar Vallabh Bhai Patel University of Agriculture and Technology, Meerut 250 110 (U.P.), India.

Datta, S. K., Singh, G., & Chakrabarti, M., (2003). *Ginger and its Products: Management of Marketing and Exports with Special Reference to the Eastern Himalayan Region.* National Consultative Meeting for improvement in productivity and utilization of ginger. Aizwal.

Etoh, T., & Simon, P. W., (2002). Diversity, fertility and seed production of garlic. In: *Allium Crop Science*. CABI Publishing, New York.

Etoh, T., (1986). Fertility of the garlic clones collected in Soviet Central Asia. *J. Jpn. Soc. Hort. Sci. 55*, 312–319.

Etoh, T., Watanabe, H., & Iwai, S., (2001). RAPD variation of garlic clones in the center of origin and the westernmost area of distribution. *Mem. Fac. Agric. Kagoshima University, 37*, 21–27.

Figliuolo, G., & Stefano, D., (2007). Is single bulb producing garlic *Allium sativum* or *Allium ampeloprasum*? *Scientia Hort., 114*, 243–249.

Figliuolo, G., Candido, V., Logozzo, G., Miccolis, V., & Zeuli, P. L. S., (2001). Genetic evaluation of cultivated garlic germplasm (*Allium sativum* L. and *A. ampeloprasum* L.). *Euphytica, 121*, 325–334.

Fritsch, R. M., & Friesen, N., (2002). *Evolution, Domestication and Taxonomy: Allium Crop Science: Recent Advances* (pp. 5–30). CABI Publishing, New York.

Gaddagimath, N. B., (1992). Studies related to genetics of economic and quality traits and exploitation of heterosis in chili (*Capsicum annuum* L.). *PhD Thesis*. University of Agricultural Sciences, Dharwad.

Giritammannava, R., (1995). Variability studies of Byadgi chili for different quantitative characters. PhD Thesis. University of Agricultural Sciences, Dharwad, India.

Goyal, R. K., & Korla, B. N., (2001). Quality variation in ginger. *Veg. Sci., 28*, 45–47.

Hiremath, N. V., (1997). Genetics of fruit yield, yield components and reaction of major biotic stresses in chili (*Capsicum annuum* L.). PhD Thesis. University of Agricultural Sciences, Dharwad.

Hsu, H. C., Hwu, K., Deng, T., & Tsao, S., (2006). Study on genetic relationship among Taiwan garlic clones by RAPD markers. *J. Taiwan Soc. Hortl. Sci., 52*, 27–36.

Huchette, O., Kahane, R., Auger, J., Arnault, I., & Bellamy, C., (2005). Influence of environmental and genetic 113 factors on the alliin content of garlic bulbs. *Acta Hort., 688*, 93–99.

Ipek, M., Ipek, A., & Simon, P. W., (2003). Comparison of AFLPs, RAPD markers, and isozymes for diversity assessment of garlic and detection of putative duplicates in germplasm collections. *Hort. Sci., 128*, 246–252.

Ipek, M., Ipek, A., & Simon, P. W., (2008). Rapid characterization of garlic clones with locus-specific DNA markers. *Sci. Hort., 32*, 357–362.

Isarel, S., & Lodha, S., (2004). Factors influencing population dynamics of *Fusarium oxysporum* f. sp. Cumini in the presence and absence of cumin crop in arid soils. *Phytopathologia Mediterranea, 43*, 3–13.

Jaleel, K., & Sasikumar, B., (2010). Genetic diversity analysis of ginger (*Z. officinale* R) germplasm based on RAPD and IISR markers. *Sci. Hort., 125*, 73–76.

Jenderek, M. M., & Hannan, R. M., (2004). Variation in reproductive characteristics and seed production in the USDA garlic germplasm collection. *Hort Sci., 39*, 485–488.

Jenderek, M. M., & Zewdie, Y., (2005). Within and between family variability for important bulb and plant traits among sexually derived progenies of garlic. *Hort Sci., 40*, 1234–1236.

Jenderek, M. M., (2004). Development of S1 families in garlic. *Acta Hort., 637*, 203–206.

Kamentsky, R., London, S. I., Bizerman, M., Khassanov, F., Kik, C., & Rabinowitch, H. D., (2003). Garlic (*Allium sativum* L.) and its wild relative from Central Asia: Evolution for fertility potential. *Proceeding of XXVIth International Horticultural Congress* (Vol. 673, pp. 83–91). Toronto, Canada. Acta Hort.

Kandiannan, K., et al., (1998). In: *First International Agronomy Congress on Agronomy.* Environment and Food Security.

Kannan, J., & Thangaselvabai, T., (2006). *J. Scott Research Forum, 3*, 16–18.

Kanthaswamy, V., et al., (1998). *South India Hort., 44*, 85–89.

Khar, A., Asha, D. A., Mahajan, V., & Lawande, K. E., (2005b). Genotype X environment interactions and stability analysis in elite lines of garlic (*Allium sativum* L.). *J. Spices Arom. Crops, 14*, 21–27.

Khar, A., Asha, D. A., Mahajan, V., & Lawande, K. E., (2006). Genetic divergence analysis in elite lines of garlic (*Allium sativum* L.). *J. Mah. Agri. Univ., 31*, 52–55.

Khar, A., Lawande, K. E., & Asha, D. A., (2008). Analysis of genetic diversity among Indian garlic (*Allium sativum* L.) cultivars and breeding lines using RAPD markers. *Ind. J. of Genet. Pl. Breed, 68*, 52–57.

Khar, A., Mahajan, V., Devi, A. A., & Lawande, K. E., (2005a). Genetical studies in elite lines of garlic (*Allium sativum* L.). *J. Mah. Agri. Univ., 30*, 277–280.

Khare, M. N., Tiwari, S. P., & Sharma, Y. K., (2014a). Disease problems in fennel (*Foeniculum vulgare* Mill) and fenugreek (*Trigonella foenumgraceum* L.) cultivation and their management for production of quality pathogen free seeds. *International Journal of Seed Spices, 4*(2). 11–17.

Khare, M. N., Tiwari, S. P., & Sharma, Y. K., (2014b). Disease problems in the cultivation of I. Cumin (*Cuminum cyminum* L.) II. Caraway (*Carum carvi* L.) and their management leading to the production of high quality pathogen free seed. *International Journal of Seed Spices, 4*(1), 1–8.

Koul, A. K., Gohil, R. N., & Langer, A., (1979). Prospects of breeding improved garlic in the light of its genetic and breeding systems. *Euphytica, 28*, 457–464.

Kumar, A., (1992). Studies on Heterosis and combining ability in F2 population in chili. *PhD Thesis*. BAC, Sabour.

Lal, G., Singh, B., Mehta, R. S., Meena, S. S., Mishra, B. K., Singh, R., Aishwath, O. P., & Sharma, Y., (2013). Scientific cultivation brings handsome return. *Indian Horticulture, 58*(6) 13–16.

Lampasona, S. G., Martinez, L., & Burba, J. L., (2003). Genetic diversity among selected Argentinean garlic clones (*Alliumsativum* L.) using AFLP. *Euphytica, 132*, 115–119.

Lankesh, R., Udaykumar, K., Srdevi, O., Laxman, C., Deepa, M., Madalageri, N. P., & Salimath, P. M., (2005). Heterosis Studies in Chili (*Capsicum annuum* L.). lant gene trait. *Peer Reviewed Open Access Journal ISS*, 1925–2013.

Lodha, S., & Mawar, R., (2014). Cumin wilt management- a review. *Journal of Spices and Aromatic Crops, 23*(2), 145–155.

Maa, B. H. I., & Klaas, M., (1995). Interspecific differentiation of garlic (*Allium sativum* L.) by isozymes and RAPD markers. *Theor. Appl. Genet., 9*, 89–97.

Mahindru, S. N., (1982). *Spices in Indian Life* (p. 26). Delhi: S. Sultanchand and sons.

Mann, J. S., & Tamil, S. M. In: *Special reference to the Eastern Himalayan Region* (pp. 82–103).

Meena, R. D., Kant, K., Sharma, Y. K., & Meena, N. K., (2013). Healthy seed spices brings more returns. *Indian Horticulture, 58*(6), 36–40.

Meena, S. S., Kakani, R. K., Singh, B., Meena, R. S., Mehta, R. S., Kant, K., Sharma, Y. K., & Meena, R. D., (2014). Ajmer Ajwain 93: An early maturing variety of ajwain developed at NRCSS for all ajwain growing areas. *International Journal of Seed Spices, 4*(2), 91–93.

Meena, S. S., Singh, B., Lal, G., Mehta, R. S., Singh, R., Aishwath, O. P., & Meena, N. K., (2013). Scientific production technology of minor seed spices brings cheers to farmers. *Indian Horticulture, 58*(6), 42–45. Kumar, A., Google book.

Muralidharan, V. K., & Velayudhan, K. C., (1983). A note on the occurrence of wild ginger in Western Ghats. *South Indian Hort., 31*, 259–260.

Nabulski, I., Safadi, A. I., Mit, B., Ali, N., & Arabi, M. I. E., (2001). Evaluation of some garlic (*Allium sativum* L.) mutants resistant to white rot disease by RAPD analysis. *Ann. Appl. Biol., 138*, 197–202.

Panja, B. N., De, D. K., Basak, S., & Chattopadhyay, S. B., (2002). Correlation and path analysis in turmeric (*Curcuma longa* L.). *J. Spices and Aromatic Crops, 11*, 70–73.

Panthee, D. R., Regmi, H. N., Subedi, P. P., Bhattarai, S., & Dhakal, J., (2006). Diversity analysis of garlic (*Allium sativum* L.) germplasm available in Nepal based on morphological characters. *Genet. Res. Crop Evol., 53*, 205–212.

Paredes, C. M., Becerra, V. V., & Gonzalez, A. M. I., (2008). Low genetic diversity among garlic (*Allium sativum* L.) accessions detected using random amplified polymorphic DNA (RAPD). *Chilean J. Agric. Res., 68*, 3–12.

Parry, J. W., (1969). *Hand Book of Spices* (Vol. 1, pp. 11–12). New York: Chemical Publishing Company Inc.

Pella, M. O., O'hearn, G. T., & Gale, C. W., (1966). Referents to scientific literacy. *Journal of Research in Science Teaching, 4*(3), 199–208.

Purseglove, J. W., (1981). *Spices* (Vol. 1, p. 1). New York: Longman Inc.

Rabinowitch, H. D., & Currah, L., (2002). *Allium Crop Science: Recent Advances* (p. 515). CABI Publishing House, UK.

Rathore, S. S., Saxena, S. N., & Singh, B., (2013). Potential health benefits of major seed spices. *International Journal of Seed Spices, 3*(2), 1–12.

Ravindran, P. N., Babu, N. K., Shiva, K. N., & Kallupurackal, J. A., (2006). *Advances in Spices Research: History of Achievements of Spices Research in India Since Independence* (p. 994). Agrobios (India), Jodhpur, Rajasthan.

Ravindran, P. N., et al., (2000). *Indian J. Areca nut Spices and Medicinal Plants, 2*, 71–78.

Ravindran, P. N., Sasikumar, B., George, J. K., Ratnambal, M. J., Babu, K. N., Zachariah, J. T., & Nair, R. R., (1994). Genetic resources of ginger (*Zingiber officinale* Rosc.) and its conservation in India. *Plant Genet. Resour. Newslett., 98*, 1–4.

Reddy, M. L. N., (1987). Genetic variability and association in turmeric (*Curcuma longa* L.). *Progressive Horticulture, 19*, 83–86.

Rosales-Longo, F. U., & Molina-Monterroso, L. G., (2007). Genetic diversity of the garlic (*Allium sativum* L.) grown in Guatemala, revealed by DNA markers. *Agronomia-Mesoamericana, 18*, 85–92.

Rosengarten, F. Jr., (1969). *Book of Spices* (pp. 24, 28, 29, 31). Wynewood: Livingston publishing co.

Sadanandan, A. K., (1993). *Proc. Nat. Sem. Potassium on Plant, Crops* (pp. 75–88). Potash Research Ins. of India, Gurgaon, Haryana.

Salvi, B. R., et al., (2000). *South Indian Hort., 48*, 16–18.

Sasikumar, B., George, J. K., & John, Z. T., (1995). A note on a ginger type collected from Western Ghat forests. *J. Spices Arom. Crops, 4*, 160–161.

Savitha, B. K., Pugalendhi, L., & Natarajan, S., (2013). Line × tester analysis to study combining ability effects in chili. *Indian J. Hort., 70*(3), 439–441.

Shashidhar, T. R., & Dharmatti, P. R., (2005). Genetic divergence studies in garlic. *Karnataka J. Hort., 1*, 12–15.

Shirsat, S., (1994). Genetic variability and divergence studies in chili (*Capsicum annuum* L.) genotypes. *M.Sc. (Agri.) Thesis*. Univ. Agric. Sci., Dharwad (India).

Simon, P. W., & Jenderek, M. M., (2004). Flowering, seed production, and the genesis of garlic breeding. *Pl. Breed. Rev., 23*, 211–244.

Singh, S. P., (2000). *Production Management of Spices*. Agrihortica publications, Gujarat.

Singh, Y., Mittal, P., & Katoch, V., (2003). Genetic variability and heritability in turmeric (*Curcuma longa* L.) *Himachal J. Agricultural Research, 29*, 31–34.

Srilakshmi, P., (2006). Genetic diversity, heritability and genetic advance studies in chili (*Capsicum annuum* L.) for quantitative and qualitative characters. MSc (Agri.) Thesis. Univ. Agric. Sci., Dharwad (India).

Status of Seed Spices Research and Development in India Balraj Singh and R. K. Solanki National Research Center on Seed Spices. Tabiji, Ajmer, Rajasthan 305 206.

Sundria, M. M., Mehriya, M. L., Rathore, B. S., & Choudhary, B. R., (2014). Cumin (*Cuminum cyminum* L.) sustainable production technology in Rajasthan. Agricultural University, Jodhpur, Rajasthan. Tetraploid clones of ginger (*Zingiber officinale* Rosc.) grown in Australia. *J. Agr. Food Chem., 54*, 29, 1414–1419.

Thangaselvabai, T., et al., (2006). *Proc. Nat. Sem. Emerging Trends in Production, Quality, Processing and Export of Spices*, pp. 28, 29.

Tiwari, S. K., (2003). Evaluation of ginger genotypes for yield and quality attributes under rain-fed and irrigated conditions. *Ann. Agr. Res., 24*, 512–515.

Vani, S. K., (2006). Studies on variability, correlation, morphological and molecular diversity in dry chilies (*Capsicum annuum* L.). *M.Sc (Agri) Thesis*. Univ. Agric. Sci., Dharwad (India).

Vijayakumar, K. R., & Mammen, G., (1990). In: *Proc. Intl. Cong. f Pl. Physiol.,* (Vol. 2, pp. 15–20). Society.

Volk, G. M., Henk, A. D., & Richards, C. M., (2004). Genetic diversity among U. S garlic clones as detected using AFLP methods. *J. Am. Soc. Hortl. Sci., 129*, 559–569.

Wohlmuth, H., Smith, M. K., Brooks, L. O., Myers, S. P., & Leach, D. N., (2006). Essential oil composition of diploid and tetraploid clones of ginger (*Zingiber officinale* Rosc.) grown in Australia. *J. Agr. Food Chem., 54*, 1414–1419.

Xu, P., Yang, C., Qu, S., & Yang, C. Y., (2005). A preliminary study on genetic analysis and purity assessment of the garlic germplasm and seed bulbs by the "fingerprinting" technique. *Acta Hort., 688*, 29–33.

Zewdie, Y., & Jenderek, M. M., (2005). Within and between family variability for important bulb and plant traits among sexually derived progenies of garlic. *Hort. Sci., 40*, 1234–1236.

Zhila, E. D., (1981). Correlations between phenotypic traits in garlic of the bolting type. *Tsitologiya-i-Genetika, 15*, 46–48.

WEBLINKS

dasd.gov.in/index.php/development-programs/958.html (Accessed on 28 November 2019).

shodhganga.inflibnet.ac.in/bitstream/10603/156/16/09_chapter2.pdf (Accessed on 28 November 2019).

CHAPTER 3

Perspective of Plantation Crops in Bihar with Special Reference to Coconut and Palmyra

RUBY RANI,[1] VISHWA BANDHU PATEL,[2] and H. P. MAHESHWARAPPA[3]

[1]*Department of Horticulture, Bihar Agricultural University, Sabour, Bhagalpur, Bihar, India*

[2]*Division of Fruits and Horticultural Technology, Indian Council of Agricultural Research–Indian Agricultural Research Institute, New Delhi, India*

[3]*AICRP, Palms, CPCRI, Kasaragod, Kerala, India*

ABSTRACT

Plantation crops are a group of commercial crops cultivated extensively over a large area of tropics and subtropics zones. Plantation crops comprise a large number of crops like coconut, areca nut, oil palm, cashew, tea, coffee, rubber, and cocoa. Their total coverage is comparatively less as compared to the other fruit crops and mostly grown by the farmers with smallholdings. These crops play an important role owing to their export potential values, domestic requirements, employment generation, and poverty alleviation, especially in the rural sector. Plantation crops have a unique role in the national economy owing to the source of basic raw materials for a number of industries and form the mainstay of the agrarian economy of the nation and thus considered as the lifeline of many states of India.

3.1 AREA AND DISTRIBUTION OF PLANTATION CROP IN INDIA

The crop-wise area, production, and productivity of major plantation crops in India during 2012–13 and 2013–14 are listed in Table 3.1.

Coconut is grown in about ninety countries in the world covering about 12 million hectares and about 75% of total area and production comes from South East Asia like the Philippines, Indonesia, and India. About 90% of the total Coconut cultivated area under in India lies in the southern states of the country, i.e., Tamil Nadu, Kerala, Telengana, Karnataka, and Andhra Pradesh. Similarly, the area under areca nut in the world is about 0.702 million hectares with an annual production of 0.854 million tonnes. The major countries producing areca nut are India, Bangladesh, China, Indonesia, Sri Lanka, and Myanmar. India ranks first in both areas (49%) and production (50%) of areca nut and the major growing states are Karnataka, Tamil Nadu, Kerala, Maharashtra, West Bengal Assam, and Meghalaya. On the other hand, India's share in cocoa production is only 0.3% of total world production. The state like Kerala, Karnataka, Tamil Nadu, and Andhra Pradesh produces cocoa. The total cocoa production in the country is 15,133 tonnes covering an area of 71,335 hectares.

TABLE 3.1 Crop-Wise Area, Production and Productivity of Major Plantation Crops in India During 2012–13 and 2013–14

Crops	2012–2013			2013–2014		
	Area	Production	Productivity	Area	Production	Productivity
Coconut	2137	15,609	7.3	2140	14,912	7.0
Cashew Nut	992	753	0.8	1011	753.0	0.7
Areca nut	446	609	1.4	452	622	1.4
Cocoa	66	13.0	0.2	71	15	0.2

Area in '000 ha, Production in '000 MT and Productivity = MT/HA.

The area under coffee plantations in India is about 400.0 thousand hectares as in 2015–16 with the productivity of 876 kg/ha. Major coffee growing area is concentrated in the southern states of Karnataka (54.95%), Kerala (21.33%), and Tamil Nadu (8.18%). The coffee industry involved a large number of enterprises involving 280,241 coffee growers of which 99% are small growers.

India is the largest producer and consumer of tea in the world where it is produced in around 563.98 thousand hectares of area in the states such as Assam (304.40 thousand hectares), West Bengal (140.44 thousand hectares), Tamil Nadu (69.62 thousand hectares) and Kerala (35.01 thousand hectares). The tea industry is India's employs over 3.5 million people across 1,686 estates and 157,504 smallholdings.

India contributes about 0.2% in the total world produce with 1% of total acreage in the world. Presently area under oil palm is around 200,000 hectares and with production of 70,000 tones a year. Largest area is under Andhra Pradesh including Telangana. The other oil palm producing states and UTs include Andaman and Nicobar Islands, Andhra Pradesh, Chhattisgarh, Goa, Gujarat, Karnataka, Kerala, Maharashtra, Mizoram, Odisha, Tamil Nadu, and Tripura.

Palmyra Palm: *Borassus flabellifer* L. or commonly called Taad or Tarh in Bihar. It is native to South and Southeast Asia, in the Indo Malaya ecozone. The palm tree of the Sugar palm group, found from Indonesia to Pakistan in India, Bangladesh, Cambodia, China South-Central and South-east Asia. The name Palmyra Palm has derived from the Portuguese word of palmeira. Toddy Palms are very common in India especially in Andhra Pradesh, Telangana, Tamil Nadu, Orissa, West Bengal, and Bihar. The total population of palmyra in India is about 102 million and about 50% of the total population is in Tamil Nadu. In Bihar, the total population of palmyra palm is estimated to be about 92 lakh. The districts with a larger Palmyra population are Gaya, Nawada, Nalanda, Aurangabad, Bhagalpur, Banka, Samastipur, and Muzaffarpur.

3.1.1 STATUS OF PLANTATION CROPS IN BIHAR

Plantation crop requires moderate climate and do not thrive well under extremes of climate. An average temperature around 25–35°C and humidity above 60% are considered ideal for the plantation crops. It requires less variation in diurnal temperature and well-distributed rainfall throughout the year. Bihar is situated between 25 to 27° North latitude; the climatic condition of Bihar is of mostly semiarid sub-tropical that experiences moderate rainfall, hot and dry summer and cold winter. This region being close to Tropic of Cancer experiences tropical climate during summer. Average temperature is 35–40°C throughout the summer season May and June are the hottest months of the year and when the maximum temperature goes up to 45C. The average minimum temperature during the coldest month of December and January goes down to 5 to 10°C. Bihar gets its maximum rainfall during South-West monsoon season which prevails from June to September. The natural precipitation ranges between 990 mm and 1700 mm, average annual rainfall being 1205 mm. Most of the annual rainfall received between the month of May and October. Thus the climate of Bihar is not very much suitable for plantation crops cultivation. However, crops like coconut, palmyra palms, betel vine, tea are cultivated in some of the area in the state.

The climate of Zone II commonly known as Kosi zone consisting the districts like Kisanganj, Araria, Purnea, Katihar, Eastern part of Madhepura,

Saharsa, and some parts of Bhagalpur has been found to be suitable for coconut cultivation (See map-1). The zone experiences an average minimum temperature of 8.8C and maximum temperature of 33.8°C. This zone comprises a network of rivers like Bagmati, Kosi and their small tributaries, thus humidity is quite high in this zone as compared to other areas of the state.

3.1.2 POTENTIAL AREA OF COCONUT CULTIVATION IN BIHAR

The total area is about 15,166 ha under coconut in the state with annual nut production of 123,755 MT (Database, 2005–06, Ministry of Agriculture, Bihar). Coconut planted in theses area are mostly of tall type mostly planted as stray, backyard or bund plantation. However few orchards in Katihar and Purnea districts have been established also (Figure 3.1).

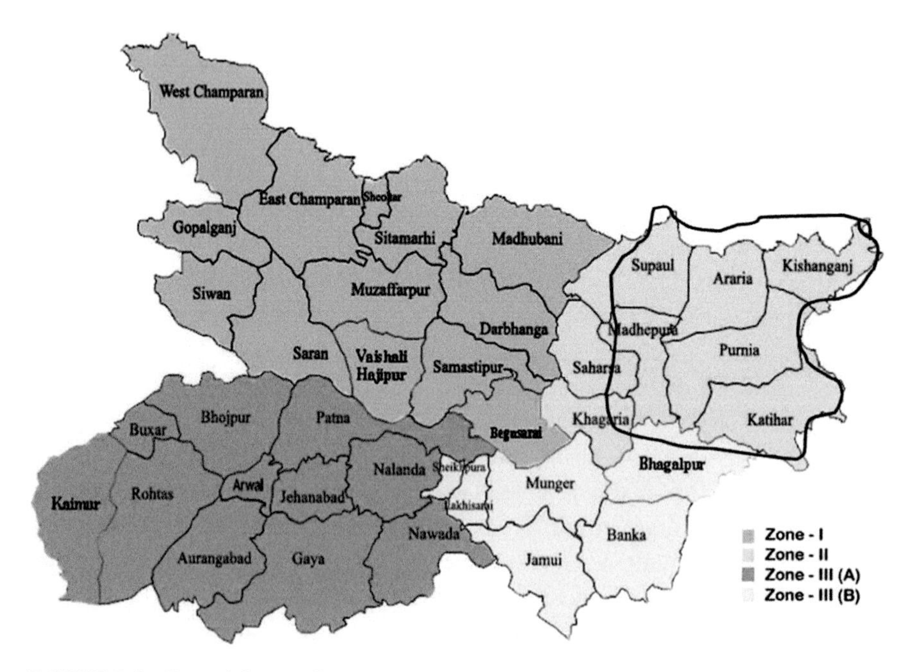

FIGURE 3.1 Potential area of coconut cultivation in Bihar.

Coconut plantation at large scale in the state was done during late nineties. Most of the plantations were of tall type. Growth of palms and yield potential of theses coconut palms were good. But in due course of time yield reduced probably due to lack of technology know-how of cultivation practices and maintenance of the palm. But well-maintained palms are still bearing up to more than 120 nuts per year. The yield of coconut reduced drastically during 2008 and in 2011 when minimum winter temperature went down to 7–10°C that prolonged up to 10 to 15 days. Even though, bearing of nut in some palms is quite good, i.e., 80–100 nuts per year.

TABLE 3.2 District Wise Population of Palm Tree in Bihar

Name of District	Palmyra Palm	Date Palm	Coconut	Total
Patna	413,292	71,373	3546	488,211
Nalanda	774,252	22,422	2730	799,404
Bhojpur	146,876	16,504	1776	165,156
Buxar	15,492	854	469	16,815
Rohtas	59,039	3238	286	62,563
Bhabhua	29,726	6596	269	36,591
Gaya	14,574,10	353,037	2134	1,812,581
Jahabnabad	334,618	6011	4089	344,718
Arwal	182,254	7419	1172	190,845
Nawada	865,097	115,456	2898	983,451
Aurangabad	400,461	106,142	1379	507,982
Saran	352,708	117,015	12,914	482,637
Siwan	172,445	65,542	5326	243,313
Gopalganj	139,112	46,629	4564	190,305
Muzaffarpur	399,867	551,370	12,850	964,087
East Champaran	212,836	120,607	10,048	343,491
West Champaran	50,510	18,738	2852	72,100
Sitamarhi	77,192	98,346	5956	181,494
Shivhar	14,922	18,077	912	33,911
Vaishali	356,617	240,378	10,282	607,277
Darbhanga	154,182	213,236	18,834	386,252
Madhubani	58,988	80,073	29,004	168,065
Samastipur	485,425	405,662	18,061	908,848
Begusarai	145,343	105,912	8342	259,597
Munger	62,668	26,106	10,578	99,352
Lakhisarai	95,415	29,122	2795	127,332
Shekhpura	267,526	32,436	1098	301,060
Jamui	149,506	209,577	5713	364,796
Bhagalpur	508,245	210,553	18,652	737,450
Banka	647,886	592,130	15,216	1,255,232
Saharsa	67,764	41,255	20,662	129,681

TABLE 3.2 *(Continued)*

Name of District	Palmyra Palm	Date Palm	Coconut	Total
Supaul	6409	9919	49,047	65,375
Madhepura	13,129	12,728	34,353	60,210
Purnea	56,799	31,617	39,509	127,925
Kishanganj	677	666	4610	5953
Araria	3849	3989	17,290	25,128
Katihar	18,738	9262	18,860	36,860
Khagaria	22,398	9568	7238	39,204
Total	9,219,373	4,009,565	396,314	13,625,252

Department of Industry, Govt. of Bihar, 2016.

Thus at present total population of coconut in the state is about 4.0 lakh with the maximum number in Supaul followed by Purnea, Madhepuera, Katihar, Madhubani, and Bhagalpur.

3.2 COCONUT RESEARCH IN INDIA

The coconut palm is known as "Kalpavriksha: the tree of heaven" as each part of the plant is used in various ways. It provides food, fuel, drink, fiber, feed and timber that are used to prepare various edible and non-edible items. Large populations in rural India either directly or indirectly depend on coconut for their livelihood and economic security.

3.2.1 GERMPLASM COLLECTION AND CONSERVATION

Research on coconut was started in 1916, when Coconut Research Stations at Kasaragod and Nileshwar was established in India to increase the nut size and yield per plant later, production of short stature varieties and resistance to different stresses was also considered under crop improvement program. Introduction of germplasm of coconut in India began during 1923–24 and diverse types improved genotype from countries like Indonesia, Philippines, Fiji, Malaysia, Sri Lanka, and Vietnam and also from Bengal, Laccadives and Andamans (under the colonial empire). The introduced genotypes were planted at the Coconut Research Station, Pilicode, for their evaluation and selection. Progenies of these were later replanted in 1940 at the Coconut Research Station, Kasaragod. In addition, 16 'Presidency varieties,' from different regions of the country were also planted at Pilicode No 1 station during June–July 1923. Further, germplasm collection was intensified in 1952 again and the first indigenous germplasm survey and collection was started in 1958.

The systematic collection, conservation, characterization of coconut diversity was initiated when the Central Plantation Crops Research Institute Kasaragod, in 1970 field gene bank was established thee to maintain the collected germplasm. Now, the worlds largest germplasm collection of coconut comprising 401 accessions (269 indigenous and 132 exotic) is being maintained in the Institute and also acts as National Active Germplasm Site for Plantation crops and National Gene Banks for cocoa, coconut and areca nut.

3.2.2 VARIETIES AND HYBRIDS

After conservation and evaluation of germplasm at different regions of the country, coconut varieties were identified and released for commercial cultivation. Basically, coconut cultivars are categorized into two groups, i.e., tall and dwarf.

3.2.2.1 TALL CULTIVARS

Tall coconut varieties are the most common. The life span of the tree is from 60 to 80 years and the height may go up to 15 to 18 m, produce good quality copra with high oil content as compared to dwarf cultivars. The different cultivars of the talis are generally named after the major place of growing The Tall cultivars commonly grown in India are the West Coast Tall (WCT), Tiptur Tall (TPT) and East Coast Tall (ECT). The tall varieties developed after improvement is Chandrakalpa or Lakshadweep Ordinary (LCT), VPM-3 (Andaman Ordinary), Philippines Ordinary (Kerachandra), and Aliyar Nagar 1 (ALR 1).

Aliyar Nagar 2 (ALR 2), Benavali Green Round (Pratap), Kerakeralam, KeraSagara (Seychelles), Philippines Tall (Chandrathara), Assam Tall (Kamaroopa), KalpaPratiba, Kalpadhenu, and Kalpa Mitra.

3.2.2.2 DWARF CULTIVARS

These are small stature plants with less life span as compared to tall. The general height is of 5–8 m, start bearing at the age of 3–4 years and have

40–50 years of life span. The nuts are smaller with soft copra and low in oil content. These varieties are mainly grown for tender water nuts The Dwarfs varieties available in India are Chowghat Orange Dwarf (COD), Chowghat Green Dwarf (CGD), Malayan Yellow Dwarf (MYD), Kenthalli (KTOD), and Gangabondam (GBGD). The improved dwarf varieties are Strait Settlement Dwarf Green, and Kalpa Raksha.

3.2.2.3 HYBRIDS

Efforts have been taken to improve the coconut for their yield, copra content, oil content, tender coconut water, biotic and abiotic stress. The improved hybrids are Chandralaksha (LCT × COD), Kerasankara (WCT × COD), Chandrasankara (COD × WCT), Lakshaganga (LCT × GBGD), Keraganga (WCT × GBGD), VHC 1 (ECT × MGD), VHC 2 (ECT × MYD), VHC 3 (ECT × MOD), Anandaganga (ADOT × GBGD), Kerasree (WCT × MYD), Kerasoubhagya (WCT × SSAT), KalpaSamrudhi, Gadavani Ganda (ECT × Gangabandom), and KalphaShankara (CGD × WCT).

3.2.3 CROP IMPROVEMENT UNDER OF ALL INDIA COORDINATED RESEARCH PROJECT ON PALMS

The all India Coordinated Coconut and Arecanut Improvement Project was conceptualized in the year 1970 and started functioning from the year 1972 with 12 centers located in eight states. It was renamed as "All India Coordinated Research Project on Palms" during 1986. Four oil palm centers were added during 1990 and two centers of palmyrah were added during 1995. The mandate of this project was to conserve and utilize the available genetic resource of palm crops and to improve them for higher yield and quality.

3.2.3.1 VARIETIES/HYBRIDS RELEASED

The following varieties/hybrids have been developed by the AICRP on palms for cultivation in different agroclimatic regions (Table 3.3).

About 112 ecotypes of coconut have been collected from different centers of AICRP, Palms located in different states like Assam, W.B, Chhattisgarh, Karnataka, Odisha, Tamil Nadu, Andhra Pradesh, and Maharashtra for desired traits and conserved for further evaluation.

TABLE 3.3 Varieties/Hybrids Developed by the AICRP on Palms for Cultivation

Varieties Released by AICRP	Hybrids Released by AICRP	Varieties/Hybrids Released Jointly by AICRP on Palms and CPCRI)	
Kamrup,	Konkan Bhatye Coconut Hybrid-1	Kalpa Dhenu	Kalpa Samrudhi
Gauthami Ganga	Godavari Ganga	Kalpa Pratibha	Kalpa Jyohti
Kera Bastar	Vasishta Ganga	Kalpa Mitra	Kalpa Surya
Kalyni Coconut-1	Kalpa Ganga	Kalpataru	Kalpa Sreshta

3.2.4 IMPROVEMENT THROUGH BIOTECHNOLOGY INTERVENTIONS

The protocol has been developed for embryo culture. Recently the opportunity of exchange of germplasm of coconut has been provided in the form of embryo through *in vitro* germination of zygotic embryo. The embryo culture protocol developed by ICAR-CPCRI was first utilized during 1994 for transferring gene. Pacific Ocean accession maintained at World Coconut Germplasm Center, Andaman Island to the mainland. Embryo rescue of rare type has also been achieved. Cryopreservation technique for conserving embryo and pollen has also been developed, where pollen can be stored without losing its viability up to four years.

3.3 PRODUCTION TECHNOLOGY REFINEMENTS

The ultimate goal of any crop research is improving productivity and quality. The research work on coconut for standardization of production technology was started in 1917, in the three stations: near the village of Nileshwar, an existing garden in the village of Kudluandin and Kasaragod (where the present CPCRI is located). The important findings as a result of outcomes of coconut research on different aspects of production technology are as under:

3.3.1 SEED NUT COLLECTION AND PLANTING

Research findings reflected that seed nuts collected during April–July germinated in less than 125 days after sowing with significantly more number of leaves, larger leaf area and wider girth at collar indicating more vigorous in growth. Whereas those collected during September–January took 160–180 days for germination. Considering all the growth characters, May and June were identified as the ideal time for sowing the seed nuts.

3.3.2 PLANTING MATERIAL

Unavailability of good quality plants is a major bottleneck in coconut development. If the seed nuts and seedlings are of poor quality, the plantation will become uneconomic and loss of time and money to the grower. Being a cross-pollinated plant coconut does not breed true to type as mother palm. Thus the selection of seed nuts and seedlings is more important. Thus, the quality of seed nuts and seedlings should be improved by a series of selections at different stages.

3.3.2.1 MOTHER PALM SELECTION

An ideal mother palm should have an age of 20 years or above, yielding more than 80 nuts per palm per year and free of any disease. A minimum of 30 leaves at the crown and nut weight more than 600 g with copra weight 150g and above is required for a mother palm. Nowadays, the Coconut Development Board is plays a major role for providing quality planting material through their DSP Farm (Demonstration and Seed Production Farm).

3.3.3 WATER MANAGEMENT

Proper tilling is very important for soil aeration and moisture conservation. Thus the experiments to demonstrate the importance of tilling the garden was started in 1919 and the research on water management was started in the late 1920s. The irrigation with 45 liters of water at 4 days interval has been found to be better in coconut. It was reported that irrigation with seawater did not leave any residual harmful effects. During the dry months irrigations of 60 mm depth at IW/CPE ratio 0.15 (average interval of 17 days) was better than 40 mm and 20 mm depth given at the IW/CPE ratios of 1.0, 0.75, or 0.5.

3.3.3.1 DRIP IRRIGATION AND FERTIGATION

Coconut gardens are commonly irrigated by flood-or basin method, having irrigation efficiency of only 30 to 60% due to the wastage of water, labor, and energy. It was standardized that WCT palms irrigated with 20 mm water/irrigation at IW/CPE ratio of 1.0 gave the highest yield. For humid tropics, it is sufficient if we irrigate at the rate of 66% of the open pan evaporation through drip in the basin area (1.8 m radius) of the palm based on the monthly mean evaporation, irrespective of the varieties. Irrigation once in four days with 45 l water to young palms planted in laterite soil resulted in better growth and vigor. Around 50% of the chemical fertilizer and the cost on labor could be brought down apart from increasing the yield by 20% following fertigation technology.

3.3.3.1.1 Findings on Drip Irrigation Trials in Coconut Under AICRP Palms

It has been found that irrigation efficiency can be increased when four drippers of discharge rates of 30 l/h is used placing at equal distance around the trunk is used at 8 days interval. Experiments were conducted at different centres of AICRP, Palms to estimate the water requirement through drip irrigation and the findings are as given in Tables 3.4 and 3.5.

TABLE 3.4 Technology Developed at CRS, Aliyarnagar, Tamil Nadu

Months	Quantity of Water to be Applied (lit/day) in the Area of Different Level of Water Availability		
	Adequate	**Moderate**	**Low**
Feb–May	65	45	22
Jan, Aug, and Sep	55	35	18
June, July, Oct–Dec	45	30	25

All India Co-ordinated Research Project on Palms: The national network for location-specific research, Vistas in Coconut Research, CPCRI-100 Years of scientific excellence, (2016). pp. 131–133.

TABLE 3.5 Water Requirement at Different AICRP Centers Using Drip Irrigation

Centers	Months	Quantity of Water (lit/day)	Months	Quantity of Water (lit/day)
Veppankulam	Mar-Sep	80	Oct–Feb	50
Arsikere	Feb-May	65–75	Jun–Jul	40–50
Ratnagiri	Oct-Jan	30	Feb–May	40

Annual Report AICRP, Palms , 2016 & 2017.

The experiment conducted on fertigation at Aliyarnagar, Veppankulam, Mondouri and Kasaragod Centers under AICRP, Palms recommended @ 50% recommended dose (RD) of fertilizer (RDF for improving the nut and copra yield per palm. Whereas, the experiment conducted at Ambajipeta, Ratnagiri, Arsikere Centers observed the application of 75% RDF through fertigation for higher nut and copra yield.

3.3.4 ORGANIC FARMING OF COCONUT

The first attempt was taken in the 1920s to increase the organic matter content in garden soil. Cow dung was used initially as organic manure. Later, green manuring like use of cow gram, glyricidia, *Calopogonium muconoides, Crotolaria retusa, Tephrosia purpurea, Crotolaria striata, T. candida,* and *Indigofera parviflora. Calopogonium muconoides* was suggested as in situ green manure crop and cowpea as cover crop. Glyricidia is suggested growing as a boundary crop with prunings of the plants once in three months to use the leaves for green manuring. Coconut husk burial was also used as a technology to enrich the soil with organic matter. The supplementation of organic sources with inorganic fertilizers (blending) is beneficial to coconut during the early establishment period.

3.3.4.1 BIO-FERTILIZER RECOMMENDATION

Application of mixture of 50 g of *Azospirillum*, 50 g of phosphobacteria (or) 100 g *Azophos,* and 50 g of VAM in sufficient quantity of compost or FYM near feeding roots once in 6 months/palm starting from planting have been found very beneficial.

3.3.4.2 ORGANIC RECYCLING

The crops like sunn hemp, Calapagonium, or Daincha may be sown as green manure in place of compost and plowed *in situ* at the time of flowering.

3.3.5 INTER-CULTURAL OPERATION WEED MANAGEMENT

The inter-space in the coconut garden has to be plowed twice in a year in June-July and December-January to keep the field weed-free and efficient use of resources.

3.3.6 *INTER CROPPING*

Inter/mixed crops should be selected according to the age and growth habit, climatic requirement, demand of the crop, irrigation facilities and soil type.

3.3.6.1 *MULTIPLE CROPPING SYSTEMS FOR DIFFERENT REGION*

The essentiality of increasing productivity from unit area. Efficient resource utilization was felt long back and screening for suitable intercrop in coconut was initiated in the early 1920s. Later in the 1970s and systematic research was initiated to grow different vegetables, flowers, medicinal plants and perennial crops as intercrops in coconut garden. Vegetable crops like brinjal, chili, tomato, etc. were grown successfully in coconut gardens. Crops like cocoa, banana, pepper, medicinal and commercial flowers were found to be suitable for growing as mixed crop in coconut garden. The results indicated that the income from coconut garden can be doubled by adopting a cropping system approach.

It was estimated that young-bearing palms (spacing 7.5 m^2) and adult palms permit less than 20% incident radiation to reach the ground, while the middle-aged palms allowed about 30% light and pre-bearing and old palms permit up to 80% light. Thus, intercropping should be selected as per the interspace available in the coconut garden.

3.3.6.2 *COCONUT-BASED CROPPING SYSTEM FOR DIFFERENT AGROCLIMATIC ZONE OF INDIA DEVELOPED UNDER AICRP, PALMS*

By seeing the success of multi-story cropping in coconut, coconut-based high-density multispecies cropping system model was established under AICRP, Palms which aims at developing a coconut-based cropping system which is self-sustaining and produces maximum biomass and returns with least inputs for the different agro-climatic zone of the country. The research over years has resulted in identifying crops that can be grown as inter/mixed crops in coconut gardens (Table 3.6).

3.7 STATUS OF COCONUT RESEARCH IN BIHAR

AICRP on Palms (Coconut) was initiated at Bihar Agricultural University (BAU), Sabour during on 6th March, 2009 to promote coconut in Bihar. The ongoing research programmers are given in subsections.

TABLE 3.6 Location-Specific Coconut-Based Cropping Systems

Cooperating Center's	Recommended Intercrops
Aliyanagar (Tamil Nadu)	Coconut + Banana + Turmeric + EFY + Tapioca + Bhendi + Cocoa
Veppankulam (Tamil Nadu)	Coconut + Black pepper + Banana + EFY + Coriander + Cocoa
Mondouri (West Bengal)	Coconut + Black pepper + Pineapple + Banana + Lime
Bhubaneshwar (Odisha)	Coconut + Banana + Tuberos (In Young Coconut Garden)
Jagdalpur (Chhattisgarh)	Coconut + Black Pepper + Carnation + Chrysanthemum
Kahikuchi (Assam)	Coconut + Black Pepper + Turmeric + Assam lemon + Banana
Konkan region of Maharashtra	Coconut + Turmeric + Banana + Pineapple + Tapioca + Nutmeg + Cinnamon
Arsikere (Karnataka)	Coconut + Banana + Drumstick + French bean + Ladies finger + Redgram
	Coconut + Banana (High Water Requirement)
	Coconut + Annual Drumstick (Medium Water Requirement)
	Coconut + Red Gram (Low Water Requirement)
Navasari (Gujarat)	Coconut + Turmeric
Sabour (Bihar)	Coconut + Guava + Banana + Turmeric + Cowpea + Pea + Mustard + Elephant Foot Yam

3.7.1 *CROP IMPROVEMENT*

A. Gen.1: Conservation and evaluation of coconut genetic resources in different agro-climatic regions

Experiment 1: Collection, conservation and evaluation of location-specific germplasm:

In Bihar, more than seven germplasm were identified from different coconut growing area of Bihar during 2012–13 to 2016–17. Seed nuts of three accessions collected during 2012–13 from Sabour and Naugacchia block were transplanted and named as SBRC-1 and SBRC-2 and SBRC-3. Four more accession identified and collected during 2014–15 and 2015–16 (Figure 3.2).

FIGURE 3.2 Different coconut growing area of Bihar.

Experiment 2: Evaluation of released varieties in coconut.

- 15 varieties namely Assam Green Tall (AGT), PHOT or Philippines Ordinary Tall (Kera Chandra), Arasampatti Tall (Aliyarnagar Tall), ADOT or Andman Ordinary Tall (IND 018), ECT (CPR 509), Sakhi Gopal Tall (IND 041), Pratap, Zanzibar Tall (IND037), LCOT (Chandra Kalpa), TPT (KalpaTharu), MYD (KalpaRaksha), GBDB (Gangabondam Green Dwarf), MOD or Malayan Orange Dwarf (IND 048), COD or COD (IND007), Gonthembilli Tall (IND 051), KGD (Kamarupa Green Dwarf) are under evaluation at Sabour.
- On the basis of data collected on vegetative growth parameters var. COD, MYD, ECT, Shakhi Gopal Tall, Kera Baster, Kera Chandra, and Chandra Kalpa have been found better for agro-climate of Bihar.

The growth pattern of all the varieties was recorded over the years at three months interval in all varieties. The average data reflected that 60–68% of total vegetative growth takes place during June to November. The extent of vegetative growth in terms of plant height was minimum (10–14%) during December to February. It was found that difference in day-night temperature also played important role in growth parameters of coconut in Bihar region (Figure 3.3).

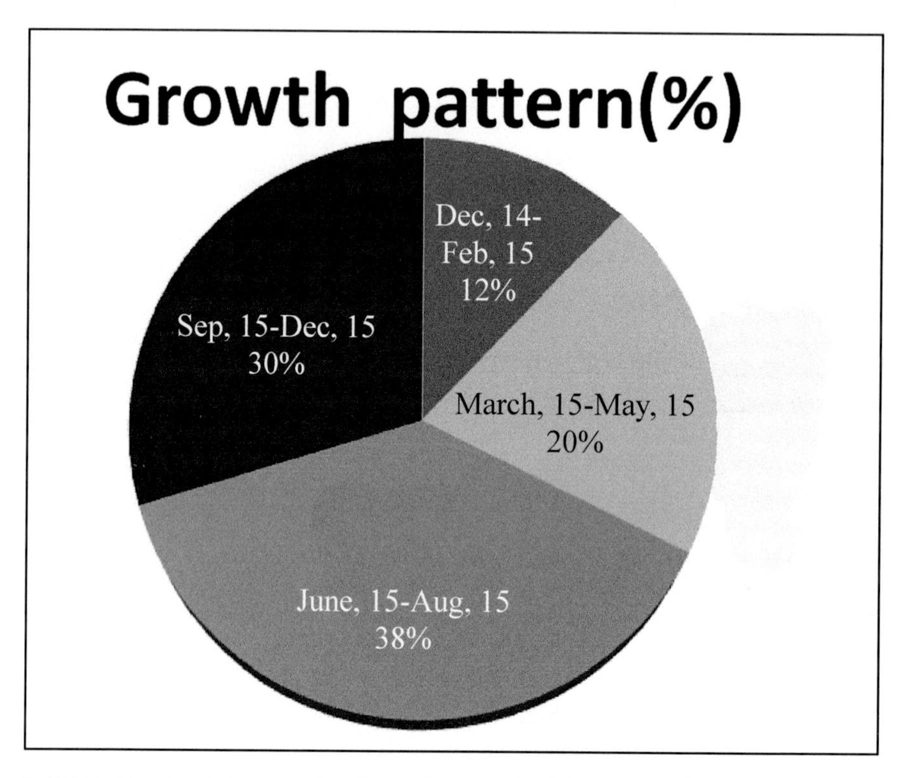

FIGURE 3.3 Vegetative growth pattern of coconut in different time of year.

B. Gen. 3: Establishment of mother blocks and production of quality planting material in coconut

Activity 1: Demonstration of released coconut varieties in different agro-climatic regions.

- Seed nuts of ten improved verities / hybrids under evaluation namely Konkan Bhyte Hybrid-I, CCS-4 (Kalpa Pratibha), Kahi Kuchi Hybrid-1, CCS-5 (Kalpa Mitra), Kera Bastar, CCS-6 (Kalpa Dhenu), Gauthami Ganga, CCS-7 (Kalpa Raksha), Kalyani coconut-1, CCS-9 (Kera Keralam) were planted and maintained in the field since 2012–13.
- On the basis of data collected on vegetative growth parameters var. Kera Bastar, Kalyani Coconut-1, Kalpa Pratibha were found better

for agro-climate of Bihar. Kahi Kuchi hybrid-1 was not found suitable for this region as this variety was found highly susceptible to cold weather.

3.7.2 CROP PRODUCTION

AGR.5 A: Studies on fertilizer application through micro-irrigation technique in coconut.

The significant influence of fertigation levels was recorded on plant growth of palms in terms of plant height, plant girth, number of functional leaves and leaf charters parameters. The experiment was laid in RBD comprising six treatments, i.e., 25%, 50%, 75% and 100% RDF through drip fertigation, 100% RDF as basin application and treatment with no fertilizer application as control. Fertigation was started at three and a half years age of palm in variety Sakhi Gopal in six equal split doses from October to May.

The highest plant height of 515.50 cm was recorded in T_5 (100% RDF fertigation) with 18.39% annual increase which was at par with T_4 (75% RDF through fertigation) Likewise the maximum collar girth (188.0 cm), number of functional leaves (26.50/plant) and annual leaf production (11.0 per plant) was recorded in T_5 (100% RDF fertigation) followed by T_4 (fertigation at 75% RDF) having collar girth of 179.00, 24.8 number of functional leaves and annual leaf production of 11.8. The plant under T_5 (100% RDF fertigation) were earliest to flower that took only 62.5 months to flower that was at par with T_6 (100% RDF though basin application).

Fertigation levels also influenced leaf nutrient status of palms having maximum nutrient status in terms of percent of N, P, and K in T_5 (100% RDF fertigation) followed by T_4 (75% RDF through fertigation) and T_6 (100% RDF though basin application.

AGR.10: Development of coconut-based integrated cropping system models for different agro-climatic regions:

The recommended coconut-based integrated cropping system models (Figure 3.4) for Bihar region is: Coconut + Guava + Banana + Turmeric + EFY + Cowpea + Pea + Mustard.

FIGURE 3.4 Recommended coconut-based integrated cropping system models.

3.8 COCONUT-BASED CROPPING SYSTEM IN BIHAR

On the basis of findings of the project following knowledge has been generated (Figure 3.5):

1. Kahi Kuchi Hybrid-1 is not suitable for growing in the Bihar region.
2. Manuring before winter, i.e., in September–October and water sprinkling during morning and in evening hours reduced frost injury in coconut seedlings.
3. Pomegranate should not be planted as an intercrop in a coconut-based cropping system for the Bihar region.
4. Seedling plants should be protected from frost during winter by covering it with transparent polythene sheets and thatching should be done to protect the young seedlings during summer (Figure 3.6).

Following technology developed by other centers of AICRP, Palms (Coconut) were transferred to farmer's field:

1. Rejuvenation of old and senile orchard by proper manuring, irrigation, and IPM.
2. Log tapping with toddy, root feeding with monocrotophos and use of pheromone trap to control red palm weevil.
3. Integrated disease management for managing bud rot disease.

FIGURE 3.5 Coconut-based cropping system in Bihar.

FIGURE 3.6 Seedling plants.

3.9 STATUS OF PALMYRA PALM

The palmyra palm is botanically known as *Borassus flabellifer* L to family Arecaceae. It is considered as a minor tropical fruit. India has the maximum palmyra having a total palm population of about 122 million (Vengaiah et al., 2012). It is a very hardy crop that can withstand extremes of climatic situations. It grows very tall ranging between 30–60 meters height and has great economic potential. All the plant parts of palmyra are used right from trunk, root, young tuberous seedlings, fruits, fruit sap, leaves, trunk; roots are used for making different edible and non-edible products. Despite, somewhat less attention has been given for the development of Palmyra palms. There is a large population of Palmyra in Bihar. Palmyra can be grown in uncultivable wastelands, farm field boundaries, bunds, roadsides, housing colonies without much care for livelihoods and for aesthetic look and ambiance.

The slow-growing palmyra has very long juvenile phase and takes from 12 to 20 years to bear. The palms are almost free from any insect and disease infestations that can tolerate prevailing extreme high and low temperatures as well as waterlogged and drought conditions. Mahatma Gandhi told the Palmyra palm an antidote to poverty.

3.9.1 SCOPE OF PALMYRA IN BIHAR

There is a rich genetic diversity of palmyra in Bihar. Diversities are with respect to plant height (dwarf and Tall), fruit size (small to large), fruit shape (oval, round), fruit color (blackish, orange, dark brown, yellowish), and a number of endosperm per fruit (one to four). Although no database is available regarding genetic diversity of palmyra in the state. Palmyra cultivation is usually done by poorest sector of people in the state mainly for toddy tapping, immature embryo (kowa), and ripened fruits. Pulp of ripened fruits consumed as raw pulp or for making different types of products. *Spongy haustorium* that forms during germination of seed nut is spongy and sweet is also used to eat as fresh. Germinating plumule is used as raw and in cooked form in the state. But no processes products are made from palmyra in the state. There is much scope of palmyra in the state. However, their use is restricted due to lack of value addition for both edible and non-edible products. By establishing palmyra palm-based processing industries a significant chunk of human labor may be engaged for their livelihood security. But no database is available on the quality of fruits of palmyra in the state.

Mat, baskets, etc., are made from leaves of palmyra in the state in rural areas that need to strengthen in the form of a cottage industry.

Plants are dioecious in nature, i.e., male and female flowers are borne on the separate palm. Both male and female trees produce spikes of flowers but only the female plant bears fruits. However, both trees are used to tap toddy. The male plant that is commonly known as *Shishua* in local language in Bihar is commonly used for toddy tapping, and female plant that is locally called "Falua" is mainly used for fruit production. Neera tapping in Bihar starts in the month of April and continues up to July. The average neera production varies from 300–450 liters per palm/year in male plants (Shishua) and about 200 liters palm/plant in female plants (Falua). Fruit production in female palm (Falua) varies from 100 to 350 nuts/palm/year. Fruits mature in the month of August-September.

Palmyra is referred as versatile tree of immense use to mankind with more than 500 uses including various edible and nonedible products. Thus, the existing population of Palmyra in the state is needed to be exploited for their economic use.

3.9.2 PALMYRAH PRODUCTS

Neera is the topmost economic produce of Palmyrah. It is a good source of minerals like phosphorus, calcium, and iron. Vitamins like A. citric acid, Niacin, Thiamin, and Riboflavin are present in neera. Neera acts as laxative and diuretic.

3.10 RESEARCH STATUS OF PALMYRA AT NATIONAL LEVEL

3.10.1 CROP IMPROVEMENT

The systematic work for crop improvement on palmyra has started very recently since 1995 under the All India Coordinated Research Project on Palms. The germplasm collection and maintenance has been started at Horticultural College and Research Institute, Killikulam Tamil Nadu, and the Horticultural Research Station, Pandirimamidi, Andhra Pradesh. Different types of germplasm have been identified from different states of India like Tamil Nadu, Orissa, Andhra Pradesh, West Bengal, and Bihar. The plants of collected germplasm are being maintained at a different center. A total of 265 germplasm are being maintained at HCRI, Killikulam till date.

In an experiment performed to evaluate the physicochemical constituent of the fruit of Palmyrah palm available in Bhagalpur district. These different types of fruits were collected from different locations of Bhagalpur district and evaluated for their quality parameter. Amongst 15 accessions collected, average fruit weight of different accession of Palmyra ranged from 1192 to 1480 g. Maximum pulp weight was recorded 590 g. Highest TSS was recorded 15.96°Brix. Fruit has been found to be a good source of antioxidant and carotenoids also. These findings can be utilized to exploit this crop at commercial level to achieve livelihood security in the state.

3.10.2 CROP PRODUCTION

There is no organized plantation of palmyra in India and thus systematic production of palmyra is not done. Work of reducing juvenile phase has been started using different growth regulators. There is no specific production technology of palmyra palm.

3.11 POST-HARVEST TECHNOLOGY

3.11.1 NEERA COLLECTION AND STORAGE

Tapping of Neera is done in both male and female trees of palmyra. The flowering of the female palms is seasonal and comes to flowering at 10–12 years under ideal growth conditions. Tapping is done normally for a period of 90–130 days from February to May.

3.11.2 FINDINGS OF EXPERIMENTS CONDUCTED UNDER AICRP, PALMS

The yield of sap varies with climatic factors, i.e., temperature and relative humidity along with genetic variability in palms. A male palm can yield maximum of 5 liters per day and female palm up to 12 liters per day.

Application of ethereal and citric acid on cut portion of spathe almost double the yield. The Time gap between two cuts should not be more than 24 hours, if so the spathe do not yield neera as it dries completely. Shape of cut surface influence the yield of inflorescence sap. Yield of Neera for male palms 30° angle of cut gives more yields followed by 45°.

1. CPCRI Method of Collection (Figure 3.7): In this method, cocosap chiller is connected directly sliced spathe. Coco sap chiller is portable device characterized by hallow PVC pipe of one end expanded into box shape contained ice cubes. The sap collected by this method can be kept as such without fermentation up to 6 hours.

FIGURE 3.7 CPCRI method of neera collection.

Inflorescence sap using CPCRI method remained in acceptable condition under ambient conditions is 3 hours, under refrigerated condition up to 10 hours, there is no significant change. The shelf life can be extended beyond one week using freezing conditions.

Neera can be stored at room temp for 8 hours by adding 200 ppm KMS, stored up to 15 days in the refrigerator and stored up to 2 days by heating at 90°C for 15 min.

Neera can be stored under the refrigerated condition for 60 days by using heating for 5 min under 90°C followed by microfiltration.

2. Kowa or Nungu or Ice Apple (Endosperm): The clear, crunchy yet jelly-like sweet young seed from the Palmyra Palm is known as Ice Apple (Figure 3.8). The young plants are eaten as a vegetable or cooked and made into meal. The fruits can be eaten raw or cooked; the sweet sap can be eaten or concentrated into palm sugar.

FIGURE 3.8 Palmyra palm.

Tender fruits are to be harvested between 60–70 days to get soft ice apple or kowa or nungu. Kowa fruits are to be harvested when the epicarp near the perianth region is light to yellowish-green in color. It is also observed that at the stage of 70 days from the initiation flower was good and more pulp as compared to others for tender nut and at the age of 80 days have more the pulp was hard and suitable for jam and jelly preparation only.

Physical and chemical properties of tender fruit has also been estimated was measured and average weight of tender fruit of after flower initiation 75 days, various from 0.625 kg to 1.25 kg, diameter of the tender fruit various from 33 cm to 45 cm, height of the ender fruit various from 13 cm to 15 cm and average weight of kowa is 75 g to 80 g and about 25% in tender fruit and tender fruits mostly 3 seeded with equal distribution and rarely two seeded and single-seeded.

The tender fruit endosperm treated with sugar syrup concentration of 50°Brix under refrigerated condition was found to be acceptable in sensory qualities up 60 days and 180 days at –4°C to –5°C except slight color change on storage.

Standardization and commercialization of inflorescence sap extraction and inflorescence sap based products (Jaggery, Palm sugar and Candy) has been done. Methods of making jaggery, palm sugar and palm candy has also been standardized.

3. Ripe Fruits: Ripe Palmyrah fruits give a sweet aroma and they drop off from the tree when fully ripe (Figure 3.9). The mesocarp is fleshy and fibrous. The fruit is roasted in fire and consumed. Several value-added products can be prepared from the fleshy mesocarp after removing the fiber. Palmyra pulp is used in the preparation of palm leather, RTS beverage, concentrated pulp, jam and bakery, and confectionary items.

FIGURE 3.9 Ripe Palmyrah fruits.

4. Non-Edible Products: Leaves are used for thatching roofs, screening as fence, as mats, baskets, fans, hats, umbrellas, buckets, sandals and as writing material and as organic fertilizer in their farms and field. Tough and long fiber extracted from petiole is used for making of ropes and fancy articles which are in good demand. Dried up leaf petioles are also in use for making of trellis for use as fence and it further serves as a fire wood.

Palmyrah trunks are used either as live poles in construction of thatch sheds, or as a timber in replacement of wooden poles. The trunk of the tree is hallowed, and is directly used as boat for travel and fishing in many parts of the tropics.

5. Palmyra Germplasm in Bihar (Figure 3.10)

FIGURE 3.10 Palmyra germplasm in Bihar.

Thus, the existing population of Palmyra in the state is needed to be exploited for their economic use.

KEYWORDS

- **coconut-based cropping system**
- **CPCRI method of neera collection**
- **palmyra palm**
- **seedling plants**

REFERENCES

Annual Report. All India Coordinated Research Project on Palms, rear, 2015–2018.

Chowdappa, P., John, G., Muralikrishna, H., & Rajesh, M. K., (2016). *Vistas in Coconut Research* (pp. 23–86). Book CPCRI-100 Years of scientific excellence. ISBN 13-978-81-932263-1-5.

Indian Horticulture Database, (2014). *NHB, Ministry of Agriculture, Government of India* (p. 85). Institutional Area, Sector-18, Gurgaon-122 015.

Vengaiah, P. C., Murthy, G. N., Prasad, K. R., & Kumari, K. U., (2012). Post-harvest technology of Palmyra (*Borassus flabellifer* L.) present practices and scope. *International Conference on Food Processing by Omics Group*. India.

CHAPTER 4

Overview of the Floriculture Sector: Performance, Problems, and Prospects

PARAMVEER SINGH, SANGEETA SHREE, and AJAY BHARDWAJ

Department of Horticulture (Vegetable and Floriculture), Bihar Agricultural University, Sabour, Bhagalpur, Bihar, India

ABSTRACT

Floriculture industry provides a fertile ground for development and commercialization of floricultural crops (bedding, flowering, foliage or indoor plants; cut greens, cut flowers, etc.), landscape industry, essential oils manufacture industry, confectioneries, perfumes and medicines. Floriculture industry is the lucrative business of India, because of its diverse utilization, satisfying aesthetic needs of the people, generating more employment, guaranteed higher returns and more earning of foreign exchange to the growers of country. Due to economic reform and liberalization policies and modified export-import policies have given a boost to this industry. After liberalization, this sector identified as a sunrise business by the Government of India and accorded this sector 100% export-oriented status. Increasing demand and high return per unit area as compared to other agricultural commodities have dared growers to take the floriculture sector. India achieved significant development in ornamental plants and flower production, particularly foliage plant and cut flowers, which have good export potential. Production of commercial floriculture has been progressively increasing with increased use of protected structure, improved varieties and production technology and accessibility of airfreight or cooling systems. India has better scope in the future as there is a shift in trend towards tropical flowers and this can be lucratively exploited by India with enormous of diversity in Indigenous flora.

4.1 INTRODUCTION

Floriculture sector is a subdivision of horticulture dealing with growing of showy plants and flowers, gardens and floriculture industry, altogether comprise floral industry (Khan and Fazili, 2015). Floricultural crops include bedding, flowering, foliage or indoor plants; cut greens, cut flowers, etc. Ornamentals are mostly herbaceous in nature. Of which, bedding and garden plants are annuals or perennials in nature. These are grown in cell packs (in flats or trays), pots and hanging baskets, usually inside a protected structure and sold for gardens and landscaping purposes. Petunias, geraniums, impatiens are the largely selling bedding plants.

Nowadays, floriculture or ornamental industry is becoming a blossomy business in the world. This sector is classified as (i) tuberous roots, tubers and bulbs, (ii) live plants, (iii) flower bud and cut flowers (fresh or dried, tinted, lightened, steeped or prepared), and (iv) foliage and parts of plant (except flowers bud and flower) of shrubs, trees, ornamental plants, grasses, mosses and lichens (fresh, dried, tinted, lightened, steeped or otherwise prepared). These are suitable for ornamental purposes and bouquets preparation. Demands for floricultural commodities are progressively growing in the domestic market as well as international markets.

Floriculture industry is the lucrative business of India, because of its diverse utilization, satisfying aesthetic needs of the people, generating more employment, guaranteed higher returns and more earning of foreign exchange to the growers of country. More specifically, aromatic flower and plant parts are being utilized as raw materials in essential oils manufacture industry, confectioneries, perfumes and medicines for direct utilization by the peoples.

Initially, floriculture is started in England and afterwards popularized in the world. Floriculture is the sunshine sector with its large genetic diversity in India. Flowers or ornamental crops are cultivated in both conditions, i.e., Open field as well as protected conditions in India. Currently, there are more focused on floriculture crops growing in protected conditions which are having more profit margin.

Flowers are the beautiful and loveliest things on earth. They immediately advise beauty because they are linked with things that offer happiness and joy. People have customarily used of flowers for expressing their innermost feelings to God, presenting to the dearly loved ones or praise someone and versifying any believable emotion. Flowers are cultivated in India from ancient times. Since immemorial times ornamental plants have been an important part of life with flowers, gardens and ornamental horticulture which are being noted in most of our historical references like the Ramayana and Mahabharata. The roman soldiers were wore Garlands which were made with olive leaves whereas; blossoms of lotus decorated the Egyptian royalty. Flowers were grown for aesthetic purposes as also for their scent, perfumes, and medicines. Changing lifestyle of persons has led to the commercialization of flower, ornamental and aromatic plants cultivation nowadays. The vast demand of ornamental plants and flowers coinciding with different occasions has led to expansion of market for ornamental plants and flowers.

Now, various flowers are cultivating for domestic requirements and export by the farmers. Flowers and ornamental plants are now touching long distances because of the accessibility of airfreight or cooling systems. Due to economic reform and liberalization policies and modified export-import policies have given a boost to this industry. After liberalization, this sector identified as a sunrise business by the GoI and accorded this sector 100% export-oriented status. Mounting demand and high return per unit area as compared to other agricultural works have dared growers to take the floriculture sector. The mounting demand of floricultural product has also increased due to increased buying capacity of middle-income class, fast urbanization, availability of more information Technology Units, Tourists, Hotels and Temples, improvement in individual Incomes and GDP, change in lifestyle and social standards of the peoples, greater consciousness among the public to ameliorate the waning environment and economic upliftment of individual people.

India achieved significant development in ornamental plants and flower production, particularly foliage plant and cut flowers, which have good export potential. Floriculture sector is imperative from the economic point of view. Production of commercial floriculture has been progressively increasing with increased use of protected structure employing polyhouse, shade net house, greenhouse, etc. Commercial production of ornamental plants and flowers in India offered a chance for rural development due to its higher profits per unit area and the new employment opportunities. India has the capacity to accomplish the gap between demand and supply of floricultural products as international demand is rising at a fast rate. India is enriched with varied agro-climatic environment and different soil types, rich water supply, sufficient technical workforce, low charge of labor, developed markets, well-developed communication and transportation facilities, growing trend in protected cultivation, utilization of cool chain facilities and progressive farmers organization offer wide opportunities for cultivating various flower and ornamental plants around the year.

Cut flower farming received importance in the globe since the early twentieth century, mainly after the Second World War. So many changes happened in this sector from the production to storage, classification, and marketing of the same to the end consumers. Modern technologies and practices are being utilized in the production to consumption for cut flower trade (Kalmegh and Singh, 2016). Due to all of this floriculture sector becoming one of the high-value agricultural trade around the globe. Worldwide trade in this industry is rising at a fast rate with an approximate yearly growth rate of about 25%. The global trade of this sector is around US$ 11 billion in which cut flowers contributed about 60% in this sector. The international exports in floriculture sector increased by twenty times from 0.5 billion to 10 billion during the year 1990 to 2010 and if it raises at this rate, its expected double by 2025 (Kalmegh and Singh, 2016). This sector could be utilized as a solution to get fast growth in economic by those nations that have different geographical conditional (Kalmegh and Singh, 2016).

4.2 GLOBAL SCENARIO OF FLORICULTURE SECTOR

More than 145 countries are associated with floriculture sector worldwide and the worldwide floriculture trade is approximate to be at US$ 70 billion at present (Misra and Ghosh, 2016). The total area occupied under floriculture is 6,20,000 ha in the world, among which Asia-Pacific occupies nearly 73% (4,53,000 ha). India covers about 51% of area under floriculture crops in Asia-Pacific region. International floriculture market is rising significantly @ of 10–15% annually. Developed countries like America, Europe, and Asia account more than 90%of the total world trade in floricultural products. The Netherlands continue leads in floriculture sector in the world, about 60% share of global floriculture exports in 2013 (Nazir, 2015). Countries viz. Netherlands, Italy, Germany, Belgium, Denmark, and the USA are major exporting countries for floriculture products. Major importing countries are Germany, France, Netherlands, USA, UK, Italy, Belgium, Switzerland, Austria, and Japan for floricultural products.

The global floriculture sector perceives fast changes due to globalization and its influence on financial growth in the different regions of the globe. Simultaneously, competition is increasing globally. Well-known traditional flowers growing countries are the Netherlands, the USA, Columbia, Japan, and Italy. Some Asian countries viz. India, China, Thailand, Bangladesh, etc., is also progressively improving floricultural production. Major ornamental plants and flower products consuming nations in the world concentrated in North America and Western Europe. Around 80% of the total flower productions of the world are consumed by Germany, USA, UK, France, Netherlands, and Switzerland altogether. Ten largest domestic markets of cut flowers are available in the world, out of ten, six markets are available in Europe viz. Netherlands, Germany, UK, France, Spain, and Italy. Other important markets are the United States and Japan, which are accounting nearly 20% each. Most recently, the Middle East and Russia have also become vital markets demonstrating fast market growth. International floriculture industry is mainly depending on the trade of cut flowers, buds of flower, cut foliage, bedding and potted plants. Most important cut flowers in international trade are rose, gerbera, chrysanthemum, lily, carnation, orchid, and anthurium (Misra and Ghosh, 2016). Worldwide total export of floricultural commodities is US$ 9,784,525,000 and the Netherlands having 47.7% share of this value. The other major exporting countries are Colombia, Ecuador, Kenya, Ethiopia, and Belgium. India having the fourteenth position in floricultural commodities export (Misra and Ghosh, 2016).

4.3 INDIAN SCENARIO

The floriculture segment in India is unorganized and done on a very small scale. Mostly smallholding farmers are engaged in the cultivation and production of ornamental plants and floricultural products in the country. These small holding growers are not aware and lacking in scientific information about effective cultivation methods of flower crops that resulted in poor production and quality products. Recently some entrepreneurs have started showing interest in the production and export of flowers in large scale and in an organized manner. But additional such steps are needed from both private entrepreneurs as well Government. Due to high competition in global market and increasing market share of floricultural products, necessary attempts requires to be taken for ensure more production of flowers and their products as per quality standards of the international market.

In India, flower cultivation was 53,000 hectares in 1993–1994, after that it has been growing with compounded annual growth rate of the 7.76% (Misra and Ghosh, 2016). Cut flower production gradually increased up to the year 2006–07; after that its production greatly rises at a Compounded Annual Growth Rate of 40.5% (Misra and Ghosh, 2016). India occupied 255,000 ha area under flower production during 2013–14 with production of 1754 thousand MT loose flowers and 543 thousand MT cut flowers (Indian Horticulture Database, 2016). This shows that production of cut flowers playing important role in development of floriculture sector in India. Presently cut flowers are greatly demanding commodity, mainly for export purpose and country has been shifting from loose flowers or traditional flowers cultivation to cut flowers in open as well as protected conditions. Different type of flowers is now being produced in huge quantity for export as well as domestic market. Country has exported 22,518.58 MT of floricultural commodities with the value of Rs. 479.42 crores to the world during 2015–16 (APEDA, 2016).

Karnataka, Maharashtra, Tamil Nadu, Andhra Pradesh, West Bengal, Rajasthan, and Haryana have emerged as major floriculture cultivation centers in India. Major cut flower cultivating states are West Bengal (33.1%), Karnataka (12.3%), and Maharashtra (10.3%). Whereas, major loose flower cultivating states are Tamil Nadu (18%), Andhra Pradesh (12.98%) and Karnataka (12%) as reported by Nazir, 2015. Floricultural commodities exports from country include fresh cut flowers to Europe, Japan, Australia, Middle East, and the USA, loose flowers to the Gulf countries, cut foliage to Europe, dry flowers to the USA, Europe, Australia, Japan, and Russia. Ornamentals and dry flowers have a great export potentiality and almost 70% of total export of floriculture commodities from India consists of dried products.

4.4 BIHAR SCENARIO OF FLORICULTURE

Horticulture has diverse components of fruit, vegetables, plantations, spices, and floriculture. All of these, vegetables and fruits division has been able to attract the early attention of farmers and entrepreneurs while floriculture sector is still fairly untouched in Bihar. The sector is still in an emerging stage of development and accounts for a negligible share in the Indian production. However, the sector has attracted attention of the major giants from the state as well as overseas market. Thus, floriculture segment offers a ray of hope to farmers of Bihar as the state has suitable soil type and favorable climatic conditions for flower production. All these favorable conditions provide an opening to the early movers in floriculture sector in respect to harvesting hidden demand in the market, simple supply chain set-up and accessible foreign markets. Floriculture sector is growing very fast as a business in all over the globe and also in India. The progressive and enterprising farmers of Bihar are fast taking up flower cultivation as an alternative to the conventional crops cultivation.

Flower cultivation in Bihar has increased recently, providing immense opportunity of employment and income in rural regions of Bihar. In Table 4.1,

TABLE 4.1 Year Wise Area and Production of Flower Crops in Bihar

Year	Area/Production	Flower Crops					
		Rose	Marigold	Jasmine (Bela)	Tuberose	Others	Total
2009–2010	Area (in hectare)	63.55	269.85	91.6	87.45	113.9	626.35
	Production (in tonnes)	80.86	4877.97	268.39	435.05	966.91	6629.18
2010–2011	Area (in hectare)	68.05	283.15	105.15	105.25	126.05	687.65
	Production (in tonnes)	86.52	5119.66	307.46	522.94	1068.23	7104.81
2011–2012	Area (in hectare)	72. 9	359.95	113.5	116.65	138.85	801.85
	Production (in tonnes)	95.14	6565.8	348.32	595.45	1210.04	8814.75
2012–2013	Area (in hectare)	73.64	314.70	113.4	110.15	132.6	744.49
	Production (in tonnes)	98.90	5603.12	317.66	535.84	1080.23	7635.75
2013–2014	Area (in hectare)	73.59	363.48	113.4	110.15	132.6	793.22
	Production (in tonnes)	98.90	6798.68	317.66	535.84	1080.23	8831.31
CAGR	Production	5.51	7.80	3.77	4.51	2.36	6.70

Source: Economic survey, 2016, GOB.

TABLE 4.2 District-Wise Acreage and Production of Flowers (Area in Hectare/Production in Tonnes)

Districts	Rose		Marigold		Jasmine (Bela)		Tuberose	
	2013–14		2013–14		2013–14		2013–14	
	Area	Production	Area	Production	Area	Production	Area	Production
Patna	4.70 (6.4)	7.87 (8)	58.00 (16.0)	1012.00 (14.9)	5.40 (4.8)	17.28 (5.4)	5.00 (4.5)	25.88 (4.8)
Nalanda	2.50 (3.4)	3.4 (3.4)	18.75 (5.2)	413.00 (6.1)	3.20 (2.8)	9.92 (3.1)	2.60 (2.4)	13.00 (2.4)
Bhojpur	2.60 (3.5)	3.48 (3.5)	19.00 (5.2)	283.00 (4.2)	3.15 (2.8)	9.45 (3)	1.80 (1.6)	9.09 (1.7)
Buxar	1.40 (1.9)	1.54 (1.6)	3.55 (1.0)	63.90 (0.9)	1.80 (1.6)	5.22 (1.6)	1.60 (1.5)	7.84 (1.5)
Rohtas	1.10 (1.5)	1.16 (1.2)	2.30 (0.6)	38.64 (0.6)	1.50 (1.3)	4.75 (1.5)	1.20 (1.1)	5.64 (1.1)
Kaimur	1.20 (1.6)	1.5 (1.5)	2.05 (0.6)	34.85 (0.5)	1.40 (1.2)	3.78 (1.2)	2.50 (2.3)	5.56 (1)
Gaya	3.20 (4.3)	4.75 (4.8)	27.5 (7.6)	512.00 (7.5)	12.70 (11.2)	25.00 (7.9)	4.85 (4.4)	22.81 (4.3)
Jehanabad	1.20 (1.6)	1.44 (1.5)	13.00 (3.6)	283.00 (4.2)	3.45 (3.0)	10.01 (3.2)	2.40 (2.2)	11.52 (2.2)
Arwal	1.30 (1.8)	1.56 (1.6)	7.75 (2.1)	135.63 (2)	3.30 (2.9)	9.57 (3.0)	1.95 (1.8)	9.56 (1.8)
Nawada	1.50 (2.0)	1.7 (1.7)	2.45 (0.7)	41.65 (0.6)	1.25 (1.1)	3.25 (1.0)	1.10 (1)	5.06 (0.9)
Aurangabad	1.50 (2.0)	1.8 (1.8)	6.25 (1.7)	217.00 (3.2)	2.40 (2.1)	8.20 (2.6)	1.35 (1.2)	6.48 (1.2)
Saran	1.30 (1.8)	1.5 (1.5)	4.4 (1.2)	77.00 (1.1)	2.30 (2.0)	6.21 (2.0)	1.60 (1.5)	7.84 (1.5)
Siwan	1.20 (1.6)	1.44 (1.5)	3.05 (0.8)	54.29 (0.8)	1.45 (1.3)	4.06 (1.3)	1.65 (1.5)	7.90 (1.5)
Gopalganj	1.25 (1.7)	1.53 (1.5)	3.50 (1.0)	60.28 (0.9)	1.50 (1.3)	4.20 (1.3)	1.45 (1.3)	7.32 (1.4)
E. Champaran	2.50 (3.4)	3.48 (3.5)	9.40 (2.6)	174.84 (2.6)	3.70 (3.3)	10.20 (3.2)	4.80 (4.4)	24.96 (4.7)
W. Champaran	2.45 (3.3)	3.33 (3.4)	8.60 (2.4)	156.52 (2.3)	2.45 (2.2)	7.60 (2.4)	3.40 (3.1)	17.34 (3.2)
Muzaffarpur	5.72 (7.8)	7.5 (7.6)	27.00 (7.4)	628.00 (9.2)	7.40 (6.5)	23.68 (7.5)	8.85 (8.0)	40.82 (7.6)
Sitamarhi	1.05 (1.4)	1.16 (1.2)	2.90 (0.8)	49.88 (0.7)	2.00 (1.8)	5.46 (1.7)	1.25 (1.1)	5.75 (1.1)
Sheohar	0.85 (1.2)	0.89 (0.9)	1.05 (0.3)	17.85 (0.3)	0.90 (0.8)	2.34 (0.7)	0.80 (0.7)	3.60 (0.7)
Vaishali	4.85 (6.6)	8.52 (8.6)	32.5 (8.9)	610.00 (9)	7.50 (6.6)	23.25 (7.3)	9.40 (8.5)	47.94 (9)
Darbhanga	1.90 (2.6)	2.38 (2.4)	6.80 (1.9)	119.00 (1.8)	2.60 (2.3)	7.28 (2.3)	1.90 (1.7)	10.03 (1.9)

TABLE 4.2 *(Continued)*

Districts	Rose		Marigold		Jasmine (Bela)		Tuberose	
	2013–14		2013–14		2013–14		2013–14	
	Area	Production	Area	Production	Area	Production	Area	Production
Madhubani	1.00 (1.4)	1.3 (1.3)	3.55 (1)	63.19 (0.9)	3.10 (2.7)	7.69 (2.4)	1.40 (1.3)	7.00 (1.3)
Samastipur	4.00 (5.4)	5.83 (5.9)	17.53 (4.8)	315.00 (4.6)	6.35 (5.6)	19.69 (6.2)	8.95 (8.1)	49.98 (9.3)
Begusarai	2.75 (3.7)	2.88 (2.9)	12.00 (3.3)	180.17 (2.7)	3.45 (3.0)	10.35 (3.3)	3.25 (3.0)	16.25 (3.0)
Munger	0.70 (1.0)	4.15 (4.2)	10.6 (2.9)	188.68 (2.8)	3.20 (2.8)	7.20 (2.3)	4.70 (4.3)	18.32 (3.4)
Sheikhpura	0.95 (1.3)	0.77 (0.8)	0.90 (0.2)	15.48 (0.2)	1.00 (0.9)	2.66 (0.8)	0.80 (0.7)	3.76 (0.7)
Lakhisarai	0.90 (1.2)	1.09 (1.1)	2.30 (0.6)	40.02 (0.6)	1.20 (1.1)	3.18 (1.0)	0.95 (0.9)	4.47 (0.8)
Jamui	1.05 (1.4)	0.99 (1.0)	1.70 (0.5)	28.90 (0.4)	0.80 (0.7)	2.08 (0.7)	0.7 (0.6)	3.15 (0.6)
Khagaria	1.05 (1.4)	1.16 (1.2)	1.95 (0.5)	33.93 (0.5)	1.30 (1.2)	3.45 (1.1)	1.30 (1.2)	6.24 (1.2)
Bhagalpur	2.7 (3.7)	3.62 (3.7)	10.95 (3.0)	202.58 (3)	6.40 (5.6)	19.20 (6.0)	4.40 (4.0)	22.44 (4.2)
Banka	0.85 (1.2)	0.98 (1)	2.10 (0.6)	36.12 (0.5)	1.35 (1.2)	3.51 (1.1)	0.80 (0.7)	3.68 (0.7)
Saharsa	1.80 (2.4)	2.23 (2.3)	7.05 (1.9)	124.08 (1.8)	2.15 (1.9)	5.81 (1.8)	2.40 (2.2)	11.52 (2.2)
Supaul	0.70 (1.0)	0.77 (0.8)	1.95 (0.5)	33.35 (0.5)	0.85 (0.8)	2.25 (0.7)	0.80 (0.7)	3.68 (0.7)
Madhepura	0.90 (1.2)	1.04 (1.1)	2.80 (0.8)	48.16 (0.7)	1.05 (0.9)	2.84 (0.9)	1.10 (1.0)	5.17 (1.0)
Purnea	1.8 (2.4)	2.39 (2.4)	7.55 (2.1)	134.39 (2.0)	2.60 (2.3)	7.28 (2.3)	3.20 (2.9)	15.36 (2.9)
Kishanganj	2.5 (3.4)	3.20 (3.2)	8.85 (2.4)	157.53 (2.3)	3.35 (3.0)	9.05 (2.9)	5.55 (5.0)	27.20 (5.1)
Araria	1.1 (1.5)	1.23 (1.2)	1.65 (0.5)	28.22 (0.4)	1.05 (0.9)	2.73 (0.9)	0.80 (0.7)	3.68 (0.7)
Katihar	2.55 (3.5)	3.34 (3.4)	10.25 (2.8)	186.55 (2.7)	2.85 (2.5)	7.98 (2.5)	7.60 (6.9)	38.00 (7.1)
Bihar	**73.59**	**98.90**	**363.48**	**6798.68**	**113.40**	**317.66**	**110.15**	**535.84**
	(100.0)	**(100.0)**	**(100.0)**	**(100.0)**	**(100.0)**	**(100.0)**	**(100.0)**	**(100.0)**

Note: Figures in parentheses denotes percentage.

the area and production of diverse flowers grown in Bihar have been presented for the period 2009–10 to 2013–14. From the table, it is evident that about 6799 tonnes of marigold, 99 tonnes of rose, 318 tonnes of jasmine (Bela) and 536 tonnes of the tuberose were produced in 2013–14 in Bihar. Over the period 2009–10 to 2013–14, the growth rate of rose and marigold was higher at 5.51% and 7.80%, respectively.

There is a large difference in the acreage and production of various flowers across the districts. It is clear from Table 4.2. For rose, the leading districts in respect of production in 2013–14 are Vaishali (8.6%), Patna (8.0%), and Muzaffarpur (7.6%). For marigold, in terms of production in 2013–14, the leading districts have been Patna (14.9%), Muzaffarpur (9.2%) and Vaishali (9.0%). For jasmine (Bela), the leading districts in terms of production in 2013–14 are Gaya (7.9%), Muzaffarpur (7.5%), and Vaishali (7.3%).

4.5 PROFITABILITY AND SCOPE IN FLORICULTURAL SECTOR

Various present and earlier studies prove that the floricultural sector is the gainful business for smallholding farmers because of they have less capacity for cultivating diverse crops. The demand of flowers and showy foliage plants has been increasing in the metro cities and towns especially from middle and higher income group peoples and hotel industry. Prices of selected flowers are fetched relatively high during the month of September to February due to various occasions (Ganvir and Patil, 2000). Deshpande and Deshmukh (2002) studied the cost of cultivation and marketing of some flower crops viz. Rose, Gerbera, Carnation, Gladiolus, and Tuberose under High-Tech cultivation. They were revealed that flowers are economically beneficial due to high-profit margin capacity in a short time, as compared to other horticulture crops like vegetables, fruits, medicinal plants, etc. It is a well-known fact that in the initial stage of flower cultivation, the expenditure on planting material is high due to all planting materials are to be procured from outside. After that propagating material like bulbs, corms, tuberous root, suckers, clumps, terminal cuttings, etc., easily becomes available from out of the previous crops. From these propagating materials could be saved an average 67 to 88% of the total working cost, as a result of which the entire floriculture unit can be treated as a profitable venture in succeeding years as reported by Deshpande and Deshmukh (2002). An experiment was conducted by Ghadge et al. (2002) to found the most appropriate and economical flower crop based cropping systems at Ahmednagar (Maharashtra). They were tried various cropping systems viz. Chrysanthemum planted in April–wheat, chrysanthemum planted April–Rabi onion and chrysanthemum planted in June–groundnut for evaluation and compared with well-established cropping system groundnut-wheat. Among the all cropping systems, the chrysanthemum planted in April–Rabi onion and marigold-onion cropping system were found to be the most economical and gainful. The economic returns obtained by these cropping sequences were Rupees 134,667.00 and Rupees 124,391.00 per hectare and B:C ratio was 5.18 and 5.21, respectively.

A variety of cut flower crops are cultivated in protected structures. Among those Roses, Gerbera, Carnations, Anthurium, Orchid, and Lilium are most common. Gerbera and Carnations can't produce in an open field. Intensive production of rose is done by using protected structures. In other words, more inputs are required under protected structures for the cultivation of roses. Obviously, the cost of cultivation per unit is higher in protected structures as compared to field conditions. Cultivation cost in protected structures is sevenfold more than in field conditions. Marketing cost is also six times more to produce cultivated in protected structures. But net profit is greater for protected produce. Floriculture sector in India achieved its great level through the use of protected cultivation technology which gives much satisfactory results to smallholding farmers in the country (Swaminathan, 2003).

An investigation was conducted by Ravinder et al. (2006) in Punjab to study contract farming in floriculture sector and they found that five companies involved in flower seed production by this farming system. They revealed that contract farming in floricultural sector had been beneficial to farmers and contracting agencies. Kale (2008) studied production cost for rose under hi-tech cultivation; she worked out cultivation cost for one hectare of rose, i.e., Rs. 147.23 lakh. Out of Rs. 147.23 lakh was spent viz. Rs. 58.87 lakh on greenhouse construction, 41.25 lakh on rose cultivation and Rs. 20.45 lakh on input used in cultivation. She also worked out the net return per flower which was Rs. 1.46, Rs. 2.46 and Rs. 4.46 for 40 cm, 50 cm, and 60 cm long stem, respectively. She concluded that net returns of the project were Rs. 51.49 lakh. After reviewed of production cost studies it observed that protected flower production venture is important in respect to farm income. It confirms by studies, flower production is getting higher earnings compared to other agricultural crops.

Pandit and Patil (2009) revealed that protected cultivation of carnation in 2300 sq.ft. area gives Rs. 3 to 5 lakh as a profit per year to the grower. Cultivation of Chrysanthemum is the easiest and cheapest flower crop which gives higher yield throughout the year (Adat, 2011). Jadhao and Bahirat and Jadhav (2011) studied cost of production, returns and profit for rose production in Satara district of Maharashtra. They were received gross value of Rs. 380,242 per hectare with the B:C ratio of 1:1.29. Bhattachrya (2013) reported that farmers shifting from conventional farming of paddy to flower crops cultivation in West Bengal.

Floricultural sector has come out as an important agribusiness. This sector provides employment opportunities and entrepreneurship in both urban areas as well as rural areas. If anyone wants to establish floriculture business NHB helps them. APEDA provides cold storage space facilities and freight subsidies to entrepreneurs. It has been proved that commercial flower cultivation has more potential than other agricultural field crops in per unit area. That's why, it is a profitable business. Floricultural sector offers a huge chance to growers in respect to income generation and empowerment. Small and marginal farmers may also utilize their small land for growing the flowers, foliage and showy plants for trade. Floricultural sector also offers careers in different segments like production, procurement, marketing, export, and research.

4.6 POTENTIALITY OF DIVERSIFICATION IN FLORICULTURAL SECTOR

4.6.1 CONVENTIONAL FLOWER CULTIVATION

Flowers grown in the open field from ancient times are known as traditional flower and growing flowers by a traditional method called conventional flower cultivation. Traditional flowers are used for social functions, public functions, worship, festivals, and individual adornments. Mostly traditional or loose flowers are used in garlands. Some traditional flowers viz. Jasmine, rose and tuberose are also used for oil extraction and perfumes production. Conventional flower cultivation is mostly done by small holding farmers. The other conventionally cultivated flowers are marigold, chrysanthemum, China aster, crossandra, barleria, nerium, cocks comb, lotus, etc. Marigold deserves special mention followed by jasmine among the loose flowers. Jasmine is known for its distinctive fragrance. It is mainly used for hair decorations by women. It is mostly grown in Tamil Nadu and Karnataka in the south and West Bengal in the east. Crossandra and aster form the rest two varieties of loose flowers; those are cultivated in wide area in certain parts of Tamil Nadu, Andhra Pradesh, Karnataka, and Maharashtra.

This segment is still unorganized and often does not get proper importance despite of its potentiality. There are urgent needs for developing high yielding varieties, throughout the year production of loose flowers viz. marigold, chrysanthemum, China aster, etc., and promotion of flowers like annual chrysanthemum, desi gulab, etc.

4.6.2 MODERN/PROTECTED CULTIVATION OF CUT FLOWERS

Flowers which are cultivated in protected structures and harvested with their long stem known as cut flower and growing system of these cut flower known as modern or protected cultivation of cut flowers. These cut flowers are used for bouquets preparation and decoration purposes. Some important cut flowers are rose, orchids, gerbera, carnation, anthurium, lilies, tuberose, gladiolus, etc. Rose is the major cut flower crop cultivated all over the country. Gladiolus is the next major cut flower crop cultivated in India. Previously gladiolus was restricted only to northeastern region. This region still continues to supply the corms to most parts of the county. But nowadays, with improved varieties and agro-techniques and better crop management, the northern parts of Delhi, Punjab, Haryana, Utter Pradesh, as well as Karnataka and Maharashtra have appeared as the major areas for gladiolus cultivation.

Tuberose is also an important and popular cut flower crop in India. It is mostly cultivated in eastern India, i.e., West Bengal, some parts of northern plains and south India. Single and double type flower varieties are available in tuberose for commercial production and both are similarly popular. Flowers of tuberose are also selling as a loose flower in some regions for making garlands and wreaths. Orchids are also an important cut flower crop in India. Orchid cultivation is basically restricted in the North-Eastern Hilly areas and some parts of the southern states viz. Kerala, Karnataka and Tamil Nadu.

Cut flower cultivation under protected structures is limited, i.e., only 5% of the total flower crop area. But its contribution to total floricultural exports is very significant. Currently, there are more than 110 export-oriented floricultural units in function. Area under protected structures has been expanded from 500 ha to around 5000 ha in recent years. Major cut flowers are rose, gerbera, carnation, etc. grown in these units, but can be diversified into anthurium, orchids, gladiolus, lilium, and tuberose due to increasing demand of tropical cut flowers worldwide. India has many advantages and great potential to boost the area under concentrated production of cut flowers and ultimately to increase the flowers and their product exports. This is possible when these units to be established in ideal locations with sound technological knowledge.

4.6.3 DRY FLOWER PRODUCTS

Mostly used flowers and plant parts for dry flowers are Helichrysum (*Helichrysum bracteatum*), Statice (*Limonium sinuatum*), Gompherena (*Gomphrena globosa*), Dahlias (*Dahlia hortensis*), poppy seed heads, roses (Rosa), Delphinium, larkspur (*Consolida ambigua*), lavender (*Lavendula augustifolia*), Jute flowers, African marigold (*Tagete serecta*), lotus pod, etc. Dry flowers comprise almost 15% of the world floriculture trade and form the major share in Indian floricultural exports. Dry flowers share is more than 60% in total floricultural products exports of India. Presently, the dry flower industry is not well organized and mostly depends on forest for plant materials and no organized cultivation of particular flowers exists anywhere in the country for dry flowers. The demand of dry flowers is rising at striking rate of 8–10% and therefore there is a vast scope available for unemployed youth, farmers and entrepreneurs in the country.

4.6.4 PRODUCTION OF FLOWER SEED

Commercial F^1 hybrid seed and open-pollinated seasonal flower seed production are considered as a rewarding business and it is popular among the farmers on a limited area. Earlier flower seed production was being done on limited scale due to limited demand as compare to other horticultural crops. Presently many companies started production of flowers seed on large scale for export to other countries viz. U.K., USA, Holland, France, Germany, Japan, etc. Area under flower seed production is nearly 600–800 ha in India. In India, major areas involved in flower seed production are Sangrur, Patiala, and Ludhiana in Punjab; Panipat and Sirsa in Haryana; Bengaluru and Rani Banur in Karnataka; Kullu Valley (H. P.); Sri Nagar Valley in Jammu and Kashmir; and Kalimpong (W. B.). Flower seed production venture offers higher returns from per unit area to the growers. Flowers seed production cost varies from Rs. 10,000 to 15,000 per hectare and generate net profit Rs. 25,000 to 75,000. There is urgent need to develop high-yielding varieties of flowers including F1 hybrids and improved agro-techniques for producing uniform flower seed according to certification standards.

4.6.5 NURSERY BUSINESS

Quality seed and propagating material is the most important goods for successful flower production. Lack of quality seed and planting material is the main barrier for not realizing the full potential of floriculture in India. Planting material of different kind's like seed of annuals, rooted cuttings, tubers, corms, bulbs, budded and grafted plants and seedlings are required for commercial flower cultivation, plants for pots and landscaping purposes. Indian flower seed and nursery industry is going enormously yearly. Several private companies involved in producing quality planting material with multinational base in India. Increasing areas under ornamental gardening, varied agro-climatic situations and accessibility of vast and economical human resource are creating immense scope for growth of nursery business in India. Nursery industry has positive influence on Indian economy in respect to income, employment generation and foreign exchange earning in global market.

4.6.6 POTPOURRI BUSINESS

Potpourri is a mixture of dried and fragrant plant parts like roots, leaves, stems, flowers, and seeds. Aromatic oils are the basis of a potpourri which is found within the plant. A significant constituent of dried flower product export includes potpourris. Nowadays, floriculture has been considered as a viable alternative for diversification in agriculture. But at present, within the floriculture sector itself so many options are available, a florist and floriculturist could take those. Essential oils and perfumery from natural sources is in great demand. Flower crops cultivated for essential oil production are limited in the country. Flower crops are grown for essential oils are included mainly jasmine, rose, tuberose, etc. Rosa damascene species is exclusively grown for essential oil extraction, attar, gulkand, rose water, rose, etc. It is mostly grown in Pushkar and Haldighati in Rajathan, Kannoj in Uttar Pradesh and some pockets of Himachal Pradesh. There is urgent need to research for development of high oil containing varieties and distillation methods for higher oil recovery. Further, identification of more aromatic plants for extraction of essential oil and standardization of production technology for these crops needs to be incorporated in the research program. Encouragement of this sector encourages auxiliary industries like making various value-added products and steam distillation units.

4.6.7 PRODUCTION OF NATURAL DYES

The natural dyes from plants were discovered a long time ago. About 450 plant species are known which could be good source of natural dyes. Natural dyes can be extracted from various parts of plant such as seeds, leaves, barks, flowers, berries, and fruits. While, many natural dyes have been replaced by synthetic dyes, but some natural dyes are still used to add color to foods, cosmetics, and fabrics. Many colors are created with flowers and flower parts. Flowers used for natural dyes are Red Rose, Hollyhock, Marigold, Chrysanthemum, Sunflower, etc. Pigments of marigold are generally used to improve the color of the eggs yolk and meat in the poultry industry. These pigments also utilized in the food and textile industry. There is a requirement to more crops identified and standardized extraction methods for full exploitation.

Technologies development in the above-mentioned areas not only improves the condition of particular floriculture sub-sector, but these provide opportunities for employment, means of income generation sources and diversification in floriculture industry.

4.7 CONSTRAINT AND CHALLENGES IN FLORICULTURE BUSINESS

Exports of floricultural commodities from India is fallen slightly in respect to value, due to increasing domestic consumption of these products at an extraordinary speed, very high import duty, perishable nature of flowers, high rates of freight, unavailability of refrigerated transportation systems. The floricultural division is also faced several challenges at the production level like quality seeds and propagating material, quality irrigation water, skilled manpower, etc. Challenges at the marketing level also faced by our exporters due to less diversification of products and differentiation, lack of defragmentation and novelty, poor quality of products and challenges related to environmental issues. Inadequate cold chain system for transportation and storage is not only affecting present produce and its marketability but also affect the future floricultural business in the country.

4.7.1 STRATEGIES FOR FLORICULTURE BUSINESS IMPROVEMENT

1. Firstly the most important requirement is to increases an integrated cold chain system for carrying produce from the growers (production site) to customers (market).

2. Developments of the non-traditional production areas may help in fulfill the growing requirements of the both domestic and global markets.
3. Developing and propagating indigenously varieties for supply of quality seed and propagating material regularly.
4. Establish training centers and organize trainings for skilled manpower.
5. To develop network of support systems involving public institutions and private organizations.
6. Reducing import tariffs with Government efforts in negotiation with preferred tariff regimes countries.
7. Periodic re-plantation.
8. Development and popularization of integrated supply chain model, i.e., integration of small and middle-class growers in large size producer supply chains.
9. Increasing the frequencies of International and chartered flights.
10. Allocation of additional cargo space for floricultural products.

4.7.2 RESEARCH WORK IN FLORICULTURE

Research work in floriculture division is being conducted by the different research organizations like ICAR and CSIR institutes, Department of Floriculture or Horticulture in SAU's, different societies and AICFIP with a network of 21 centers (15 budgetary, 4 institutional and 2 voluntary Centers). Flower crops like rose, chrysanthemum, gladiolus, orchid, tuberose, jasmine, aster and marigold have obtained larger attention of researchers. Recently, research work in floriculture division focused on improvement in flower crops, improved propagation techniques, standardization of agro techniques, measures of plant protection and management of flower produce after harvest. Earlier most of researchers focused on traditional flower crops because of major flower crops are being cultivated in open field. But presently, mass multiplication of planting materials through tissue culture and technologies of protected cultivation has also got attention. Many ornamental and cut flower species and varieties have been identified which are appropriate for cut flower and landscaping or garden display. Various varieties also developed in cut flowers which are suitable for protected structures and also for open field cultivation. In India, the private sector contribution in floriculture research activities is very negligible.

4.8 IMPROVED VARIETIES IN FLOWER CROPS

Floriculture as a commercial activity is still largely practiced on smallholdings all over India. Flower crops cultivated in 2,49,000 ha area and this large area produced 1.69 million tonnes loose flowers and 96,327 lakh cut flowers during 2014–2015 (NHB, 2015). The improved varieties of various flowers and ornamental plants developed by different institutes can meet the increasing market demand for these crops with higher quality and yield. The information regarding improved varieties in different flower crops is discussed in the following subsections.

4.8.1 ROSE

1. **Pusa Arun:** Queen Elizabeth × Jantar Mantar.
   Recommended area: Northern plains.
   Average yield: Pusa Arun variety produces 20 flowers/plant in winter season and 35–40 flowers/plant in spring season.
   Characteristics of variety: Fleshy and meteoric dark red, big sized flower, double blooms, long and strong stems. It has 38 to 40 thick petals. Flowers are gently scented and good for cut flowers cultivation and also suitable for exhibition purpose. Pusa Arun is tolerant against powdery mildew disease and red scale.
2. **Pusa Shatabdi:** Jadis × Century Two.
   Recommended area: Northern plains.
   Average yield: 20–30 flowers/plant in winter season and 35–40 flowers/plant in spring season.
   Characteristics of variety: Attractive light pink flowers, fleshy 35–40 pink color petals, moderately tolerant to leaf spot and powdery mildew diseases, suitable for cut blooms and exhibition purposes and mildly fragrant blooms.
3. **Pusa Ajay:** Pink Parfait × Queen Elizabeth.
   Recommended area: Northern plains.
   Average yield: 15–20 flowers/plant in winter season and 35–40 flowers/plant in spring season.
   Characteristics of variety: Foliage pigmented, glossy and dark pink blooms, 35–40 petals, fleshy and dark pink color petals, recurrent blooming, moderately tolerant to powdery mildew and black spot diseases, mildly fragrant blooms, suitable for cut flower and exhibition purposes.

4. **Pusa Komal:** Pink Parfait × Suchitra.
   Recommended area: Northern plains.
   Average yield: 20 flowers per plant in winter and 45 flowers per plant in spring.
   Characteristics of variety: Thornless variety, petals are red with lighter shade on the reverse side, black spot tolerant.

5. **Pusa Mohit:** Suchitra × Christian Dior.
   Recommended area: Northern plains.
   Average yield: 20 flowers per plant in winter and 45 flowers per plant in spring.
   Characteristics of variety: Thornless variety, red petals with lighter shade on the reverse side and tolerant to black spot.
   Recommended cultivation practices
   Spacing: 50–60 cm.
   Budding: February–March.
   Planting time: September–October.
   Pruning: First fortnight of 2 October.
   Fertilizer requirement: 75 g N, 125 g P O and 100 g K O/1.44 m area and 4–5 kg FYM Irrigation: As per need.
   Disease control: Spray Captan @ 0.2% for black spot.
   Insect control: Parathion for red scale and malathion @ 0.1% for aphids control.

4.8.2 MARIGOLD

A. African Marigold

1. **Pusa Basanti Gainda:** Golden Yellow × Sun Giant.
   Recommended area: Throughout India.
   Average yield: Fresh flowers 20–25 t/ha; seed 70–100 kg/ha.
   Characteristics of variety: Medium size flower, flowers in lemon yellow color, blooms in 135 to 145 days after sowing, suitable for growing in pot and garden display purposes.

2. **Pusa Narangi Gainda**: Cracker Jack × Golden Jubilee.
   Recommended area: Throughout India. Popular in southern India due to big flower size.
   Average yield: 25–30 t/ha of fresh flowers, 100–125 kg/ha of seeds.
   Characteristics of variety: Deep orange flowers with ruffled florets, blooms in 125–135 days after sowing, rich source of carotenoids

(329 mg/1000 g petals), broadly used in food, poultry industry, and pharmaceutical industries.

B. French Marigold

1. **Pusa Arpita:** Selection from heterozygous (local) population.
   Recommended area: Northern plains.
   Average yield: 18–20 t/ha.
   Characteristics of variety: Medium-sized flowers with light orange color, blooms during mid-December to mid-February in northern India.
   Recommended cultivation practices
   Seed rate: 600–800 g/ha.
   Spacing: 45 × 60 cm.
   Sowing time: Second fortnight of July to the first fortnight of August.
   Fertilizer requirement (NPK, kg/ha): 120–80–80.
   Irrigation: As per need.
   Disease control: Spray soluble sulfur solution @ 0.2% for powdery mildew and rust diseases.
   Insect control: Spray Dicofol @ 0.3% for mites.

4.8.3 GLADIOLUS

1. **Pusa Shubham:** Lucky Shamrock × Green Lilac Open.
   Suitable for: Northern plains.
   Average yield: Produces 1.6–2.3 spikes per plant, average 2 corms per plant and 20 cormels per mother corm.
   Characteristics of variety: 14 to 16 florets per spike, florets are cream to yellow color, early in flowering (72 days) compact spikes with good vase life (10 days).
2. **Pusa Kiran:** Selection from the open-pollinated population from cv. Ave.
   Recommended area: Northern plains.
   Average yield: Produces 1.9–2.7 shoots, >2 corms, and 20 cormels.
   Characteristics of variety: Produces 1.9–2.7 shoots and more than 2 corms and 20 cormels from each mother corm.
3. **Pusa Manmohak:** A selection from Mayur × Hunting Song.
   Recommended area: Northern plains.
   Average yield: Produces 2.22 spikes and 2.33 corms.
   Characteristics of variety: Mid-maturing, blooms in 100–105 days, florets in saffron red color, whitish stripes on the throat of two oppositely

located lower petals, spikes length more than 93 cm, rachis length 55 cm, 19–21 florets per spike, 5–6 florets open at a time, excellent for vase decoration, vase life is 10 days.

4. **Pusa Vidushi:** Selection from the progeny of Melody × Berlew.
   Suitable for: Northern plains.
   Average yield: Produces 2.11 spikes and 2.33 corms.
   Characteristics of variety: Plants light green in color and straight, early and mid maturing variety, blooms in 80–85 days, straight spikes, 15–16 florets/ spike, florets purplish white in color with grayed purple spots on throat base, 11 days vase life.

5. **Pusa Red Valentine:** Selection from the open-pollinated population of the 'Regency' variety.
   Suitable for: Northern Plains.
   Average yield: Produces 2.11 spikes and 2.33 corms.
   Characteristics of variety: Plants are healthy, green in color and straight reaching a height of 125 cm, each corm produces two shoots with 7–8 leaves, mid maturing variety, flowered in 95 days, long straight spikes and good length of rachis about 50–55 cm, 18–19 florets/spike, florets brick or blood-red in color and lower petals have sun ray-like small lines, 10 days vase life, produces 2.33 corms and > 28 cormels.

6. **Pusa Srijana:** Selection from the progeny of Berlew and Heady Wine.
   Suitable for: Northern Plains.
   Average yield: Produces 2.11 spikes and 2.33 corms.
   Characteristics of variety: Very long sturdy spike (115 cm) with good rachis length (> 56 cm), 16–20 florets, more than 9 days vase life, produces 2.88 corms and 49.78 cormels per plant, medium to late flowering hybrid, takes 107 days to first floret opening after planting, floret color is red-purple group (72B) (inner two petals are dark/pink with one white stripe on center and outer petals are light white at base) with a vase life of 7 days.

7. **Pusa Unnati:** A selection from the progeny of Berlew and Heady Wine.
   Recommended area: Northern Plains.
   Average yield: Produces 2.11 spikes and 2.33 corms.
   Characteristics of variety: Medium long sturdy spikes (>85 cm), rachis length more than 49 cm, 15–17 florets, vase life (9 days), good multiplier produces 3.10 corms and 27.44 cormels from each mother corm, early flowering hybrid, first floret opened in 73.22 days, dark pink

or mauve color floret (purple group, N-78B), best for kitchen garden, garden display and landscaping also.

Recommended cultivation practices:
Requirement of corms for planting: 1.5 lakh/ha.
Spacing: 60 cm.
Sowing: October to November.
Fertilizer requirement (NPK, 2 g/m): 25–16–25.
Irrigation: As per need.
Disease control: Spray of Captan @ 0.2% for black spot.
Insect control: Mixing of Thimet 10 G granules @ 20–25 kg/ha in soil at land preparation to control chafer beetle and 0.2% spray of Metacid-50 for control of aphids and thrips.

4.8.4 CHRYSANTHEMUM

1. **Pusa Anmol:** Mutant of cv. Ajay.
   Recommended area: Hills and plains to get off-season blooms.
   Average yield: 100–150 flowers/plant.
   Characteristics of variety: It is bushy variety with yellowish pink flowers, thermo-photo insensitive variety, blooms thrice in a year, i.e., October–November, February–March and June–July, flowers in 85–100 days, ideal for loose flowers, vase life on the plant is 20–22 days.
2. **Pusa Centenary:** Gamma-ray induced mutant of cv. Thai Chen Queen.
   Recommended area: Hills and plains to get off-season blooms.
   Average yield: 10–12 standard flowers/plant.
   Characteristics of variety: A vigorous variety that produces very big size yellow flowers, blooms in 100–110 days after transplanting, suitable for cut flowers, vase life on plant is 20–22 days.
3. **Pusa Aditya:** Seedling of cv. Jaya.
   Recommended area: Northern Plains.
   Average yield: 300–400 flowers/plant (Spray type).
   Characteristics of variety: The Plants are bushy (30–35 branches) and medium in height (55–60 cm) with moderate spread (45–50 cm). The variety is a spray type that produces star-shaped semi-double flowers that resemble to gazania flowers. Flowers are yellow in color (5 A) at the periphery and orange red color (45A) in the center. The florets are spatulate with distinct keel. This is suitable for garden display purposes.

4. **Pusa Chitraksha:** A seedling of cv. Lal Pari.
Recommended area: Northern Plains.
Average yield: This is spray type variety, 400–500 flowers/plant.
Characteristics of variety: Plants are bushy with 24–30 branches, plant height is 60–65 cm with best spread of 60–65 cm, spray type variety, flower deep magenta in color, florets are spatulas in shape with magenta color (59A), disc florets are yellow (12A) that provide good contrast, suitable for pot culture and good for garden display purpose.

5. **Pusa Sona:** A seedling of cv. Sadbhawana.
Recommended area: Northern Plains.
Average yield: 200–300 flowers/plant (Spray type).
Characteristics of variety: Bushy type plant (20–25 branches), dwarf in height (25–30 cm) with an excellent spread (50–55 cm). Spray type, produces single flower, flowers yellow in color (8A), disc florets are also yellow (12C), early variety (20 days early flowering), suitable for pot mums.

6. **Pusa Kesari:** A gamma-ray induced mutant of Thai Chen Queen.
Recommended area: Northern Plains.
Average yield: 30 flowers/m^2.
Characteristics of variety: The plants are tall in height (65–70 cm) with good spread (60–65 cm). The flowers are semi-double and big in size (9–10 cm in diameter). It is red in color, red-colored ray florets (171A) appeared as chimers, suitable for pot culture and cut flower.

7. **TQP-06-01:** An induced mutant of Thai Chen Queen.
Recommended area: Northern Plains.
Average yield: 30 flowers/m^2.
Characteristics of variety: It is a gamma-ray induced pink-colored mutant of cv. Thai Chen Queen, which is orange in color. Pink colored ray florets (65D) appeared as chimeras which were used for *in vitro* regeneration to establish as a new variety. The plants are medium in height (50–55 cm) with good spread (60–65 cm). The semi-double flowers are big (7–8 cm diameter). The variety is suitable as a cut flower and pot culture.

Recommended cultivation practices:
Spacing: Line to line 30 cm and plant-to-plant 30 cm.
Time of cutting: May-June.
Pinching: After 30 days of planting.
Transplanting: First fortnight of August month.

Fertilizer requirement (NPK, kg/ha): 200-100-100.
Irrigation: As per need.
Insect control: Aphids-control by spray of Rogor (30%) @ 2 ml/l water.

4.8.5 TECHNOLOGY FOR DRYING FLOWERS

Drying of Chrysanthemum: Drying technology for chrysanthemum varieties like Gauri, Vasantika, and Jayanti has been standardized. Fresh flowers are inserted in silica gel and dried at 45°C for 48 hours in hot air oven which was ideal for Vasantika, Gauri and Jayanti.

4.8.5.1 CROP-SPECIFIC DRYING METHODS FOR ANNUAL FLOWERS

1. **Crop Drying method**
   - Calendula: Press drying; Flowers drying in the microwave oven for 90 seconds.
   - Marigold: Press drying; Drying of flowers in microwave oven for 120 seconds.
   - Larkspur: Press drying; Drying at 40 C in hot air oven for 24 hours.
   - Pansy: Press drying; Drying of flowers in microwave oven for 100 seconds.
   - Poppy: Press drying; Drying of flowers in microwave oven for 90 seconds.
2. **Initiatives Taken at B.A.U., Sabour**
   i. High value flower crops and their varieties tried in protected cultivation/ polyhouse. The findings in brief are given hereunder.
      - ➢ In Orchid, cultivar D. Sonia produced maximum number of flower sticks and gave more earnings (Agenda Notes, 12th RCM, Rabi, 2016).
      - ➢ In Anthurium, the cultivar Xavia followed by Angel performed better with respect to growth, quality and yield of spikes (Agenda Notes, 12th RCM, Rabi, 2016).
      - ➢ In gerbera, 10 varieties viz. Laura, Szantal, Delfin, Newada, Olympia, Kormoran, Partrizia, Rock, Feliks, Samuraj were evaluated. Among all of these varieties patrizia recorded maximum number of flowers per plant with longest stalk length (Singh, 2017).

ii. Identified marigold lines suitable for summer cultivation. The findings in brief are given hereunder.

   - An experiment was carried out to evaluation and screening of twenty genotypes during summer season (March, 2014) from which three promising lines were identified. Maximum yield was recorded in BRM 113 (461.93 g) followed by BRM 714 (438.70 g) and BRM 414 (412.30 g). High variability among the genotypes was observed in traits viz. Days to flowering, height of plant, spread of plant, length of stalk, secondary branches, flowering duration, number of flowers/plant and yield/plant which can be helpful for future breeding program for marigold improvement. Varietal differences were found significant for these traits (Annual Report BAU, 2014–15).

3. Ongoing Research Projects:

i. Selection of suitable vegetable and flower cultivars for protected cultivation in Bihar (Figures 4.1–4.6).
ii. Identification of marigold lines suitable for summer cultivation.
iii. Value addition of flowers through dehydration techniques.

FIGURE 4.1 Chrysanthemum in polyhouse.

FIGURE 4.2 Anthurium var. Xavia in shade net house.

FIGURE 4.3 Anthurium var. Angel in shade net house.

FIGURE 4.4 Dendrobium var. Singapore White.

FIGURE 4.5 Dendrobium var. Sonia.

FIGURE 4.6 Gerbera cultivation in polyhouse.

KEYWORDS

- **chrysanthemum**
- **floriculture**
- **flower crops**
- **natural dyes**
- **polyhouse**
- **potpourri business**

REFERENCES

Adat, S. S., (2011). Chrysanthemum cultivation. *Baliraja Magazine* (pp. 82–85).

Agenda Notes (p. 2). 12th RCM, Rabi 2016, Directorate of Research, BAU, Sabour.

Annual Report BAU, (2014–2015). *Sabour*, p. 47.

APEDA, (2016). *Ministry of Agriculture, Government of India, New Delhi.* http://agriexchange.apeda.gov.in/indexp/genReport_combined.aspx (Accessed on 21 November 2019).

Bahirat, J. B., & Jadhav, H. G., (2011). To study the cost returns and profitability of rose production in Satara district, Maharashtra. *The Asian Journal of Horticulture*, *6*(2), 313–315.

Bhattacharya, R., (2013). Behavioral, education, economics, business and industrial engineering. *International Journal of Social*, *7*(6), p. 1398.

Deshpande, S. D., & Deshmukh, C. M., (2002). Hi economics of production and marketing of selected floriculture plants under Hi-tech growing. *Journal of Maha. Society of Agril. Economics,* 42.

Ganvir, & Patil, (2000). Marketing of selected flowers in GalveKadi (pune market). *Maharashtra Economic Society Journal*, 24, 25.

Ghadge, H. L., Mokate, A. S., Deshmukh, P. H., & Pawar, V. S., (2002). Production potential and economics of floriculture-based crop sequences in irrigated areas of Maharashtra. *Indian Journal of Agronomy*, *47*(4), 499–503.

Kale, M. S., (2008). *Gulab Phulanche Utpadan ek Vyasti Adhyayan*. Dissertation submitted to T.M.V., Pune.

Kalmegh, S., & Singh, N., (2016). Review of floriculture as a promising industry for marginal farmers in Maharashtra. *International Journal of Scientific and Technology Research*, *5*(7), 141–144.

Khan, D., & Fazili, A. I., (2015). A SWOT analysis of floriculture industry in Kashmir. *Abhinav International Monthly Refereed Journal of Research in Management and Technology*, *4*(12), 1–4.

Misra, D., & Ghosh, S., (2016). Growth and export status of Indian floriculture: A review. *Agricultural Reviews*, *37*(1), 77–80.

National Horticulture Board Database, (2015). Available from: http://nhb.gov.in (Accessed on 21 November 2019).

Nazir, M., (2015). *Floriculture-Scenario*. Posted on 30/12/2015 by Daily excelsior, http://www.dailyexcelsior.com/floriculture-scenario (Accessed on 21 November 2019).

Pandit, S. N., & Patil, M., (2009). Studied the cultivation of carnation in polyhouse with organic farming in Pune district. *Baliraja J.*, 20–22.

Ravinder, S., Dhaliwal, H. S., & Joshi, A. S., (2006). Contract farming of floriculture in Punjab-problems and prospects. *Journal of Ornamental Horticulture*, *9*(3), 15.

Singh, P., Bhardwaj, A., Kumar, R., & Singh, D., (2017). Evaluation of gerbera varieties for yield and quality under protected environment conditions in Bihar. *Int. J. Curr. Microbiol. App. Sci.*, *6*(9), 112–116.

Swaminathan, M. S., (2003). An action plan for agriculture for coming 25 years. *Report of Maharashtra Council of Agriculture Education and Research*, pp. 30–39.

PART II

Natural Resource Management

CHAPTER 5

Climate Change and Natural Resource Management

SUBORNA ROY CHOUDHURY and SYED SHERAZ MAHDI

Department of Agronomy, Bihar Agricultural University, Sabour, Bhagalpur, Bihar, India

ABSTRACT

Climate change now becomes a serious concern for agricultural production and its sustainability. Natural and anthropogenic moderation causes irregularities in the normal climatic phenomenon, thus creating a severe threat to human survival. Higher number of natural calamity phenomenon like every year occurrence of heals and cold waves, tropical cyclone, flood, drought, etc., are apprehending the climate change phenomenon. Several scientific technologies have been developed to cope up with the ill effect of the climate change. Site specific management of natural resources like water, nutrient; tillage management practices, development of new varieties are considered as an adaptive strategy to sustain agricultural production under changing climatic scenario.

5.1 CLIMATE CHANGE

Global warming and climate change are the most widely used terms since last three decades and they are often used synonymously. But the term 'climate change' broadens its impacts from every sphere. It includes changes in the pattern for an extended period of temperature, precipitation, or wind. It may occur either instantaneous natural factors (slow changes in the Earth's orbit around the sun) or faulty human activities. In India, climate change is explicated by the concurrent occurrence of disasters like drought (65% Indian landmasses are prone), flood (12% area), and tropical cyclones (~8%). As a corollary, climate change put forward its destructive impacts on livelihoods viz., socio-economic, cultural, ecological systems, infrastructural, etc., and posing severe risks to the development of the developing country. Since inceptions its impacts are observed at multiple levels from the global to the local; thus, its mitigation strategy should be local to global.

5.2 CLIMATE CHANGE AND INDIA

Although, India is an accelerated growing economy country in the world, but it ranked fourth-largest emitter of greenhouse gases (GHG), contributed about 5.8% of total global emissions. In the last two decades, the GHGs emission rate has been enhanced by 67.1% (1990–2012) and projected to be 85% by 2030 under a business-as-usual scenario. India is designated as the most vulnerable country in perspective of climate change impacts due to huge population pressure, acute dependency on agriculture and excessive utilization of natural resources. Cereal productivity is likely to decrease up to 10–40% by 2100 and severe loss is expected in the winter season. There are several evidences of negative impacts on wheat and rice productivity in the major growing areas of India due to temperature hike, water stress and reduction in a number of rainy days. Model projects a significant decrease in cereal production by the end of this century (Mahdi et al., 2015). However, climate change behavior and its consequent impacts are different in different parts of the country. For example, Parts of western Rajasthan, Southern Gujarat, Madhya Pradesh, Maharashtra, Northern Karnataka, Northern Andhra Pradesh, Utter Pradesh, Bihar, and West Bengal are expected to be more vulnerable in respect of extreme events.

5.3 CLIMATE CHANGE AND BIHAR

Bihar state is usually vulnerable to hydro-meteorological natural disasters, with north Bihar being highly flood-prone and south Bihar being highly drought-prone. The northern and southern parts of Bihar were assessed to be most vulnerable regions in the Indo-Gangetic Plains (IGP). Twenty-eight districts have been categorized under medium to very high degree of vulnerability owing to high exposure, high sensitivity and low adaptive capacity of people (Sehgal et al., 2013). Last 45 years trend analyzed weather attribute data of different Agro-climatic zones of Bihar revealed the signs of climate change induce variability in temperature and rainfall events (intensity, frequency and duration), which projects an increasing trends in maximum and minimum temperatures by 2–4°C and irregular distribution (–25 to + 30% deviation) in the monthly rainfall pattern that may cause large implications on the agricultural production, food security and livelihoods. Weather-related extremes (heatwave, cold waves, floods, droughts, cyclones, etc.) have become a recurrent phenomenon, which affects more than 45% of the state's geographical area (Economic Survey, 2014). Frequent disasters are often attributed to the state's ever-increasing population (103.8 million), its growth rate (25.42%/decade) and density (1102 people per sq. km), which has made the region second most populous state in India (Economic Survey, 2014).

Climate change may be ascribed with the aberrations in the climatic normals and Bihar is a live example within effect especially on water availability. The mean annual rainfall in Bihar is 1120 mm, but its magnitude varies from 1000 mm in the western and south-western Bihar to 2000 mm in the extreme eastern and northern Bihar and 33% of the State receives <750 mm rainfall annually. Thus, Bihar vulnerable with two opposite natural calamities; southern part of Bihar by drought and northern part by flood. Although 35% area in Bihar (northeastern part) receives mean annual rainfall once in 4–5 years since last two decades. The major implications for changes in the rainfall pattern threaten crop protection and food security. Lack of state level climate models and/or vulnerability studies, as well as low community awareness, the state like Bihar is potentially more delicate and vulnerable towards the changing climate.

5.4 TRUTHS OF CLIMATE CHANGE

The IPCC postulated that if the average surface temperature is rised by 1.4°C to 5.8°C in the next 100 years, sea levels will expected to rise by 9–88 cm depending partially on the future GHG emissions rate. This situation will also aggravate tropical cyclones intensity, higher wind speeds, and rainfall, as well as the intensity of other extreme events of weather, including storms. The ranges of projected temperature and sea-level rises reflect different assumptions about future global GHG emissions and, to a lesser extent, uncertainties associated with earth's circulation system. Despite so many international efforts to reduce global emissions it is continued to grow, particularly in the developing countries. There is also growing concern that biological feedback, such as melting of permafrost, emissions from peat bogs, may cause higher emissions. These raises trigger the possibility of global warming, causing the release of biologically fixed carbon as carbon dioxide. Thus carbon dioxide concentration becomes doubled by the end of this century from pre-industrial era.

5.5 CAUSE OF CLIMATE CHANGE

Excavating intensely into the roots of climate change it becomes fairly clear that there are enough scientific evidences that high concentrations of GHG in the atmosphere, due to human activities, are intensifying the natural "greenhouse effect" thus increasing the Earth's temperature, ultimately changing the climate and direct or indirect effect on agriculture. Agriculture accounts for approximately 10–12% of total global anthropogenic of GHG emissions. People usually release chlorofluoro carbons into the atmosphere, thus degrading stratospheric ozone and increasing biologically harmful ultraviolet (UV) radiation that reaches earth's surface. Methane (CH_4) and nitrous oxide (N_2O) is primarily responsible for global warming, because their global warming potential (GWP) are 25 and 298 times greater, respectively, than that of CO_2 over a time span of 100 years (IPCC, 2007). Global warming may distort global carbon cycle, thereby structures and functions of ecosystem. GHGs emission from agricultural land depends on soil properties as well as environmental factors which get influenced by the management practices and global warming respectively (Roy Choudhury et al., 2015a).

5.6 IMPACT OF CLIMATE CHANGE

5.6.1 MAJOR CEREALS

Heat stress caused significant reduction in phenological dates and final yield of wheat. The results obtained from the research conducted in BAU, Sabour, exhibited that the days taken to attain anthesis, maturity and yield were less under elevated temperature of 5°C and above 5°C, when compared to phenological dates and yield of wheat crop exposed no heat stress (Normal). It can be further inferred that among the different crop growth stages, a thesis to milk stage was most sensitive to heat stress, where physiological maturity got reduced to 9 days and yield was reduced to 29.05% (Agricultural Technologies Inventory, 2016). The grain yield gets declined at elevated temperature significantly than the normal. When the temperature rises beyond 5°C above normal the grain yield was lowest to the tune of 3660 kg/ha, followed by 5°C with 3965 kg/ha compared to normal 5363 kg/ha. Rice yield reduction due to elevated temperature was 26.0 and 31.75% for 5°C and above 5°C treatments respectively (Research Farm, BAU, Sabour). Similarly based on another model it was revealed that with 2°C increase in temperature during panicle initiation to anthesis stage in rice resulted in higher grain yield reduction for 30th June transplanting compared to 15th July transplanting. The impact of temperature rise during a thesis period of wheat was more for timely sown conditions compared to late sown conditions (Agricultural Technologies Inventory, 2016).

5.7 NATURAL RESOURCE MANAGEMENT TO OVERCOME THE IMPACTS OF CLIMATE CHANGE

It was found that GHG emissions from the agriculture sectors contribute over 30% of the current annual total emissions (deforestation and forest degradation 17.4%, agriculture 13.5%). Soils significantly contributed towards GHG emission into the atmosphere. Many methods have been tested to reduce GHGs emissions from agricultural field as well as mitigate the impact of climate change. Some important methods for reducing the GHG emission are described in subsections.

5.7.1 SITE SPECIFIC NUTRIENT MANAGEMENT TECHNIQUES

Application of fertilizer as per the requirement of crop improves the fertilizer use efficiency due to application of lesser amount of fertilizers. As a consequence, less accumulation of nutrients within the crop root zone reduces the emission of GHG particularly nitrous oxide where nitrogen is added as per the requirement of the crop. Numerous studies have revealed the impact of chemical fertilizers on CH_4 and N_2O emissions which depend on rate, type, and mode of fertilizer application. Application of ammonium nitrate exhibits competitive inhibition to nitrate reduction caused reduced methane emission. Under field conditions, it has been found that the application of sulfate-based fertilizers such as $(NH_4)_2$ SO_4 and $CaSO_4$ have reduced CH_4 emission, while application of K_2HPO_4 enhanced the CH4 emission. It has been reported that application of undecompost organic manure increases the methane flux. On the other hand, application of well-decomposed vermicompost (Research Farm, BAU, Sabour) and the use of sulfur amendment has been found to be very effective for reducing methane flux from rice fields.

5.7.2 METHOD OF FERTILIZER APPLICATION

It is also important in the mitigation of GHG. It has been found that the application of urea in plow layers gives less emission of N_2O than band application. The addition of phosphorus and liming materials can also affect nitrous oxide evolution from the soil. However, phosphorus induced emissions are higher than those obtained with lime. Plant uptake of fertilizer N can be improved by various methods as deep placement and placing fertilizers in the band.

5.7.3 TILLAGE MANAGEMENT

Optimizing tillage practices, i.e., no tillage or minimum tillage (Research Farm, BAU, Sabour), bed planting and reducing compaction also help in the mitigation of nitrous oxide and methane flux from the soil. Direct sowing of rice recorded lower methane flux as compared to mat type paddy transplanter and manual transplanting (Research Farm, BAU, Sabour).

5.7.4 WATER MANAGEMENT

Out of several factors affecting GHG emission, water management plays a very crucial role. Various technologies viz., drip irrigation, sprinkler irrigation, use of soil moisture sensor to predict irrigation requirements have immense potential in increasing water and N-use efficiencies. It has been found that water regime in irrigated rice fields with higher water percolation and poor water supply often results into multiple aerations, which has direct bearing on CH_4 emissions. Therefore, water management practices might be one of the most promising measures to reduce CH_4 emission from rice fields. Keeping the rice field moist rather than flood or continuously saturated, thereby minimizing anaerobic conditions, and improving root growth and diversity of aerobic soil organisms, helps in mitigation of climate change.

5.7.5 USE OF NITRIFICATION INHIBITORS

N_2O is formed as a gaseous intermediate during oxidation of ammonia (NH_3), i.e., nitrification and during reduction of nitrates (NO_3), i.e., denitrification by soil microbes participating in the N-metabolism cycle. Paddy fields are considered a major source of N_2O emission as flooded paddy soils have both an aerobic surface layer and an anaerobic sub-surface layer where N_2O can be produced via nitrification of ammonium to nitrate and denitrification of accumulated nitrate respectively with the aid of specific microbes. Thus in order to control N_2O emission from soil into the atmosphere, use of specific microbial inhibitors, which can inhibit or slow down N_2O formation, can prove to be a potential management strategy for safeguarding our environment from the harmful effects of N_2O increase in our atmosphere. Thiourea had shown an extremely positive effect on plant growth as well as yield but its inhibitory effect on N_2O emission was not found to be substantial in comparison to other treatment. However, the application of Dicyandiamide (DCD) was found to very effectively control N_2O emission rate from the soil while also maintaining the yield at a considerable higher level.

5.7.6 SEED INOCULATION

Seed inoculation with stress-tolerant strain of *Pseudomonas putida* helped sorghum and pearl millet seedlings survive at 500C up to 21 days, whereas the controlled seedlings could survive only up to 10 days.

5.7.7 USE OF MICROBES

One of the key emerging technologies to reduce GHGs emissions from rice fields is the use of zymogenous bacteria, acetic acid and hydrogen producers; methanogens, methane oxidizers, etc. in rice, which will help in maintaining the soil redox potential in a range where both N_2O and CH_4 emission are low.

5.7.8 SELECTION OF SUITABLE VARIETIES

The emission of methane and nitrous oxide gases can be mitigated by choosing the appropriate variety that can help in reducing the ill impacts of climate changes. The nitrous oxide flux from five different varieties of vegetable pea and concluded that variety VL-7 is most suitable for reduced N_2O emission from the field grown with vegetable pea. Crop cultivars with low emission, high N use efficiency, faster decomposition of residues and less water demanding have potential to reduce nitrous oxide emission from soil.

5.7.9 CLIMATE RESILIENT VARIETIES

Utilization of tolerant genetic resources for direct use or for development of new resistant variety or hybrid is the most important mitigation approach. Use of tolerant rootstocks for grafting scions of otherwise good quality but susceptible genotypes can be an efficient way to cope up with the vagaries of climate change. Grafting of fruit trees has been common for centuries, but it is relatively new in vegetable farming. The rootstock is selected based on its ability to enhance vigor and marketable yield and resisting the relevant climatic stress. In tomato *Solanum pimpinellifolium* resistant to heat stress, *S. cheesmaniae, S. chilense, S. pimpinellifolium, S. penellii, S. lycopersicum var. cerasiforme, S. peruvianum* resistant to drought stress, *S. cheesmaniae, S. chilense* resistant to salinity stress, *S. habrochaites* is resistant to Chiling stress and in brinjal, *Solanum incanum* resistant to drought stress, *S. torvum, S. sisymbrifolium* resistant to heat stress can be efficiently used as rootstock for grafting.

5.7.10 APPLICATION OF HORMONES

The adversity of temperature (both high and low) leads to flower drop resulting in no fruit set in vegetables. Foliar application of para-chlorophenoxy acetic acid (PCPA) 50–100 ppm at the flowering stage triggers the fruit set at low and high temperature as found in tomato. Besides, tomato fruit setting has also been reported to be promoted by gibberellic acid (GA) at low concentration (20 ppm). The application of synthetic auxin 4-CPA (4-chloro phenoxy acetic acid) reduced pre-harvest fruit drop and increased the number of fruits per plant and yield.

5.8 AGRONOMIC PRACTICES

Mulching is another management aspect that can confront the environmental vagaries. Mulching improved the growth of brinjal, okra, bottle gourd, round melon, ridge gourd, and sponge gourd compared to the non-mulched controls. Straw mulch lowers temperature in summers. Silver mulch reflects back heat. In winters, black polythene mulch increases soil temperature, thus enhancing germination and root activity. Studies of AVRDC (1990) have shown that dark-colored plastic mulch in combination with rice straw was effective for vegetables, which prevents injuries due to direct sunlight and maintains a conducible temperature for crop growth Mulching also reduces fruit cracking and unmarketable fruits numbers.

5.8.1 *PROTECTED CULTIVATION*

Protected structures such as polyhouses, greenhouses and shade nets can be very effective in the changing climate scenario. Polyhouses are less prone to waterlogging higher yields of crops can be obtained. Low/medium cost polyhouses could raise temperature in the winters by 6–7°C and satisfactory crop growth can occur. Shade nets and shade net houses can be effectively used to reduce high temperature stress. Polyhouses with cooling system and/ or temperature control can also be the means of crop cultivation under high temperature stress.

5.8.2 *PLANTING METHODS*

Planting vegetables in raised beds can improve the ill effects of flooding during the rainy season and ultimately enhance vegetable yield. Raised bed planting facilitates good drainage and confiscates anoxic stress.

5.9 APPROACHES OF BIHAR AGRICULTURAL UNIVERSITY (BAU), SABOUR

5.9.1 *AGRONOMIC APPROACHES*

Good agronomic practices, efficient nutrient and weed management, and plant protection measures would have predominant influence on the productivity of different crops.

5.9.2 *CROP MANAGEMENT*

Direct seeded rice (DSR) followed by zero till wheat, under rice-wheat cropping system which can augment annual GWP by reducing 62% of CH_4 emissions through moist irrigation or alternate wetting and drying in rice (Research Farm, BAU, Sabour). Therefore, DSR followed by zero tillage technology may play a major role in making Bihar self-sufficient in cereals. Effort should be given to provide an enabling environment for the accelerated spread of this technology as an important element of sustainable intensification in Eastern India (Agricultural Technologies Inventory, BAU, Sabour).

5.9.3 NUTRIENT MANAGEMENT

50% reduction in Nitrogen application and incorporation of bio-fertilizers, brown manuring and residue mulching with or without FYM in soil helps in maintaining organic carbon, soil temperature, reduction in NO_3 loss, improvement in soil micro flora, etc. were found important strategies to mitigate the ill-effects of climate change in agriculture (Research Farm, Sabour and Patna). In an another experiment regarding precision Nitrogen management it was found that SPAD based real-time N management was effective in saving of nitrogenous fertilizer up to 24% in rice and 27% in wheat (Research Farm, BAU, Sabour) without deteriorating the grain yield over conventional fixed time N management in the cropping system (Agricultural Technologies Inventory, BAU, Sabour).

1. **Conservation Agriculture:** Seed-cum-fertilizer drill under no-till condition increased yield, provided soil N for subsequent crops, enhanced organic matter content of the soil and microbial biomass, reduced moisture loss, saved energy as compared to conventional tillage (Research Farm, BAU, Sabour).
2. **Adjustment in Sowing Dates:** Among the different dates of sowing for wheat, 15th November sown crop recorded the highest yield and 15th July transplanting for rice (Research Farm, BAU, Sabour). Delayed transplanting of rice on 30th July and 15th August lead to yield decrement of 51 and 55 kg/ha/day respectively compared to 15th July transplanting of rice. Delayed in sowing from 15th November, yield reduced by 6, 15, and 32% which accounted for yield decrease of 17, 23 and 33 kg/ha/day for 30th November, 15th and 30th December sowing respectively (Agricultural Technologies Inventory, BAU, Sabour).
3. **Application of Synthetic Compounds:** The extent of increment of yield was found to be around 10–11% on average as a result of foliar spray of synthetic compounds (KNO_3 and $Cacl_2$) (Research Farm, BAU, Sabour). The treatments recorded significantly higher yield also significantly reduced electrolytic leaf leakage and Proline content in flag leaf both at anthesis and grain filling stage as well as maintained a higher level of chlorophyll content. Thus these treatments significantly improved high-temperature stress characteristics of late sown wheat (Agricultural Technologies Inventory, BAU, Sabour).

5.10 NANO-TECHNOLOGY

Nanotechnology seems to play pivotal role in agricultural sciences in various sub-disciplines (Increasing input use efficiency, draught stress alleviation, novel agrochemicals formulations, increasing shelf life of harvested fruits and vegetables and production of quality planting materials, etc.) in changing climatic scenario. Specific application of nanotechnology in climate change alleviation is being mentioned in the following sections like, alleviation of draught: Novel superabsorbent nanopolymeric hydrogel having equilibrium water absorbency (WA) (ranging from 350 g g^{-1} to 500 g g^{-1}) is good materials under moisture stress condition. Products are biodegradable, can withstand high salinity level (100 Mm NaCl solutions), input use efficiency: Nanofertilizers (P, K, Zn, Fe, etc.) with higher use efficiency and greater diffusion rate in rhizosphere as well as within plant system would bring cost-effective nutrient formulation in changing climate scenario. Mode of nutrient transport via diffusion is temperature and moisture controlled which could be effectively controlled by nano-formulations, agrochemical formulation: Shift in pest dynamics and population is anticipated in changing climate scenario. Novel nanopesticides (Nano silver, cu, and chitosan) with minimal dose, more biodegradability and cost effectiveness would of great interest in changing climate scenario.

Carbon sequestration: Nano titanium dioxide (TiO_2) particles are reported to increase carbon sequestration potential in soil by escalating degree of aromatization and polycondensation of carbon. Increased C storage would facilitate in combating global warming, photosynthetic radiation use efficiency: Silicon nanowires, a solar cell for Mg and Ti nanoparticles that increases photosynthetic radiation use efficiency. Silica nanoparticles as a tool for fluorescence collection efficiency enhancement would be of great use in changing climate scenario. Temperature sensors for managing radiation use efficiency would be of good use, increasing shelf life: nanosilicon, silicon dioxide, and nanopolymeric coating materials would improve shelf life of fruits and vegetables in changing temperature scenario. Nanobiosensors and nanosensors for intelligent packing materials development and determination of contaminants in packaged food materials are of great help. Nevertheless agricultural application of nanotechnology is its infancy stage. Systematic scientific researches on long-term basis need to be done following interdisciplinary approach with due consideration in nanotoxicology for its wide scale application.

1. **Breeding Approaches:** Efforts are being made for crop improvement by large scale screening of the accessions/germplasm lines and

by using various traditional and non-traditional methods resulted in number of good improved crops varieties.

2. **Sabour Surbhit (Rice):** It can perform well under limited irrigated condition. It is the semi-dwarf variety with long and super fine slender grain and strong aroma with yield potential of 30–35 quintal per hectare and suitable for irrigated medium up land and medium land situation of Bihar (Agricultural Technologies Inventory, BAU, Sabour).
3. **Sabour Ardhjal (Rice):** It can save about 50% of water requirement for rice. It is an aerobic rice variety with long slender grain suitable for upland and medium land under non-puddled, direct seeded and non-flooded irrigated condition with intermittent irrigation. The yield potential is 50–55 quintal per hectare (Agricultural Technologies Inventory, BAU, Sabour).
4. **Sabour Deep (Rice):** This variety is very much suitable for contingent plan under late arrival of monsoon. It is an early maturing, semi-dwarf variety with long slender grain with yield potential of 40–45 quintal per hectare and suitable for irrigated medium up land and medium land situation of Bihar (Agricultural Technologies Inventory, BAU, Sabour).
5. **Sabour Samriddhi (Wheat):** It is tolerant to numerous emerging diseases like, Leaf Blight, Brown Rust and major insect-pest, Suitable for timely sown irrigated condition (TS- IR), 120–125 Days Maturity, Medium Height, Lodging resistance, Non Shattering, Fertilizer responsive, having good Chapatti Making Quality, Yield 45–50 Q/Ha (Agricultural Technologies Inventory, BAU, Sabour).
6. **Sabour Nirjal (Wheat):** It is lodging resistance and low water required variety. Suitable for timely sown rainfed condition (TS-RF), 125–135 Days Maturity, Medium Height, Long Earhead, Non Shattering, Fertilizer responsive, Tolerant to Leaf Blight, Brown Rust and major insect-pest, Good Chapatti Making Quality, Yield 25–30 Q/Ha (Agricultural Technologies Inventory, BAU, Sabour).
7. **Sabour Shreshtha (Wheat):** Suitable for late sown irrigated (LS-IR) condition, 105–110 Days Maturity, Semi Dwarf, Lodging resistance, Non-Shattering, Fertilizer responsive, Tolerant to Leaf Blight, Brown Rust and major insect-pest, Good Chapatti Making Quality, Yield 43–45 Q/Ha (Agricultural Technologies Inventory, BAU, Sabour).
8. **DHM 117 (Maize):** It is tolerant to major evolving diseases and insect-pests. Medium duration hybrid recommended for Kharif and rabi season, Yield: 95–100 q/ha, Grain color orange, Plant stay green at harvest (Agricultural Technologies Inventory, BAU, Sabour).

9. **SHM 1 (Maize):** It is tolerant to major evolving diseases and insect-pests. Early duration (72–75 days) hybrid, Recommended for Kharif season, Yield: 55–60 q/ha, Grain color yellow (Agricultural Technologies Inventory, BAU, Sabour).
10. **SHM 2 (Maize):** It is tolerant to newly emergent diseases and insect-pests. Heat stress tolerant medium duration (100–105 days) hybrid Recommended for spring season, Yield: 65–70 q/ha, Grain color orange yellow (Agricultural Technologies Inventory, BAU, Sabour).

In order to improve the present cultivar and enhancing the yield, BRC-1 and BRC-2 (Chickpea), BRM-1 and BRM-2 (green-gram), BRP-1 and BRP-2 (an early and extra-early cultivar of Pigeon pea) have been identified by the breeders which have potential to cope in several biotic and abiotic stresses. At present, multi-locational trails is going on for these identified varieties and will be released in near future.

Therefore, there have been only limited studies on the impacts of climate change at state level of Bihar. The identification of suitable response strategies is crucial to develop sustainability in agriculture. The important mitigation and adaptation strategies required to cope with anticipated climate change impacts is the resilient management of natural resources which include adjustment in sowing dates, breeding of plants that are more resilient to variability of climate, and improvement in agronomic practices. The proper understanding of climatic conditions and efficient utilization of natural resources are, therefore, of great concern for the improvement and sustainability of agricultural production by maintaining the food security vis-à-vis environmental safety with high economic yield and lowering the GHG emission.

KEYWORDS

- **direct seeded rice**
- **gibberellic acid**
- **global warming potential**
- **greenhouse gases**
- **harmful ultraviolet**
- **para-chlorophenoxy acetic acid**

REFERENCES

Agricultural Technologies Inventory, (2016). An ISO 9001: 2008 Organization, Bihar Agricultural University, Sabour-813210, Bhagalpur (Bihar).

AVRDC, (1990). *Vegetable Production Training Manual* (p. 447). Asian Vegetable Research and. Development Center. Shanhua, Tainan.

Committee on Earthquake Engineering Research, (1982). "*Earthquake: Engineering Research-1982* (p. 266)." National Research Council. National Academy Press, Washington, DC.

Economic Survey of India, (2014). India Today.in New Delhi.

IPCC, (2007). Climate change 2007: Impacts, adaptation and vulnerability. In: Parry, M. L., Canziani, O. F., Palutikof, J. P., Van Der Linden, P. J., & Hanson, C. E., (eds.), *Contribution of Working Group II to the Fourth Assessment Report of the Intergovernmental Panel on Climate Change* (p. 976). Cambridge University Press, Cambridge, UK.

IPCC, (2013). Summary for policymakers. In: Stocker, T. F., Qin, D., Plattner, G. K., Tignor, M., Allen, S. K., Boschung, J., Nauels, A., Xia, Y., Bex, V., & Midgley, P. M., (eds.), *Climate Change 2013: The Physical Science Basis*. Contribution of Working Group I to the Fifth Assessment Report of the Intergovernmental Panel on Climate Change. Cambridge University Press, Cambridge, United Kingdom and New York, NY, USA.

IPCC, (2014). Climate change 2014: Impacts, adaptation, and vulnerability. In: Field, C. B., Barros, V. R., Dokken, D. J., Mach, K. J., Mastrandrea, M. D., et al., (eds.), *Part A: Global and Sectoral Aspects. Contribution of Working Group II to the Fifth Assessment Report of the Intergovernmental Panel on Climate Change* (p. 1132). Cambridge University Press, Cambridge, United Kingdom and New York, NY, USA.

Jain, S. K., (1983). "*Analytical Models for the Dynamics of Buildings*." EERL 83–02. (PB-84-161009).

Mahdi, S. S., Dhekale, B. S., Roy, C. S., Bangroo, S. A., & Gupta. S. K., (2015). "On the climate risks in crop production and management in India: A review". *Australian Journal of Crop Science, 9*(7), 585–595.

National Disaster Management Authority. Government of India. Website: ndma.gov.in.

Roy, C. S., Chowdhury, R. M., & Brahmchari, K., (2015a). "Approaches to reduce greenhouse gases from rice-rice cropping sequence in new alluvial zone of West-Bengal". *Ecoscan., 9*(1&2), Supplement on Rice.

Roy, C. S., Chowdhury, R. M., & Brahmchari, K., (2015b). "Moderation of green-house gas emission from rice based cropping system to combat the effect of climate change". *Research on Crops, 16*(3).

Sehgal, V. K., Singh, M. R., Chaudhary, A., Jain, N., & Pathak, H., (2013). *Vulnerability of Agriculture to Climate Change: District Level Assessment in the Indo-Gangetic Plains*.

Singh, C. S., (2004). "Earthquake resistant building design." *M.Tech Thesis*. NIT Patna, Patna University, Patna (India).

Singh, C. S., (2013). "Earthquake resistant design of concrete gravity dam." *PhD Thesis*. NIT Patna, Patna University, Patna (India).

CHAPTER 6

Soil Carbon Sequestration: With a Particular Reference to Bihar

RAJEEV PADBHUSHAN,[1] ANUPAM DAS,[1] and SWARAJ KUMAR DUTTA[2]

[1]*Department of Soil Science and Agricultural Chemistry, Bihar Agricultural University, Sabour, Bhagalpur, Bihar, India*

[2]*Department of Agronomy, Bihar Agricultural University, Sabour, Bihar, India*

ABSTRACT

Anthropogenic influences especially deforestation, fossil fuel and crop residue burning have contributed a significant enhancement of carbon dioxide concentration (CO_2) in the atmosphere, thus become one of the lead causes of global warming. Agriculture and allied sectors contributed global warming through emitting methane, carbon dioxide, and nitrous oxide into the environment and deforestation, intensive cultivation, enteric fermentation, and indiscriminate inorganic fertilization are the paramount processes for those emissions (IPCC, 2007). As a consequences of this ill effect, sea level rises, natural calamities viz., drought, flood, tropical storms, etc., become more regular, erratic rainfall distribution, disruption of ecological balance occur that also adversely affect agricultural production sustainability and soil health. There is a crucial need for some feasible strategies to mitigate this alarming situation.

6.1 CARBON SEQUESTRATION

To overcome the consequences of GHGs emissions mainly carbon-rich gases need to be trap in the various ecosystems that can overall control the ill-effects of climate change. The emission can be reduced by retarding at their source or buried into the ocean, trap trough the terrestrial environments (vegetation, soils, and sediments) and geologic transformations. A negative balance between emissions and the uptake of CO_2 can able to reduce global warming. The trapping of the carbon in the ecosystem for a longer period time eventually control the adverse affect on the atmosphere is termed as "carbon sequestration." The feasibility of different aspects of carbon sequestration under a diversified environment and their consequences has been evaluated by a number of researchers working throughout the globe.

6.1.1 TYPES OF CARBON SEQUESTRATION

The carbon sequestration is broadly classified into three groups:

1. Chemical transformation;
2. Engineering techniques; and
3. Terrestrial sequestration.

Atmospheric CO_2 is converted into mineral carbonates. Chemical transformation is a stable solution for sequestering atmospheric CO_2 but is cost involvement renders its wide spread applicability. Engineering techniques of carbon sequestration are carbon dioxide buried into the soil or ocean where it can reside for a longer period of time. The most important carbon sequestration having agricultural significance is terrestrial carbon sequestration. It is also referred as "biological sequestration." It includes forest, wetlands, grasslands and resource conservation practices that enhance the storability of carbon in the system. The reduction of gas emissions through minimizing agricultural tillage practices and subduing wildfires is also important part of terrestrial carbon sequestration. Conversion of land use pattern, i.e., restoration of wetlands, degraded lands, etc. may increase carbon sequestration, enhance ecosystem livelihood and recreational potential.

6.1.1.1 SOIL CARBON SEQUESTRATION

Soil carbon sequestration is an important component of terrestrial carbon sequestration. Building healthy soil maintenance and management of carbon sources are required. Agricultural practices through organic amendments addition, conservation agriculture, and crop residue management retain carbon in the soil. Trapping of carbon for longer period is only possible through the above agricultural practices on continuous basis. The phenomenon of this carbon trapping through long-term management practices in the soil is called "Soil Carbon Sequestration." Many researchers in the world are engaged in defining the management practices that enhance the carbon retention and sequestration in the soil. The practices of the carbon farming studied on soils and crop production seem very effective to some extent by capturing carbon in soil to combat this climate change.

6.1.1.2 SOIL CARBON SEQUESTRATION AND SOIL HEALTH

Soil quality/health is a key driver of agricultural productivity, environmental resilience, and human food security worldwide. Healthy soils also provide numerous environmental co-benefits to society, including soil carbon sequestration and reduced risk of flooding, erosion, pest and plant disease outbreak at various scales. Soil quality and soil health are interchangeable terms include physical, chemical and biological environment as a whole. Modern intensive cultivation causes decline in the quality of soil. The reason behind this deterioration is decline in the soil carbon stock due to mismanagement followed in modern cultivation. This causes degradation of soil and decline in microbial population. The studies have also reported that conventional cultivation was more beneficial for sustained soil health. Long-term soils studies have shown that improvement in trapping of carbon in the soil due to recommended management practices results into improved soil health. Tropical and subtropical climate is characterized to higher organic matter decomposition rate due to persisting high temperature for long duration. This may causes carbon losses from the soil and overall impact the carbon stock. The decline in carbon content influences the nutrient availability in the soil.

Building healthy soil maintenance and management of carbon sources are required. Agricultural practices through addition of organic sources, conservation agriculture, and crop residue management retain carbon in the soil. Trapping of carbon for longer period is only possible through the above agricultural practices on a continuous basis. The phenomenon of this carbon trapping through long-term management practices in the soil is called "Soil Carbon Sequestration." Many researchers in the world are engaged in defining the management practices that enhance the carbon retention and sequestration in the soil. The practices of the carbon farming studied on soils and crop production seem very effective to some extent by capturing carbon in soil to combat this climate change.

The importance of soil organic matter to soil fertility, crop productivity and terrestrial cycling of carbon, nitrogen and other nutrient is well known. Organic matter addition in the soil enhances not only soil health but also work as an important source to maintain both active and passive pools of carbon. Studies have advocated that increase in amount of aggregate associate carbon and passive pools of carbon as a result of organics addition enhance the carbon stock in soil. Many long-term field trials have shown that organic amendments increase macroaggregates and their associated carbon along with the amount of structurally trap carbon from the added biomass and mitigate the soil-based eco-climate. Now a day to take sustainable crop production and soil health management of organics has been prioritized. Really this carbon farming practice is better option to control global warming impacts that is mainly due to release of carbon-based greenhouse gases (GHG) in the atmosphere.

The concept of conservation agriculture also works on the same theory of sustainability. It is practiced to foster natural ecological processes to increase yield and environmentally sustainable. Minimal soil disturbance, crop rotation, and permanent soil cover are the basic principles on which conservation agriculture stand. This carbon farming practice on long-term capture carbon and protect the soil health. Minimal soil disturbance can be done by following minimal tillage practice, because tillage expose the soil carbon to the environmental oxidation and this eventually emitted CO_2, a GHG. It has been observed that just plowing soil only one year can also negate the carbon accumulation and if continues this process for many years reduces the overall carbon storage potential of the soils. Thus, the idea of "no-till" practice has been found more practical for overall enhancement in the carbon storage potential on longer period of use.

Bare soils adversely affect the soil quality. That is why cover cropping practice is being followed under this agricultural production system. Permanent cover crop has the potential to increase net carbon in the soil. The field research has proved that if one-fourth of the cropland globally followed by cover crop practice-could offset approximately 8% of the total agricultural emissions. This can be better starting appeared for soil-based climate mitigation. Agricultural emissions make up approximately 11–12% of the overall global emissions. The practice of cover cropping is creating difference in these total agricultural emissions that would really source generation for sustained environment.

Agricultural constraints like nutrient losses, low nutrient use efficiency and intensive environment pollution is real culprit for sustained agricultural production system. Crop residue management is better option to overcome these constraints. Under this carbon farming practice residue of the harvested crop is allowed to incorporate in the soil and thus retaining nutrient and biomass in soil. Overall, increases in the soil carbon content. Residue management is also providing diverse and positive effect on soil health.

6.2 MITIGATION OPTION FOR CLIMATE CHANGE

Climate change is real threat to food security and livelihoods for the population. Rising global temperature, increased climate variability and extreme weather events impact the crop production and its sustainability. Combating climatic change in present scenario is a big challenge for the scientific community and it can be possible through the soil management practices. "Carbon farming" practices are one of the better options to reverse the climate change impact. Under this concept management practices are followed to build carbon stock in the soil and draw down carbon in the atmosphere through improved soil management that stabilizes the climate require taking carbon out of the atmosphere.

Agriculture and climate change are the interrelated processes, both of which take place globally. Agriculture emits GHG that contributes to the climate change mainly through four ways viz., deforestation, cultivation, enteric fermentation and inorganic fertilization. According to IPCC, 2007 these agricultural phenomenon releases GHG like methane, carbon dioxide and nitrous oxide abundantly. The energy conversions from grain to meat also contribute towards climate change. Different energy-intensive practices from farm to the food end also contribute towards climate change measurably. Many researchers globally provided quantitatively measure emissions from the different sources. Smart cultivation practice can provide a pathway for combating the ill effects of climate change and keep the environment safe for sustainable life for future generation.

Global warming is a global problem persisting from almost a century. Carbon sequestration can help to control this problem by sequestering carbon-rich compositions in the soil and protect the environment from the greenhouse effect. It also helps by improving productivity, water quality, and restoration of degraded soils and ecosystems and soil-based climate change mitigation.

Soil carbon sequestration is a conduit among the three global issues—climate change, desertification, and biodiversity conservation. But there is a tradeoff relationship between enhanced carbon stock and methane flux. Soil carbon sequestration capacity can be improved by some recognized management practices like reduced or zero tillage, residue retention, use of organic manure, etc. But soil carbon sink capacity depends on its saturation deficit. Saturation deficit depends on some intrinsic properties of soil like nature and amount of clay, moisture and thermal regime, texture, etc.

6.3 CARBON SEQUESTRATION POTENTIAL

Carbon sequestration capacity in a soil can be viewed from three levels-potential, attainable and actual. The potential carbon sequestration capacity for soils can be approximately equated to the initial value observed under historic land use prior to any change in the site management. The attainable soil carbon sink capacity refers to the amount which can be achieved following the best management practice over a long period of time and would hardly approach more than 50% of the potential capacity. The actual soil carbon sink under common land use will be usually low.

Soil is the second-largest sink for the sequestering carbon from the atmosphere and its sequestration rate can be enhanced with the adoption of appropriate management practices. On study over carbon stock and rising greenhouse emissions, the usefulness of soil as a carbon sink and source of solution has been observed. There is a dichotomous philosophy remains behind the carbon sequestration, either upholds native soil organic carbon (SOC) concentration or protect the same from microbial degradation. The management strategies that satisfy either or both the options will work for long run and mitigate the problems of climate change. The phenomenon is very specific to the state and origin, thus it become a matter of big puzzle for scientific community. Uniform recommendation is very scanty throughout the globe. Conservation of SOC not only mitigate the effect of climate change but also provide some auxiliary benefits that improves soil physical (e.g., soil structure, porosity, etc.), chemical (e.g., available nutrient content) and biological environment for plant growth.

Carbon retention potential and carbon sequestration potential are two similar words used as misnomer. Carbon retention potential is the capacity of a system to store carbon for short duration while this storage of carbon is done by the system for long duration is the carbon sequestration potential. Agricultural system as a part of the terrestrial ecosystem has the potential to sequester carbon but its magnitude varied with the soil carbon concentration, bulk density, and depth of soil.

6.4 ORGANIC FARMING AND SENSITIVE INDICATORS OF CARBON

Organic farming is a type of farming system which has gained importance under present scenario worldwide. For human health prospective as well as soil health this system is much outfitted. The concept of organic farming has provided a suitable pathway for sustainable crop production. The practices of organic farming provide invaluable insight in improving the soil fertility and its productivity for sustainable agriculture. The basic perspective for application of organic amendments in the soil is required to maintain organic carbon stock and the microbial activity. In India, Sikkim state has shown the way related to organic farming to the country.

Liability of carbon, carbon management index, sustainability index and sensitivity index are the important sensitive indicators to study the stratification SOC and its influence on diverse management practices. Liability of carbon is explained using graded use of sulfuric acids. The use of this acid in the soil will extract different amount of carbons. On calculation finally various labile and non-labile form of carbon is obtained. The loss of carbon from a small pool, i.e., labile pool is more critical than loss of carbon from a large pool, i.e., non-labile pool. Carbon management index compares the changes that occur in carbon pools as a result of agricultural practices with respect to natural ecosystem as reference. Other carbon indicator is sensitivity index which is used to compare the magnitude of changes in different carbon pools compared to stable reference soils. Sustainability index is an indicator represents the set of requirement that is felt appropriate to evaluate the sustainability of particular farm. Studies have shown that these parameters improve the carbon pools on addition of recommended management practice. The effects of manures and fertilizers on carbon pools and on sensitive indicators have been studied extensively and positive impact on soil.

6.5 SOIL CARBON SEQUESTRATION AND BIHAR

Bihar is one of the eastern states of the country India. A various type of soils are present in different part of the states. Alluvial and red soils are prominent in the state. Rice, maize and wheat are the main field crops covering larger area of the state. Soil carbon sequestration Studies have been implemented under different land use systems agroforestry and under various long-term experimentations running under the jurisdictions of Bihar Agricultural University (BAU), Sabour. Some of the important findings of these works have been briefed below.

A survey on the carbon status of the red soil under different land use was carried out in the year 2014. In surface soil, the highest SOC was recorded in forestland use (0.75%) followed by orchard land (0.51%), grazing land (0.45%) and cultivated land (0.39%), whereas in subsurface soil the highest SOC was observed in forest land (0.46%) followed by orchard and grazing land (0.33%) and the lowest in cultivated land (0.29%) (Figure 6.1). The highest SOC stock was observed in forestland use (61.37 Mg ha^{-1}) followed by orchard land (53.81 Mg ha^{-1}), grazing land (53.47 Mg ha^{-1}) and cultivated land (52.54 Mg ha^{-1}) (Figure 6.2). The present study showed that the decline in soil carbon and its stock from natural land use to anthropogenic land use. This might be the mismanagement practice followed due to excessive cultivation. The modification of artificial land use to back natural land use by establishing forestry in the region improves the sequestration of carbon and thus helps to mitigate the climate change.

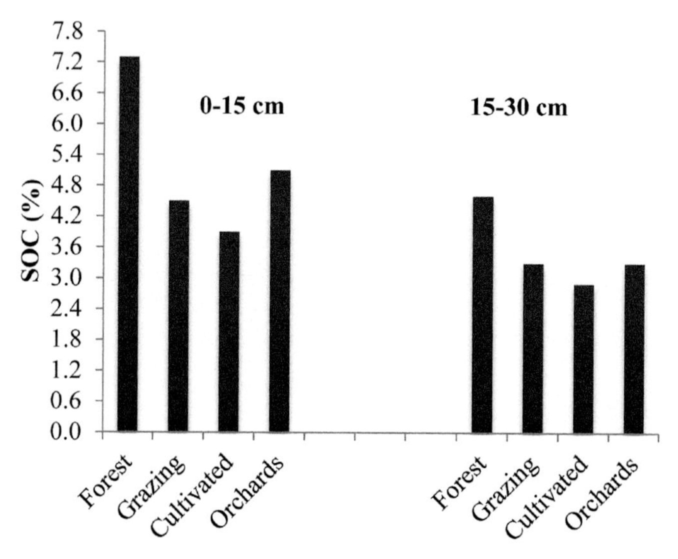

FIGURE 6.1 Effect of land use on soil organic carbon (SOC) and different land use systems in the red soil.

A long-term organic farming experiment was initiated in alluvial soil under scented rice-potato-onion cropping system and on 9th cycle increase in SOC, aggregate associated carbon and passive pools of carbon was observed. The increase in the amount of carbon on long-term organic amendments suggested improvement of soil carbon sequestration. At a soil depth 0–10 cm, SOC content was increased by 9–14%, with the application of sole organic amendments (viz., FYM + VC + NC, FYM + VC + NC + PSB, FYM + BFN + BM + PSB and NPK + FYM) over sole mineral fertilizer treatment (NPK). Subsurface soil (10–20 cm) also showed similar pattern alike surface soil (0–10 cm). (Figure 6.3) (Padbhushan et al., 2016b).

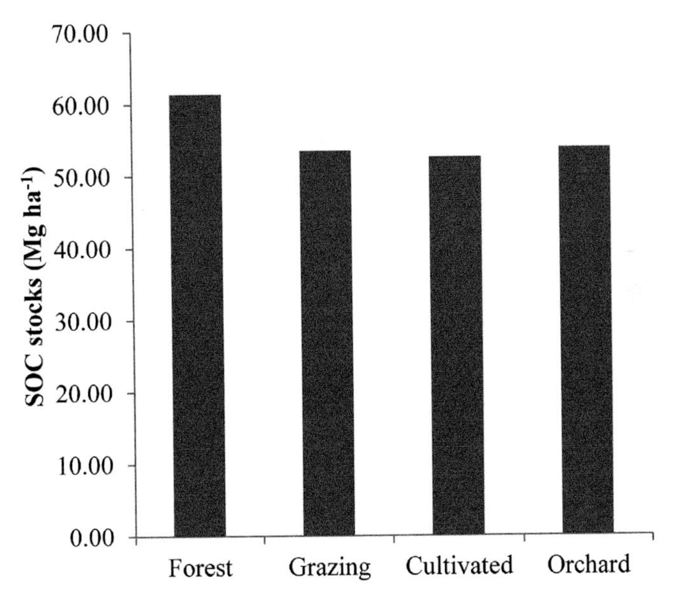

FIGURE 6.2 Soil organic carbon stocks (Mg ha^{-1}) under different land uses in the red soil of Banka district.

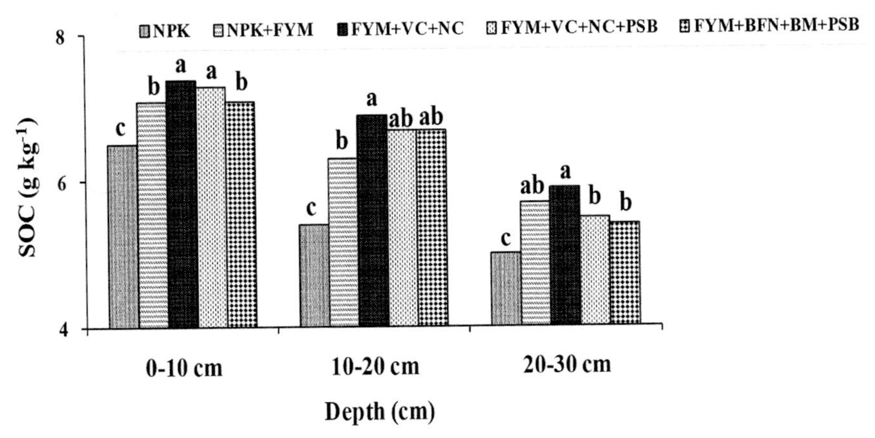

FIGURE 6.3 Effect of long-term use of fertilization on soil organic carbon (g kg^{-1} soil) in soil depths (0–10 cm, 10–20 cm and 20–30 cm) under scented rice-potato-onion cropping system.

Aggregate associated carbon in >5000 μm soil aggregate fraction was 69–133%, higher in case of FYM + VC + NC + PSB as compared to NPK in soil depth 0–30 cm, respectively. Similarly, carbon concentration in soil fraction 2000–5000 μm was 29–55%, 1000–2000 μm was 35–92%and 500–1000 μm was 62–126% higher in FYM + VC + NC + PSB as compared to NPK in the above-mentioned depths, respectively (Table 6.1). Long term addition of organic amendments exhibit a higher redundancy to microbial degradation, organic amended treatments retained higher carbon in different aggregated size classes over sole mineral fertilizer. High content of polysaccharides (cellulose and hemicelluloses) in organic amendments could lead to the production of higher carbon (Padbhushan et al., 2016a).

Organic amended treatments had more humic acid and fulvic acid as compared to blanket recommended treatments. Treatment FYM + VC + NC + PSB had the highest humic acid (6.05 g kg^{-1}) and fulvic acid (2.83 g kg^{-1}) followed by FYM + BFN + BM + PSB>FYM + VC + NC>NPK + FYM, while the lowest humic acid (3.70 g kg^{-1}) and fulvic acid (1.96 g kg^{-1}) was obtained in the treatment NPK. The soils containing more amounts of these acids can be considered sequestering more carbon and ultimately declining the impact of global warming (Table 6.2).

A study was conducted to reveal the long term influence of organic amendments on native carbon pools under alluvial soil of Bihar and it was observed that organic amendments enhanced the liability of carbon by 54.5–77.3% as compared to chemical fertilizer applied alone in soil depth 0–10 cm. The amount of carbon pools decreased on increasing soil depth (0–10 cm > 10–20 cm > 20–30 cm). In the soil depth 0–10 cm, carbon management index values were 1.80–1.62 times more in organic manure added treatments as compared to chemical fertilizer applied alone. Sensitivity index of labile carbon for different treatments showed positive values, which indicates positive impact of the management practices on soil organic matter content and on the soil quality. The management practices are approaching towards natural ecosystem (Padbhushan et al., 2015).

Some experiments have also been conducted on carbon sequestration potential in diversified cropping system and greenhouse emissions through the particular cropping system and reported the context of climate change well furnished. These studies have provided detailed information about the soil carbon sequestration and also quantified greenhouse emissions from alluvial soils of Bihar. In the western part of the state where crop residue burning in the crop field is a big challenging issue. This has ill effects on soil as well as climate. The government has banned the crop residue burning and suggested them to go for crop residue management. Because of this

TABLE 6.1 Effect of Long-Term Use of Fertilization on Water Stable Aggregates Organic Carbon (g kg^{-1}) in Soil Depths (0–10 cm, 10–20 cm and 20–30 cm) Under Scented Rice-Potato-Onion Cropping System

Soil Depths (cm)	Soil Fractions (μm) Treatments	>5000	2000–5000	1000–2000 WSAOC	500–1000 (g kg^{-1})	250–500	100–250	Mean
0–10	NPK	3.6^{c}	4.7^{a}	4.9^{c}	2.9^{a}	3.8^{b}	5.5^{a}	4.2
	NPK + FYM	5.9^{b}	6.2^{b}	6.7^{b}	4.3^{a}	4.1^{b}	5.6^{a}	5.5
	FYM + VC + NC	8.2^{a}	7.1^{c}	8.9^{a}	4.5^{a}	4.4^{ab}	5.9^{a}	6.6
	FYM + VC + NC + PSB	8.4^{a}	7.3^{c}	9.4^{a}	4.7^{a}	5.1^{a}	5.8^{a}	6.8
	FYM + BFN + BM + PSB	6.3^{b}	6.1^{b}	6.5^{b}	3.7^{b}	4.2^{b}	5.6^{a}	5.4
	Mean	6.5	6.3	7.3	4.0	4.3	5.7	
	LSD (P = 0.05)	0.9	0.8	0.9	0.5	0.7	NS	
10–20	NPK	3.1^{c}	3.1^{b}	4.8^{c}	1.9^{b}	5.6^{c}	3.5^{b}	3.7
	NPK + FYM	4.3^{b}	3.3^{b}	5.5^{b}	2.1^{b}	6.5^{ab}	3.6^{ab}	4.2
	FYM + VC + NC	6.2^{a}	4.9^{a}	6.2^{a}	4.2^{a}	6.6^{ab}	3.7^{ab}	5.3
	FYM + VC + NC + PSB	6.6^{a}	5.1^{a}	6.5^{a}	4.3^{a}	6.8^{a}	3.9^{a}	5.5
	FYM + BFN + BM + PSB	4.7^{b}	3.1^{b}	5.2^{bc}	2.2^{b}	6.1^{bc}	3.7^{ab}	4.2
	Mean	5.0	3.9	5.6	2.9	6.3	3.7	
	LSD (P = 0.05)	0.7	0.6	0.6	0.4	0.8	NS	
20–30	NPK	2.3^{b}	2.4^{b}	3.6^{b}	1.4^{c}	3.6^{c}	0.8^{a}	2.4
	NPK + FYM	2.7^{b}	2.7^{ab}	3.8^{b}	2.1^{b}	4.2^{ab}	0.8^{a}	2.7
	FYM + VC + NC	3.6^{a}	3.0^{a}	4.9^{a}	2.7^{a}	4.5^{a}	0.9^{a}	3.3
	FYM + VC + NC + PSB	3.9^{a}	3.1^{a}	5.0^{a}	2.7^{a}	4.8^{a}	0.9^{a}	3.4
	FYM + BFN + BM + PSB	2.9^{b}	2.8^{ab}	3.7^{b}	1.8^{b}	3.7^{bc}	0.9^{a}	2.6
	Mean	3.1	2.8	4.2	2.1	4.2	0.9	
	LSD (P = 0.05)	0.4	NS	0.6	0.3	0.6	NS	

*Values followed by different letters in the same column and LSD are significantly different at P = 0.05.

management practice the carbon retention in the soil has improved and improved the carbon sequestration potential of the system.

TABLE 6.2 Effect of Different Treatments on Soil Humic Substances (g kg^{-1} soil) in Long-Term Organic Farming System

Passive Pools	Humic Acid	Fulvic Acid
Treatments		
NPK	3.70[a]	1.96[a]
NPK + FYM	4.56[b]	2.72[c]
FYM + VC + NC	5.51[c]	2.80[c]
FYM + VC + NC + PSB	6.05[d]	2.83[c]
FYM + BFN + BM + PSB	5.60[c]	2.41[b]

*Values followed by different letters in the same column.

Some of these studies in the state showed both conditions of recommended management practices and mismanagement anthropogenic activities. The climatic condition of the Bihar is subtropics. Thus prolonged high temperature aggravated the organic matter decomposition rate. Decompositions of these native organic matters decline the content and storability of soil carbon for longer period. This impacts the soil fertility of the region and climate change.

KEYWORDS

- **biological sequestration**
- **carbon dioxide**
- **carbon sequestration**
- **fulvic acid**
- **humic acid**
- **soil organic carbon**

REFERENCES

Amarjeet, K., (2016). *M.Sc. Thesis*. Bihar Agricultural University, Sabour (Bhagalpur), 813210.

Eric, S., Robert, B., Stephen, F., Robert, G., Jennifer, H., Yousif, K., Larry, T., & Mark, W., (2016). *Edited by Dale, L. Simmons Graphic Design by Christine, T. Mendelsohn Figure*

4 by Eric, A. Morrissey and Sean Brennan For additional information about this report, please contact Eric Sundquist (esundqui@usgs.gov). For additional information about the U.S. Geological Survey, please visit our web site at: http://www.usgs.gov/ (Accessed on 21 November 2019).

IPCC, (2007). *Fourth Assessment Report Climate Change 2007-Impacts*. Adaptation and vulnerability.

Léopold, B., Rebecca, C. C., Ritt, K., Sara, L., & Hortencia, R., (2016). *Soil Health and Carbon Sequestration in US Croplands: A Policy Analysis Prepared for: Natural Resources Conservation Service (NRCS) of the United States Department of Agriculture (USDA) and the Berkeley Food Institute (BFI) By: Goldman School of Public Policy*. University of California, Berkeley.

Rajeev, P., Anupam, D., Rajiv, R., Rajendra, P. S., Anshuman, K., & Rajesh, K., (2016a). Long-term organic amendment application improves influence on soil aggregation, aggregate associated carbon and carbon pools under scented rice-potato-onion cropping system after the 9th crop cycle. *Communications in Soil Science and Plant Analysis*. doi: 10.1080/00103624.2016.1254785. Online Published.

Rajeev, P., Rajiv, R., Anupam, D., & Rajendra, P. S., (2015). Assessment of long term organic amendments effect on some sensitive indicators of carbon under subtropical climatic condition. *The Bioscan., 10*(3), 1237–1240.

Rajeev, P., Rajiv, R., Anupam, D., & Rajendra, P. S., (2016b). Effects of various organic amendments on organic carbon pools and water stable aggregates under a scented rice-potato-onion cropping system. *Paddy and Water Environment*. doi: 10.1007/s10333-015-0517-814: 481-489.

Rattan, L., et al., (2004). *Climate Change and Food Security*. doi: 10.1126/science.1097396 Science 304, 1623.

CHAPTER 7

Agromet Advisory Services: Tool to Mitigate the Effect of Extreme Weather Events

SUNIL KUMAR

Department of Agronomy, Bihar Agricultural University, Sabour, Bhagalpur, Bihar, India. E-mail: sunilkumaragromet@gmail.com

ABSTRACT

Climate change has increased the extreme weather events in many parts of the world. Situation is adverse for the agriculture sector which is the back bone of the country like India. Some efforts has been initiated to combat the effect of climate change. But for farmers, it is very less and they have not adapted all the recommendations due to several reasons. Extreme weather events like extreme heat and cold, drought, extreme heavy rainfall, storms, hail etc. has been increased in its frequency and it's duration by climate change. Agromet Advisory Services (AAS) initiated by India Meteorological Department in collaboration with National Centre for Medium Range Weather Forecast (NCMRWF) and State Agriculture University (SAUs) showed the path to support efforts made by other organizations and Government. Farmers are given weather forecast with agro advisory for 5 days in advance. Weather parameters like maximum temperature, minimum temperature, rainfall, wind direction, wind speed, maximum relative humidity, minimum relative humidity are forecasted and are provided to the farmers through different media at district level and now it has been started at block level. Feedback of the farmers are also taken to improve the forecast and agro advisory. It has improved the productivity of the crop and has decreased the loss due to different aberrant weather situation not favorable for agriculture. It will also be helpful to increase the income of farmers by reducing the cost of cultivation. Further improvement is needed for weather forecast which may be increase its accuracy and duration will be certainly helpful to diminish the ill effect of extreme weather and climate change.

7.1 INTRODUCTION

Agriculture is the backbone of the Indian economy and several factors affect its success or failure. Weather and climate are very important factors which effect on each phase of growth and development of plant. Variation in the weather or climate during crop season, like delay in the monsoon, extreme rainfall, flood, droughts, long spells of very high or very low temperatures may affect the crop growth as well as quantity and quality of the yield. Yield loss may be reduced by proper management in real-time by planning in advance based on timely and accurate weather forecasts. The weather-based agro advisory includes best-suited crops in the region based on expected weather and actual weather condition which may be affecting regular farm works of farmers. Weather forecast and agro advisory try to reduce the losses and risks, reduces the cost of cultivation by saving irrigation cost, labor cost and in total increases the agricultural production. Rathore et al. (2001) discussed properly about how the weather forecast by the National Center for Medium Range Weather Forecast (NCMRWF) would be issued for different location for five days in advance. Damrath et al. (2001) reported that statistical interpretation methods maybe helpful to increase the reliability of the rainfall forecast. Climate change, which is one of the major challenges of the 21st century, is tackled by Agromet Advisory Services (AAS). The Gramin Krishi Mausam Sewa (GKMS) is the new name of AAS in India and it intends to link the farmers in respond of climate change at root level. GKMS aims to create efficient, effective and synergistic linkages to improve the delivery of these services to the farmers. GKMS operates under the guidance of India Meteorological Department (IMD), which is serving for Indian farmers to enhance their livelihoods. It directly addresses the needs of Indian farmers, contributing to sustainable growth in and transformation of Indian agriculture by providing effective advisory services. It gives the opportunities for AAS as well as a great challenge to adapt in response to climatic and other drivers of change, and begins to outline the possible roles and characteristics of adaptive AAS.

Day-by-day increase in uncertainties of weather and climate is giving a big threat to food security of the country and is becoming obstacle for farmer's empowerment by taking decision in agricultural risk management. In coming decades, there is high probability that the combination of long-term variation of climate and greater frequency of extreme weather events will likely to have adverse impacts on the agricultural production.

Agro-meteorological service provided by IMD is an innovative step to disseminate the weather information and based on weather forecast, crop/ livestock management strategies and farming operations are advised to increase the crop/livestock production by providing real time crop and location specific agrometeorological services with outreach to village level. It has a potential to change the face of the country in terms of food security and poverty alleviation. It has tried to make a tremendous difference to the agriculture production by taking the advantage of favorable weather and minimizes the adverse impact of adverse weather. A mechanism has been developed to integrate weather forecast and to prepare agromet advisories which contribute significantly to increase the farm productivity and is trying to solve the problem of food security.

7.2 AGROMET ADVISORY BULLETINS

Medium range forecast information and advisories at district level help to maximize output and to reduce crop damage or yield loss. It also helps farmers to plan in advance for pesticide applications, irrigation scheduling, disease, and pest outbreaks and many more weather-related farm-specific operations. These may include cultivar selection, their dates of sowing/planting, dates of intercultural operations, dates of harvesting and post-harvest operations. It helps to increase profits by consistently delivering weather information, analysis and decision support for farming operations.

7.3 POPULATION TREND

Globally, sustainable long-term food production is needed to be doubled in order to meet the basic needs of increased urban population. Demographic changes include not only urbanization, but also increasing migration and seasonal mobility. Climate change, increasing population and food security, among other factors, are highlighting the importance of the provision of different ecosystem services and the finite nature of these land resources. To avoid expansion into other ecosystems, agricultural productivity will have to increase, while minimizing the associated environmental damage and with net reductions in GHG emissions from food production and postharvest activities (World Bank, 2010a). This concept of increasing output but using less resources and reducing the environmental impact has been described by UNEP (2011) as 'decoupling.' Climate change is likely to alter countries' comparative advantages in agriculture, and thereby alter the pattern of international trade (Tamiotti et al., 2009).

The visions and policies are fairly consistently based on the premise of increasing agricultural productivity to drive economic growth and poverty reduction. This is being implemented alongside major statements regarding food security. Current policies are generally supportive of agricultural practices that focus on increasing short-term production (e.g., expansion of agricultural land, increasing mechanization, increasing use of fertilizer and other inputs).

7.4 CHARACTERISTICS OF AAS AND EXPLORATION OF 'ADAPTIVE' ATTRIBUTES

Over many decades in India, regular use of old ICTs (such as telephone or television) was rare for large sections of the rural populations. The most striking was radio, which quickly became widespread due to the easy availability and of low cost. GKMS have maintained communication departments and produced regular radio programs on agricultural topics for broadcast to rural populations, often through state-owned radio stations. In some cases, educational videos or TV programs were also produced via mobile audio-visual vans. The content was largely created and controlled by the AAS organizations and targeted at the farmer recipient. Since the turn of the millennium, vary fast growth of private mobile-phone ownership and use in both rural and urban areas, increasing access to TV and video-screening facilities, and digital filming apparatus (cameras, mobile phones), and the more recent spread of internet access in towns and even into smaller towns and canters via mobile net services, have offered a whole new world of opportunity for multi-directional communication. Interestingly, in our experience, even while AAS has embraced new participatory approaches such as farmers field schools and farmers participatory research to mobilize communities and to harness complementary contributions from researchers, farmers and AAS staff for innovation, AAS in general seems to have been relatively slow to explore opportunities for revolution in multi-stakeholder information sharing, knowledge creation and advocacy activities offered by combinations of new and old ICTs. A major challenge for ICTs in AAS vis-à-vis climate change issues will be to develop ICTs as multi-way platforms and break with the unidirectional communication traditions of the past. It is not only AAS staff who are in need of information and perspectives about climate change science and their expression in their local environment. Researchers and official meteorological stations are one source of these, but both AAS leaders and researchers also need to access and learn from the experience of farmers, frontline AAS staff and other sector staff living and working in the focus areas. Adaptive capacity varies widely between individuals and communities, due to differences of access and control of assets and the environment in which people are living. In order to strengthen adaptive capacity, GKMS needs to be able to recognize these differences and develop strategies to address them. Self-organization is a key element of adaptive capacity. Climate change has emerged only recently as a critical issue and so most AAS individuals would have received little specific training in relation to climate change in their formal training. This is starting to change, but many AAS actors have limited capacity to actively seek and use new knowledge and information.

7.5 AGROMET ADVISORY METHODS

In moving towards adaptive AAS, the advisory methods used are critical. In dealing more explicitly with climate change and other uncertainty, methods need to emphasize such aspects as: strengthening the capacity of clients (rather than delivering messages), and enabling clients to experiment and use climate information, strengthening the self-organization of farmers, enhancing local-level innovation, improving links between research and extension, and considering the content of advice in relation to what is appropriate to the context (e.g., balancing production-innovation, growth, and climate resilience).

7.6 GRAMIN KRISHI MAUSAM SEWA (GKMS)

The Agromet advisory service based on a medium-range weather forecast has been made operational by NCMRWF. NCMRWF has proposed to establish 130 agro meteorological field units (AMFUs) in all the agro-climatic zones, which cover the entire country (Figure 7.1 and Table 7.1). The agro advisory bulletins are prepared by the panel of scientists based on the "Package of practices recommendations developed by the concerned University for the Respective Zones. The essence of the agro advisory is to make the farmers aware of farm operations for sustainable agricultural production based on weather forecast. Weather forecast for 5 days district wise is received regularly twice a week on Tuesday and Friday. For e.g., during the year 2015–16,

102 weather forecasts for 365 days were received. Based on total forecasts received 102 Agromet Advisory Bulletins were prepared and circulated among 50 selected farmers of different villages of the region. Weather forecasts and agro-advisories were also communicated telephonically to some of the nearby farmers of Bhagalpur, Banka, Munger, Jamui, Lakhisarai, Sheikhpura Patna, Gaya, Aurangabad, Jahanabad, Nawada, Kaimur, Rohtas, Bhojpur, Buxar, Arwal and Nalanda districts of agro-climatic zone III A and III B. For immediate benefit of the local farmers in 100 km radius, All India Radio, Bhagalpur broadcasted Agromet Advisory bulletins in between 18.30 and 19.00 hours on every Tuesday and Friday of the week as their regular feature. Similarly different Newspapers like "Hindustan," "Danik Jagaran," "Dainik Bhaskar," "Prabhat Khabar," "Rashtriya sahara," and "Nayi baat" published weather forecasts and Agro-advisories on regular basis. Weather forecasts and agro-advisories were also sent by email/ faxed to Annadata Programme of ETV Bihar, Sahara TV Bihar, Sadhna TV Bihar for telecast, for the benefit of farmers of this state. Agro advisories were also sent by email to IMD, Pune and uploaded on website of the Agromet Division of IMD and Website of Bihar Agricultural University (BAU), Sabour. Among the officials, Agromet Advisory bulletins were sent to Joint Director of Agriculture and all District Agriculture Officers of different districts of zone III A and III B, Project Directors of ATMAs of Bihar, Block Agriculture Officers, for communication among the farmers through Village Extension workers or local medias. The college also organized Kisan Mela and Kisan Gosthi in which the visiting farmers were made aware about this service. Weather forecast and agro advisories were also sent through SMS on mobiles of Farmers of zone III A and III B on regular basis. Till date 220 SMS had been sent and more than 10 million farmers had been benefited and 5,34,736 Farmers from 17 Districts of Bihar has been registered and are getting SMS at a time (Figures 7.2 and 7.3).

FIGURE 7.1 Agro climatic zones of Bihar.

TABLE 7.1 Agro Climatic Zones of Bihar and AMFUs of GKMS

S. No.	Agro Climatic Zone	Districts	Agromet Field Unit (AMFU)
1	North West Alluvial Plain Zone (Zone-I)	W Champaran, E Champaran, Gopalganj, Vaishali, Sitamarhi, Muzaffarpur, Siwan, Darbhanga, Samastipur, Sheohar, Begusarai, Madhubani, Saran.	Pusa (Samastipur), RAU, Pusa
2	North East Alluvial Zone (Zone-II)	Saharsha, Purina, Katihar, Supaul, Khagaria, Madhepura, Kishanganj, Araria.	Agwanpur (Saharsha), BAU Sabour
3	South Bihar Alluvial Zone (Zone III A and B)	IIIA: Bhagalpur, Sheikhpura, Lakhisarai, Jamui, Munger, Banka. IIIB: Bhojpur, Buxar Aurangabad, Nalanda, Jahanabad, Gaya, Nawada, Patna, Kaimur, Arwal Rohtas.	Sabour (Bhagalpur), BAU Sabour

7.7 OBJECTIVES OF THE PROJECT

The first objective of the project is to pass on weather data biweekly to NCMRWF, Delhi and DDGM (Agrimet) India IMD, Pune from 1.8.2007. The second objective is to prepare Agromet Advisory Bulletin on the basis of forecast received from Meteorological center Patna after modification of the forecast. Dissemination of the Agromet Advisory Bulletins (prepared biweekly) is the another objective through All India Radio, different T.V. channels, Newspapers, IMD website and also through personal contact to the selected farmers of zone III A, III B of Bihar. Regular collection of feed-back from the selected farmers of these zones is also the duty of AMFUs. At meanwhile verification of the reliability of weather forecasts using actual observation recorded in the observatory is also one of the objectives of the project. At the first time, the preparation of Agromet Advisory Bulletin was started weekly from April 1997 and from 1998 it was started biweekly preparation of bulletin.

7.8 STATUS OF DISTRICT AGROMET ADVISORY SERVICES (AAS) BY FARMER'S PORTAL

There is a problem in preparing the District Agromet Advisory bulletin through the farmers portal. It needs fast internet connectivity for online preparation. So, it could not be prepared online through the farmers portal. In Hindi, it needs to improve the software so that some mistake in the name of place or crop will be improved. For example, at our Center, it has not been written correct in heading. In Hindi bulletin, Sabour is not written in right way and it was informed to the office by mail. For one district Arwal, moderated weather forecast is not given. There is no facility directly to upload all the pdf files on IMD website and on our University website. The very important fact is that, there is no weather data for all districts to fill up the table for all districts regarding previous week weather condition.

ग्रामिण कृषि मौसम सेवा
बिहार कृषि महाविद्यालय
बिहार कृषि विश्वविद्यालय, सबौर, भागलपुर (बिहार)

बुलेटिन नं0–59
दिनांक –07/08/15

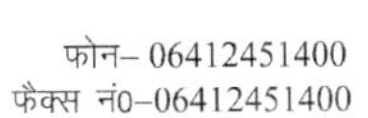

फोन– 06412451400
फैक्स नं0–06412451400

पटना जिला के लिए कृषि मौसम सलाह बुलेटिन
(17– 21 अक्टूवर 2015)
भारत मौसम विभाग द्वारा प्राप्त आगामी पॉच दिनों का मौसम पूर्वानुमान

दिनांक	17 अक्टूवर	18 अक्टूवर	19 अक्टूवर	20 अक्टूवर	21 अक्टूवर
वर्षा	0	0	0	0	0
बादल	साफ	साफ	साफ	साफ	साफ
अधिकतम (डिग्री से0 ग्रे0)	32	32	30	31	30
न्युनतम (डिग्री से0 ग्रे0)	22	22	21	22	21
आद्रता (प्रतिशत)	50-82	50-72	50-72	52-80	52-80
हवा की गति(कि0मी0 / घ.)	4	3	4	3	3
हवा की दिशा	दक्षिण पश्चिमी	दक्षिण पश्चिमी	दक्षिण पश्चिमी	दक्षिण पश्चिमी	दक्षिण पश्चिमी

कृषि मौसम सलाह (मौसम पूर्वानुमान के अनुसार)

धान– धान में अगर गंधी बग का प्रकोप हो तो क्युनालफास दवा 1.5 प्रतिशत अथवा मिथाइल पाराथियान 2 प्रतिशत या फॉलीडाल धूल 25 किलोग्राम प्रति हे0 की दर से प्रयोग करें। धान की दुग्धावस्था में ब्रुप्रोफेजिन 1.5 मि0 ली0 प्रति ली0 पानी के साथ ब्राउन प्लान्ट हॉपर से बचने के लिए प्रयोग करें। खेत में 5 सेमी0 पानी लगा रहने दें।

सब्जीयॉ – अच्छी तरह तैयार खेत में फूलगोभी की रोपनी करें तथा पिछात किस्मों के बिचड़ें गिरा दें।

गन्ना – ' शरद्कालीन ईख की रोपनी 15 अक्टूवर के बाद आरंभ कर दें।

चारा फसल– रबी चारा हेतु बरसीम (वरदान– 1 और स्कावी), लूसर्न एवं जई (केंट) लगावें। बुआई के पूर्व बीजोपचार अवश्य कर लें।

पशुपालन – प्रत्येक पशु को कृमि निरोधक दवा अवश्य खिला दें। खूर और मुख पका बीमारी की रोकथाम के लिये टीके लगवये ।

मुर्गी पालन – मुर्गी घर में पर्याप्त प्रकाश की व्यवस्था करें।

फल एवं वानिकी – नये लगाये गयें बाग में पानी का निकास एवं निकाई गुडाई करते रहें। फलदार वृक्षों एवं वानिकी वृक्षों के पौधों को लगाने का कार्य जारी रखे। अगर दीमक का प्रकोप हो तो क्युनालफास दवा 2 मि0 ली0 प्रति पानी के साथ प्रयोग करें।

नोडल पदाधिकारी
Email: amfusabour@gmail.com

FIGURE 7.2 Agromet advisory bulletin for Patna district (Hindi).

GRAMIN KRISHI MAUSAM SEWA
DEPARTMENT OF AGRONOMY
BIHAR AGRICULTURAL COLLEGE,
BIHAR AGRICULTURAL UNIVERSITY, SABOUR, BHAGALPUR
(BIHAR)

Bulletin No. 72 Ph: 06412451400
Date:04/10/16 Fax: 06412451400

Fig 3: Agromet Advisory Bulletin for Patna district (hindi)

Agromet Advisory Bulletin for Bhagalpur district
(05-09 October 2016)
Weather forecast for next five days received from India meteorological department

Date	05 October	06 October	07 October	08 October	09 October
Rain fall (mm)	3	5	8	0	0
Cloud cover	Partly Cloudy	Partly Cloudy	Partly Cloudy	Partly Cloudy	Partly Cloudy
Max. Temp.(^{0}c)	33	32	32	33	33
Min. Temp. (^{0}c)	25	24	24	24	24
R.H. (%)	53-82	50-92	56-91	56-87	59-90
Wind Speed (km/hr)	4	3	5	6	8
Wind Direction	South Easterly	South Easterly	South Easterly	South Easterly	South Easterly

Agromet Advisory (Based on weather forecast)

- Farmers are advised to monitor crop regularly, if population of **stem borer** is above ETL, broadcast of Cartap 4% granules or carbofuran 3% G @ 10 kg/acre or chlorpyriphos 20 EC @ 2 ml/ litre of water for effective control is recommended.
- Present weather conditions are highly favourable for infection of **Blast** disease in rice. Early symptoms include development of eye shaped spot in the leaf sheath with light centre and dark boundaries. At later stage these spots coalesce to form big spots. If symptoms appear then, spraying of Bavistin 2 gram/litre of water 2-3 times with an interval of 10 day is advised for effective control.
- In present weather condition, constant monitoring for **brown plant hopper** (BPH) in paddy field is advised. Farmers are advised to enter into the middle of the crop field and see mosquito like insect at the basal portion of the plant. If insect population is above ETL, spray of Imidaclorprid 17.8 % SC@ 1.0 ml / 3 lit or fenobucarb 50 EC @ 1 ml/ litre or buprofezin 25 EC @ 2 ml/ litre of water is recommended.
- This is suitable time to all crops and **vegetables** for hoeing & weeding after that split dose of nitrogen should be broadcast for optimum growth. Constant monitoring of ladyfinger, brinjal and chilli crop against attack of mite and jassid is advised. To control mite, spraying of phosmite @ 2 ml per litre of water is advised while spraying of Dimethoate 30 EC @ 2 ml/liter of water is advised against jassid when sky remain clear.
- In **maize** fields, constant monitoring of stem borer should be done. If insect population is above ETL then spraying of carbrayl @ 2.0 gm/ lit. of water is recommended in clear sky.

FIGURE 7.3 Agromet advisory bulletin for Bhagalpur district (English).

7.8.1 FEEDBACK FROM THE FARMERS

To improve the quality of the AAS, regular direct interactions are being made with local farmers. The feedback is collected regularly through Kisan Chaupal (Gosthi), Kisan mela, Farmers gathering and through interaction with farmers personally. A survey was conducted to obtain the feedback from the participating farmers. Data were collected through semi-structural questionnaire and personal contact was made with the farmers for getting the feedback from the participating farmers. Out of 50 farmers, only 44 farmers replied to the questionnaire. It was seen that about 90% farmers were benefited through weather forecasts and Agromet Advisory Bulletins with respect to various agro management viz; timely preparation of land, use of suitable varieties, timely sowing of seeds, amount of water applied, timely irrigation scheduling, fertilizer application, Seed treatment with suitable chemicals, plant protection measures and its application, etc. The responses of the farmers were observed to be very positive and encouraging.

7.9 IMPACT OF AGROMET ADVISORY BULLETINS

To strengthen weather-based Agromet advisory for the benefit of farmers and to ensure food security it is felt that there is a need to study the economic impact of weather-based agro advisory which was already in operation under NCMRWF. From the study, it is clear that the economic gain by an individual farmer is considerable and it varied from 10 to 35% or more depending upon the crop and the season if one follows the biweekly agro advisory over the non-followers. No doubt, the weather-based agro advisory has an overall beneficial effect and farmers gain knowledge in addition to monetary benefit. The Agromet advisory service based on weather forewarning is an effective tool if properly and timely disseminate to the farmers. On the basis of weather forecasts received from Meteorological Center, Patna (Bihar) and Agromet Advisory Bulletins were prepared and served among the farmers through various medias as well as by direct contact with the local farmers of the region. A sample survey was made to get the feedback from the participating farmers. The views of the farmers are collected through semi-structural questionnaires regarding the benefit of Agromet Advisory Bulletins circulated, are being briefly presented below.

During the month of April, 15 there was a prediction of no rain and the same was communicated time-to-time to the farmers of this region with advise to continue their harvesting and threshing operations of rabi crops without any delay. Thus, the farmers could take advantage of clear weather and dry westerly winds in all these post-harvest operations of crops.

During the month of May, 15 there were forecast of no rain and the farmers were advised in this forecast to irrigate the summer crops and vegetables and for spray in chili. Following our suggestions from the advisories, the farmers were thus benefited by good crops and vegetables.

During the month of June, 2015 there was forecast of rain between 12–14 June, accordingly farmers were advised to keep themselves in readiness for sowing of medium duration of rice varieties in the nursery. At the same time, they were also advised to sow Kharif fodder. The rain received the above said period and the farmers were benefited by timely sowing of Kharif fodder and timely sowing of medium and long duration rice seedlings for raising seedlings.

During the month of July, 2015, there was prediction of heavy rain during 8–10 July and similar situation occurred during thc above said period. The farmers were advised to reap the benefit of predicted rain in transplanting of long and mid duration rice and sowing of short duration rice for seedlings if not done earlier. Thus most of the farmers could reap the benefit of predicted rain in transplanting of rice and also utilized rainwater in irrigation to prior transplanted rice crop.

During the month August, 2015 there was forecast of rain during 01–02 August. The same was communicated to the farmers of this locality and were advised to transplant short duration and photosensitive local tall varieties of rice seedlings, if not transplanted earlier. At the same time, they were also advised to utilize predicted rain in irrigation to transplanted rice crop and to drain excess water from Kharif maize, pulses, and oilseed crops to avoid waterlogging. Rain occurred during the period and farmers were benefited by timely transplanting of rice crop, utilized rainwater in irrigation to prior transplanted rice and saved their other Rabi crops from waterlogging.

During the month of September, 2015 there was a forecast of no rain during 9–13 Sept. 2015. Accordingly, farmers were advised to irrigate the transplanted rice crop and to maintain the water in the field due to panicle initiation stage of the crop which is very crucial period for better production.

During the month of October, 2015 there was prediction of no rain during 17 to 21 October 2015 and farmers were advised to spray Quinalphos in rice crop due to attack of Gundhi bug. The same was followed by the farmers and took the benefit of timely spraying of the insecticide in rice crop.

During the month of November and December, 15 there was forecast of no rain and the same weather situations continued during that month. The farmers were communicated about the prediction of clear weather and advised to reap the benefit of clear weather in harvesting of matured rice crop and to prepare field for sowing of rabi crops like-maize, wheat, pulses, and oilseeds. The same was followed by the farmers and took the benefit of timely sowing of the above-mentioncd crops.

TABLE 7.2 Monetary Gains Accrued to Farmers During the Year 2015–2016

Season	Crop Grown by the Farmers	Mean Productivity Realized in kg/ha		Additional Production Gains of Adoptive Farmers (kg/ha)	Price Rs./kg	Additional Income Rs./ha
		GKMS Adoptive Farmers	GKMS Non-Adoptive Farmers			
1. Kharif	Rice	3000	2760	240	12.00	2880/-
	Maize	4000	3740	260	13.00	3380/-
	Vegetables (Brinjal)	20 MT	17 MT	03 MT	15	4,50,000/-
2. Rabi	Wheat	3500	3200	300	10.00	3000/-
	Maize	5800	5510	290	14.00	4060/-
	Gram	1200	1000	200	80.0	1600/-
	Lentil	1150	900	250	70.0	1750/-
	Vegetable (Cauliflower)	21 MT	18 MT	3 MT	14	4,20,000/-
3. Summer	Moong	0880	0760	120	60.00	7200/-
	Vegetable (Okra)	15 MT	13.5 MT	1.5 MT	16.0	2,40,000/-
	Fruit-1. Mango 1. Litchi	8.2 MT	7.0 MT	1.2 MT	30	3,60,000/-
	1.	7.6 MT	6.3 MT	1.3 MT	70	9,10,000/-

During the month of January, 2016 there were the prediction of no rain during 6–10 January. This was communicated to the farmers well in advance and was advised to irrigate the crop and spray insecticides and fungicides for control of insects, pest of their Rabi, maize, potato and wheat crops. No rain was received during that said period and thus the adoptive farmers could save their crop from attack of pest.

During the month of February, 2016, there was prediction of no rain and the same was communicated time to time to the farmers of this region with advice to spray the insecticides in mustard against aphid and fungicides in pea against powdery mildew. Thus, the farmers could take advantage and sprayed the same to protect the above-mentioned crops.

During the month of March, 2016, there was prediction of no rain and the same was communicated time to time to the farmers of this region with advice to spray the insecticide in chickpea against pod borer. Thus, the farmers could take advantage and sprayed the same to protect the mentioned crop.

7.10 ECONOMIC IMPACT OF GRAMIN KRISHI MUASAM SEWA

7.10.1 CASE STUDY

GKMS is a useful information tool to the farmers of this region in deciding their planning and budgeting for weather-agro management operations to achieve maximum benefit from predicted weather forecast served to them periodically by AAS Unit comprising of eminent scientists of different disciplines of Bihar Agricultural College, Sabour.

Table 7.2 indicates that in the case of Kharif rice, maize and brinjal monetary value of saving was estimated as Rs. 2880, Rs. 3380, and Rs. 4,50,000 per hectare. The estimated loss to the non-adoptive farmers who did not care the Agro Advice rendered by GKMS, served by this center was estimated at 240, 260 and 03 MT per hectare in rice, maize, and brinjal respectively under this region.

In the case of Rabi season wheat, maize, gram, lentil and cauliflower monetary value of saving was estimated Rs.3000/-, Rs. 4060, Rs. 1600/-, Rs. 1750/- and 4,20,000/- per hectare respectively. The grain yield losses to wheat, maize, gram, lentil, and cauliflower the non-adoptive farmers who did not follow the Agromet Advisories was 300 kg, 290 kg, 200 kg, 250 kg, and 03 MT per hectare, respectively.

Similarly, in the case of summer season Moong, Okra, Mango and Litchi are cultivated and the benefit to the Growers was estimated as Rs. 7200/-,

2,40,000/-, 3,60,000/-, and 9,10,000/- per hectare to adoptive farmers of the locality.

So, all overviews it was indicated that the adaptive farmers who adapted agro advisory services got the benefit of Rs. 1600 to 4060 for cereals crop and from Rs. 2.4 lakh to 4.5 lakh and Rs. 3.6 to 9.1 lakh in vegetables and fruits crop, respectively.

7.11 VERIFICATION AND ANALYSIS OF WEATHER FORECAST

A five-day medium-range weather forecast was received from M.C, Patna on every Tuesday and Friday of the week. The data related to the weather forecast of the whole year was grouped in four distinct seasons, i.e., pre-monsoon, monsoon, post-monsoon and winter for analysis and verification. Both qualitative and quantitative verification analysis was carried out using skill score and critical values for error structure. The correlation coefficient and root mean squire error have also been worked out of all the four seasons.

7.11.1 QUALITATIVE VERIFICATION ANALYSIS

7.11.1.1 RAINFALL FORECAST VERIFICATION

For qualitative analysis verification of rainfall forecast, the skill score test has been used as suggested by NCMRWF, which are based on 2 × 2 contingencies table. The result of all the four seasons for the year 2015–16 has been presented in Table 7.3.

It is evident from the perusal of Table 7.3 that the value of ratio score was higher during pre-monsoon (94%), post-monsoon (95%), and winter (93%) season because this technique of analysis considered NN cases also. The value of ratio score during monsoon season was 68%, this clearly shows that there was better occupancy of forecast during monsoon season.

The value of threat score, which considered only YY cases, was also found maximum during pre-monsoon season (82%). During monsoon, Post monsoon and winter seasons its value observed were 53, 52 and 0% respectively.

7.11.1.2 ANALYSIS VERIFICATION OF OTHER WEATHER PARAMETERS

Qualitative analysis verification of some other weather parameters was also carried out using standard statistical procedure for all the four meteorological seasons and has been presented in Table 7.3.

TABLE 7.3 Rainfall Prediction Trends During Different Seasons at AMFU, Sabour of Bihar Agricultural College, Sabour (Year 2015–2016)

Sl. No.	Type of Skill Score	SEASON			
		Pre-Monsoon	**Monsoon**	**Post-Monsoon**	**Winter**
1.	Ratio Score	0.943	0.679	0.952	0.932
2.	Bias Score	0.921	0.604	0.523	0.677
3.	Probability of Detection	0.867	0.572	0.543	0.000
4.	False Alarm Ratio	0.015	0.036	0.000	0.032
5.	Threat Score	0.823	0.536	0.521	0.000
6.	Haidke Skill Score	0.778	0.468	0.391	–0.684
7.	Hansen and Kuipper Score	0.843	0.402	0.506	–0.036

7.12 CORRELATION-COEFFICIENT AND ROOT MEAN SQUARE ERROR (RMSE)

The perusal of correlation coefficient and root mean square errors (RMSE) data which were worked out using standard statistical procedure between weather forecast and actual weather prevailed during the same period indicated that the forecasts made by this AMFU were more or less close to correctness excluding wind direction. All observed weather parameters via; cloud cover, rainfall, wind speed, max. and min. temp, except wind direction were found in the line of forecast made in all the four seasons respectively. The RMSE values of wind direction were found too high in all the four seasons to accept any homogeneity in the predicted and observed values. The RMSE value of rainfall during monsoon season was also higher which clearly indicated that forecasts of rain were more or less correct but amount of rain predicted never tallied with observed value of rain occurred.

7.13 QUANTITATIVE VERIFICATIONS ANALYSIS

The quantitative verification analysis worked out between weather forecast made and actual weather prevailed during the same period and has been presented in Table 7.5 (a, b, c, and d). Total numbers of forecasts received during the year 2015–16 were 360 out of which 91 were during pre-monsoon season, 121 were during monsoon season, 92 were during post-monsoon season and 56 were during winter season. These forecasts have been verified by correct, usable and unusable methodology. In this procedure some limits of predicted values of different parameters as suggested by NCMRWF, were used for quantitative verification analysis. Forecast values falling within these limits are recorded as correct and usable and beyond these limit, the forecasts are rated as unusable for meteorological application (Table 7.4).

TABLE 7.4 Season-Wise Correlation Co-Efficient and Root Mean Square Error (RMSE) Value of Different Weather Parameters

Sl. No.	Weather Parameters	SEASON (2015–16)							
		Pre-Monsoon		Monsoon		Post-Monsoon		Winter	
		CC	RMSE	CC	RMSE	CC	RMSE	CC	RMSE
1.	Cloud cover	0.549	2.332	0.442	2. 778	0.5467	2.794	0.561	2.066
2.	Rainfall	0.802	4.966	0.309	15.315	0.9400	9.021	0.802	0.880
3.	Wind speed	0.720	2.863	0.771	3.749	1.1354	3.548	0.426	3.739
4.	Wind direction	0.545	98.678	0.359	82.287	0.3592	90.125	0.568	95.522
5.	Max. temp.	0.877	1.651	0.771	2.112	0.9665	1.055	0.837	2.825
6.	Min. temp.	0.870	2.378	0.230	1.825	0.9453	2.197	0.781	2.672

TABLE 7.5(a) Weather Parameters During Pre-Monsoon Seasons-2015

Sl. No.	Weather Parameters	Pre-Monsoon Season, 2015			
		Correct	Usable	Unusable	Total
1.	Cloud cover	66 (72)	12 (13)	13 (14)	91 (100)
2.	Rainfall	84 (92)	03 (3)	04 (4)	91 (100)
3.	Wind speed	45 (49)	14 (15)	32 (35)	91 (100)
4.	Wind direction	34 (37)	16 (17)	41 (45)	91 (100)
5.	Maximum Temperature	72 (79)	10 (11)	09 (10)	91 (100)
6.	Minimum Temperature	67 (73)	14 (15)	10 (11)	91 (100)

TABLE 7.5(b) Weather Parameters During Monsoon Seasons-2015

Sl. No.	Weather Parameters	Monsoon Season, 2015			
		Correct	Usable	Unusable	Total
1.	Cloud cover	88 (72)	17 (14)	16 (13)	121 (100)
2.	Rainfall	65 (53)	26 (21)	30 (25)	121 (100)
3.	Wind speed	32 (26)	23 (19)	66 (55)	121 (100)
4.	Wind direction	48 (39)	12 (10)	61 (51)	121 (100)
5.	Maximum Temperature	57 (47)	36 (30)	28 (23)	121 (100)
6.	Minimum Temperature	59 (47)	37 (32)	26 (21)	121 (100)

TABLE 7.5(c) Weather Parameters During Post Monsoon Seasons-2015

Sl. No.	Weather Parameters	Post-Monsoon Season, 2015			
		Correct	Usable	Unusable	Total
1.	Cloud cover	61 (66)	08 (9)	23 (25)	92 (100)
2.	Rainfall	88 (96)	01 (1)	03 (3)	92 (100)
3.	Wind speed	35 (38)	12 (13)	45 (49)	92 (100)
4.	Wind direction	46 (50)	19 (21)	27 (29)	92 (100)
5.	Maximum Temperature	60 (65)	27 30)	05 (5)	92 (100)
6.	Minimum Temperature	53 (58)	21 (23)	18 (19)	92 (100)

TABLE 7.5(d) Weather Parameters During Winter Seasons (2015–2016)

Sl. No.	Weather Parameters	Winter Season (2015–2016)			
		Correct	Usable	Unusable	Total
1.	Cloud cover	40 (71)	6 (11)	10 (18)	56 (100)
2.	Rainfall	46 (82)	6 (11)	4 (7)	56 (100)
3.	Wind speed	29 (52)	15 (27)	12 (21)	56 (100)
4.	Wind direction	23 (41)	17 (30)	16 (29)	56 (100)
5.	Maximum Temperature	32 (57)	11 (19)	13 (24)	56 (100)
6.	Minimum Temperature	33 (59)	14 (25)	9 (16)	56 (100)

Note: Figures in parenthesis indicate in percent.

7.13.1 CLOUD COVER

It evinced from the above tables that out of 91, 121, 92 and 56 cloud cover forecasts received during pre-monsoon, monsoon, post-monsoon and winter season, respectively, 85, 86, 75 and 82% were correct in the respective seasons. The correctness of cloud cover ranged between 75 to 86% in different seasons during the year 2015.

7.13.2 RAINFALL

Total numbers of rainfall forecasts received during Pre-monsoon, Monsoon, Post-monsoon and winter seasons were 91, 121, 92, and 56, respectively. Out of these rain forecasts in the respective seasons 95, 74, 97 and 93% were found to be correct. During the rainy season, the correctness of forecasts was only 69% while during other seasons the correctness of rain forecast were much truer because on most of the days there were neither prediction of rain nor it occurred during the said period.

7.13.3 WIND SPEED

During pre-monsoon, monsoon, post-monsoon and winter seasons, total number of wind speed forecasts were 91, 121, 92, and 56, respectively. Out of these forecasts, 64, 49, 51, and 79% were observed to be correct in the respective seasons. The percent of correctness of wind speed forecast was highest (79%) during winter and lowest during monsoon seasons (49%).

7.13.4 WIND DIRECTION

Total number of wind direction forecast received during the four meteorological seasons, i.e., pre-monsoon, monsoon, post-monsoon and winter was 91, 121, 92, and 56, respectively. Out of these wind direction forecasts, 51, 49, 71, and 71% were correct during the respective seasons, which clearly indicated that on most of the days, forecast of wind direction were beyond and specified limit of ± 30°.

7.13.5 MAXIMUM TEMPERATURES

In case of maximum temperature, out of 91, 121, 92 and 56 forecasts received during pre-monsoon, monsoon, post-monsoon, and winter seasons 82 during pre-monsoon 93 during monsoon, 87 during post-monsoon, and 43 during winter were found correct. This clearly indicated that the percentage of correct forecasts were 90, 77, 95, and 76 in the respective seasons.

7.13.6 MINIMUM TEMPERATURE

Similarly, in the case of minimum temperature forecasts, out of 91, 121, 92 and 56 forecasts received during the respective season, 81, 96, 74 and 47 were correct indicating 88, 79, 81 and 84% correctness in the respective seasons.

7.14 AGROMET ADVISORY BULLETIN BASED ON NDVI MAP

The map (Figure 7.3) is regularly prepared by India IMD and is sent at fortnightly to different AMFUs for preparation of Agromet Advisory bulletin. Agriculture vigor is observed for the period and for the area concerned. The analysis is done based on the vigor and weather. Then Agromet advisory bulletin is prepared for the next five days based on the study.

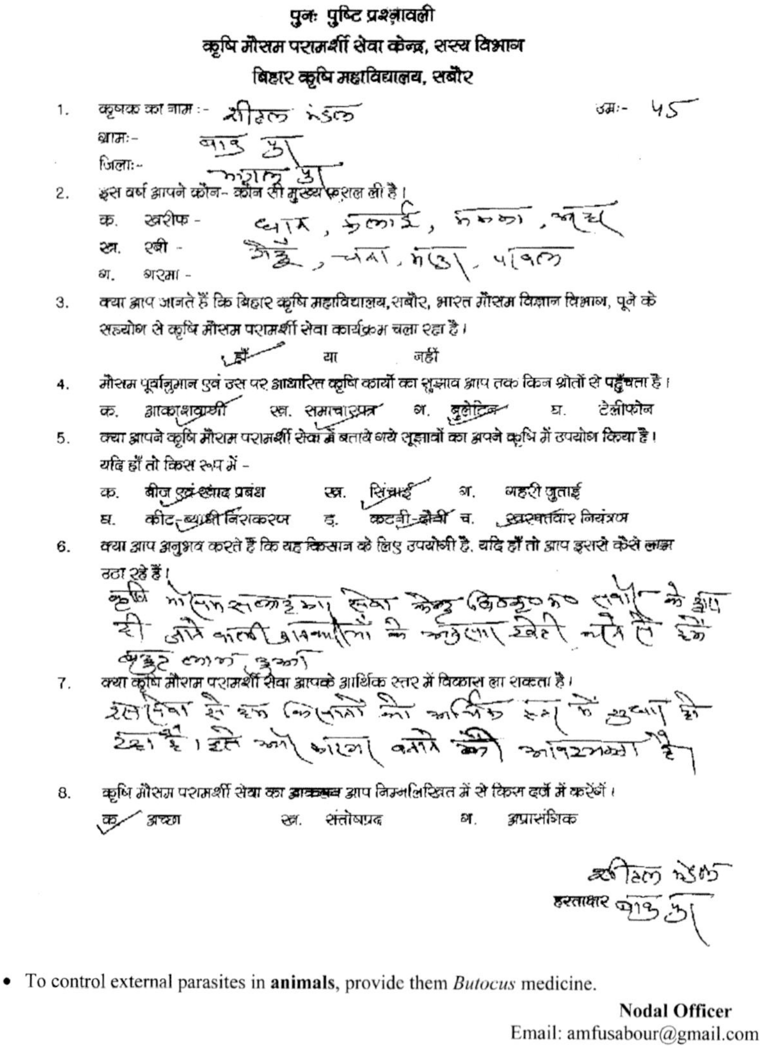

पुनः पुष्टि प्रश्नावली

कृषि मौसम परामर्शी सेवा केन्द्र, सस्य विभाग

बिहार कृषि महाविद्यालय, सबौर

1. कृषक का नाम :- शीतल मंडल उम्रः- 45
   ग्रामः- बाबू पुर
   जिलाः- भागल पुर
2. इस वर्ष आपने कौन- कौन सी मुख्य फसल ली है।
   क. खरीफ - धान, [illegible], मक्का, अरहर
   ख. रबी - गेहूँ, चना, मसूर, प्याज
   ग. गरमा -
3. क्या आप जानते हैं कि बिहार कृषि महाविद्यालय, सबौर, भारत मौसम विज्ञान विभाग, पूने के सहयोग से कृषि मौसम परामर्शी सेवा कार्यक्रम चला रहा है।
   हाँ या नहीं
4. मौसम पूर्वानुमान एवं उस पर आधारित कृषि कार्यों का सुझाव आप तक किन श्रोतों से पहुँचता है।
   क. आकाशवाणी ख. समाचारपत्र ग. बुलेटिन घ. टेलीफोन
5. क्या आपने कृषि मौसम परामर्शी सेवा में बताये गये सुझावों का अपने कृषि में उपयोग किया है।
   यदि हाँ तो किस रूप में -
   क. बीज एवं खाद प्रबंध ख. सिंचाई ग. गहरी जुताई
   घ. कीट-व्याधी निराकरण ङ. कटनी-दौनी च. खरपतवार नियंत्रण
6. क्या आप अनुभव करते हैं कि यह किसान के लिए उपयोगी है. यदि हाँ तो आप इससे कैसे लाभ उठा रहे हैं।
   कृषि मौसम सलाहकार सेवा केन्द्र बि०कृ०म० सबौर के द्वारा दी जाने वाली जानकारियों के अनुसार खेती करते हैं इससे बहुत लाभ हुआ
7. क्या कृषि मौसम परामर्शी सेवा आपके आर्थिक स्तर में विकास ला सकता है।
   इस सेवा से हम किसानों को आर्थिक स्तर में सुधार हो रहा है। इसे और ज्यादा बढ़ाने की आवश्यकता है
8. कृषि मौसम परामर्शी सेवा का आकलन आप निम्नलिखित में से किस दर्जे में करेंगे।
   क. अच्छा ख. संतोषप्रद ग. अप्रासंगिक

शीतल मंडल
हस्ताक्षर बाबू पुर

- To control external parasites in **animals**, provide them *Butocus* medicine.

Nodal Officer
Email: amfusabour@gmail.com

FIGURE 7.3 Feedback from the farmer of the Bhagalpur district.

7.15 AGROMET ADVISORY BULLETIN BASED ON THE STANDARDIZED PRECIPITATION INDEX (SPI) MAP

The map is received from IMD Pune. This is prepared based on the rainfall and its deviation from the normal of the area and the period. The map with full details of the map is sent to the AMFUs by email (Figures 7.4 and 7.5).

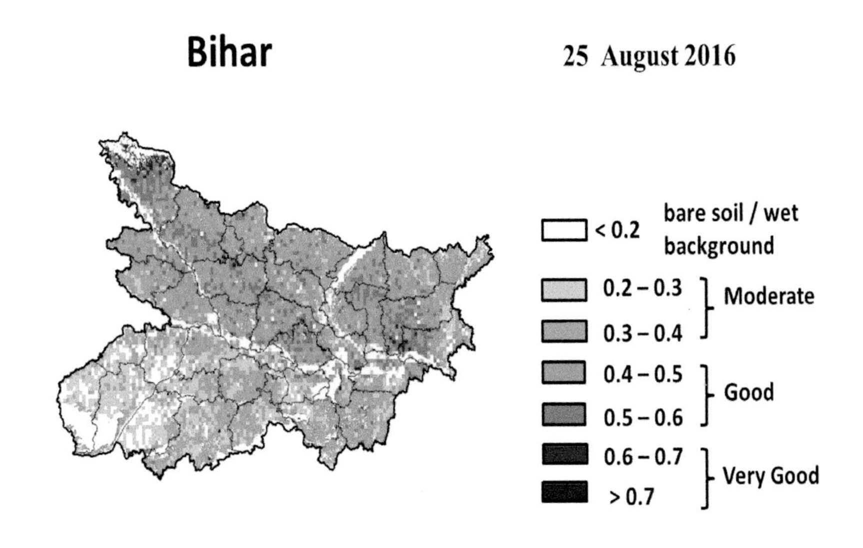

Agriculture vigour is moderate over southern parts of Bihar whereas northern part shows good NDVI values.

FIGURE 7.4 NDVI (normalized difference vegetation index) map.

7.15.1 DETAILS ABOUT THE MAP OF SPI (STANDARDIZED PRECIPITATION INDEX) (FIGURE 7.5)

- Extremely wet/severely wet conditions experienced over most parts of Uttar Pradesh, Madhya Pradesh, Maharashtra, some parts of Andhra Pradesh, Rajasthan; Karnataka; Koraput, Nawarangpur, Rayagada districts of Odisha, Monghyr, Patna, Rohtas districts of Bihar.
- Moderately/severely dry conditions experienced over East Siang, Papumpara districts of Arunachal Pradesh, Lakimpur, Sibsagar, Karimganj, Sonitpur, Bongaigaon, Tinsukia, Kamrup Metro districts of Assam and South Tripura district of Tripura.

MINISTRY OF EARTH SCIENCES
INDIA METEOROLOGICAL DEPARTMENT
HYDROMET SECTION, PUNE

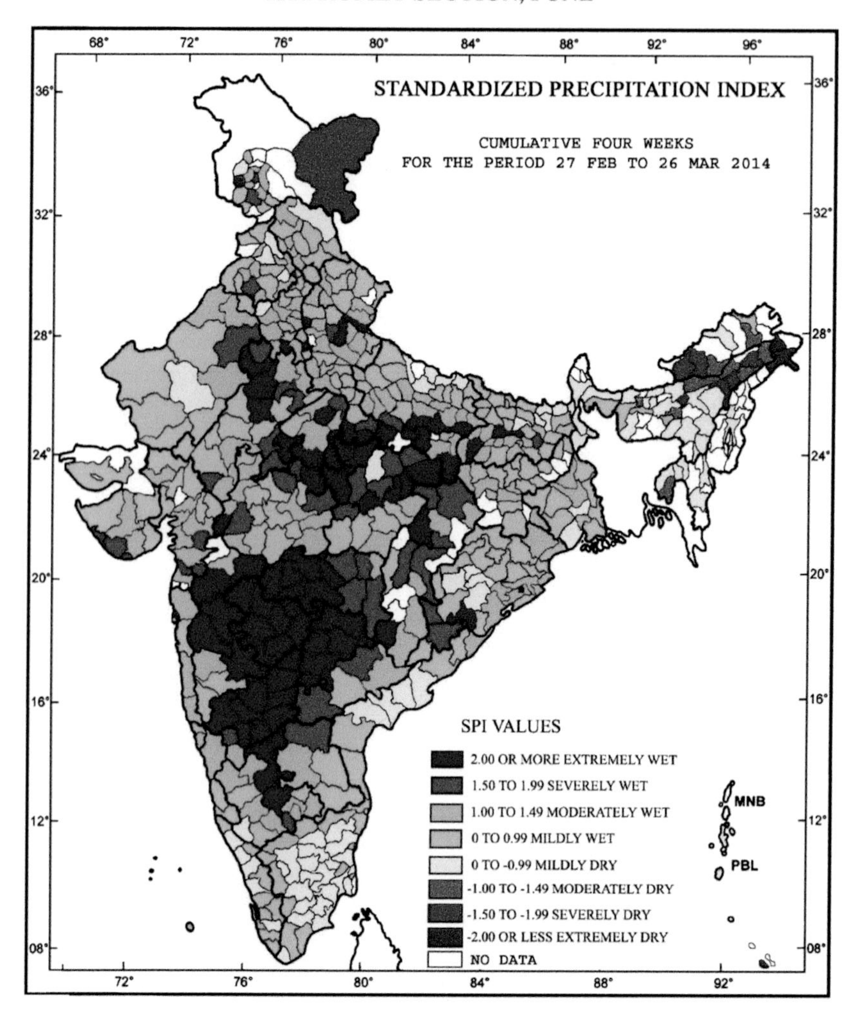

FIGURE 7.5 Standardized precipitation index for the period 27 February to 26 March 2014.

- Extremely dry conditions experienced over Changlang, East Kameng, Lohit, West Kameng districts in Arunachal Pradesh, Dibrugarh, Jorhat, Morigaon, Golaghat districts of Assam, Ladakh (Leh), Poonch districts of Jammu and Kashmir.
- Rest of the country experienced mild wet/dry conditions.

The Standardized Precipitation Index (SPI) is prepared by Hydromet section of India IMD and is sent fortnightly to every AMFUs for every districts of the Country. Based on the moisture condition, advisory regarding irrigation is given to the farmers based on the weather forecast for the next five days (Table 7.6).

TABLE 7.6 Weather Forecast for the Next Five Days Received from India Meteorological Department

Date	5 October	6 October	7 October	8 October	9 October
Rainfall (mm)	3	5	8	0	0
Cloud cover	Partly Cloudy	Partly Cloudy	Partly Cloudy	Partly Cloudy	Partly Cloudy
Max. Temp. (°C)	33	32	32	33	33
Min. Temp. (°C)	25	24	24	24	24
R.H. (%)	53–82	50–92	56–91	56–87	59–90
Wind Speed (km/hr)	4	3	5	6	8
Wind Direction	South Easterly	South Easterly	South Easterly	South Easterly	South Easterly

7.15.2 AGROMET ADVISORY (BASED ON WEATHER FORECAST)

- Farmers are advised to monitor crop regularly, if population of stem borer is above ETL, broadcast of Cartap 4% granules or carbofuran 3% G @ 10 kg/acre or chlorpyriphos 20 EC @ 2 ml/l of water for effective control is recommended.
- Present weather conditions are highly favorable for infection of Blast disease in rice. Early symptoms include development of eye shaped spot in the leaf sheath with light center and dark boundaries. At later stage these spots coalesce to form big spots. If symptoms appear then, spraying of Bavistin 2 gram/liter of water 2–3 times with an interval of 10 days is advised for effective control.
- In present weather conditions, constant monitoring for brown plant hopper (BPH) in paddy field is advised. Farmers are advised to enter into the middle of the crop field and see mosquito-like insects at the basal portion of the plant. If insect population is above ETL, spray of Imidacloprid 17.8% SC@ 1.0 ml/3 liters or fenobucarb 50 EC @ 1 ml/l or buprofezin 25 EC @ 2 ml/l of water is recommended.
- This is suitable time to all crops and vegetables for hoeing and weeding after that split dose of nitrogen should be broadcast for optimum growth.

Constant monitoring of ladyfinger, brinjal, and chili crop against attack of mite and jassid is advised. To control mite, spraying of phosmite @ 2 ml/l of water is advised while spraying of Dimethoate 30 EC @ 2 ml/l of water is advised against jassid when sky remains clear.

- In maize fields, constant monitoring of stem borer should be done. If insect population is above ETL then spraying of carbaryl @ 2.0 gm/ liter of water is recommended in clear sky.
- To control external parasites in animals, provide them *Butocus* medicine.

KEYWORDS

- **agro meteorological field units**
- **agromet field unit**
- **brown planthopper**
- **root mean square error**
- **standardized precipitation index**

REFERENCES

Damrath, U., Doms, G., Friihwald, D., Heise, E., Richter, B., & Steppeler, J., (2000). Operational quantitative precipitation forecasting at the German weather service. *J of Hydrology, 239*, 260–285.

Rathore, L. S., Gupta, A., & Singh, K. K., (2001). Medium range weather forecasting and agricultural production. *Journal of Agric. Physics, 1*(1), 43.

Tamiotti, L., Olhoff, A., Teh, R., Sommons, B., Kulaçoglu, V., & Abaza, H., (2009). *Trade and Climate Change: A Report by the United Nations Environment Programme and the World Trade Organization*. WTO, Geneva, Switzerland.

UNEP, (2011). In: Fischer-Kowalski, M., Swilling, M., Von Weizsäcker, E. U., Ren, Y., Moriguchi, Y., et al., (eds.), *Decoupling Natural Resource Use and Environmental Impacts from Economic Growth*. A report of the working group on decoupling to the International Resource Panel. United Nations Environment Programme.

World Bank, (2010a). *World Development Report 2010: Development and Climate Change*. The World Bank, Washington, DC, USA.

CHAPTER 8

Soil Pollution: Studies with a Specific Reference to Bihar

M. K. DWIVEDI, S. KUMAR, A. KOHLI, Y. K. SINGH,
SHWETA SHAMBHAVI, and R. KUMAR

*Department of Soil Science and Agricultural Chemistry,
Bihar Agricultural University, Sabour, Bhagalpur, Bihar, India*

ABSTRACT

Accelerated soil pollution that has been taking place in India during the last few decades has created environmental complexities including soil. Soil pollution adversely affects the vegetation in urban and peri-urban areas. In the past, soil research focused largely on agricultural soils but now increased research interest is being observed in urban soils too because of the significant increase of the urban population. Undisturbed or partially disturbed urban soils are indicator of environmental perturbations which are responsible for altering soil profile characteristics. Soil properties that are subject to change over relatively short time periods can be a result of natural process and pollutant deposition. The Arsenic contamination in soils of Bihar is a serious threat to human being as well as to soil flora and fauna. Study reveals that the Arsenic contamination in soils of Bihar is mainly due to excessive and irrational exploitation of Arsenic contaminated groundwater. Besides this anthropogenic activity, some geogenic activities are also responsible for Arsenic contamination in soils.

8.1 INTRODUCTION

The accumulation of any substance that makes soil, air, and waterless valuable and less desirable is known as pollution. The decline in the production of soil system due to the presence of various soil pollutants is also known as soil pollution. The term pollution is used to describe the presence of any elemental, molecular or ionic species at a concentration that has been unconsciously raised as a result of anthropogenic activities. Pollutant substances degrade the quality of the environment. Pollution adversely affects the health of human beings, animals and plants and it may be caused by a simple substance such as soluble salts or toxic substances. Pollution leads to degradation of the environment and organisms particularly the humus fraction of soil. To notice the level of pollutants or contaminants in any part of the soil, air, and water needs sensitive and suitable methods of chemical analysis. The soil, water, and air quality are being continuously degraded by construction activities, waste dumpings, and fumes, respectively. Therefore, the activities of human beings are strongly associated with pollution and its control in the environment. The possible reasons for pollution are changes in land use patterns, soil erosion, increasing salinity and sodicity, shortage of water and use of refined technologies.

The soil pollutants affect the physical, chemical and biological characteristics of the soils and results in the decline of soil fertility as well as its crop productivity. Build-up of these soil pollutants also affects the inhabiting soil flora and fauna. Persistence of pesticide residues, fertilizers, radioactive substances, industrial wastes, and plastic materials are the main contributing agents towards soil pollution. Soil pollution can lead to other pollutions like air pollution and water pollution. Strengthening of agricultural production by the use of excessive fertilizer and pesticide application, contaminated irrigation water, etc. has created the problem of soil pollution. Soil pollution can be checked by restricting the use of the above-mentioned pollutants and utilizing better agricultural production systems like organic farming and adopting other suitable production systems. Sustainable production can happen if we follow suitable and adopt such practices that protect the soil from pollutants hazard.

In any cultivated region, soil pollution is one of the main reasons for low crop productivity and soil health. Decline in soil fertility and availability of various toxic materials has affected the soil quality. Very few literatures are available to understand the impact of soil pollution on crop productivity with respect to Bihar. In this part, the information about the soil pollution with special reference to Bihar, a state of potentiality with prospect to crop production has been outlined.

8.2 SOURCES OF SOIL POLLUTION

The general soil pollutants are chemical pesticides, organic wastes, inorganic pollutants, salts, radionuclides and acid rain.

8.2.1 CHEMICAL PESTICIDES

To control detrimental organisms in every field of modern life especially agriculture, chemical pesticides are being used. Various forms of pesticides are insecticides, fungicides, herbicides, rodenticides, and nematocides. Application of these chemicals at high concentrations is detrimental but at low concentrations, they may be beneficial. Any chemical pesticide becomes pollutant if it harms beneficial soil organisms. Application of pesticides to plants, animals or other places enters the soil system through the soil-plant-atmosphere continuum (SPAC).

Following are the acceptable properties of pesticides:

1. Their residence time in the environment must be very short. Their residues must not be accumulated in the food chain. More persistent pesticides may kill non-target organisms.
2. The pesticide must not have the capability to cause cancer in animals, tissue deformation and mutations.
3. The pesticide must not spread extensively in the environment after its application.
4. Chemical pesticides should be of low cost, low corrosiveness and of low caustic in nature.

8.2.2 ORGANIC WASTES

Sewage, sludge, and effluents (organic wastes) produced from domestic or industrial activities are the major source of toxic elements. There must be technologies to ascertain the level of organic and inorganic chemicals prior to its application in agricultural fields.

1. **Garbage:** Municipal garbage after decomposition can be used as a compost to increase crop production and soil conditioner. Urban garbage contains a high amount of decomposable organic matter and can be used as composting material.
2. **Sewage, Sludge, and Effluents:** These are being generated from industrial wastewater treatment plants. Long-term use of industrial effluents in agricultural crops can build-up soil salinity since its biochemical oxygen demand (BOD) is very high.

8.2.3 INORGANIC POLLUTANTS

The maximum permissible limit (MPL) of several chemicals in soil, water, air, and food chain has been prescribed and the primary objective is to protect human and animal health. The hazardous chemicals may reach the soil through the inherent system, active volcanoes and release of metals into the environment from industries and waste disposal. The inorganic pollutants that are considered to be hazardous include Arsenic (As), Cadmium (Cd), Chromium (Cr), Copper (Cu), Mercury (Hg), Nickel (Ni), Lead (Pb), and Zinc (Zn).

8.2.4 SALTS

Soluble salts are mineral substances present in all-natural waters bodies. As water is evaporated or flows on the surface, the salt concentration increases. Salt accumulation in soil and water bodies of arid, semi-arid and coastal regions of the world has been observed. Excess salts can hinder plant growth, accelerate corrosion of metals and do contamination of drinking water. The low-cost remedial measure of salt management in the soil is the urgent need of today.

8.2.5 RADIONUCLIDES

Radioactive substances containing unstable chemical isotopes of elements and give off radiation (alpha, beta, gamma, and other rays) at some specific half-life rate are known as radionuclides. The decay time of radionuclides is in contrast to stable isotopes. These can destruct biological tissues and are carcinogenic and mutants in nature. The presence of these substances in the soil degrades its quality and adversely affects the population of flora and fauna reside in the soil. The retained residues of radioactive compounds in the soil further decline the soil characteristics.

8.2.6 ACID RAIN

Rainfall containing dissolved acids with very low pH value can be said to be acid rain. The name of acids present in the acid rain are sulfuric acid (H_2SO_4) and nitric acid (HNO_3). Carbonates can be dissolved by acid rain and kill aquatic life. It increases the toxicity of Aluminum (Al^{3+}) in the soil. In the atmosphere, Nitric acid comes from different oxides of Nitrogen accumulated in the atmosphere. The atmospheric sources responsible for soil acidification are sulfuric acid, nitric acid, hydrochloric acid, carbonic acid, sulfur dioxide, reduced sulfur compounds, oxides of nitrogen and ammonia.

8.3 METAL CONTENTS IN SOIL

During the last few decades, soil and plants are being exposed to some metals as influenced by anthropogenic factors. Biochemical and geological cycles of many heavy metals have been changed due to man-made activities. We can determine and evaluate the heavy metal content and its exposure in the environment by various methods; however, it is accepted by many researchers that, soil analysis is one of the crucial factors for the identification of environmental contamination. An estimation of the environmental threat caused by soil contamination is principally important for cultivated and uncultivated areas due to the fact that metals are harmful to human health, its lasting effect remains in soils for a reasonably long time and may shift into the food chain in considerable amounts. These days, growing concern about the studies of the value of soil and wastes scarce to dust fall deposition generated from industrial and urban actions is observed. However, little has been recognized about the accurate mechanism of transport of heavy metals into soil and the mode they are released. It could be quite endorsed to the complex organo-mineral character of soils. The variations in metal content of soils can be linked with composition, oxidation/reduction and adsorption/desorption processes, physical transport or categorization in addition to man-made metal input. The fact which should be taken into account also that, metal is associated with soil in different ways and strengths that make the analysis more difficult. Soil is a vital component of rural-urban environments where land management is the key to soil properties and quality.

Apart from mining and manufacturing activities, the use of synthetic products (e.g., paints, pesticides, batteries, industrial waste and soil application of industrial effluents) may result in heavy metal contamination of urban and rural soils. Heavy metals also occur in nature, but not often at toxic levels. Potentially contaminated soils may take place at old landfills and orchards that used insecticides containing heavy metal as an active constituent, field, and farms that had earlier applications of effluents or sewage sludge, areas in or around mining waste heaps, industrial lands where chemicals may have been deposited or in non-point source areas downwind from point source.

The surplus heavy metal build-up in soils is lethal to human beings and animals. Chronic introduction of heavy metals over a longer period of time leads to transfer in the food chain. Severe poisoning from heavy metals is uncommon (but possible) through intake or skin contact. Chronic troubles linked with continuous and long-term use of heavy metals are Lead (Pb, Atomic No. 82), Cadmium (Cd, Atomic No. 48) and Arsenic (As, Atomic No. 33). Frequent problem causing cationic metals (positively charged metallic elements) in soil are Mercury (Hg, Atomic No. 80), Cadmium, Lead, Nickel (Ni, Atomic No. 28), Copper (Cu, Atomic No. 29), Zinc (Zn, Atomic No. 30), Chromium (Cr, Atomic No. 24) and Manganese (Mn, Atomic No. 25). The most familiar anionic compounds (negatively charged metallic elements) are Arsenic, Molybdenum (Mo, Atomic No. 42), Selenium (Se, Atomic No. 34) and Boron (B, Atomic No. 5).

8.4 TRACE ELEMENTS AS SOIL CONTAMINANTS

It is discernible from the literature that, during the last two decades, the trace elements have been considered as environmental contaminants, even more than organic chemicals. The problems associated with man-made environmental pollution of metal sources have been creating a matter of concern in the densely populated cities of India. In this view, industrial and agricultural practices, in particular, are answerable for extensive contamination of the environment in many places. Therefore, the impacts of this pollution on the associations between animals and human health and exposure to such elements via air, water, and food, is a vital area of environmental study (Fifield and Haines, 1995).

Human activities globally are extremely changing the distribution and characteristics of the world's forests (Noble and Dirzo, 1997). In fact, human ecological studies and the promising field of forest history progressively showed that human influences have long been noticeable in many forests (Baiee, 1989; Thirgood, 1989; Denevan, 1992; Lepofsky et al., 1996; Roosevelt et al., 1996; Schnieder, 1996; Kirby and Watkins, 1998; Agnoletti and Anderson, 2000).

In China, environmental pollution has been growing for the last decades. High atmospheric emission of sulfur has been, and still is, of major concern (NEPA, 1997). Heavy metals contents in Chinese cultivated soils have been studied by many researchers, but small information exists on heavy metal contamination in forest soils.

The core origin of air pollution is fuel combustion through any of the sources. In India, 25% of the total energy (of which 98% comes from oil) is consumed by transport sector only, which is reported to be contributing more than 50% of air pollution in most of the metro cities and in some cases, it was even up to 80%. A study in 2001 showed that, air pollution contribution of transport sector was about 72% in Delhi and 48% in Mumbai.

One of the anthropogenic sources of heavy metals exist in various industrial point sources (Nilgun et al., 2004).

Hydrospheric heavy metals are important because they act together with soil/sediment samples of geological origin and further can influence biological processes. Ingole and Bhole (2000) revealed that plants, especially aquatic species, can accumulate heavy metals and act as indicators of the condition of the water environment in which they are located.

Substantial amounts of lead have accumulated in soils all over the world due to man-made activities in the last few decades. This metal is highly lethal for human and animals. So recognizing and characterizing its performance in soils is essential. Lead forms strong complexes with organic matter therefore, it often suffers almost entire retention within forest soils (Wang et al., 1995). Many heavy metals are biogenic elements, i.e., they occur in inadequate amounts in living organisms and play definite roles in them. However, higher concentrations can cause serious problems. In recent years, due to anthropogenic activity, some heavy metals accumulate in upper soil layer, enter into the food chain and affect human health.

All over the world, there is a long custom of intensive farming within urban and peri-urban areas (Smit et al., 1996). However, most of these peri-urban lands are contaminated with pollutants including heavy metals such as Cu, Zn, Pb, Cd, Ni, and Hg. These metals are generated mostly through industrial effluents, sewage, and sludge, vehicular emission, diesel generators and application of pesticides in agriculture. This loading of heavy metals frequently leads to deterioration of soil health and food chain contamination mainly through the vegetables grown on such soils (Jackson and Alloway, 1992; Rattan et al., 2002).

The emission pathways of metal pollutants into the atmosphere are of very different types viz. volcanic activity, agricultural emission, soil erosion and man-made. In other terms, pollutants are emitted from natural and anthropogenic sources. An accurate evaluation of natural source strength is pretty difficult but also important, as for many elements, natural emissions exceed those from anthropogenic sources. Among the natural sources of trace metals, the windblown dust or dust fall and volcanic eruptions are considered as the most important (Thakur et al., 2002).

Today's lead (Pb) and cadmium (Cd) pools of Swedish forest soils are mostly originating from anthropogenic sources (Andersson et al., 1992; Johansson et al., 1995). In southern Sweden the Pb pool of the top soil has increased during 2000 years between 5 and 10 times the background level (Johansson et al., 1995). While iron crusts (ferricretes) are pervasive in western and Central Africa, those are characterized by a contrasted seasonal climate; they generally not seen in tropical rain forest regions, due to distorted climatic conditions.

8.5 EFFECTS OF SOIL POLLUTION

The main reason for soil pollution is human activities. Contaminates adversely affect the health of soil directly and indirectly. The soil pollution interrupts the natural environment and affects the ecological balance. The nutrient supplying capacity of soil declines abruptly and in turn affects sustainability. The soil pollution influences the plant growth and nutrient stock poorly and leads to accumulation of toxic material in the soil. Once toxic materials enter in the plant system, they disturb the different physiological and biochemical pathways. Consuming contaminated food material cause ill effects in the human and the animals. Several infectious diseases in human and animals are results of these contaminants. The health hazards are great concern for us. Soil pollution results in deterioration of soil characteristics and disrupts the arrangement of soil particles (soil structure). Alteration in soil structure causes death of many beneficial microorganisms. It affects the life of larger predators (e.g., birds) that do environmental cleaning to supporting life system on this planet. Soil pollution also affects the water bodies. For example, the use of excessive nitrogenous and phosphatic fertilizer in the soil washes by running water and gets accumulated in water bodies (eutrophication). These fertilizers affect the quality of the water and growth and development of such organisms that cause pollution of water systems.

The cleanliness of the environment is directly related to the state of the soil. Release of toxic gases in the environment that pollute the air which is inhaled by humans and animals for their survival is due to anthropogenic soil contamination. People residing near the polluted land are prone to many severe diseases like cancer, leukemia, kidney and liver damage, etc. People of those regions face the problem of migraines, nausea, fatigue and skin diseases. These problems are caused by direct poisoning of soil pollution.

The long term effects of soil pollution have become more important all over the world that has affected the life of the entire living society residing on the earth. Higher dose of heavy metal in the potable and edible food material enters into food chain and finally does detrimental problems. Remediation of soil pollution is a big challenge and requires considerable awareness and attention of society. Biological soil remediation is one of the techniques which can be utilized on mass scale in getting rid of soil pollution.

8.6 BIOLOGICAL SOIL REMEDIATION

Not like the physical remediation techniques, biological techniques are usually performed in-situ. These techniques comprise of microbial remediation, phytoremediation, fungal remediation and composting. Further, the descriptions are not broad, as we obtain to emphasize the major benefits and drawbacks of each technique and provide a relationship among the techniques. Widespread literature on these techniques and methodologies are available for further understanding.

8.6.1 MICROBIAL REMEDIATION

Remediation through microbes refers to the utilization of microbes in degrading contaminants into a reduced toxic form. This technique can be very effective in the treatment of polycyclic aromatic hydrocarbons or PAH's, pesticides and Polychlorinated biphenyls or PCB's. The normal cost is relatively low and timeframe is little. However, there is a likelihood of enhanced toxicity of definite metals.

8.6.2 PHYTOREMEDIATION

Phytoremediation is the method of using plants to extract contaminants or to degrade/decompose them in the soil. As in the case of microbial remediation, the cost is little. However, the time frame can be longer than quite a few years. Efficiency of soil to meet agricultural standard varies because one species of plant is characteristically used on one type of contaminant, potentially leaving behind a variety of contaminants. Hence, the contaminated plants used for extraction must be disposed off.

8.6.3 FUNGAL REMEDIATION

Remediation through Fungi refers to the use of distinct types of fungi to degrade contaminants. This commercial technique can be available only after its development.

8.6.4 COMPOST REMEDIATION

Remediation through compost involves the addition of compost materials to the soil. This is inexpensive and rapid to do. However, it is not an accurate remediation method, as the contaminants normally remain together in the soil. Further, the compost addition can, on the other hand, is used to make a raised bed, in which the plant roots may not have contact to contaminated soil.

8.6.5 NON-REMEDIAL OPTIONS

Other than these remediation techniques, soil contamination issues can be dealt with the other options also. The options involve growing the produce in a detached container or raised bed above the contaminated soil. It can be used in an endeavor to avoid plant roots from attaining the contaminated soil. Likewise, techniques such as aquaponics are an additional method to prevent growing directly in the soil. Alternately, for highly contaminated soil, one may consider for another portion of land for the garden. As expected, the health risks of growing food grains in contaminated soil can be noteworthy.

8.7 SOIL POLLUTION IN BIHAR

Soil is one of the most important resources of a nation. It is the gift of nature of immense value. The most common use of the word soil is in the sense of a medium in which plants grow, although it has a different connotation at different time and place, and for persons engaged in different professions. Almost all the economic activities are directly or indirectly dependent on soil. Thus soil is the backbone of agricultural and industrial development. Soil has a number of characteristics, which may be regarded as the aggregate of the physical, chemical and biological properties. The Bihar plane, in most part, consists of a wide alluvial layer of drift origin.

Human efforts for the better living standards and pollution of the environmental are the two sides of the same coin. Considering rapid industrialization, consequent urbanization and ever-increasing population, the basic services of life, i.e., water, air, and soil, are being populated constantly. Industrial complexes have become the center of environmental pollution. When changed in its quality and composition as affected by anthropogenic activities, air may be regarded as polluted. Since the atmosphere has a considerable absorptive capacity, the release of fewer amounts of pollutants into the air does not lead to any severe effects.

Various industrial installations form the stationary sources of the urban air pollution viz. asphalt plants, brick chimney plants, boiling and heating installations, cement manufacturing, fertilizer manufacturing, mineral acid manufacturing, paper and pulp manufacturing, thermal and nuclear power plants, sewage treatment plants and engineering workshops, etc. The mobile sources of air pollution are automobiles such as cars, scooters, motors, trucks, and buses moving on the urban roads. North Bihar's Saran district has a variety of temporary as well as permanent brick chimney plants adjoining the district headquarter Chhapra town (Srivastava and Singh, 2012). All the brick chimneys emit CO_2, SO_2, etc., which directly or indirectly interrelate with the soil constituents. The interaction of the dust smoke with soil may bring about physical and chemical changes of the top fertile layer of soils.

Groundwater contamination with natural Arsenic and Fluoride is posing a serious problem to the soil as well as the environment and causing serious health hazards to human beings, soil flora and fauna. This is because the source of Arsenic and fluoride contamination in soils is unexceptionally contaminated groundwater (Table 8.1). It is only due to the excessive and irrational exploitation of contaminated groundwater, that there occurs a substantial build-up of Arsenic and fluoride in soils. In India, this problem is mainly concerned in the states like Bihar, UP, Chhattisgarh and some parts of Jharkhand. Arsenic contamination reaches the soil due to various geogenic and anthropogenic activities and it pollutes the soil badly. More than 10 million people in the rural part of Bihar are exposed to high levels of Arsenic through naturally contaminated drinking water sources. Arsenic levels in the water go beyond the permissible limits in some areas. To the inhabitants, Arsenic contamination in soil and food chain could be a potential threat. People suffering from Arsenicosis symptoms and supposed to Arsenic induced cancers were observed in the state. The state needs ample Arsenic mitigation policies and decision-making tools to help prioritize the areas. For assessing and mapping the vulnerability to groundwater Arsenic contamination, extent, distribution and its possible mitigation options a decision-making tool such as a composite vulnerability framework would be the absolute necessity.

Shallow depth hand pump water samples collected from some villages in the Ganga basin in Kahalgaon. Nathnagar, Sultanganj, Pirpainti and Sabour block of Bhagalpur district were found to be contaminated with Arsenic in quantities more than the permissible limit (> 50 ppb). The maximum fluoride content of water samples was found to be 3.61 ppm in Kola Khurd village of Jagdishpur block which was substantially higher than the permissible limit of 1.50 ppm. Fluoride contamination was also detected in some villages of Shahkund and Nathnagar blocks. Vegetable crops were found to contain more Arsenic than cereals, pulses and oilseed crops in the Arsenic contaminated areas. No Arsenic and fluoride contamination was detected n water samples collected from a depth of more than 130 ft. In the studied areas, the Nathnagar block showed maximum Arsenic content (460 ppb). Some villages of Kahalgaon, Pirpainti, Sabour and Sultanganj blocks also exhibited Arsenic content beyond the permissible limit of 50 ppb. Arsenic content in soils of different blocks of Bhagalpur varied from non-detectable to 3.88 ppm whereas the content in plant parts of cereals, tubers, bulbs, pulses, oilseeds, and vegetables are presented in Table 8.2.

TABLE 8.1 General Information on Groundwater of Bhagalpur

S. No.	Parameters	Values
1	EC (dS m^{-1})	0.07–1.3
2	pH	6.8–8.35
3	Total dissolved solids (ppm)	163–788
4	Total hardness as $CaCO_3$ (ppm)	61–505
5	Calcium (ppm)	11–172
6	Magnesium (ppm)	7.3–81
7	Sodium (ppm)	14–189
8	Potassium (ppm)	0.18–4.0
9	Chloride (ppm)	10.6–200
10	Carbonate (ppm)	0.12–0.91
11	Bicarbonate (ppm)	102–742
12	Sulfate (ppm)	<3.0
13	Fluoride (ppm)	0.0–3.61
14	Arsenic (ppm)	0.0–0.46

Experimental studies conducted at Bihar Agricultural University (BAU), Sabour, Bhagalpur for minimizing Arsenic hazards in rice have suggested that the Arsenic accumulation in grain, straw, and roots of rice could be reduced with the application of organic materials such as FYM, rice straw, green manure. The same results were found to be true when studies were upscaled on farmer's fields in the villages of Nathnagar and Pirpainti blocks. The incorporation of organic manures minimized the Arsenic accumulation in rice grain and straw and also sustained the soil health against Arsenic contamination.

TABLE 8.2 Arsenic (As) Content of Various Crops Grown in Contaminated Areas

Crops	Range of as Content (Parts Per Billion)	Mean Content of as (Parts Per Billion)
Wheat	43–76	57
Maize	38–85	74
Potato	119–156	131
Brinjal	75–108	84
Cauliflower	51–129	81
Coriander	15–47	35
Beans (Seem)	18–36	27
Onion	154–255	180
Tomato	20–61	35
Mustard	23–81	50
Linseed	18–52	32
Lentil	25–48	33
Green gram	21–41	34

8.8 WET AND DRY POLLUTANT DEPOSITION ON SOIL

The wet removal process is helpful only during the monsoon period (June-September) when about 90% of the annual rainfall occurs in India. During the rest of the year, dry conditions prevail which determine the atmospheric deposition chemistry in India. Ambient concentration and atmospheric reactions are restricted by the continuous input of dust particles suspended in the air which are contributed by soil suspension for the duration of dry weather conditions. Hence, dustfall deposition is a significant removal method in India as it provides an extremely good sink for acidic gaseous pollutants covering the earth's atmosphere (Kulshrestha et al., 2003).

Saxena et al. (1997) considered that besides wet deposition, dry deposition is another major atmospheric removal process of both gases and particulates to the earth's surface. The dry deposition of minute acidifying substances containing SO_4^{2-}, NO_3^- and NH_4^+ contribute to the total acid input to ecosystems. For hefty particles containing base cations, the understanding of deposition is important for the understanding of through fall measurements, nutrient cycling, and assessment of significant critical loads.

Dry deposition of airborne pollutants contributes importantly to the atmospheric load of ecosystems and is studied intensively. The dry deposition process is influenced by numerous chemical, physical, and biological aspects of the atmosphere, the deposited substance, and the surface structure (Sehmel, 1980; Hosker and Lindberg, 1982). Factors influencing the rate of dry deposition may have different effects of the deposition of particles and gases. Differences in factors influencing deposition may occur within small distances and within short periods of time. Forest edges provide a situation where many factors regulating deposition are changing within very small distances. Wiman and Agren (1985) showed in their model studies that, the higher the wind speed at the forest edge increased the dry deposition of particles.

8.9 CONCLUSION

Leaving crop residues on the soil and incorporating it into the soil reduces erosion and increases soil organic matter. The introduction of organic matter into the soil also makes compaction less likely. Crop rotation is a valuable way to improve soil fertility, reduce erosion and control pests. With respect to organic farming, there has been a point of view both for and against. Critics argue that organic farming cannot produce the amount of food required for today's population and in certain conditions only it is economically viable. However, a group of organic farming feels that if the unseen costs of soil erosion and pollution are taken into account, it is a feasible approach. One more way to reduce pollution impacts on soil is via use of integrated pest management (IPM). This is a technique that uses a complete thought of all the natural aspects of a crop and the meticulous pests to which it is vulnerable to establish pest control strategies that uses no or few pesticides.

KEYWORDS

- **biochemical oxygen demand**
- **compost remediation**
- **fungal remediation**
- **integrated pest management**
- **maximum permissible limit**
- **soil-plant-atmosphere continuum**

REFERENCES

Andersen, M. K., Refsgaard, A., Raulund-Rasmussen, K., Strobel, B. W., & Hansen, H. C. B., (2002). Content, distribution, and solubility of cadmium in arable and forest soils. *Soil Science Society of America Journal, 66*, 1829–1835.

Banat, K. M., Howari, F. M., & Al-Hamad, A. A., (2005). Heavy metals in urban soils of central Jordan: Should we worry about their environmental risks? *Environmental Research, 97*, 258.

Bruemmer, G. W., Gerth, J., & Herms, U., (1986). Heavy metal species, mobility and availability in soils, Z. Planzenernaehr. *Bodenk, 149*, 382–398.

Davies, T. D., & Mitchell, J. R., (1983). Dry deposition of sulfur dioxide onto grass in rural eastern England with some comparisons with other forms of sulfur deposition. In: Pruppacher, H. R., Semonin, R. G., & Slinn, W. G. N., (eds.), *Precipitation Scavenging, Dry Deposition and Resuspension* (pp. 795–806). Elsevier, New York.

Dudka, S., Piotrowska, M., Matysiak, Z., & Witek, T., (1995). Spatial distribution of trace metal concentrations in arable soils and crop plants of Poland. *Polish Journal of Environmental Studies, 4*, 9.

Fifield, F. W., & Haines, P. J., (1996). *Environmental Analytical Chemistry*. Blackie Academic & Professional, Glasgow.

Goyal, S. K., Ghatge, S. V., Nema, P., & Tamhane, S. M., (2006). Understanding urban vehicular pollution problem vis-a-vis ambient air quality - case study of a megacity (Delhi, India). *Environmental Monitoring and Assessment, 119*, 557–569.

Grzebisz, W., Cieśla, L., Komisarek, J., & Potarzycki, J., (2002). Geochemical assessment of heavy metals pollution of urban soils. *Polish Journal of Environmental Studies, 11*, 493.

Hosker, Jr., R. P., & Lindberg, S. E., (1982). 'Review: Atmospheric deposition and plant assimilation of gases and particles.' *Atmos. Environ., 16*, 889–910.

Ingole, N. W., & Bhole, A. G., (2000). Bio-accumulation of chromium, nickel and zinc by water hyacinth. *Pollution Research [Pollut. Res.], 19*(4), 575–583.

Jackson, M. L., (1973). *Soil Chemical Analysis*. Prentice-Hall of India Pvt. Ltd., New Delhi.

Kumar, R., Rani, A., Kumari, K. M., & Srivastava, S. S., (2005). Atmospheric dry deposition to marble and red stone. *J. Atmos. Chemistry, 50*, 243–261.

Mehra, R. K., (2016). Soil and environment. In: *Text Book of Soil Science* (pp. 456–484). ICAR New Delhi.

NEPA and the National People's Congress, (1997b). *The Handbook for Environmental Laws and Enforcement*. Beijing: China Environmental Science Press.

Nilgun, G., Omar, A., & Gurdal, T., (2004). In vestigation of soil of multi- element composition in Antalya, Turkey. *Environmental International, 29*, 631–640.

Noble, I. R., & Dirzo, R., (1997). Forests as human-dominated ecosystems. *Science, 277*, 522–525.

Omar, A., & Al-Khashman, (2004). Heavy metal distribution in dust, street dust and soils from the workplace in Karak Industrial Estate, Jordan. *Atmospheric Environment, 38*, 6803–6812.

Rattan, R. K., Datta, S. P., Chandra, S., & Saharaan, N., (2002). '*Heavy Metals in Environments-Indian Scenario.*' (Vol. 47, pp. 21–26, 29–40). Fert. News.

Saari, A., Martikainen, P. J., Ferm, A., Ruuskanen, J., De Boer, W., Troelstra, S. R., & Laanbroek, H. J., (1997). Methane oxidation in soil profiles of Dutch and Finnish coniferous forests with different soil texture and atmospheric nitrogen deposition. *Soil Biol. Biochem., 29*(11/12), 1625–1632.

Saikh, H., Varadachari, C., & Ghosh, K., (1998). Changes in carbon, nitrogen and phosphorus levels due to deforestation and cultivation: A case study in Simlipal National Park, India. *Plant and Soil, 198*, 137–145.

Sami, M., Waseem, A., & Akbar, S., (2006). Quantitative estimation of dust fall and smoke particles in Quetta valley. *J. Zhejiang Univ. Science B., 7*(7), 542–547.

Saxena, A., Kulshrestha, U. C., Kumar, N., Kumai, K. M., Praksh, S., & Srivastava, S. S., (1997). Dry deposition of sulphate and nitrate to polypropylene surfaces in a semi-arid area of India. *Atmospheric Environment, 31*(15), 2361–2366.

Smit, J., Ratta, A., & Bernstein, J., (1996). *Urban Agriculture: An Opportunity for Environmentally Sustainable Development in Sub-Saharan Africa*. Building.

Srivastava, K. P., & Singh, V. K., (2012). Impact of air pollution on pH of soil of Saran, Bihar, India. *Research Journal of Recent Sciences, 1*(4), 9–13.

Thakur, M., Deb, M. K., Imai, S., Suzuki, Y., Ueki, K., & Hasegawa, A., (2004). Load of heavy metals in the airborne dust particulates of an urban city of Central India. *Environmental Monitoring and Assessment, 95*, 257–268.

Tuzen, M., (2003). Determination of heavy metals in soil, mushroom and plant samples by atomic absorption spectrometry. *Microchemical Journal, 74*, 289.

Wang, E. X., & Benoit, G., (1996). Climate change and agriculture in China, *Environ. Sci. Technol., Environ. Sci. Technol., 30*(7), 2211–2219.

CHAPTER 9

Nanotechnology in Agricultural Science

NINTU MANDAL,[1*] KASTURIKASEN BEURA,[1] and ABHIJEET GHATAK[2]

[1]*Department of Soil Science and Agricultural Chemistry, Bihar Agricultural University, Sabour, Bhagalpur, Bihar, India*

[2]*Department of Plant Pathology, Bihar Agricultural University, Sabour, Bhagalpur, Bihar, India*

[*]*E-mail: nintumandal@gmail.com (N. Mandal)*

ABSTRACT

Nanotechnology, the science as well as the art of manipulating matter at atomic or nanoscale (1 nm = 10^{-9} m) is cutting edge technologies in agricultural sciences. Nanotechnological interventions in increasing input use efficiencies, decontamination of toxicants, manipulating genetic materials, increasing shelf life of harvested produce and production of quality planting materials may lead to sustainable agricultural production. Nanoscience and nanotechnology unit of BAU, since its inception is engaged in developing novel agrochemicals with an intelligent delivery system as well as human resource development (UG and PG teaching and research in Nanotechnology in Agriculture). Novel nanopolymeric hydrogel, nanoformulation of P, Zn, Fe, chitosan, Ag, and Cu have been developed and evaluated under laboratory and greenhouse conditions. Field evaluation of novel nanomoleclues for benefit:cost ratio and multi-location field trials are underway.

9.1 INTRODUCTION

Nanoscience is the study of matter at atomic or nanometer (1 nm = 10^{-9} m) scale. Nanotechnology is the designs, fabrication and control structure for specific applications. According to US EPA (the United States Environment Protection Agency, 2007), nanoparticles are substances that are less than 100 nm in size in more than one dimension. Such particles are minerals that are as small as roughly 1 nm and as large as several tens of nanometers in at least one dimension. Limiting size in one, two, or three dimensions results in a nanosheet (e.g., Vernadite), a nanorod (e.g., Palygorskite), or a nanoparticles (Ferrihydrite), respectively.

9.1.1 BRIEF HISTORY OF NANOTECHNOLOGY

Richard P. Feynman in 1959 suggested that it would be possible to build machines small enough to manufacture objects with atomic precision. His talk "*There's Plenty of Room at the Bottom*" is widely considered to be the foreshadowing of nanotechnology. In 1990, a team of IBM physicists revealed that they could write (arrange) the letters IBM using 35 individual atoms of Xenon.

In 1992, a book entitled *Nanosystems: Molecular Machinery, Manufacturing and Computation* was published by Eric Drexler where he outlined a way to manufacture extremely high-performance machines out of molecular carbon lattice (diamondoid).

Around the year 2000, federal funding for nanotechnology in the United States began with the National Nanotechnology Initiative (NNI). NNI defined nanotechnology as dealing with materials with sizes between 1 and 100 nanometers exhibiting novel properties.

The Government of India, in May 2007 launched a mission on nanoscience and nanotechnology (nano mission) with an allocation of Rs. 1000 cores.

Nanosynthesis involves two fabrication techniques, i.e., top-down approach and bottom-up approach. The first approach is working with macroscopic materials and coming gradually to nanoscale, relatively easy but there may be imperfection in the crystals. The second one, i.e., bottom-up approach is working with atomic level particles and gradually build-up of nanoscale materials. It is time taking, but more perfection on the structure is obtained. There is a summary of these two techniques.

9.2 SYNTHESIS AT NANOSCALE

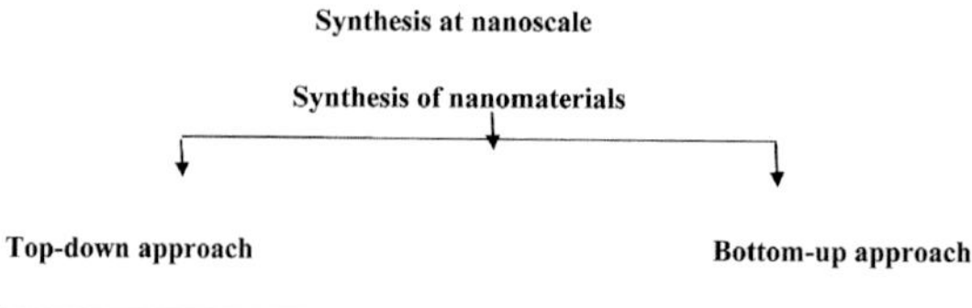

Processes	Materials synthesized
Arc discharge method	Fullerenes and carbon nanotubes
Laser ablation	Broad range including
Ball milling	Composites and mixtures of elemental powders
Inert gas condensation	Oxides, alloys and semiconductors

Processes	Materials synthesized
Homogenous nucleation	Noble metal particles, gold nanoparticles
Chemical vapour deposition	Widely used to produce CNTs, fullerenes and boron nanotubes
Molecular beam epitaxy	Compound semiconductor and thin films
Sol-gel synthesis	Colloidal and oxide nanoparticles
Hydrothermal synthesis	Elemental nanopowders and mostly oxides
Microwave method	Oxide nanoparticles including TiO_2

9.3 CHARACTERIZATION TECHNIQUES

9.3.1 NANOSCALE CHARACTERIZATION TECHNIQUES

The development and continual refinement of nanoscale techniques have allowed for much of the revolution in nanoscience and nanotechnology. Wide array of microscopes such as scanning electron microscopy (SEM), transmission electron microscopy (TEM), atomic force microscopy (AFM) and epifluorescence microscopy are used for characterization at nanoscale. Analytical methods ranging from atomic absorption (AA) and inductively coupled plasma (ICP) emission spectrophotometry (ICP) to extended x-ray absorption fine structure (EXAFS) and x-ray absorption near edge structure (XANES) spectroscopies are crucial for understanding material composition and structure.

9.4 APPLICATION OF NANOTECHNOLOGY

Nanomaterials have a wide potential in agricultural science. Several fields of potential nanotechnology Increase in input use efficiency (Fertilizers, agrochemicals), controlled release agrochemical formulations, biosensors in precision agriculture and so on (Figure 9.1).

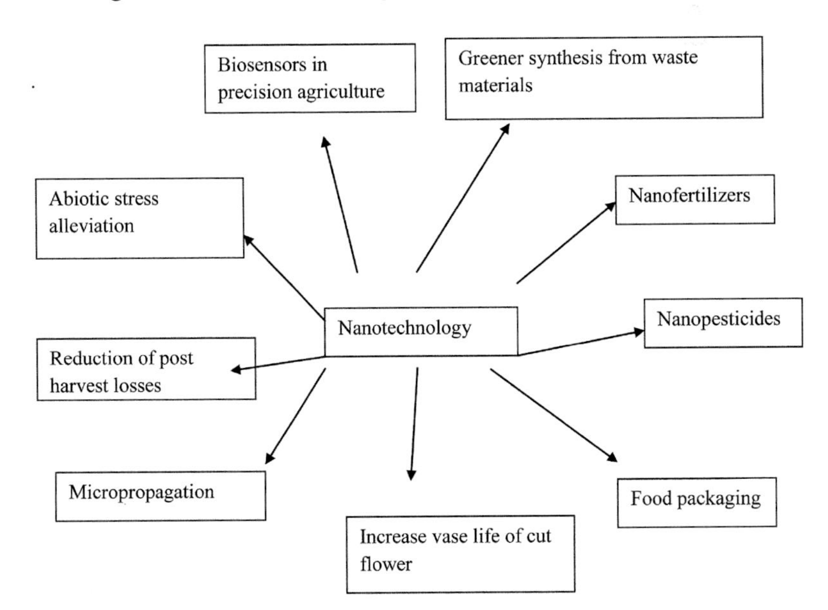

FIGURE 9.1 Application of nanotechnology in agriculture.

Horticultural science also having immense application of nanomaterials *viz* increase of vase life of cut flowers, food packaging, biosensors, control release fertilizers, and pesticide formulations, micropropagation, as well as horticultural waste materials, can be a good source for nanoparticle synthesis for other applications.

9.4.1 AGRICULTURAL APPLICATIONS OF NANOTECHNOLOGY: GLOBAL SYNTHESIS

9.4.1.1 NANOCLAY POLYMER COMPOSITES (NCPC): NOVEL FERTILIZER FORMULATIONS

Slow-release carrier of urea fertilizer was developed by aqueous solution copolymerization of partially neutralized acrylic acid and acrylamide in presence of kaolin nanopowder. The release of urea could be affected by many factors such as contents of acrylamide, crosslinker, kaolin, the neutralization degree of acrylic acid, temperature, pH and ionic strength of release medium (Liang and Liu, 2007).

In a greenhouse experiment, Sarkar et al., 2013 reported that Olsen P, $CaCl_2$-extractable P and mineral N content in soils increased significantly due to the addition of fertilizer as NCPC (Nanoclay polymer composite) as compared to conventional fertilizer. Moreover, nutrient availability in soils receiving lower doses of fertilizer as NCPC was statistically at par with the higher dose of conventional fertilizer.

NCPCs based controlled release nutrient formulation was reported by Sarkar et al. (2013). Among various types of clays, smectite-dominated clay had the highest hydration and distention among the selected clays (Kaolinite and Mica); this resulted in more crosslinking in the composite, which in turn, resulted in the lowest equilibrium water absorbency (WA) and nutrient-release (N and P) rate of the corresponding NCPC. Smectitic types of clay minerals are reported to be exfoliated into the galleries whereas mica and kaolinite, intercalation occurred.

Superabsorbent polymer with slow-release phosphate fertilizer (SAPSRPF) was reported by Zhan et al., (2004). The experimental result for phosphate release in the SAPSRPF about 26.5% of phosphate in SAPSRPF was released out within 24 h; on the 72nd hour, the total amount of phosphate released out was about 47%. The phosphate released out in this period mainly existed in the SAPSRPF surface layer, because it would go into water without diffusing as soon as it was dissolved, so the release rate was fast, and on the (672th hour) 28th day, the total amount of phosphate released out had gradually increased to 79%.

However, the application of NCPCs in the field have met some problems because most of these superabsorbent are based on pure poly (sodium acrylate) and then they are too expensive and not suitable for saline containing water and soils (Mohan et al., 2005). Recently there have been many reports on introducing inorganic clays into pure polymeric superabsorbent to improve swelling property, hydrogels strengths, and reduce production cost (Wu et al., 2003).

Nanomaterials based controlled release nutrient formulations were reported recently (Sarkar et al., 2013; Mandal et al., 2015). Nanoclay polymeric based controlled release formulation for fertilizers to supply N, P had been attempted (Sarkar et al., 2013) and under pot experimentation reported to give promising results. Zincated nanoclay polymer composites (ZNCPC) as controlled releaser Zn formulation was reported to enhance Zn uptake, P uptake in rice and also stimulated soil microbial activities in terms of enhanced soil dehydrogenase, acid and alkaline phosphatase activity (Mandal et al., 2015).

Mosanna and Behroztar (2015) reported that the use of zinc nano-chelate application had positive effect on yield and yield components. Based on the results, soil application of nano-chalate zinc had the highest plant height.

Prasad et al., (2012) was observed that application of nano ZnO to Peanut resulted to increases pod yield per plant was 34% higher compared to bulk $ZnSO_4$. Consequently, and with the foliar application of nano-scale ZnO particles at 15 times lower dose compared to the chelated $ZnSO_4$ recommended and we recorded 29.5% and 26.3% higher pod yield, respectively, compared to chelated $ZnSO_4$.

The effect of NCPC combination increases the availability nutrient in soil and enhanced the microbial activity due to high organic matter decomposition and mineralization (Mukhopadhyay et al., 2015).

ZnO-NPs enhanced growth of mung bean and chickpea (*Cicer arietinum*) seedlings at low concentrations (Mahajan et al., 2011). They reported that mung bean seedlings, the best growth response for root (a 42% increase in length or 41% in biomass) and shoot (98% in length or 76% in biomass) was observed at a concentration of 20 mg L^{-1} over those of the control; 1 mg L^{-1} caused significant increases in root (53% in length or 37% in biomass) and shoot (6% in length or 27% in biomass) growth for the chickpea seedlings. However, the decline in growth rates of roots and shoots were observed beyond these optimal concentrations.

Starch (by 1.1–1.6 times), glutelin (by 0.9–2 times), and Zn (by 1.7–2.5 times) content increased by application of ZnO-NPs in the harvested cucumber fruits (Zhao et al., 2013, 2014). Further, there were no adverse effects of ZnO-NPs observed on any growth-related parameters. 2 mg L^{-1} of ZnO-NPs enhanced root elongation of germinated radish (*Raphanussativus*) and rape (*Brassicanapus*) seeds over those of the control (deionized or DI water only), while 2 mg L^{-1} of metallic Zn-NPs improved growth of ryegrass (*Loliumperenne*) seedlings (Lin and Xing, 2007).

9.5 SUPER ABSORBENT HYDROGEL

Super absorbents are three-dimensionally cross-linked hydrophilic polymers capable of swelling and retaining huge volume of water in swollen state. Recently, research on the use of superabsorbent as water management materials for agricultural and horticultural application (Mohana et al., 2003) and as slow-release carrier of agrochemicals has attracted great attention (Shavit et al., 2003; Sarkar et al., 2013). The optimized combination of superabsorbent and fertilizers may improve the nutrition of plants and concomitantly mitigate the environmental impact from water-soluble fertilizers, lowered evaporation losses and reduced frequency of irrigation (Li et al., 2005: Chatzoudis and Valkanas, 1995).

Equilibrium WA in distilled water showed that for polyacrylamide/clay (PAM/clay) composites incorporated with the same clay content was in the order PAM/vermiculite > PAM/mica > PAM/attapulgite ≈ PAM/kaolinite > PAM/Na^+ -montmorillonite and they also proposed that this might be due to the hydration and distension difference of these clays (Zhang and Wang (2007). For example, Na^+ -montmorillonite clay had the highest hydration and distention among the clays selected, that resulted more crosslinking in the composite which in turn resulted the lowest equilibrium WA of the corresponding PAM/clay composites. Equilibrium WA of PAM/clay composites in distilled water decreased with increasing clay content from 5 wt% to 40 wt%. Decreasing tendency of WA with increasing clay content might be due to the fact that the clays could react with acrylamide and acted as crosslink points in the corresponding polymeric network and their observations were in conformity with the study of other researchers (Lin et al., 2001; Li et al., 2004).

Moisture release curve (expressed in terms of difference in moisture held at zero pressure and a particular pressure) indicates that soil amended with 0.75% NSAPC (Novel Nano-Superabsorbent Composite) led to maximum improvement in its water release property. As expected, 0.75% P-gel amended soil-less medium exhibited most superior moisture release pattern (Singh et al., 2011).

SAPSRPF was prepared by esterification of polyvinyl alcohol (PVA) and phosphoric acid (H_3PO_4), which was a slow-release fertilizer as well as superabsorbent polymer was reported by Zhan et al., 2004. The product's WA was about 480 times its own weight if it was allowed to swell in distilled water at room temperature for 24 h.

9.6 BAU INITIATIVES IN NANOTECHNOLOGY

Nanoscience and Nanotechnology unit was established at BAU, Sabour (O.O. No: 40/Registrar/BAU., Sabour dated 30.07.2015) under the department of Soil Science and Agricultural Chemistry, BAC, Sabour.

9.6.1 EDUCATION

- The unit is offering one UG course (ASS 301 Introduction to Nanotechnology) and one PG (MSc, Ag, and PhD) (SOILS 516 Introduction to Nanotechnology) course.
- Innovative teaching methodologies including models, small projects, multimedia projector are being used.

9.6.2 RESEARCH PROJECTS

- Development of multinutreints NCPCs fertilizer formulation for cereals.
- Development of partially acidulated nano-rock phosphate formulation for rice wheat cropping system.
- Novel antimicrobial polymeric coating materials for increasing shelf life of fruits.
- Novel nanobiomolecules for controlling pathogenic fungi.

9.6.3 ACHIEVEMENTS

- **Novel Superabsorbent Hydrogel:** Novel superabsorbent hydrogel development protocol has been designed having equilibrium WA (ranging from 230 g g^{-1} to 500 g g^{-1}) (Table 9.1). Products are biodegradable, can withstand high salinity level (100 Mm NaCl solution). These are being processed for patent filing and subsequently will be licensed for commercial production.

TABLE 9.1 Equilibrium Water Absorbency of Novel Superabsorbent Hydrogel

Combination	Water Absorbency ($g\ g^{-1}$)
Polymer Iymer I	550,500,000
Polymer II	450
Polymer III	350
Polymer IV	275
Polymer V	250
Polymer VI	230
LSD ($P<0.05$)	15

- **Novel Nano Fe Fertilizer:** Novel nanoformulation of Fe has been developed; protocol has been standardized in the laboratory. Products content Fe ranging from 6% to 10%. Fe use efficiency under pot experiment goes high (25%) in comparison with conventional fertilizer ($FeSO_4$. 7 H_2O) having use efficiency of 1–5%. Novel Fe nanofetilizers seem to be a promising technology for increasing Fe content in cereal grains which will in the future eliminate Fe malnutrition. These products will be submitted for patent and in due course will be released for commercial production.
- **Novel Nano P Fertilizers**: Protocol for development of novel nano P fertilizers from indigenous rock phosphate (Udaipur, Purulia, and Mussourie) has been developed. Watersoluble P content in nanoformulations goes high (Figures 9.2 and 9.3) as compared to rock phosphate. P use efficiency in nano P goes high up to 30% in pot experiment as compared to conventional (DAP, SSAP) P fertilizers (use efficiency 15–20%). These products will be submitted for patent and in due course will be released for commercial production.

9.7 SYNTHESIS, FORMULATION AND EVALUATION OF NANO ZN FERTILIZERS UNDER RICE RHIZOSPHERE

Various nano Zn formulations (ZNCPC and Nano ZnO) were evaluated in comparison with $ZnSO_4$. $7H_2O$ (Conventional Zn fertilizers) under rice-rhizosphere. There was significant increase in DTPA ZN content under NCPC based Zn formulations (Table 9.2). Apparent Zn recovery went as high as 25.56% under NCPC Zn formulation (Table 9.3) followed by nano ZnO spray (15.25%).

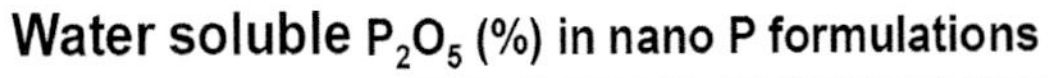

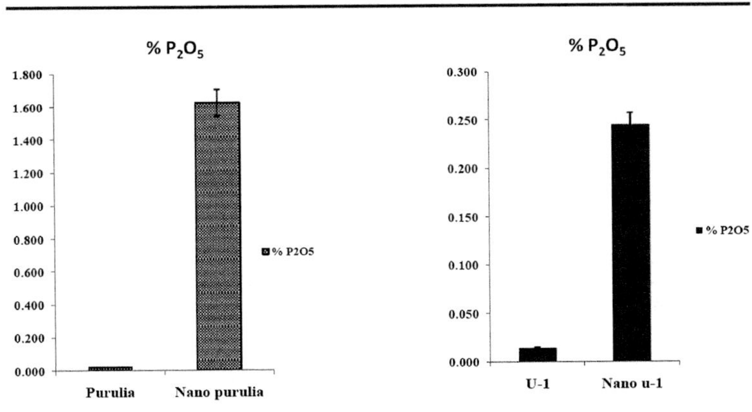

✓Purulia nano formulation demonstrated 81 times greater WSP content comparing to that of Purulia RP. Whereas, the same for that of U-I is 24 *times* and U-II was 13 times.

FIGURE 9.2 Watersoluble P content in Rock phosphate and nanoformulations.

Water soluble P_2O_5 (%)

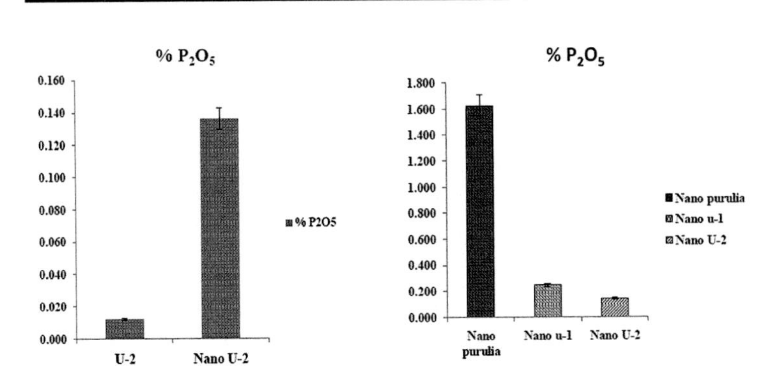

✓Among the three, Purulia nano formulation is giving best result in reference to water soluble P content.

✓Low grade P containing RP (e.g. Purulia RP) can be successfully used for maximizing P use efficiency.

FIGURE 9.3 Watersoluble P content in various nanoformulations.

TABLE 9.2 Effect of Different Inorganic and Bio-Inoculating Treatments on DTPA Extractable Zn in Soil (mg kg^{-1}) of Rhizosphere Soil at Different Growth Stages of Rice Plants

Treatment	S_1 Maximum Tillering	S_2 Panicle Initiation	S_3 Harvesting	Mean
T_1: Control	0.50	0.73	0.64	0.62
T_2: RDF ($ZnSO_4$, $7H_2O$) soil application	0.76	1.05	0.98	0.93
T_3: $ZnSO_4$, $7H_2O$ (2%) + Lime spray	0.58	1.59	1.54	1.24
T_4: RDF + *Azospirillum brasilense*	2.65	3.52	3.46	3.19
T_5: Nano Zn spray at recommended dose	0.64	1.39	1.43	1.15
T_6: NCPC* based Zinc	2.63	3.60	3.50	3.24
Mean	1.02	1.98	1.92	1.73
Initial value				0.48
Particulars		$SE_{m(\pm)}$	CD at 5%	CD at 1%
Stage (S)		0.05	0.13	0.18
Treatment (T)		0.06	0.19	0.25
Interaction (S × T)		0.11	0.32	0.43

TABLE 9.3 Effect of Various Nano Zinc Sources with Zinc Mobilizer on Apparent Zinc Recovery (%)

Treatment	Apparent Zinc Recovery (%)
T_1: Control	-
T_2: RDF ($ZnSO_4$, $7H_2O$) soil application	2.23
T_3: $ZnSO_4$, $7H_2O$ (2%) + Lime spray	3.14
T_4: RDF + *Azospirillum bracilense*	10.76
T_5: Nano Zn spray at recommended dose	15.25
T_6: NCPC* based Zinc	25.56

However, these products need to be evaluated in field experiments at multi-location trial.

9.8 MULTINUTREINT NANOCLAY POLYMER COMPOSITE FORMULATION (MNCPC) FOR CEREALS

Nanotechnological intervention in rhizosphere is of utmost importance. Novel NCPCs based multinutreint formulations have been developed for cereal crops. Product contains N, P, K, and Zn (%): 34, 6.67, 12 and 1 respectively (Maize) product contains N, P, K, and Zn (%) 36, 6, 6 and 1 (Rice). Laboratory released study (Figure 9.4) revealed that MNCPCs had a controlled release of nutrients as compared to conventional fertilizer molecules.

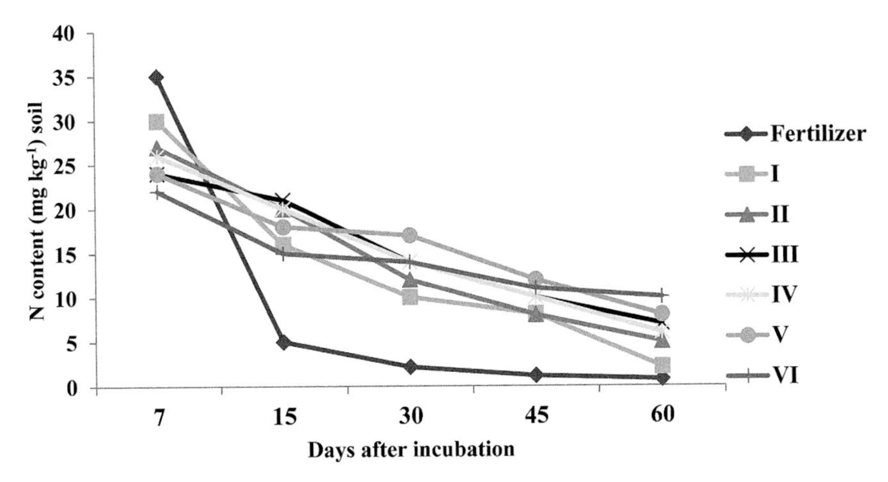

FIGURE 9.4 Release of N over incubation period in laboratory experiment.

9.9 NOVEL NANOFORMULATIONS FOR CONTROLLING PATHOGENIC FUNGI

Novel chitosan nanoformulations has been synthesized in the laboratory. Ag and Cu based nanocomposites have been developed in the laboratory. These have been tested *in vitro* on *Sclerotium rolfsii*. These formulations under controlled condition giving promising results in terms of germination of sclerotia (Table 9.4; Figure 9.7), radical growth (Figure 9.5) and status of scerotia germination (Figure 9.6).

TABLE 9.4 Sclerotia Germination of *Sclerotium Rolfsii* Upon Dipping of in Colloidal Nanoformulation of Chitosan

Dipping Period in Hour	Status of Germination					
	Day 1	Day 2	Day 3	Day 4	Day 5	Day 6
Mock*	-	+	+	+	+	+
0	-	+	+	+	+	+
4	-	-	-	+	+	+
6	-	-	-	+	+	+
8	-	-	-	+	+	+
24	-	-	-	-	+	+
28	-	-	-	-	-	+
30	-	-	-	-	-	+
32	-	-	-	-	-	-

*Mock: water dipping for 0 h.

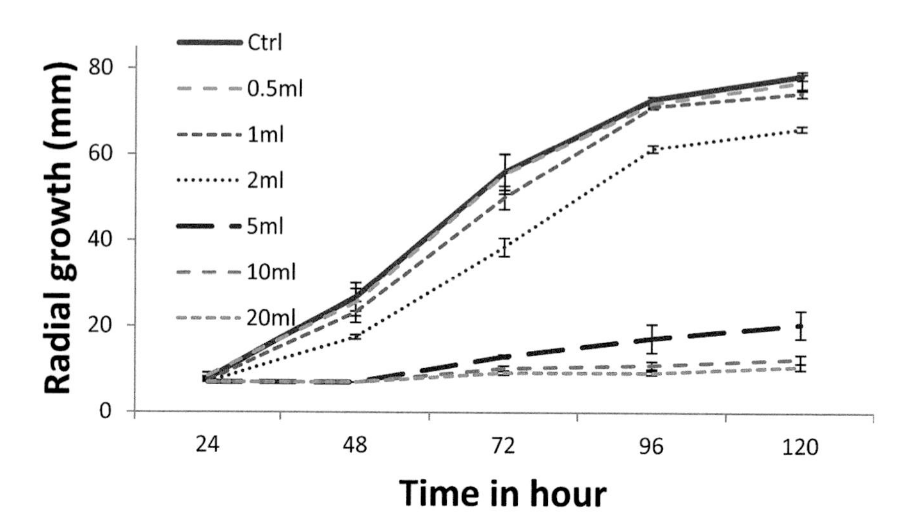

FIGURE 9.5 Effect of concentration of chitosan nanoformulation on radial growth of *Sclerotium rolfsii*.

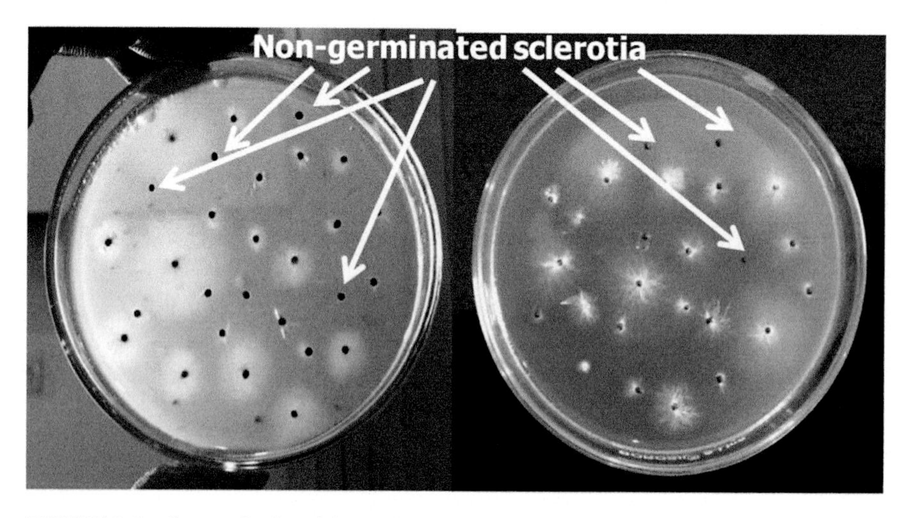

FIGURE 9.6 Status of sclerotial germination upon treatment of chitosan colloid.

9.10 FUTURE PERSPECTIVES

Efficacy of novel nanoformulations developed need to be evaluated under field experimentation in long-term basis at multi-location trial.

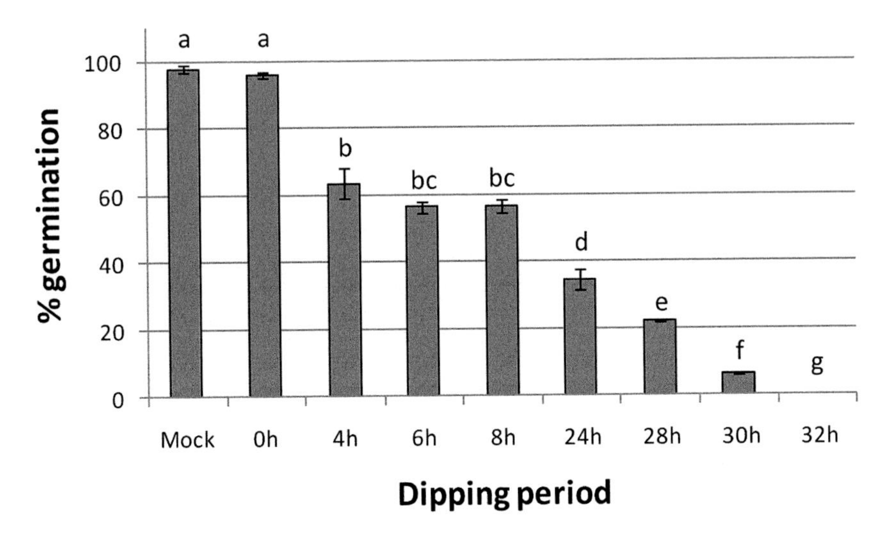

FIGURE 9.7 Germination percent of sclerotia.

KEYWORDS

- **atomic absorption**
- **atomic force microscopy**
- **farmers adaptability**
- **human resource development**
- **inductively coupled plasma**
- **nanotechnology**

REFERENCES

Lin, D., & Xing, B., (2007). Phytotoxicity of nanoparticles: Inhibition of seed germination and Influence of CeO_2 and ZnO nanoparticles on cucumber physiological markers and bioaccumulation of Ce and Zn: A life cycle study. *Journal of Agriculture and Food Chemistry, 61*, 11945–11951.

Linng, R., & Liu, M., (2007). Synthesis of polymer–montomorillonite nanocomposites by in situ intercalative polymerization. *Journal of Applied Polymer Science, 106*, 3015–3030.

Mahajan, P., Dhoke, S. K., & Khanna, A. S., (2011). Effect of nano-ZnO particle suspension on growth of mung (Vigna radiata) and gram (*Cicer arietinum*) seedlings using plant agar method. *Journal of Nanotechnology,* 7. Article ID 696535.

Mandal, N., Datta, S. C., Manjaiah, K. M., Dwivedi, B. S., Kuramr, R., & Aggarwal, P., (2016). Evaluation of zincatednanoclay polymer composite (ZCNPC) in releasing Zn, P and effect on soil enzymatic activities in wheat rhizosphere. *Journal of Agricultural and Food Chemistry.* Accepted manuscript jf-2016–03946r.R.

Mohana, R. K., Padmanabha, R. M., & Murali, M. Y., (2003). Crystallization, properties, and crystal and nanoscale morphology of PET–clay nanocomposites. *Polymer International, 52*, 768–781.

Sarkar, S., Datta, S. C., & Biswas, D. R., (2014). Synthesis and characterization of nanoclay-polymer composites from soil clay with respect to their water-holding capacities and nutrient-release behavior. *Journal of Applied Polymer Science.* doi: 10.1002/app.39951.

Wu, J., Wei, Y., Lin, J., & Lin, S., (2003). Preparation and properties of organosoluble montmorillonite/polyimide hybrid materials. *Polymer, 44*, 6513–6525.

Zhan, F., Liu, M., Guo, M., & Wu, L., (2004). Preparation of superabsorbent polymer with slow-release phosphate fertilizer. *Journal of Applied Polymer Science, 92*, 3417–3421.

Zhao, L., Peralta-Videa, J. R., Rico, C. M., Hernandez-Viezcas, J. A., Sun, Y., & Niu, G., (2014). $CeO_2$and ZnO nanoparticles change the nutritional qualities of cucumber (*Cucumis sativus*). *Journal of Agriculture and Food Chemistry, 62*, 2752–2759.

Zhao, L., Sun, Y., Hernandez-Viezcas, J. A., Servin, A. D., Hong, J., & Niu, G., (2013). Root growth. *Environmental Pollution, 150*, 243–250.

CHAPTER 10

Endomycorrhizal Fungi: Phosphorous Nutrition in Crops

MAHENDRA SINGH, RAJIV RAKSHIT, and KASTURIKASEN BEURA

Department of Soil Science and Agricultural Chemistry, Bihar Agricultural University, Sabour, Bhagalpur, Bihar, India,
E-mail: m.singh30648@gmail.com

ABSTRACT

Phosphorus (P) is one of the most important plant growth nutrient which is immobile in soil. In soil, it occurs in three forms namely, soluble and insoluble inorganic phosphorous and organic phosphorus. Uptake of phosphorus from soil solution is increased by endomycorrhizal fungi in addition to increase plant root surface area. They are commonly found in association with approximately 85% of angiosperm plants. The enhancement in growth of plants is mainly attributed to uptake of less mobile nutrients such as P, Zn, Cu, etc., from the soil. The other advantageous properties are their role in the plant growth hormone production, greater capability to endure water stress and synergistic association with agriculturally beneficial soil microorganisms. It is believed that mycorrhizal fungi secret some phosphatases enzyme which also increased the availability of P. The field studies have shown that inoculation with efficient mycorrhizal fungi not only improves growth and yield of crop plants while it reduces the application of phosphatic fertilizer. Advantages of endomycorrhizae are attaining through the application of suitable and efficient AM fungi inoculums and augment resident mycorrhizal fungal actions in the soil through manipulating agricultural practices in support of these inoculums.

10.1 INTRODUCTION

Mycorrhiza is a mutualistic symbiosis between certain groups of soil fungi and most plant root systems (Hata et al., 2010). The most important assistance of mycorrhiza is the enhanced growth of plants which is mostly due to improvement in phosphorus nutrition. Various mechanisms (e.g., exploration of large soil volume, faster movement of mycorrhizal hyphae and solubilization of soil phosphorus) are responsible for increasing the uptake of phosphorus by mycorrhizal plants. Non-nutritional benefits to plants, such as changes in water relations, phytohormone levels, carbon assimilation, secretion of enzymes, increased microbial count in soil, etc. have also been reported, but they are difficult to interpret (Andrade et al., 1998).

Phosphorus is one of the major essential macronutrients which limit plant growth owing to its low bioavailability in soils (Feng et al., 2004). Improving plant acquisition of P from soil is an obvious alternative for the management of those low P soils (Zhu et al., 2003). It is commonly known that *arbuscular mycorrhiza* (AM) fungi, they act as a direct link between soil and roots, AM fungi help plants to capture water and nutrients (especially P) from the soil, and in return, the plant provides the fungus with relatively constant and direct access to carbohydrates (Smith and Read, 2008), which are translocated from their source to root tissue and on to fungal partners.

Soil enzymatic activities regulate the various indices of soil fertility, soil productivity and soil quality (Busto and Perez-Mateos, 1997). AM fungi can increase soil enzyme activities, such as phosphatase dehydrogenase, urease, protease and β-glucosidase (Caravaca et al., 2004). Mar Vazquez et al., (2000) reported mycorrhizal colonization induced qualitative changes in the microbial population and enzyme activities in the rhizosphere of maize plants. Rao and Tak (2001) found that mycorrhizal fungal inoculation resulted in enhanced plant growth, total uptake of N, P, and many other nutrients, activities of dehydrogenase, phosphatases and nitrogenase in the rhizosphere in gypsum mine spoil. Due to the energy-intensive manufacture of chemical fertilizers, use of microbial inoculants to supplement a part of phosphorus requirement has been attaining enormous significance.

10.2 TYPE OF MYCORRHIZAE

There are two main kinds of mycorrhiza- the ectomycorrhizae (ECM) and the endomycorrhizae. In the ECM (also called ectotrophicmycorrhize), the fungus completely encloses each feeder rootlet in a sheath or mantle of hyphae. The hyphae or germ tube of spore of mycorrhizae get penetrate only between the root cortex cells. In endomycorrhizae fungi, it does not form a hartig net which is an external sheath but lives inside the cells of root and develop structure namely vesicles and arbuscules. Hence, it is established straight connections among the cells of the root and the nearby soil of roots.

The details of beneficial effects of both ecto- and endomycorrhizae on plant growth is presented here.

10.2.1 ECTOMYCORRHIZAE (ECM)

ECM are generally formed by hundreds of different fungal species belong to class Basidiomycota and Ascomycota. Most of the fungi forming ECM with trees are basidiomycetous belonging to Hymenomycetes and Gasteromycetes (families: Agaricaceae, Tricholomataceae and Sclerodermataceae, etc.) and some belonging to class Discomycetes, Plectomycetes and Zygomycota. The well- known genera in ECM are *Pisolithus*, *Thelephora*, *Scleroderma*, *Cenoccocum*, *Boletus*, *Laccaria*, etc. An ectomycorrhizal root increases the surface area of root hairs and is sheltered by a stratum or sheath of fungal which more or less looks like host plant tissue.

10.2.2 ENDOMYCORRHIZAE

Endomycorrhiza is an obligate symbiontic association between fungi and roots of plants. About 80 fungal species form endomycorrhiza which are mainly the members of Zygomycota. The hyphae of endomycorrhizae enter the cortical cells of feeder roots of nearly all the cultivated plants and many forest trees, shrubs, and wild herbaceous plants. Endomycorrhiza forms large vesicles and Arbuscules in their host cells and because of this they are also known as vesicular arbuscular mycorrhiza (VAM).'

10.2.3 ARBUSCULAR MYCORRHIZA (AM)

AM occurs in roots of most of angiosperms, pteridophytes and bryophytes, although absent in plants which form only ectomycorrhiza (Pinaceae, Betulaceae) or two other specific types of endomycorrhizae of Ericales and Orchidales. AM develops special characteristics structures called arbuscules and vesicles. The Arbuscules help in the transfer of nutrients from a distant location in the soil to the root system.

10.3 ENDOMYCORRHIZAE FUNGI AND PHOSPHOROUS NUTRITION

Uptake of phosphorus from the soil is performed by mycorrhizal fungi in addition to increasing the surface area of plant roots. Although there are various types of mycorrhizal fungi, but here AM fungi are discussed. AM fungi are omnipresent, which are present in most of the soils. They belong to the phylum Glomeromycota, which has three classes (Glomeromycetes, Archaeosporomycetes, and Paraglomeromycetes) with five orders (Glomerales, Diversisporales, Gigasporales, Paraglomerales, and Archaeosporales), 14 families and 26 genera (Stermer, 2012). AMF are not host-specific although evidence is growing that certain endophytes may form a preferential association with certain host plants (Rivera et al., 2007 and Bagyaraj, 2011).

A pot experiment was conducted at BAU, Sabour by Singh et al. (2015) which revealed that the higher P uptake was recorded with the application of *Glomus mosseae*, in both straw (0.065 g plant^{-1}) and grain (0.12 g plant^{-1}). However, *Glomus coronatum* was quite competitive with *Glomus mosseae* with respect to the uptake of P in straw and grain of maize. *Glomus mosseae* also gave significantly higher P uptake by straw when compared with all treatments except *G. coronatum.* Application of *G. coronatum* significantly increased P uptake in straw by 53.65, 96.87, 21.14, and 231.57% higher than *G. intraradices, G. margarita, G. decipiens* and local, respectively. A similar trend was observed in the uptake of P by grain. Application of *G. decipiens* significantly increased straw-P uptake by 62.5173.68 and 550% over *G. margarita,* local species and control, respectively. Similarly, application of *G. decipiens* numerically increased the grain-P uptake by 11.49, 49.23, and 155.26% over application of *G. margarita,* local species and control, respectively.

10.4 INITIATIVE TAKEN AT BAU

10.4.1 COLLECTION OF LOCAL AM FUNGI INOCULUM

We have collected root samples from different crops from different locations of Bihar. Analysis of root samples for mycorrhizal root colonization was done by following standard methodology. The maximum root colonization was observed under the litchi root from Muzzafarpur, Bihar and lowest in the Brinjal roots from the Halsi, Lakhisarai, Bihar, India (Table 10.1 and Figure 10.1).

TABLE 10.1 Native Status of AM Fungi Colonization of Various Crops of Bihar

Sl. No.	Location	Crop	pH	Colonization (%)
1	Rajpur, Sabour	Maize	5.9	60.34
2	University Orchard	Litchi	7.8	65.33
3	Chandeli, Sabour	Wheat	8.2	55.00
4	Chandeli, Sabour	Maize	8.3	60.34
5	Suryagarh, Lakhisarai	Potato	6.9	44.50
6	Saidpur, Lakhisarai	Coriander	7.75	35.55
7	Saidpur, Lakhisarai	Tomato	8.35	39.00
8	KVK Halsi, Lakhisarai	Brinjal	8.1	34.66
9	KVK Halsi, Lakhisarai	Gram	8.2	40.55
10	KVK Halsi, Lakhisarai	Potato	7.5	50.55
11	Lakhisarai	Gram	8.15	35.80
12	Fatehpur, Lakhisarai	Lentil	7.8	49.88
13	Katoria, Banka	Gram	6.5	47.56
14	KVK Banka	Pigeonpea	7.5	40.00
15	Satisthan, Tarapur	Sugarcane	7.3	45.76
16	Satisthan, Tarapur	Gram	7.9	50.55
17	MurarPur, Bhagalpur	Potato	8.1	44.70
18	Maize (Vill- Rajpura, Sabour)	Litchi	7.8	50.34
19	Litchi orchard (RAU, Pusa)	Litchi	8.5	80.66
20	Litchi orchard (NRC, on Litchi, Muzaffarpur)	Litchi	8.2	85.22

10.5 CONCLUSION

It has been observed from the investigation that the maximum root colonization was found under the litchi root from Muzzafarpur and lowest in the Brinjal roots from the Halsi, Lakhisarai. The application of *Glomus mosseae* increased phosphorous uptake by grain and straw when compared with uninoculated treatment.

FIGURE 10.1 Collection of VAM inoculum from Litchi orchard, BAU, Farm, RAU, Pusa and NRC on Litchi, Muzafarpur.

KEYWORDS

- **arbuscular mycorrhiza**
- **ectomycorrhizae**
- **glomalin**
- **maize**
- **mycorrhiza**
- **phosphorous**

REFERENCES

Andrade, G., Linderman, R. G., & Bethlenfalvay, G. J., (1998). Bacterial associations with the mycorrhizosphere and hyphosphere of the arbuscular mycorrhizal fungus *Glomus mosseae*. *Plant Soil, 202*, 79–87.

Bagyaraj, D. J., (2011). *Microbial Biotechnology for Sustainable Agriculture, Horticulture and Forestry*. New India Publishing Agency, New Delhi.

Busto, M. D., & Perez-Mateos, M., (1997). Stabilisation of cellulases by cross-linking with glutaraldehyde and soil humates Extraction of humic-fl-glucosidase fractions from soil. *Biores. Technol.*, *60*, 27–33.

Caravaca, F., Alguacil, M. M., Azcon, R., Díaz, G., & Roldan, A., (2004). Comparing the effectiveness of mycorrhizal inoculation and amendment with sugar beet, rock phosphate and *Aspergillus niger* to enhance field performance of the leguminous shrub *Dorycnium pentaphyllum* L. *Appl. Soil Ecol.*, *25*, 169–180.

Feng, K., Lu, H. M., Sheng, H. J., Wang, X. L., & Mao, J., (2004). Effect of organic ligands on biological availability of inorganic phosphorus in soils. *Pedosphere, 14*, 85–92.

Hata, S., Kobae, Y., & Banba, M., (2010). Interactions between plants and arbuscular mycorrhizal fungi. *Int. Rev. Cell Mol. Biol.*, *281*, 1–48.

Mar Vazquez, M., Cesar, S., Azcon, R., & Barea, J. M., (2000). Interactions between arbuscular mycorrhizal fungi and other microbial inoculants (*Azospirillum, Pseudomonas, Trichoderma*) and their effects on microbial population and enzyme activities in the rhizosphere of maize plants. *Appl. Soil Ecol., 15*, 261–272.

Rao, A. V., & Tak, R., (2001). Influence of mycorrhizal fungi on the growth of different tree species and their nutrient uptake in gypsum mine spoil in India. *Appl. Soil Ecol.*, *17*, 279–284.

Rivera, R., Fernandez, F., Fernandez, K., Ruiz, L., Sanchez, C., & Riera, M., (2007). Advances in the management of arbuscular mycorrhizal symbiosis in tropical ecosystems. In: Hamel, C., & Plenchette, C., (eds.), *Mycorrhizae in Crop Production* (p. 151). Haworth Food & Agricultural Products Press, New York.

Singh, M., Beura, K., Pradhan, A. K., Rakshit, R., & Lal, M., (2015). Ability of arbuscular mycorrhiza to promote growth of maize plant and enzymatic activity of an alluvial soil. *Journal of Applied and Natural Science, 7*(2), 1029–1035.

Smith, S. E., & Read, D. J., (2008). *Mycorrhizal Symbiosis*. Academic Press, London, UK.

Zhu, Y. G., Smith, F. A., & Smith, S. E., (2003). Phosphorus efficiencies and responses of barley (*Hordeum vulgare* L.) to arbuscular mycorrhizal fungi grown in highly calcareous soil. *Mycorrhiza, 13*, 93–100.

CHAPTER 11

Water Management in Horticultural Crops

K. KARUNA and ABHAY MANKAR

Department of Horticulture, Bihar Agricultural University, Sabour, Bhagalpur, India

ABSTRACT

Agricultural practices like soil management, nutrient management, irrigation management, disease and pest management, etc., all are helpful in sustainable water availability to the plants. Water is an important input for sustainable production of agricultural produce. Water is an important constituent of food production in plants and most of these food materials are converted into end products i.e. yield. The availability of fresh water also decreases day by day and efficiency of irrigation is also low. Only 65 per cent of irrigation water is utilized by the plants. Due to lack of surplus fresh water for irrigation as well as climatic changes it is need to increase efficiency of irrigation water to the plant by using advanced method of irrigation technique to cope up unavailability of irrigation water and climatic changes. Another way for maximum utilization of water in plant is partial root drying. It is most applicable in dry areas. Under this condition half part of the root zone is irrigated and left half part. Water percolated from irrigated part to other parts of the root zone and make available to the plants.

11.1 INTRODUCTION

Agriculture always have implicated economic risks due to almost every farming depends on weather and after long research it may show the impact of weather-related problems on plant and correlate the hazards with climate to minimize them in a particular environment. In the early days, farmers showed that when water applied externally in the field it can minimize drought effect on the crop. Water management is the major concern that enhances efficiency and yield of horticultural produce. Under horticultural crops vegetables (except potato) and ornamentals are irrigated while fruits and other tree crops are predominately rainfed. This can be responsible for the poor productivity of many horticultural crops. According to Ryan and Pitman (1998) crop once needed to require water to its root zone, supplied it for a long period of time and again dispose further unwanted water. It shows the lack of water management practices in farmer's field. Long time of watering can cause negative environmental effects in the root zone of any crops, particularly deposition of salts (Gardner, 1993; Van Schilfgaarde et al., 1974).

Water is stored in soil which helps in continuous release of available water during the growing period crops. The microbial activities are maximized in the presence of optimum available water and good aeration. The fertilizer compounds are converted into available forms and dissolved into water. The nutrients are being absorbed by the roots through the water solution due to water potential gradient of plant. Water is essentially required for physical, chemical and biological activities of soil and plant, in addition to plant metabolic activities and ET requirement. Hence, it is life sustaining and renewable source. It is a prime importance for plant constituent. A plant life has been defined as a 'supported column of water' because its physiological processes occur in solutions when huge quantities of water are present. Adequate, timely and assured availability of irrigation is critical to agriculture, horticulture and plantation crops for ensured yield. Water management is important to reduce the wastages of water. It is our need to enhance water use efficiency (WUE) and ensure uniform water distribution. Water acts as important part of all living beings because its role as solvent for synthesis of proteins, nucleic acids, and other metabolites. Water influences photosynthesis, transpiration, respiration, translocation, absorption and use of mineral nutrients and cell division besides some other metabolic processes. All these above physiological activities depend on water absorption rate and release of water from the plant. Vapor pressure gradient in atmosphere and leaves helps transpiration rate. Dry matter accumulation depends on these physiological activities which ultimately relates to productivity of crops. For this reason, actively growing plants as the case of all the vegetable crops need to maintain liquid phase continuity from soil water through its vascular system and all the way to evaporative sites in leaves. So, water management in annual crops especially vegetables and flowers are very essential for good yield and quality. Vegetables liked by the people who have qualities like crispness, tenderness, succulence, and flavor. All these parameters depend on proper supply of water at right stages. Texture of vegetables also determined by cumulative effect of turgor pressure, cell wall characters and tissue structure.

11.2 IMPORTANCE OF WATER FOR PLANT LIFE

1. Water is the prime constituent of every cell.
2. Water is a way of transpiration of nutrients from roots to shoots and vice-versa.
3. Water is the source for many metabolic reactions.
4. It acts as solvent as well as medium for biochemical reactions in all living beings.
5. It regulates temperature inside the plants.
6. It actively takes part in photosynthesis and hydrolytic process.
7. Water imparts turgidity to cells and helps them to maintain their form and structure.
8. Gain and loss of water from cells and tissue is responsible for other movements in plants like opening of stomata, nocturnal folding of leaves in some plants, opening and closing of flowers in certain plants and sensitivity of leaflets to touch as in *Mimosa pudica*.
9. Cell elongation phase in growth is dependent on water.

Therefore, water for plants is important as a solvent, constituent, reagent as well as maintainer of cell turgidity.

11.2.1 FACTOR INFLUENCING WATER REQUIREMENT

1. **Soil Moisture and Humidity:** When soils are wet and also follow the heavy rain or irrigation, soil solution gets diluted and creates anaerobic condition in the basal part of the plant and therefore more water is required by the plant. In other words, under dry condition water requirement by the plant increased. If the atmospheric humidity is high, the plant required less water because of low transpiration.
2. **Fertility Level:** Water requirements of plants are considerably lower on fertile soils than on unfertile lands. Therefore, adequate manuring results in efficient water use by the plant.

3. **Cultivation:** Water requirement of plant is materially reduced by cultivation because it increases moisture content of soils and the supply of available plant nutrients by providing good aeration. It also depends on type of crops, age of plants, geometry of plants, etc.
4. **Light:** Water requirement of plants increases in exposed situations than in protected situations and thus provision of windbreaks is essential to reduce transpiration loss.

These factors show that in scientific water management practices, it is essential to maintain optimum soil nutrient status as it reduces water requirements.

11.3 SYSTEMS OF IRRIGATION

A major factor attributed to lower water productivity in inefficient irrigation systems. The overall efficiency of canal system is about 30–40% while that of ground water-based systems is about 60%. The contribution of the losses during supply of water is approx 25–40%. With proper designing, these losses may be minimized to 5–20%. This may be done from both flood irrigation methods and pressurized irrigation methods. Pressurized irrigation system can be extensively used for improving irrigation efficiency and enhancing water productivity (Anonymous, 2002; Singh et al., 2000; Choudhary and Kumar, 2005). Pressurized irrigation system is the efficient method of irrigation for fruit and plantation crops where it saves 30–70% supplied water and increases yield by 25–80%. The present status of area covered by drip irrigation is over 3.0 lakh ha which is about 0.5% of net irrigated area and major chunk of area is in few states viz. Andhra Pradesh, Maharashtra and Karnataka. However, now the area growing fast in pressurized irrigation system because of many development schemes are running at national level such as National Horticulture Mission (NHM), Rashtriya Krishi Vikas Yojana (RKVY), etc.

1. **Basin System:** This irrigation method is commonly used in India. It is made around the plants which cover a small area. The size of the basin depends on the age and canopy spread of the tree. The depth of basin varies from 7–15 cm from the surface of the ground. A gentle slope is provided from the stem of the plant to the outer side of the basins.

The periphery of basin increases as the size of the plants increased. In very old orchards, this system cannot be suitable as the root system would have gone for beyond the canopy spread of the trees and the entire orchard may be necessary to irrigate.

Merits:

i. Useful for new orchard;

ii. Limited area is wetted which economize water.

Demerits

i. Highly costly;

ii. High loss of water by evaporation and seepage.

2. **Flood System:** When the land is flat, and water supply is ample, this system is preferred. For this, the entire area is divided into convenient beds which are flooded for irrigation. This system is also generally used in vegetable crops and flowers.

Merits:

i. Useful for old orchard;

ii. Suitable for interculturing.

Demerits:

i. High loss of water through seepage and evaporation.

3. **Furrow System:** This system is practiced in orchards in Western countries. Furrows are made between the tree rows at one or both directions of the row. Its number depends on the requirement of the plants as well as age of the plants. The most suitable furrow structure is about 60–90 m length and 45 cm width and 15 cm deep. Size of furrow varies on the basis of soil type and topography of the field. For example, in highly sloppy land the furrow length will be reduced. Furrows should be shallow to assure quick spread of irrigation water of targeted area.

Merits:

i. Water losses are lesser in comparison to basin system;

ii. Water distribution is more uniform.

Demerits:

i. It requires proper land leveling;

ii. Less costly.

4. **Ring System:** In this irrigation method water is supplied in a ring around the tree. This irrigation method is highly suitable for *citrus species* because, water is not allowed to touch the trunk of the tree thereby reducing the chances of collar rot of the tree which is susceptible from this disease. Ring size should be increased as the tree grows.

   Merits:
   i. Useful for establishment of new orchard;
   ii. Limited area is wetted to economize water;
   iii. Less risk of spread of diseases.

   Demerits:
   i. Interculture is difficult;
   ii. Labor intensive;
   iii. High loss of water through seepage and evaporation.

5. **Check Basin System:** In this method, an irrigation channel is prepared in between two rows of trees. The channel is then jointed through square or rectangular shaped beds passing through a number of trees. These beds are uniform in size, level inside and are connected together by irrigation channels. This irrigation method can be suitable for many crops, soils, and farming practices. The supply of water is done by direct or cascade. It is expensive and needs abundant water to sink upon finely pulverized soil and this has an adverse effect on soil texture.

   Merits:
   i. Simple layout.

   Demerits:
   i. Interculture is difficult;
   ii. Loss of water due to seepage and evaporation;
   iii. Risk of stagnation of water;
   iv. Labor intensive;
   v. Distribution of water is less uniform.

6. **Trickle/Drip Irrigation System:** This system has originated in Israel and has accepted internationally and popular as an efficient method of irrigation having maximum water economy. This is highly water use efficient (WUE) method of irrigation with very less requirement of irrigation water. Especially in arid region, this method of irrigation is very beneficial technique of irrigation. In this system plants

watering with a rate equivalent to its consumptive use so that plants would not experience any moisture stress.
Water-soluble fertilizers may be applied along with irrigation water through the drip system. The non-soluble manures and fertilizers are also placed near the drippers where irrigation water is supplied. The plants are irrigated daily by this method as per requirement.

Merits:

i. After installation maintenance is simple.
ii. Uniform supply of water.
iii. Water use efficiency is high.
iv. No wastage of water by evaporation, percolation, leaching or run-off.
v Fertilizers, insecticides, fungicides may be supplied along with irrigation water.
vi Eliminate soil erosion.
vii Improve quality and productivity of crops.
viii Problem soil may be managed more effectively.

Demerits:

i. High initial investment.
ii. Frequent clogging of emitters and drippers. It might be due to accumulation of salt, algae or other waste materials.
iii. Skilled man power is needed for design of this system, maintenance as well as operation.
iv. Threat of all kinds of pollution due to use of nonbiodegradable plastic materials.
v. Lack of training for efficient maintenance of this irrigation system.

11.4 HOW TO INCREASE EFFICIENCY OF DRIP IRRIGATION SYSTEM?

The initial establishment cost of this irrigation system is higher which restricts its faster area expansion. This system further enhancing productivity of any crops by increasing the yield level through better management of other inputs. Fertigation is one such method, in which fertilizer is applied along with this system. The work conducted under AICRP on water management has noted that fertigation improves the yield by 20–50%; it economizes fertilizer use by 20–25%. The fertilizer use may be further economized by differential dose of fertilizer during different growing stage of plants instead of applying it in equal splits during that period. Singandhupe et al., (2005) found that differential fertilizer application with reduced fertilizer dose during initial crop stage and higher dose during maximum vegetative growth period and reproductive phase of tomato reduces fertilizer requirement by 23% due to fertigation.

Horticultural crops have an advantage of wide spaced and therefore establishment cost of this irrigation system is less. Under horticultural crops especially fruit crops the space between rows either remains fallow or used for growing rainfed crops and therefore not utilizing the maximum potential of that space. A suitable modification in irrigation system may ensure that intercrop is also irrigated efficiently without any significant additional investment. A study at Water Technology Center for Eastern Region (WTCER), Bhubaneswar has found that by providing microtubes in its system, the banana rows and intercrop of turmeric can be irrigating efficiently without hampering the irrigation potential of banana (Mohanty et al., 2006). The productivity of banana increased from 61.25 t/ha to 68.35 t/ha with an additional yield of turmeric 4.1 t/ha.

Partial root drying (PRD) is also an important irrigation technique that tends to increase WUE in crops. In this technique, half portion of the root system is irrigated while the other half portion allowed drying out. Alternate irrigation through drip of partial root zone maintained hot pepper yield with up to 40% saving of irrigation water (Kang et al., 2001). Irrigation amounts were reduced by 28 and 12% in a pear orchard using fixed partial root-zone irrigation and alternate partial root irrigation as compared to conventional flood irrigation (Kang et al., 2002).

Drip irrigation along with mulching enhancing its effectiveness by preventing loss of moisture by evaporation. Various organic mulches like straw, hay, organic manure, tree leaves, etc. and inorganic mulches like alkathenes are available.

1. **Sprinkler Irrigation:** This system of irrigation can be beneficial for almost all types of crops on plain as well as all types of topography and soil. This system can be designed as per the necessity of the crops. The design of this irrigation system can help to meet out the demand of water by the crop at a particular growth stages. This system is very beneficial in arid areas because lack of sufficient irrigation water in that area.

   The different parts of this system are power generator, pump, pipeline and sprinklers. Power generator used in this system may be mechanical or electrical. There are many types of pump used to lift the water. (a) Centrifugal pump used to lift the water up to 4.5–6.0 m. (b) A turbine pump useful if water is very deep. (c) A piston pump used to spray liquid fertilizers as well as irrigation water. The nozzles used in this system can be one or two, stationary or revolving and are mounted on pole pipes which are attached to the laterals. Every sprinkler head supplies water in a circular area. Spread of water depends on size of nozzle and water pressure. Water spread in a circular area and its diameter depends on size of nozzle and water pressure.

Merits:

i. A fixed amount of water may be applied.
ii. Application of fertilizers, herbicides, and pesticides can be applied easily through this method of irrigation.
iii. It is a portable system and can be used anywhere as per need.
iv. It is a suitable method of irrigation where water is scanty.
v. It removes run-off and erosion of soil.
vi. Lands with irregular topography can be irrigated with a low leveling and disturbance of the surface soil.
vii. Very high water supply efficiency.
viii. Water loss due to deep percolation almost totally avoided and economize the water loses.

Demerits:

i. Freshwater is required to avoid choking of nozzles.
ii. Pipe systems spread on the field may interfere with farm operations, movements of animal and farm implements.
iii. Highly costly for its installation.
iv. Operating costs also higher.

2. **Quality of Irrigation Water:** It is an important component for drip irrigation. It should preferably be normal for best growth and development. Quality of normal water depends upon nature and quantity of salt present in it. Generally, water contains soluble salt of magnesium, calcium, sodium in the form of ions of chloride, sulfate, bicarbonate, and carbonate. Besides, these elements the heavy metals are also available in water which causes detrimental effect on the irrigation system as well as plants also. For this purpose borehole water sources have pure water than other kinds. The quality of irrigation water mainly depends on two aspects, i.e., microbiological and chemical characteristics of water. Other factors also affect quality

of irrigation water such as presence of pollutants like silt particles, peat, moss, etc. The chemical characteristics of irrigation water are pH, electrical conductivity, chloride level, carbonate, bicarbonate, alkalinity, sulfate level, iron, and other elements.

11.5 TIME AND FREQUENCY OF IRRIGATION

There are several devices useful to determine time as well as frequency of watering. Irrigation is done to supplement the deficits of water in soil. Irrespective of soil, irrigation also depends on the intensity of rainfall over an area. In other ways, similar soils under different atmospheric conditions require varying irrigation practices, depending upon not only the rainfall, but also on humidity and temperature.

1. **Wilting of Leaves:** Generally, the growers irrigate their orchards when the leaves start showing wilting. The other visual plant symptoms are change in foliage color, leaf angle, reduced growth, etc. This is not a correct measure, as even before the leaves start wilting, there are much damage is done by moisture stress. Curling of immature leaves is first visible sign of stress. For scheduling irrigation, however, measurement of relative water content (RWC) in plant is essential as practical one.
   Moisture available in the plant may be expressed by RWC in leaves. It can be measured by drying in oven at 60–70°C and expressed in percent on dry weight basis. Several other indirect plant measurements, such as leaf thickness, trunk diameter, and fruit measurements also indicate water available in the plants.
2. **Soil Tensiometers**: It can be established at the specified depth in root zone of the orchard to show the soil moisture tension. Tension is the rigidity with which moisture is hold by the soil. Trees cannot absorb soil moisture as the tension of soil moisture increases. This is very useful in coarse textured soil because most of the usable soil moisture is at lower tension.
3. **Gypsum Blocks:** It is fitted in the orchard at several places and at different depths. These have wires and the electrical resistance is measured by using a moisture meter. Higher the electrical resistance, lesser is the available soil moisture.
4. **Soil Signs:** It indicates the quantity of soil moisture present as related to its texture. In this method soil samples of root zone and try to make

a ball by squeezing a handful of soil. Throw the ball above and catch it and on that basis of symptoms, determine the percentage of soil moisture. Field could be irrigated at 50–70% depletion of soil moisture.

Soil Moisture (%)	Coarse Soil	Medium Soil	Fine Soil
50–70%	Tendency to form a ball	Ball can be formed by pressing causes slight bending	Ball can be formed by pressing and ribbon can be formed by pressing between thumb and forefinger

11.6 ANNUAL WATER REQUIREMENT OF SOME IMPORTANT HORTICULTURAL CROPS

Crops	Water Requirement (mm)
Banana	1200–2200
Bean	300–500
Cabbage	380–500
Citrus	900–1200
Cotton	700–1300
Grape	500–1200
Groundnut	500–700
Maize	500–800
Onion	350–550
Pea	350–500
Pineapple	700–1000
Potato	500–700
Sorghum/millets	450–650
Soybean	450–700
Sugarbeet	550–750
Sunflower	600–1000
Tomato	600–800

11.7 EFFECT OF WATER ON CROP GROWTH AND YIELD

Optimum soil moisture requirement of every crop during the different growing period varies. Every stage during growing period are very sensitive to soil moisture and these stages are also called moisture sensitive stage. The term critical stage is usually known as the most sensitive stage at which shortage of water in any crop can cause economical loss to the growers. Insufficient irrigation during moisture sensitive stage can reduce the productivity and after that supply of water cannot recover the yield of the crop. Vegetables are highly susceptible to water stress because it has very weak root system. Vegetables require more water because it transpire and utilize high amount of water during its life time. If water scarcity prevailed at any critical growth stage cause reduction in yield.

11.8 EFFECT OF WATER ON CROP STAGES

Young and delicate plants require frequent supply of water. In initial growth period plant grow at faster rate. For this all physiological and biochemical processes in the plants are put up in a faster rate. This process leads to demand of water supply liberally to maintain the active vegetative growth period and physiological activities in the plants. The annual crops like vegetables and flowers causes serious loss of yield when water stress occurred at any critical stages and this stage differed from crop to crop. Stress of water at any critical stages causes loss in yield and ultimately reduces the profit of the growers. So this is the proven fact that every crop require sufficient water supply in whole life for better growth and yield. Apart from this, plants also require some extent of water stress for certain periods to increase productivity. This period may be identified to save water, particularly in that area where water scarcity prevailed.

11.9 IRRIGATION/FERTIGATION STUDIES

11.9.1 EFFECT OF FERTIGATION ON VEGETATIVE PARAMETERS

According to Ramana et al. (2014) who obtained highest plant growth parameters, i.e., plant height (3.09 m) and canopy volume (20.9 m^3) when higher dose of NPK were applied in sweet orange. Kachwaya and Chandel (2015) reported that higher doses of NPK (150, 100, 120 kg/ha.) through fertigation increased growth parameters like maximum fruit length (42.49 mm), fruit breadth (31.74 mm) and fruit weight (19.87 g) in strawberry cv. Chandler.

Ahmad et al. (2010) also observed that vegetative growth was positively related to the amount of nitrogen applied through drip fertigation in sweet cherry. The findings in relation to canopy volume was found agreement with results of Menino et al. (2003), who found that application of nitrogen at higher rate, i.e., 720 g N leads to the greatest tree canopy volume in Valencia trees. The highest value of LAI may be due to the effect of nitrogen, the findings in relation to LAI by nitrogen application were supported by Chatterjee, 2013. Experiment on citrus was conducted at BAU, Sabour, Bihar, India which revealed that the application of fertigation with NPK at the dose of 120% RDF (360:108:108 g/plant/year) showed that increase in plant height (36.75 cm), trunk girth (4.67 cm), canopy volume (1.83 m^3) leaf area index (5.51) and growth of current season shoot (9.42 cm) of the plant were more responsive to higher dose of fertigation, i.e., T_1 followed by T_2 - 100% RDF (300:90:90 g/plant/year).

11.9.2 EFFECT OF FERTIGATION ON PHYSIOLOGICAL PARAMETERS

The data RWC and internal CO_2 concentration of leaves were found statistically non-significant by different level of fertilizer treatments in citrus. Nitrogen is an essential part of many compounds, including chlorophyll, nucleic acid, proteins and enzymes and found necessary for plant growth and development (Sah et al., 2014). The pyrole rings of chlorophyll arise after prior combination of amino acids, glycine, and succinic acid. Nitrogen application increases chlorophyll content, which results in production of photosynthates. Intrigliolo et al. (1992) found that continuous fertigation significantly improved physiological plant status like photosynthesis rate, transpiration rate and stomatal conductance in citrus tree. (Leuning, 1995) found that photosynthetic rate, required supply of CO_2 and stomatal conductance of leaves are highly correlated to light intensity, which subsequently increase with higher leaf area. Leaf area increase can be due to the positive effect of nitrogenous fertilizer on many important plant structures, genetic and metabolic compounds in plant cells (Don, 2001). Nutrient application through fertigation resulted in maintaining optimum moisture and nutrient reserve in soil and hence, responsible for better physiological activity in citrus plant whereas, under water deficit condition plant close their stomata to prevent dehydration which affects both transpiration and photosynthesis in citrus (Medina et al., 1999). The leaf RWC was found non-significant effect in different treatments which might be due to light, temperature and stress conditions during the period of investigation.

11.9.3 EFFECT OF FERTIGATION ON YIELD

The investigation was carried out at RAU, Pusa, Samastipur, Bihar, India on pointed gourd to show the effect of fertigation along with mulch on yield. The treatment has 100%, 80%, 60% fertigation through drip and 100% fertilizer application as farmer's practices, i.e., flood irrigation with and without plastic mulch. The highest fruit length (10.55 cm), width (3.96 cm) and mean fruit weight (38.50 g) were recorded with 100% fertigation with mulch. Maximum fruit weight per vine (6.31 kg) and yield (15.78 tons/ha) were noted with 100% fertigation along with mulch. Though, it was statistically at par with 80% fertigation with mulch. The yield enhancement due to fertilizer application as well as soil moisture maintenance in the root zone might be the reason of increase in yield. Thus 80% fertigation with mulch was found to be most effective treatment with 20% fertilizer and water-saving 29.50% increase in yield as compared to the traditional irrigation practices.

11.10 IN-SITU, EX-SITU WATER HARVESTING, AND THEIR CONSERVATION

Rainwater, either store in *in-situ* or *ex-situ* rainwater harvesting, which is very beneficial for the crops to irrigate at critical growth periods of both rainfed and irrigated areas. It is the prime limiting production factor in rainfed horticulture whereas in irrigated horticulture it supplements water supply and helps to minimize pressure on applied irrigation water. Therefore, conservation and storage of rainwater in subsurface and surface storages and its efficient use is the challenging issues for stabilizing horticulture production. Water harvesting and integrated watershed management have shown promising response especially annual and rainfed horticultural crops.

KEYWORDS

- **partial root drying**
- **relative water content**
- **soil signs**
- **vegetative parameters**

REFERENCES

Ahmad, M. F., Samantan, A., & Jabeen, A., (2010). Response of sweet cherry (*Prunus avium*) to fertigation of nitrogen, phosphorus and potassium under Kerawa land of Kashmir alley. *Indian J. Agric. Sci.*, *80*(6), 512–516.

Anonymous, (2002). *Biennial Report of National Network Project on Drip Irrigation for Perennial Horticultural Crops*.

Barnes, J. D., Balaguer, L., Manrique, E., Elvira, S., & Davison, A. W., (1992). A reappraisal of the use of DMSO for the extraction and determination of chlorophylls a and b in lichens and higher plants. *Environ. Experi. Bot.*, *32*, 85–100.

Chatterjee, R., (2013). Physiological attributes of tomato (*Lycopersicon esculentum* Mill.) influenced by different sources of nutrients at foothill of eastern Himalayan region. *J. Appl. Natur. Sci.*, *5*(2), 282–287.

Choudhary, M. L., & Kumar, R., (2005). Role of efficient water management technologies. *Journal of Water Management, 13*, 73–78.

Choudhary, R., (1984). The effects of selective combination of N, P, & K on growth, yield and quality of sweet orange (*Citrus sinensis,* Osbeck) cv. Mosambi. *M.Sc. Thesis*. Rajendra Agricultural University, BAC, Sabour, Bihar.

Don, E., (2001). *Efficient Fertilizer Use Nitrogen* (4th edn., pp. 66–84),. IM Global. III inois, USA.

Gardner, W. R., (1993). The future of irrigated areas. In: Buxton, D. R., Shibles, R., Forsberg, R. A., Blad, B. L., Asay, K. H., Paulsen, G. M., & Wilson, R. F., (eds.), *International Crop Science* (Vol. 1, pp. 97–99). Crop Sci. Soc. Amer., Inc., Madison, Wis.

Ghosh, S. P., Verma, A. N., & Govind, S., (1981). Rejuvenating citrus orchards in north-eastern hill region. *Indian J. Hort.*, *26*(1), 20–24.

Haynes, R. J., (1985). Principles of fertilizer use for trickle irrigated crops. *Ferti. Res.*, *6*, 235–255.

Intrigliolo, F., Coniglione, L., & Germana, C., (1992). Effect of fertigation on some physiological parameters in Orange trees. In: Tribulato, E., Gentile, A., & Refergiato, G., (eds.), *Proc. Inter. Soci. Citricul.* (Vol.2, pp. 584–589). Acireale, Italy.

Jacob, A., & Uexkull, H. V., (1958). *Fertilizer Use Nutrition and Manuring of Tropical Crops*. Centered "Etude de Azote, Hannover.

Kachwaya, D. S., & Chandel, J. S., (2015). Effect of fertigation on growth, yield, fruit quality and leaf nutrients content of strawberry (*Fragaria* × *ananassa*) cv. Chandler. *Indian J. Agric. Sci.*, *85*(10), 1319–1323.

Kang, S. Z., & Zhang, J., (2004). Controlled alternate partial root-zone irrigation: Its physiological consequences and impact on water use efficiency. *Journal of Exp. Bot.*, *55*(407), 2437–2446.

Kang, S. Z., Hu, X. T., Goodwin, I., Jerie, P., & Zhang, J., (2002). Soil water distribution, water use and yield response to partial rootzone drying under flood irrigation condition in a pear orchard. *Sci. Hort., 92*, 277–291.

Kang, S. Z., Li, Z. J., Hu, X. T., Jerie, P., & Zhang, J., (2001). An improved water use efficiency for hot pepper grown under controlled alternate drip irrigation on partial roots. *Sci. Hort., 89,* 257–267.

Koo, R. C. J., (1981). Results of citrus fertigation studies. *Proc. Florida Stat. Hort. Soci., 93*, 33–36.

Lekvinadze, P. A., (1972). The effect of different superphosphate rates on some quality indices in satsumas. *Subtropicheskie Kul'tury, 5*, 83–88.

Leuning, R., (1995). A critical appraisal of a combined stomatal-photosynthetic model for C3 plants. *Plant Cell Environ.*, *18*, 339–355.

Medina, C. L., Machado, E. C., & Gomes, M. M. A., (1999). Stomatal conductance, transpiration and photosynthesis Valencia orange under water deficiency. *Rev. Bras. Physiol. Veg.*, *11*, 29–34.

Menino, M. R., Corina, C., Amarilis, V., Victor, V. A., & Jose, B., (2003). Tree size and flowering intensity as affected by nitrogen fertilization in non-bearing orange trees grown under Mediterranean conditions. *J. Plant Physiol.*, *160*, 1435–1440.

Mohanty, S., Srivastava, R. R., Behera, M. S., & Singandhupe, R. B., (2006). Evaluation of intercrops in drip irrigated banana. *Journal of Agricultural Engineering, 43*(4), 15–21.

Nirgude, V., Karuna, K., Mankar, A., Kumar, V., & Spandana, M., (2016). Response of NPK fertigation on pheno-physiological status of *Citrus sinensis* Osbeck cv. Mosambi under high density planting. *J. Appl. & Nat. Sci., 8*(3), 1306–1311 (2016) 1311.

Panse, V. G., & Sukhatme, P. V., (1967). *Statistical Methods for Agricultural Workers*. Indian Council of Agricultural Research, New Delhi.

Ramana, K. T. V., Lakshmi, L. M., Gopal, K., Krishna, V. N. P. S., Lakshmi, T., Sarada, G., Gopi, V., & Sankar, T. G., (2014). Nitrogen and potassium based fertigation response on plant growth, yield and quality of sweet orange (*Citrus sinensis* Osbeck) cv. Sathgudi. *Research and Reviews: J. Agric. Allied Sci.*, *3*(3).

Rani, R., Nirala, S. K., & Suresh, R., (2012). Effect of fertigation and mulch on yield of pointed gourd in calcareous soil of North Bihar. *Environment and Ecology, 30*(3A), 642–645.

Roose, M. L., Col, D. A., Atkin, D., & Kuper, R. S., (1986). Yield and tree size of four citrus cultivars on 21 root stocks in California. *J. Amer. Soci. Hort. Sci.*, *114*, 135–140.

Ryan, W., & Pitman, W., (1998). *Noah's Flood: The New Scientific Discoveries About the Event That Changed History*. Simon & Schuster UK Ltd., London.

Sah, H., Pratibha, Kumar, R., & Topwal, M., (2014). Response of NPK on growth, yield and quality of oriental pear: A review. *Indian Hort. J.*, *4*(1), 01–08.

Shirgure, P. S., (2012). Micro-irrigation systems, automation and fertigation in Citrus. *Scientific J. Rev.*, *1*(5), 156–169.

Shirgure, P. S., (2013). Yield and fruit quality of Nagpur mandarin (*Citrus reticulata* Blanco) as influenced by evaporation based drip irrigation schedules. *Scientific J. Crop Sci.*, *2*(2), 28–35.

Singandhupe, R. B., Edna, A., Mohanty, S., & Srivastava, R. C., (2005). Effect of fertigation on field grown tomato (*Lycopersicon esculentum* L.), *Indian J. Agric. Sci., 75*(6), 329–332.

Singh, H. P., Samnel, J. C., & Kumar, A., (2000). Micro-irrigation in horticultural crops. *Indian J. Hort., 45*(1), 37–44.

Singh, R. M., Singh, D. K., & Rao, K. V. R., (2010). Fertigation for increased crop yield and fertilizer saving. *Agric. Engi. Today*, *34*(2).

Solaimalai, A., Baskar, M., Sadasakthf, A., & Subburamu, K., (2005). *Fertigation in High Value Crops-A*.

Van Schilfgaarde, J., Bernstein, L., Rhoades, J. D., & Rawlins, S. L., (1974). Irrigation management for salt control. *J. Irr. Drain. Div. Amer. Soc. Civil Eng., 100*, 321–338.

Wassel, A. H., Ahmed, F. F., Ragab, M. A., & Ragab, M. M., (2007). Response of Balady mandarin trees to drip irrigation and nitrogen fertigation II-Effect of nitrogen fertigation and drip irrigation on fruit setting, number and quality of fruits of Balady mandarin trees (*Citrus reticulata*). *Afric. Crop Sci. Conf. Proc.*, *8*, 513–522.

Weatherley, P. E., (1950). Studies in the water relations of the cotton plant. I. The field measurement of water deficits in leaves. *New Phytol.*, *49*, 81–87.

CHAPTER 12

Orchard Floor Management

MANOJ KUNDU

Department of Horticulture, Bihar Agricultural University, Sabour, Bhagalpur, Bihar, India, E-mail: manojhorti18@gmail.com

ABSTRACT

Fruit crops are very slow growing in nature. Under the conventional planting system, fruit plants are normally planted at a wider distance which results in the creation of a significant amount of vacant land in fruit plantation for the entire economic life of the fruit plant. In general, these vacant lands become the habitat for different orchard flora which ultimately reduces the amount of quality harvest in terms of yield of the main crop and leads to huge economic losses to the growers. Therefore, maintenance of the orchard floor from the very beginning of orchard establishment is the key operation as it will improve the fertility status of orchard soil. It can be done by adopting different techniques such as the application of organic fertilizers, planting cover crops, sod culture techniques, etc. which will helps to prevent the topmost soil layer of the orchard to expose directly to the atmosphere. In addition, all these techniques help to incorporate organic matter to the soil of the orchard. Further, scientific management of orchard floor would also suppress the weed growth around the trunk of main fruit crop and deliver enough space for different cultural operations in entire orchard floor including spraying of different nutrients, pesticides, fungicides and harvesting of the produce. In addition, short duration vegetables like tomato, brinjal, legumes, cole crops, etc., has the potentiality to grow as intercrops at the vacant space in-between two rows of the main crop which ultimately gives the extra profit to the growers particularly during the pre-flowering stage of fruit crops. Further, a growing short duration intercrop may also reduce the load of weed population on the orchard floor. However, the cultivation of intercrops in the vacant space of the orchard should be done in such a way so that these intercrops would not have any harmful effect on main fruit crops particularly for nutrient competition, competition for natural resources, different biotic and abiotic factors, etc.

12.1 OBJECTIVES OF ORCHARD FLOOR MANAGEMENT

While managing the surface of any fruit plantation, the management of both tree rows as well as alleyways is most important. Covering of alley row by using grasses is the general practice while the three rows remain clean without any vegetation (Figure 12.1). The main objective of floor management involves the suppression of weed population in the plantation, maintaining soil fertility status, minimizing the death of beneficial insect populations to retain the ecological balance, and optimum utilization of natural resources like land, light, air, water, etc. However, during maintenance of the orchard floor, care must be taken in such a way that it ultimately reduces the soil erosion and compaction. In addition, a proper floor management system always gives economic benefits to the orchardist for different objectives. Generally, grassy alleyways reduce the erosion of surface soil of the plantation and also reduce the compaction of soil in sub layer while vegetation-free clean tree rows minimize the competition between main crop and other vegetation on the floor especially for natural resources and inputs.

FIGURE 12.1 Grass alleyway in the cherry orchard.

12.2 BENEFITS OF PROPER ORCHARD FLOOR MANAGEMENT

1. Unutilized space of the plantation can be unitized for growing different short duration crops as intercrop which will ultimately give some additional economic return to the growers.
2. Good floor management system will increase the soil fertility of the plantation particularly by growing cover crop or intercrops. Further, it also maintains the soil moisture status of the plantation by hindering sunlight to come in direct contact with the surface soil.
3. Cover crops and intercrops are helpful to reduce erosion of top soil especially during rainy season.
4. Weed growth can be suppressed by planting cover as well as inter-crops which ultimately helps in the reduction of the weed population in the plantation and also protect the main fruit crops particularly at the early stage of their growth from different biotic and abiotic stresses like wind damage, cold injury, hot stress, etc.

12.3 FACTORS AFFECTING ORCHARD FLOOR MANAGEMENT

12.3.1 WEED POPULATION

Any unwanted plants naturally grown with the main crop in the orchard is known as weed. Weed population is very harmful for the main crop as they create completion with the main crop not only for natural resources but also for the other inputs and ultimately reduces the growth as well as economic return of quality produce in optimum quantity of the leading crop. Further, they create suitable micro-climate for different insect-pest and micro-organisms for the entire life span of the main fruit crop of the plantation. Hence, it is the urgent need for the orchardist to keep the surface of the orchard clean without any weed growth, which ultimately will reduce the competition of weed population with the main crop for any inputs and natural resources and also reduce the problem of biotic and abiotic stresses.

12.3.2 SOIL STABILIZATION

During establishment of an orchard, it is essential to keep in mind that the soil of the plantation is not very compact and it is free from any hardpan or compact layer in the subsurface soil as these hardpan or compact layer in the sub-surface soil can hamper the root growth particularly during the early stage of plant growth. Therefore, tillage or plowing of surface soil of the plantation once in a year is essential with the dual benefit of reduction of weed growth, exposing the eggs or larvae of different pests, improving soil aeration process by loosening of soil. However, soil erosion is one of the key drawbacks under such condition particularly in sloppy land. Thus, the cultivation of cover crops or intercrops in between the tree rows is the commonly adapted practice in the fruit plantation. Further, some compact and stabilized floor is also required in the orchard for smooth conduction of various intercultural operations like pruning, thinning, hoeing, spraying of plant growth regulators (PGRs), micro-nutrients, pesticide, fungicides, harvesting, and transportation of fruits, etc. Alleyway space has compact and stabilized soil which supports all these operation and also maintains the soil properties in the plantation.

12.3.3 IRRIGATION WATER

The need of water to irrigate a particular orchard is partially influenced by the intensity of orchard floor management, adopted in that plantation. Proper maintenance of ground floor of the orchard which includes cover crop and intercrop cultivation may require some additional amount of irrigation water for optimum growth and nutrition of those cover crop and intercrop along with the main fruit crop. To maintain the grassy alleyway, it is essential to supply irrigation water on the entire orchard floor. The need of water also varies with the nature of grasses selected for the alleyway. For the slow growing grasses, it require to supply comparatively smaller amount of water for irrigation as compared to fast growing one or broad leaved plants which require significantly higher more amount of irrigation water for their growth and development. However, the conventional clean weed-free strip and grass alleyway significantly minimize the competition with the leading fruit crops of the plantation for different inputs. But continuous cultivation under clean tilled condition may cause evaporation losses and with the increase of time span, the structure of soil of the orchard may start to disrupt leading to the formation of dust. Therefore, mulching can be done over the tilled portion of the orchard floor to prevent the surface evaporation losses, particularly in the water scarcity area.

12.3.4 PEST MANAGEMENT

Management of pest population is another key component to improve the productivity of a fruit orchard. The approach to control the pest population in a healthy orchard includes the maintenance of insect pest diversity below economic threshold level. Generally, the insect population increases at the faster rate at the ground storey of the orchard as it remains covered with green growth of the main crop especially during the initial stage of the orchard establishment and with the population of cover crops as well as intercrops particularly after the completion of juvenile phase of the main crop. Leaf-feeding arthropods have the natural habitat at the ground storey of the plantation. During mowing of ground floor vegetation or ground floor weeds, these insect pests migrated to the standing fruit trees of the plantation resulting higher infestation on the standing fruit crops. However, the infestation can be reduced by avoiding mowing operation during the month coincides with very hot and dry condition. Further, mowing operation should also be avoided when the population dynamics of the target pest remain very high on the orchard floor. In addition, pest population on the orchard floor can also be reduced by mowing the ground floor vegetation on alternate row, so that the pest population can only migrate from one row to other rows, covered with green vegetation instead of their migration on the standing fruit trees of the orchard.

12.4 ORCHARD FLOOR MANAGEMENT SYSTEM

The important feature of an ideal orchard floor would be easy in maintenance, facilitation of optimum growth and development of the main fruit crops, prevention of the completion among the main trees of the orchards with other flora and fauna; especially for water or nutrients, reduction of soil erosion, maintenance of fertility status of the soil orchard, etc. However, the amalgamation of all these characteristics in a particular orchard under a single floor management frame is not always possible but rather stability among all these factors should achieve.

Different approaches are there to manage the orchard floors under different situations with some specific advantages. These are:

- Weed control;
- Grass alleys;
- Solid vegetation covers;
- Clear cultivation;
- Mulching;
- Cultivation of intercrop;
- Fruit-based cropping system.

12.4.1 WEED CONTROL

Control of weed population is a unique challenge in fruit plantation due to very high cost involvement weed control process. Organically weed can be controlled through cover crop cultivation, uprooting of weed population, mulching to smoother weeds, application of organic herbicides and application of flamers for burning or cooking of weeds.

Minimum competition for weed growth is a key objective for the entire fruit growers of the world, especially during the initial phase of orchard establishment when the main crops of the plantation are very young. Weed competition during that stage can reduce the growth of the main fruit crop up to 60–70% which ultimately may shorten the developmental process of bearing shoots of main crops in the upcoming, resulting huge economic losses. Cultivars which are dwarf in nature or grafted on dwarf rootstocks are highly susceptible to adverse effect of weed population. Age of the main fruit crops (vegetative growth phase), system of irrigation, nature of weed are the important factors playing significant role to increase the population dynamics of different weeds in the fruit plantation.

Any unwanted plants, grown simultaneously with the main fruit crops of the plantation and create completion with the main fruit crops particularly for different natural resources and other inputs- resulting huge economic loss for the orchardist is known as weed. Even a crop, planted by the orchardist for getting extra economic benefits, can also be classified as weed if they competing with main crop for all those in puts. Among different types, weeds which are perennial in nature are highly devastating as they create completion with the main fruit crop both during vegetative growth phase and fruiting season. However, weed growth particularly during winter season not create any serious issues for the orchardist as the main fruit crops remain in dormant condition during that particular time period. However, care should be taken in such a way so that the weed population is suppressed well before the completion of dormancy of the main fruit crops. Besides, weed growth during that time can also minimize the soil erosion from the ground floor of the plantation. However, before the resumption of active growth of main crops in the following spring, mowing of the weed population followed by leveling them over the uncovered surface soil of the plantation is significantly effective to increase the organic matter content of the orchard soil which in turn may improve the productivity of quality produce of the plantation.

Among the different techniques to suppress the weed population in a perennial plantation, use of flamer over the weed vegetation on orchard floor, particularly over the grassy weed is a vital and improved technique. Propane burning technique is commonly adopted for this purpose which includes the use of one or two metal flame orifices, connected with propane tanks to initiate the heating process on the ground floor around the trunks of the fruit tree. A safety switch is attached with the propane tank. Pulling of this switch at moderate speed helps to burn the emerging weeds before attaining a height of more than 2 inches. However, the major drawback of this technique involves the chance of injury on the main crops and the damage of irrigation pipes and trellis materials. Further, fire hazards may also create from the dry leaves of the weeds.

12.4.2 SOLID GRASS COVER

Solid grass cover is being currently used by the orchardists throughout the world to manage the orchard floor properly. This technique is highly effective on erosion prone area specifically on steep slopes. Here grasses or green vegetation are permitted for their natural growth over the entire orchard floor as solid vegetation cover and mowing is done when the grasses attain a certain height. Apart from the conservation of soil moisture and maintenance of soil temperature, this method can also increase the organic matter content in the orchard soil significantly.

12.4.3 CLEAN CULTIVATION

This system of floor management is more adaptable in the arid region of the country. Shallow tillage is the widely adopted technique to make the entire orchard floor weed free. However, application of herbicides also has the potentiality to make the ground floor of the plantation clean without any weed population. This technique of clean cultivation is specifically adopted in area experiencing with frequent radiative spring frosts as clean floor radiates significantly higher amount of heat back into the atmosphere than the cover crops. However, implementation of shallow tillage technique has certain limitations which include damage on the feeder roots, present at the sub-surface soil resulting the reduction of tree vigor.

12.4.4 GRASS ALLEYS

Adaptation of grassy alleys in between tree rows with vegetation-free strips within the row is one of the key techniques to manage the ground floor of any perennial plantation. This technique is advantageous over both solid grass cover and clean cultivation. Under this technique, herbicides are normally applied to develop vegetation free strips around the tree within the rows. However, the cultivation of cover crops is followed in the alley in between tree rows. The vegetation free strips within the tree rows help to grow the roots of the main fruit crops freely without any competition with weeds or grass sod for nutrients, water, and other inputs. However, the grass alley in between tree rows of the orchard prevents the surface soil from its erosion. Further, it also helps to maintain the moisture content of the soil, improve the soil fertility status and also provide the necessary space for different intercultural operations through mechanical aids. Some commonly used grasses for orchard alleyways are Creeping red fescue, Perennial rye, Kentucky bluegrass, and Orchard grass.

12.4.5 GROWING A NITROGEN-RICH COVER CROP IN THE ALLEYWAY

Normally, grasses are allowed to grow on the alleyway of any perennial plantation because it has the potentiality to prevent the erosion of surface soil and also maintain the moisture content in the soil significantly. However, instead of allowing the grasses in the alleyway, the cultivation of leguminous crops is more beneficial as it can give additional income to the growers along with the main crops at the same time and also helps to improve the health of the orchard soil by adding additional nitrogen to the soil. Growing of leguminous crops followed by mowing and keeping those plants on the same row will band the extra amount of nitrogen near the root zone of the main crop. Apart from these, the presence of leguminous crops in the alleyway of an orchard will significantly increase in arthropod population within the orchard which ultimately will maintain the ecological balance between the insects and predators of the orchard.

12.4.6 MULCHING

Besides the retention of moisture content in the soil, mulching prevents weed population to grow around the tree basin. Effective mulches include straw, sawdust, dry leaves, pruned materials of fruit trees, shredded new print, etc., being an important tool of mulching, paddy straw significantly suppress weed population to grow on the orchard floor; however, it is advisable to take care during incorporation of paddy straw as mulching material so that the weed seeds cannot be introduced the plantation with the straw. Further, wood chip mulch is another very effective tool that not only prevents the weed growth but also having positive impact on vegetative and reproductive growth of standing orchard crop resulting higher return of quality fruit. Shredded new print or paper prevents the light to transmit underneath. Therefore, when this Shredded new prints or papers are used for mulching, it prevents the seedlings growth effectively by avoiding light penetration underneath (Figure 12.2). However, the major limitation of using paper mulch includes its frequent application at regular interval. Another useful mulching material is the weed fabric. It is made up of water permeable woven plastic which helps to penetrate water to underneath soil; however, prevent the light to penetrate. The added advantage of using it as mulching material is its longer durability as can last for more than fifteen years but the initial cost of this material is relatively high (approx. \$3000 acre^{-1}).

FIGURE 12.2 Plantation under clean cultivation (left) and under paper covering (right).

In addition, living mulches can also be used for the management of orchard floor. Among different living mulch different leguminous crops, turf grasses, clovers are very promising. However, addition of this type of mulch materials are highly beneficial particularly in the orchards which are already in bearing stage and requirement of different inputs for initial growth and development of main orchard crops are not too critical.

12.4.7 CULTIVATION OF INTER CROP

Intercropping, an effective approach for orchard floor management has commonly been practiced by orchardists in the entire world since time immemorial. When more than one intercrops are grown together, it is very important to make sure the availability of adequate space and other inputs to avoid any competition among the crops for the required inputs. Cultivation of cover crops depends on several factors namely: plant density, maturity dates of intercrops, plant architecture, spreading of root system of individual crops, etc.

12.5 OBJECTIVES OF INTERCROPPING SYSTEMS

Cultivation of intercrop in between the vacant space of two standing fruit crops within the orchard was initiated with the idea to earn some extra profit from those intercrops when the fruit plants are in the juvenile stage. Further, it also insures the profit of the growers when the main fruit crops fail to produce economic return due to any natural calamities or any reasons. However, the cultivation of intercrops in the perennial plantation has the following objectives:

1. Proper utilization of all the natural resources, artificially supplied inputs.
2. Insurance against the failure of main crop due to any natural calamities or any other factor.
3. Increase the cropping intensity per unit land area.
4. Higher profit gain.

12.5.1 PRE-REQUISITES OF FRUITFUL INTERCROPPING SYSTEM

- The peak nutrient and water demanding period of intercrop and main fruit would be different from each other.
- Competition for light among intercrop and main crop would be as minimum as possible.
- The maturity period of intercrop and main crop would be differed by minimum one month.

12.5.2 SELECTION OF INTERCROPS

- Proper selection of intercrops depends on various factors such as land availability, soil fertility status, input availability, etc. Under the availability of heavy inputs, intercrops help to return a higher profit for growers especially at the early age of the orchard. However, in the area having very high land and labor cost, this practice of intercropping within the perennial plantation may not give profitable return.
- For well-established fruit orchard, shade tolerant crops like ginger, turmeric, yam are advisable to grow while legumes, vegetables are recommended for young orchards.
- Short statured fruit crops with short juvenile period viz. papaya, banana, lemon, lime, guava, phalsa, pineapple, and strawberry can also be cultivated as an intercrop in the vacant space of tall statured fruit orchard. However, a thorough analysis is requiring selecting a fruit crop for this purpose, so that root system of such intercrop may not compete with root system of main crop particularly for space, water, and nutrients.

The following fruit crops are most suitable as intercrop for different fruit orchard:

Main Crops	Intercrops
Mango	Pineapple, guava, papaya, lime, phalsa, custard apple, dragon fruit, apple ber
Litchi	Guava, papaya, lemon
Sapota	Papaya, pineapple, guava, custard apple, dragon fruit
Guava	Papaya, pineapple, phalsa, apple ber
Mandarin/sweet orange	Papaya, pineapple, guava, custard apple
Date palm	Mandarin, sweet orange, sapota, guava, mango
Aonla	Lime, phalsa, guava, papaya

12.6 FACTORS AFFECTING CHOICE OF INTERCROPS

1. The selection of intercrop mainly depends on the edaphic and climatic suitability of certain regions or areas. For a particular locality, crops having wide adaptability to the climate and soil condition of that particular locality should have been selected. Under poor soil condition, selection of leguminous crop is recommended as intercrop.

2. Avoid the crops as intercrop which are heavy feeder for water and nutrient because they create competition with main crops for water and nutrients which ultimately will reduce the yield potential of main crop. Therefore, it is advisable to select those crops as intercrop which has less nutrient and water requirement and critical requirement period not overlaps with critical requirement period of main crops.
3. For the orchards having well connectivity with the retail market, situated in a big city or town, it is beneficial to grow different high valued vegetable crop as intercrops; however, for distance market, crops like potato, onion, garlic is preferred to cultivate as they having longer storage life in comparison to other perishable vegetables.
4. At the early stage of any orchard, vegetable crops that require maximum sunlight for their growth and development should be selected as intercrop; while, in old plantation, it is always recommended to keeps shade loving plant as intercrop. However, the selection of field or grain crop as an intercrop is not profitable as they cause severe loss of nutrients and water from the soil of the plantation.
5. Selection of root crops (particularly for seed production) as intercrops within the established fruit orchard is not preferable as the entire seed development period may coincide with the peak period of fruit growth and development of main orchard crop resulting heavy exhaustion of resources from the orchard soil by intercrop.
6. Tall and erect statured plants should be avoided as intercrop as they may affect the growth and developmental process of the main crop. Further, creeper or vine type vegetables like cucurbits should also be discouraged as intercrops.
7. Crops which are highly susceptible to different pest and diseases and particularly those intercrops having the same host range with main fruit crops for different pest and diseases, should be avoided from grown together.
8. Considering leguminous crops as cover crops within the fruit orchard is advisable as they facilitate the process of fixation of atmospheric nitrogen into the soil of the plantation resulting in improvement of soil health. Green gram, black gram, cowpea, etc., are recommended as rainy season cover crops while chickpea, faba bean, pea, and mustard are suitable as winter season cover crops.
9. If the texture of orchard soil is light in nature, then adaptation of sunhemp in the form of cover crop is advisable while in heavy textured soil, growing of *Dhaincha* is preferable to enrich the fertility status

of orchard soil. Further, it also helps to avoid soil erosion problems. They are normally sown prior to the start of monsoon and after two months, they are incorporated in orchard soil through plowing.

12.7 CROP COMBINATION

Lots of research work has been conducted in the entire world regarding the selection of suitable intercrops in a fruit orchard particularly for the early phase of orchard growth to exploit the profit of the growers as maximum as possible. Usually the shot duration legumes such as gram, mung, pea, beans; field crops viz. potato, maize; vegetables viz. Chilies, carrot, cole crops, radish, okra, leafy vegetables, onion, etc. are most suitable as intercrop in a fruit orchard (Figure 12.3) although, their selection ultimately depends on farmers choice. Besides, turmeric, ginger, and sweet potato also have the potentiality to give better yield under the condition of partial shade.

Various crop combinations for interspaces of fruit orchard are as follows:

- Cabbage (Oct-Feb.); Cowpea (March-June); Methi (July-Sept.).
- Okra (Sept.-Dec.); French bean (Jan-March); Tomato (April-July).
- Cauliflower (Aug.-Nov.); Potato (Dec.-March); Dolicos (April-July).
- Cabbage (Sept.-Dec.); Brinjal (Jan.-May); Cucumber (June-Aug.).
- Palak (Nov.-Jan.); Cluster bean (Feb.-May); Brinjal (June-Oct.).

12.8 FRUIT CROP BASED CROPPING SYSTEM

Fruit crop based cropping system having three or more different types of crops with different morpho-phenological features, growing simultaneously at different tiers of the same land which ultimately increase the use efficiency of natural resources resulting increased overall productivity. Increasing cropping intensity of a particular fruit orchard is the key objective of fruit crop based cropping system. To develop such cropping system, growers usually prefer to plant some short stature, quick-growing fruit trees with very short juvenile period at middle tier and bigger canopy sized fruit crops as the main orchard crop of the plantation while legumes, vegetables, flowers or fodder for ground tier crop. In a widely spaced orchard of mango, litchi or jackfruit, fruit crops like papaya, guava, citrus, phalsa can be planted as filler crop in close spacing in between row of main orchard crop (mango, litchi or jackfruit) (Figure 12.4). From the second year onward, these filler trees will start to produce and will give economic yield up to 8–10 years. But the root system of these filler crops will not occupy the space within the root zone of main orchard trees. Further, it also includes the adaptation of different annuals as ground storey crop. Thus, this system comprises the cultivation of perennial and annual plant species simultaneously as different components on the same land to facilitates the maximum utilization of space in four dimensions (length, width, height, and depth) leading to maximum economic return from the system.

FIGURE 12.3 Intercropping under litchi plantation.

12.8.1 BENEFIT OF FRUIT BASED CROPPING SYSTEM

1. Fruit trees are efficient enough to provide higher economic return even under stressed growing conditions prevailing under the upland situations than the other annual crops.
2. It increases the productivity through efficient utilization of air space which remains unutilized in conventional cropping system.

FIGURE 12.4 Multi-tier cropping system with mango as main crop while guava as filler crop.

3. This fruit-based cropping system utilize the natural resources like soil, water, space and environment in more efficient way without any wastage of natural resources. In case of annual crops, 74% of roots do not go beyond 50 cm soil depth whereas in perennial crops, top 50 cm depth of soil remains almost devoid of feeder roots. Feeder roots of most of the perennial fruit crops confined within 50 to 150 cm soil depth with the average spreading of in 2–3 m radius around the tree base. Further, during the initial 10 years of orchard establishment, the canopy of the main crop utilizes only 20–30% of land space. Therefore, only 25–30% of total orchard land is utilized by the main fruit crops and they draw nutrient only from 50–150 cm soil depth. Fruit-based multistoried cropping system offers ample scope for optimum utilization of natural resources above and below the ground level by different component of the system such as main crop, filler crop and intercrop. The differential root distribution pattern of different component crops of the system helps in the effective exploitation of nutrient, water, light and air as a whole resulting higher profit gain. Moreover, it increase organic carbon content in the surface soil by the decomposition of fallen leaves from the fruit trees resulting increased biological activity in the system leading to environmental stability in the rhizosphere.

4. The summation of productivity of the component crops can increase the total productivity of the land where poor soil fertility coupled with low water holding capacity of the soil contributes towards low productivity of any crop in this region.
5. Higher labor requirement per unit area of multitier system contribute towards creation of job opportunities at site.
6. High return per unit area under upland conditions is the ultimate result of fruit-based system.

12.8.2 COMPONENTS OF FRUIT BASED CROPPING SYSTEM

The fruit-based cropping system is a system having at least three different types of crops for different height viz. main crop (upper storey), filler crop (middle storey) and intercrop (ground storey). By efficient utilization of natural resources like water light, from different heights below and above the ground level, respectively, this self-sustainable system can increase the cropping intensity and productivity many folds comparing to conventional farming system. The three important components under this system are as follows:

- **Main Crops:** Fruit crops with relatively larger canopy structure, long juvenile phase with long economic life are mainly considered as the main crop in fruit-based cropping system. They are planted at wider spacing as compared to spacing maintained under conventional system, so that the filler and intercrops can easily be accommodated in between the main crops. The canopy of this category of plants utilizes the top most layer of the multitier system during its entire economic life. However, during the initial 10 years of orchard establishment, they utilize only 25–30% of total orchard space and the area coverage by this crop increase thereafter with 100% utilization only after 20–25 years of planting.
- **Filler Crops**: Short stature, prolific bearing plants having short juvenile phase and relatively shorter economic life are recommended as filler crop under fruit-based cropping system. The main objective of planting this category of plant is to gain some income during the early stage of the plantation until the main crops not overcome their juvenile phase by optimum utilization of unused land or space. After 8–10 years of orchard age, when the main crop start to produce economic yield, this filler crops utilize the middle layer of the multitier system. They are normally planted at closer spacing within the row of main crops.

- **Intercrops:** The lower most layer of the multitier system is occupied by intercrops. Generally, they can be short duration annuals particularly legumes, old seeds, vegetables or flower crops. They utilize the entire ground space which remains unutilized after planting the entire main and filler crops. The main objective of planting this category of plants in a multitier system is to gain some additional income from the unutilized land and resources. Moreover, they also help to suppress the weed population during the early stage of the plantation and also may improve the fertility status of orchard soil. However, after 20–25 years plantation, when the entire orchard space covered with the canopy shade of main orchard crop, it is too difficult to cultivate intercrops in the ground storey except some shale loving crops like ginger, turmeric, foot yam, etc.

12.9 THE PROCEDURE FOR LAYING OUT OF FRUIT CROP BASED CROPPING SYSTEM

While planning for establishing fruit-based cropping system, selection of proper combination of fruit crops is the key factor. The choice mainly depends on edaphic factors, climatic condition, availability of different input facilities like irrigation facility, nutrient availability, infra-structure facility and socio-economic requirements in that area. During planting under fruit-based orchard, crops having relatively bigger canopy structure like mango, litchi, aonla, jackfruit, sapota should be consider as the main crop whereas short-statured crops such banana, guava, lemon, lime, custard apple, papaya, etc. should be selected as filler. However, for the ground storey, different leguminous crops, oil seeds and fodders are most suitable. Apart from these, high valued flower and vegetable can also be selected as the suitable ground storey crops. During the first ten years age of orchard, legumes like pegionpea, horse gram, black gram, cowpea, French bean; oilseeds like soybean, groundnut and fodder like guinea grass, hybrid Napier, etc. are desirable for Kharif planting while vegetables or flowers crops recommended for rabi planting at the ground storey of fruit-based cropping orchard. After 10 years, cultivation of shade-tolerant crops like foot yam, turmeric or ginger is preferable.

Selection of suitable varieties of different component crops is the key for improving productivity under multi-tier cropping system.

Suitable varieties of different fruits crops are as follows:

- **Mango:** Amrapali, Mallika, Bombay Green, Dashehari, Chausa.
- **Litchi:** Shahi, Ajhauli, Rose Scented, Swarna Roopa, China.

- **Aonla:** Narendra Aonla-7, Kanchan.
- **Sapota:** PKM-1, Kalipatti, DHS-1, Cricket Ball.
- **Jack fruit:** SwarnaPoorti, SwarnaManohar.
- **Guava:** L-49, Allahabad Safeda, ArkaMridula.
- **Custard apple:** Balanagar, ArkaSahan.
- **Papaya:** Pusa Dwarf, CO-1, PusaNanha, CO-6.
- **Lime:** Kagzi.
- **Lemon:** Assam lemon.
- **Banana:** Dwarf avendish, Basrai, Grand Nine.

The following steps are followed for laying out of fruit crop-based multi-tier system:

1. **Marking of Pits:** Marking of pits is done at 5 × 5 m spacing accommodating 400 plants per ha. Different markers are used to mark the pits for main and filler plants. The pits for main plants are marked at a spacing of 10 × 10 m whereas the pits of filler plants are marked at a distance of five meters between the rows and five meters within the plants of the main crop. Such layout accommodates 100 main plants and 300 filler plants per ha.
2. **Pit Preparation:** Summer season is the best time to prepare the pit. Pit size varies with the canopy structure of the fruit crop. In general, pit of 1 m × 1 m × 1 m (length × width × depth) is optimum for the crops having bigger size canopy such as mango, litchi, apple, pear, etc.; however, pit of 60 × 60 × 60 cm size is optimum for the crops having smaller canopy size such as citrus, guava, pomegranate, etc. During pit preparation, surface soil is kept separately from the soil of lower level and the dug-up pits are left exposed during entire summer season to minimize the load of harmful pests in the pit.
3. **Pit Filling:** Beginning of monsoon is the optimum time to carry out operation of pit filling. At the time of pit filling, well decomposed, dry FYM @ 15–20 kg, karanj cake @1 kg, single super phosphate @ 300 g is mixed with previously separated surface soil of the pit and put on the empty pit first.
4. The planting is done after 1–2 rains when the soil gets properly settled.
5. **Intercropping:** For intercropping; five-meter wide inter-row spacing is used. Care should be taken to keep an area of about 4 ft^2 around the fruit trees.

List of recommended intercrops under fruit crop based cropping system

Crop	Sowing/ Planting Time	Planting Distance (cm)	Fertilizer (kg/ha)			Recommended Cultivars
			N	P_2O_5	K_2O	
Okra	June-July	40 × 20	120	80	60	Arka Anamika, ArkaAbhay, ParbhaniKranti
Foot yam	May	75 × 75	125	50	120	Gajendra, Santragachi
Sweet potato	May-June	60 × 20	75	50	75	Gauri, Sankar, Pusa Safed, Sree Bhadra
Turmeric	May-June	40 × 20	80	60	60	Suguna, Sudarshana, Suvarna, Rajendra Soniya
Ginger	May-June	40 × 20	80	60	60	Suprabha, Suruchi, Surbhi

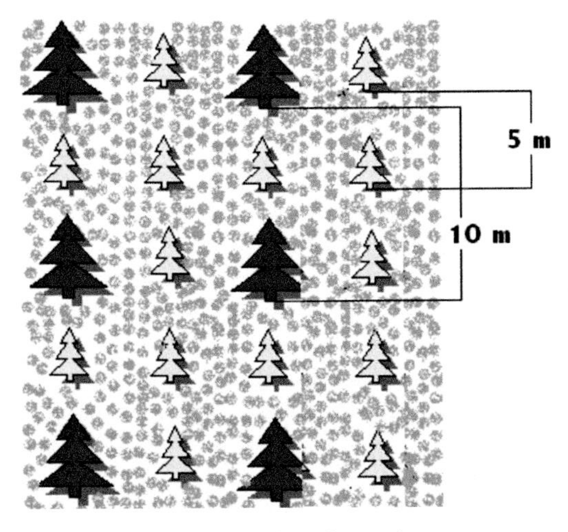

Layout of fruit crop based cropping system.

6. **Aftercare of Young Fruit Crops:** Proper care of young fruit plants is essential for developing healthy and productive plants. After planting, soil application of Chloropyriphus (0.3%) is done for management of white ant. Winter protection is essential for plants up to 2 years of age to minimize mortality of young plants due to winter injury particularly in crops like litchi. Covering of young plants with a thatch prepared by paddy straw or locally available grasses during the winter months. Besides, mulching by paddy straw or locally

available grasses is general practice during summer to retain soil moisture around root zone of young fruit crops. However, irrigation of these young fruit crops at weekly interval during winter and two times per week during summer.

7. **Canopy Management:**
   a. **Young Fruit Plants:** To develop proper framework of main fruit crop of the plantation, training of young plants should be done from the first year of planting. All the growing branches should be headed back above the height of 80 cm from ground level before the start of rainy season. The shoot emerged out from the headed back portion should also remove after retaining only 3–4 well-spaced healthy shoots oriented towards four direction of the main trunk. These healthy shoots ultimately develop the initial framework of the plant as primary branches. In the following year, 3–4 healthy shoots should be selected as secondary branches after heading back of primary branches before the onset of monsoon following the same techniques of previous year. The operation should be repeated on the secondary branches during third year to develop the tertiary one. Branches growing inwards or towards the center of main fruit plants must be removed.
   Heading back of filler crops (guava, lemon, lime, custard apple) to develop proper framework should be done at 30–40 cm above the ground.
   b. **Bearing Plants:** In the case of bearing plants, the criss-cross and overcrowding branches, unproductive and dead branches are removed. Opening the center of the canopy by removal of the branch covering the center of the canopy at the top from its origin helps in minimization of pest incidence ingrown up plants. In case of a fruit which bears on current season growth like litchi, regular pruning of shoot is done immediately after harvesting.

8. **Fertilizer Application:**
   - Around 30 kg well rotten, dry FYM, 2.0 kg karanj cake, 100 g urea, 300 g S.S.P. and 80 g MOP per year is optimum for main crop, up to 2–3 years of age. With respect to micronutrients, each plant is given 10 g each of zinc sulfate, manganese sulfate, boric acid per annum per year age. With increasing age, the dose of the fertilizers is increased at the same ratio till 10 years of age. This

means a 10-year-old fruit tree will require 3–4 kg karanj cake, 10 kg FYM, 1 kg urea, 3 kg SSP, 0.8 kg MOP, 100 g each of zinc sulfate, manganese sulfate, boric acid, copper sulfate per plant per annum. After ten years, the dose is applied at a constant rate.

- A dose of 1.5–2.0 kg karanj cake, 100 g urea, 300 g S.S.P., 80 g M.O.P. and 10 g each of micronutrients like zinc sulfate, manganese sulfate, boric acid with 25–30 kg farmyard manure per plant in every year is optimum for filler crops at the age below 2 years. With increasing age, the dose of chemical fertilizers is increased proportionately till 5 years. Thus, the manurial requirement for a fully developed 5 years old fruit plant will be 30–40 kg farmyard manure, 3 kg karanj cake, 500 g urea, 1.5 kg SSP kg S.S.P., 0.40 kg M.O.P. and 50 g each of zinc sulfate, manganese sulfate, boric acid per plant per year. After 5 years, the dose is applied at a constant rate.
- Half the dose of fertilizer along with full quantity of manure and cake may be applied in June. The remaining quantity of fertilizers can be given in September. Light irrigation is required thereafter. In the case of fruit like litchi, aonla, the second split is applied during March. Application of agricultural lime @ 3–4 kg per tree once in 3 years has been found beneficial in acid soils.

9. **Water Management:**
   - ✓ At early stage of orchard, irrigation at 15–20 days interval in winter and 8–10 days interval in summer is preferable.
   - ✓ Coving the surface soil of the orchard with mulching just after completion of monsoon help to conserve the soil moisture for maximum time period.
   - ✓ In the bearing orchard of litchi, irrigation is not desirable between the month of December to fruit set.
   - ✓ Mango needs frequent irrigation during fruit development period.
10. **Disease and Pest Management:** Under clean cultivation, the probability of disease and pest infestation on main crop remains very low as the entire floor of the plantation remain in clean condition in such situation. However, the probability of infestation under cover cropping or intercropping system is too high although they automatically suppressed weed growth on the floor of the plantation. By increasing the humidity and preventing proper air circulation, cover cropping or intercropping system create the micro climate favorable for different pest and diseases particularly during reproductive phase of main fruit

crop and also during the monsoon time, resulting heavy loss in fruit set and ultimately economic loss from the main fruit crop. Therefore, adaptation of optimum plant-protection measures starting from the orchard establishment is the urgent need of the hour to avoid the harmful effect of various pest and disease infestation problem on the main fruit crops, particularly undercover cropping or intercropping system. Opening of central leader of main fruit crop during initial year of orchard establishment is helpful to allow ample amount sun light to reach to the floor of the plantation, preventing the micro climate to develop. Further, regular pruning of main fruit crop each and every year is also essential to avoid the problem of disease and pest infestation on orchard trees.

12.10 RESEARCH WORK DONE AT BAU, SABOUR

A significant research work was carried out at BAU Sabour for managing the floor of strawberry plantation. Strawberry plants are generally herbaceous and smaller in size, so the fruits touch the ground during its growth and development. To avoid the direct contact of strawberry fruits with soil and also to maintain its quality, mulching is very essential tool of strawberry cultivation. Mulches maintain the soil moisture and soil temperature regime. It also controls weeds and deters the migration of insect-pests. Different organic and synthetic mulches are widely used by the growers based on the availability of raw material and climate condition of that region. Black plastic was found beneficial as mulch material for improving plant growth especially in winter season, conserves moisture and keeps the field weed free. Plastic mulch improves the uptake of minerals and water by increasing soil temperature. It keeps the fruits clean and protects them from any infections by avoiding direct contact from soil. For that U-V stabilize black polyethylene mulch of 25 micron thick and 1–1.2 m width was used in the plantation. Thereafter, well leveled raised bed of 30–35 cm height and 1.0 m width was prepared. Drip system was installed well before covering the bed with plastic mulch. Thereafter, surface of bed was covered with that plastic mulch in such a way that ensures close contact of mulch with surface soil of the bed. Side of those mulches is covered with the soil to fix it on bed tightly. Thereafter, holes of small size are made on the mulched plastic according to the spacing of planting. It was followed by planting of strawberry saplings on those holes. It was found that the plastic mulch improves plant growth, enhances early flowering and fruit setting. Further, it was also recorded that the covering of strawberry beds with plastic mulch significantly increase the yield (380 g/plant) and TSS of the fruit (11.8°B). The maximum number of marketable fruits with average berry weight of 14.21 g was also obtained under black mulch treatment than mulching with straw and without mulch. Plastic mulch produces clean fruits without the occurrence of grey mold infestation.

In addition, percent of marketable fruits under plastic mulch condition was about 80% whereas straw mulch produces only 67% marketable fruits. Further, no weed problem was occurred in the plastic mulched plot. Fertilizer and water use efficiency (WUE) was also increased under plastic mulch besides an increase in overall yield and marketability of fruits.

Fruits without mulch.

Plants under straw mulch.

Fruits under plastic mulch.

KEYWORDS

- **canopy management**
- **crop combination**
- **cultivation**
- **fruit-based cropping system**
- **intercrops**
- **pest management**

CHAPTER 13

High-Density Planting in Fruit Crops for Enhancing Fruit Productivity

V. K. TRIPATHI,[1] SANJEEV KUMAR,[2] VISHAL DUBEY,[3] and MD. ABU NAYYER[4]

[1]*Department of Horticulture, C.S. Azad University of Agriculture and Technology, Kanpur, U.P., India, E-mail: drvktripathicsa@gmail.com*

[2]*U.P. Council of Agricultural Research, Gomti Nagar, Lucknow, U.P., India*

[3]*Department of Applied Plant Sciences (Horticulture), B.B.A. University, Lucknow (U.P.), India*

[4]*Department of Agriculture, IIAST, Integral University, Lucknow (U.P.), India*

ABSTRACT

Planting of fruit trees rather at closer spacing than the recommended one within unit land area, using certain special techniques with the sole objective of obtaining maximum crop yield per unit area without sacrificing quality is often referred as "high density planting." In high density planting (HDP) trees are planted very close together. Plant density may varies with the region, species to be grown, crop variety, rootstock used, cost of planting material, labor charges, assumed return from the orchard and techniques adopted for different horticultural operations for a crop. During last four decades, the importance of high density planting for higher productivity of fruit crops have been realized and now it has become one of the most successful tools of the Hi-Tech horticulture ensuring efficient use of land, water, nutrients and solar radiation with higher production per unit area. HDP offers early cropping and higher yields, improved fruit quality, reduced labor costs, enhanced mechanization in production with efficient use of different production resources leading to higher income per unit area. It can be achieved by the use of dwarfing rootstocks and genetically dwarf cultivars available in different crops, proper and timely use of growth retardants and mechanical methods for plant size control along with using incompatible rootstocks according to the situation. The various researches carried out in different parts of the country it has been found that under HDP of mango 1600 plants/ha, in guava 2222 plants/ha, in citrus up to 3000 plants/ha, in papaya 6400 plants/ha, in banana 4500 to 7000 plants/ha and in ultra-high density guava 5000 plants/ha were found beneficial for getting higher yield with good returns.

13.1 INTRODUCTION

Generally, most of the fruit crops are cultivated using a traditional planting system under which it is difficult to achieve desired level of production, because large trees provide low production per unit area and need higher labor inputs. Early height control and tree management are not commercially practiced in India. Hence, there is over riding need to improve the existing planting system. With ever-increasing land cost, and the need for early returns on invested capital, in the recent years, there is worldwide trend towards planting of fruit plants at closer spacing which leads to high-density planting (HDP) of plants.

Planting of fruit trees rather at closer spacing than the recommended one within unit land area, using certain special techniques with the sole objective of obtaining maximum crop yield per unit area without sacrificing quality is often referred as "HDP." HDP with some modifications suited to respective fruit crop could greatly enhance productivity with enhanced land use efficiency and long-term benefits. First time in records, high-density orchards were planted in Europe at the end of 1960s. In HDP trees are planted very close together. The number of trees may ranges from 500 to 1,00,000 per hectare depending upon the crop, variety, and region. The exact limit of plant density to be termed as high density is not well defined. On the basis of number of trees planted in one-hectare area, HDP system may further be divided into the following three groups:

- Medium-high density: 500–1,500 plants per hectare;
- Optimum-high density: 1,501–10,000 plants per hectare; and
- Ultra-high density: 10,001–1,00,000 plants per hectare.

Plant density may vary with the region, species to be grown, crop variety, rootstock used, cost of planting material, labor charges, assumed return from the orchard and techniques adopted for different horticultural operations for a crop. As the plant population density increases, the cost of plantation in an unit area increases drastically.

The underlying principle of HDP is to ensure the maximum utilization of area, labor, and solar radiation to earn maximum possible return per unit of inputs and resources during the initial years of planting by accommodating more number of plants per unit area. Higher and quality production is achieved from the densely planted orchard through judicious canopy management and the adoption of the suitable plant training system. Considering the soil conditions, plant geometry and manipulation in the spacing higher production and productivity can also be achieved.

Canopy management for proper light distribution is key factor in accommodating higher number of productive trees per unit area. However, unlike temperate fruits, where tree management technologies have been developed and refined for over a century, the same tools and experiences are lacking for tropical and subtropical fruits especially mango, guava, aonla, citrus, etc. However, tree management techniques for few subtropical and tropical fruit crops have been developed and are being used in different parts of the country, which can be adopted after certain modifications in different growing regions.

Advantages: The main advantages of HDP are:

- It resulted early cropping and higher yields per unit area.
- It improve fruit quality.
- It reduced labor costs.
- It reduced cost of production.
- Enhanced mechanization in production.
- Make efficient use of different production resources.
- Make maximum utilization of land and space of orchard.
- Efficient use of nutrients, water, and solar radiation.
- Possible to plant more trees per unit area leading to higher income.

1. **Characteristics of HDP System:** In HDP, plant canopy designing is very important. The plant should have:
   - Maximum number of fruiting branches and minimum number of structural branches.
   - The plants should be trained in central leader system of training in which main branch/stem is surrounded by nearly horizontal fruiting branches.

- The branches should be arranged and pruned in such a way that each branch casts a minimum amount of shade on the other/lower branches.
- The height of plants should be one and half times of its diameter at the base.

2. **Methods of HDP:** In HDP, plants are planted at close spacing's, which in turn is made possible through control of plant size or by planting in a system which accommodates more number of plants per unit area. Manipulation/control of plant size is an important pre-requisite for success of HDP in any fruit crops. Commonly the following methods are applied to control the size/vigor of the plants and to make suitable them for HDP.

   A. **Use of Dwarfing Rootstocks:** Rootstocks play an important role in controlling plant size. Some rootstocks causes' vigorous growth and some dwarfing effects on the scion cultivars. In temperate fruits, a large number of dwarfing rootstocks are available, whereas, these are lacking in tropical and subtropical fruits. Since rootstock trials in fruit trees take 15 to 20 years to draw conclusions, hence there is an urgent need to develop rapid screening technique for identifying dwarfing rootstocks. The dwarfing rootstocks of some of the fruit crops are given below:

Crop	Dwarfing Rootstocks
Mango	Vellaicolamban, Kurukan, Totapari Red Small, Olour, Moovadan
Citrus	Thomasville, Citrangequate, Feronia, Troyer Citrange, *Fortunella* spp., Cuban Sheddock and *Severinia buxifolia*
Ber	*Zizyphus rotundifolia, Zizyphus nummularia*
Guava	*Psidium pumilum, P. fridrichsthalianum*
Apple	M9, M26, M27, MM106, MM109, MM111, Bud9, Bud146, P22, Ottawa 3
Pear	Quince A, Quince C, Adam
Peach	Siberian C, St Julien X, *P. besseyi,* Rubira
Plum	Pixy
Cherry	Colt, Charger, Rubira

   B. **Use of Genetically Dwarf Cultivars:** The availability of genetically dwarf cultivars provides a great scope in the success of HDP. Such varieties are limited in number which is available only in few crops which are given below:

Crop	Genetically Dwarf Cultivars
Mango	Amrapali
Litchi	Shahi, Early Bedana, Purbi, Rose Scented
Banana	Dwarf Cavendish, Basrai, Williams, Rasthali, Poovan, Robusta, Lacatan
Papaya	Pusa Nanha, Pusa Dwarf
Mandarin	Kawano Wase, Satsuma Wase
Sapota	PKM-1, PKM-3
Pineapple	Kew
Apple	All Spur group cvs. (Starcrimson, Redspur, Goldspur, Wellspur, Hardispur, etc.)
Cherry	Compact Lambert, Meteor, North Star
Peach	Red Heaven
Citrus	Meyer Lemon, Ponderosa Lemon

C. **Use of Growth Retardants:** Chemical growth retardants are effectively used in many fruit crops in combination with pruning to control tree size and inducing the early flowering but there are many concerns about the future of these treatments due to quarantine and residue restrictions. Most widely used growth retardants are SADH or Alar (1000–2000 ppm), CCC or Cycocel or Chlormequat (1500–3000 ppm), Paclobutrazol or Cultar (1000–2000 ppm), TIBA (50–100 ppm) and Maleic Hydrazide (1000–1500 ppm). They also help in the early attainment of physiological maturity.

D. **Use of Mechanical Methods**: Among the production packages, tree canopy management, especially size control, has become a priority for the orchardist due to the demands imposed by modern markets in terms of production costs, yield and fruit quality. An interest in mechanical tree size control (hedging, topping, training, pruning and tree thinning) has increased because of the trend to high-density orchards on standard root-stocks. This may be the only practical method of tree size control of inherently large trees in HDP. Tree height control is accomplished by removal of the multiple central leaders within the canopy of the young tree. The dominant, up word shoot or shoot from each growth umbel (whorl of new shoots) is removed and when there are parallel branches within the canopy, the more dominant upright shoots are removed. The height management should begin in the first year. Single stemmed trees are headed between 40 and 70 cm depending on the desired height of the lowest scaffold limbs.

Out of various training systems adopted in apple spindle bush, dwarf pyramid and espalier; in peach cordon; in grape head system are found to control the growth and size of the plants. Different training methods and pruning in any form has a dwarfing effect in fruit plants. In slow-growing plants, proper pruning and training respond more favorably in maintaining a given shape and size of the plant without sacrificing yield. Plant size control through pruning is limited mostly in crops like grape, apple, peach, plum, cherry, etc.

Pruning and hedging practices should be based on our knowledge of the factors in tree physiology and development *viz.*, length and number of growth flushes, nutrient distribution, light intensity, bloom characteristics, climatic information, carbohydrate and nitrogen metabolism and location and status of stored food at pruning time.

In absence of a dwarfing rootstock, the practical approach would be to plant at optimum densities for early production and to thin the orchard (tree thinning) just prior to crowding. A care should be taken so that planting distance remains workable before and after thinning.

Under HDP, plant height should not be greater than twice the width between the outer perimeters of the tree canopies between rows for interception of optimum light. In the orchards, in which spacing between plant in narrow, hand or mechanical pruning must be practiced to prevent overcrowding of branches and kept them more exposed to sunlight. For this pruning should be performed immediately after harvesting of fruits.

E. **Use of Incompatible Rootstock:** The use of graft incompatible scion and stock also found to induce dwarfness in the composite plant, though it is not commercially exploited. In ber, *Z. rotundifolia* and *Z. nummulania* induce dwarfness in cultivated varieties due to graft incompatibility, which can be utilized for HDP of fruit crops.

F. **Use of Cultural Methods:** Cultural methods such as regulated irrigation, nitrogen supply, root restriction, spacing arrangement, poorly adopted rootstock, and high budding can be used to a limited extent under special circumstances to control tree height in HDP.

3. **Performance Under HDP:** Under HDP increase in yield per unit area has been reported in many fruit crops (Bose et al., 1999). Performances of some subtropical and tropical fruit crops under HDP are given here as under:

Crop	Variety	Spacing (m)	No. of Plants/ ha	Yield (t/ha)
Mango	Amrapali	2.5 × 2.5	1,600	22.20
Banana	Robusta	1.0 × 1.0	10,000	99.90
	Robusta	2.4 × 1.8	2,315	41.40
	Basrai	1.5 × 1.5	4,444	78.00
	Nendran	1.2 × 1.2	7,000	72.12
Papaya	Pusa Nanha	1.25 × 1.25	6,400	103.60
	Babaco	1.0 × 1.25	8,000	320.00
	Coorge Honey Dew	1.2 × 1.8	4,629	146.00
Pineapple	Kew	0.22 × 0.60 × 0.75	63,758	118.80
	Kew	0.25 × 0.35 × 0.75	72,727	106.90
Mandarin	Kawano Wase	2.0 × 1.5	3,333	43.33
	Satsuma Wase	1.0 × 1.0	10,000	60.00

4. **Constraints in High-Density Planting (HDP)**
   - Lack of standardized production technology in various fruits crops under HDP.
   - High initial establishment cost of orchard.
   - Lack of promising dwarfing rootstocks in many fruit crops.
   - Incidence of some diseases in HDP, e.g., Sigatoka leaf spot and Finger tip in banana.

13.2 MEADOW ORCHARDING

This is an ultra-high density or super-intensive system of fruit growing. In this system, plants are planted at densities of 30,000 to 1,00,000/ha (Sharma et al., 1998). The term 'meadow' stands for a tract of grassland, in which fruit could be harvested by moving off the treetop mechanically similar to moving of the grassland. This system is mostly followed in apple production, where plants are induced to form fruit buds in their first year of planting with plant bio-regulators treatments. In the second year, plants flower and produce fruits, after which they are cut back to a stump from which a new shoot is regenerated to repeat the biennial cycle. It is envisaged that ultimately fruit harvesting is carried out by moving off the orchard (hence the term 'meadow') with some form of combined harvester which could separate the fruit from the shoots.

Apart from the high yields which are theoretically possible, the attractiveness of such a system lies in the possibility, it offers for the complete mechanization of both pruning and harvesting, which are the two most labor-intensive operations in the traditional fruit orchard. Orchard soil is not cultivated but maintained in weeds free condition by the use of simazine (1 kg/ha, twice a year). In this system single stem is allowed to bear fruits.

13.2.1 ESTABLISHMENT OF MEADOW ORCHARD

For the quick establishment of a meadow orchard, plant cut back maiden trees at a spacing of 35 × 40 cm, if planted in autumn, such trees can be induced to form fruit bud in their first year of growth. In another method, rootstocks are planted at a spacing of 30 × 45 cm in autumn and bud them *in-situ* in the summer of the following year.

In their first year of growth, when the plants are about 60 cm (2 ft) in height, growth retardant such as SADH (2000–2500 ppm) is sprayed, which causes a cessation of vegetative growth and encourage the development of fruit buds in the axils of leaves. In the second year (fruiting year), early in the season repeated application of SADH 2000–2500 ppm is also done to reduce vegetative growth and June drop of fruits.

Fruits could be harvested by a machine that would gently comb the apples off the trees, hopefully with less bruising than is caused by shake-catch methods of harvesting. At the end of the season, the trees should be pruned down to centimeters of the graft union by means of a cutter bar mechanism, and at the same time apply a fungicidal wound dressing to the cut surface of the stumps. In the next season, the side shoots should be removed from the stumps.

The meadow orchard system may not be beneficial to those fruits which are difficult to propagate by cuttings because of high initial cost of establishment as well as the problems of sprouting in rootstocks.

13.3 HIGH-DENSITY PLANTING (HDP) IN MAJOR SUBTROPICAL AND TROPICAL FRUITS

Generally, planting densities are decided after considering vigor of tree, prevailing climatic conditions, soil fertility, water availability and market requirements. Some of the examples, which indicate the superiority over traditional planting systems with regard to productivity, have been summarized below:

1. **Mango:** Despite the rich genetic diversity prevailing, the productivity of mango in India is very low as compared to other mango producing countries of the world. Among the several reasons, lower plant population per unit area arising from the crop geometry leading to sub-optimal exploitation of different resources appears to be the key factor contributing to low mango productivity in the country. The conventional planting adopts 10–12 m apart (100–69 plants/ha) which is proved less efficient for higher productivity.

For getting higher production and ensure optimum interception of light in HDP, plant height should also control. It should not be greater than twice the width between the outer perimeters of the tree canopies between the rows. For this immediately after harvesting, light pruning should be followed.

HDP makes maximum use of land to achieve high yields from the orchard areas. Pandey and Majumdar (1988) grafted *in-situ* rootstock seedling grown at 2.5 × 2.5 m spacing, with Amrapali scion and found fruit yield of 11.5 tonnes in the fourth year with the density of 1600 plants/ha. The productivity increased further reaching 24.46 t/ha in the eleventh year with an average yield of 12.2 t/ha.

Ram and Sirohi (1988) planted Dashehari at 12 × 12 m (69 plants/ha) and 2.5 × 3.0 m (1333 plants/ha) spacing and found that growth of Dashehari trees in HDP in terms of scion length, main stem diameter and circumference increased progressively up to the tenth year, when branches started touching to each other, dehorning of 50% branches after fruit harvest was done in the eleventh year and another 25% was done in the twelfth year. Dehorning of touching branches at this stage brought a reduction of 37% in scion length and 59% in circumferences. The yields also increased progressively, both in low as well as is HDP of Dashehari and the average yield per tree were also similar under both the densities up to 10 years. Thus the yield increased in high-density plantation was because of increase in number of trees per unit area giving 2.4 t/ha in fifth year, which further increased to 18.6 t/ha at the end of tenth year as compared to only 0.2 t/ha in the fifth year and 16 t/ha in tenth year in low density orcharding. Thus for high-density orcharding an irregular bearing Dashehari and regular bearing Amrapali is highly suitable with light to heavy pruning followed by fertilization after fruit harvest.

In India, some efforts were made at IARI, New Delhi, GBPUAT, Pantnagar and HARP, Ranchi which indicated positive results in Dashehari

(1333 plants/ha) and Amrapali and highlighted the need of better orchard management for sustained productivity. Studies at CISH, Lucknow have indicated that sustained yields of about 14–16 t/ha can be obtained by adopting medium density of planting in mango with crop geometry of 5 × 5 m (400 plants/ha) for cv. Dashehari (Mishra, 2013).

2. Guava: It is one of the most suitable crops for HDP as it bears fruits on current season growth and responds significantly to pruning. HDP along with pruning have given encouraging results in guava. Guava planted at 6 × 6 m spacing (277 plants/ha) responds very well to pruning. CISH started HDP at a spacing of 3 × 1.5 m (2222 plants/ha), 3 × 3 m (1111 plants/ha) and 3 × 6 m (555 plants/ha) along with pruning for management of tree size and improving the fruiting potential of guava trees. Eight years after planting, the highest fruit yield (159.39 kg/plant) was recorded from the trees spaced at 3 × 6 m followed by 124.12 kg/plant under 6 × 6 m spacing. Trees spaced at 3 × 6 m had heavier fruit weight. HDP coupled with canopy management produced 47.1 t/ha at 3 × 6 m as compared to 6 × 6 m spaced trees (28.53 t/ha). This technology has become popular among the farmers across the country.

In ultra-high-density guava orcharding, production starts from the very first year of planting and the productivity is higher with superior fruit quality besides ease of tree management. Ultra-HDP of guava accommodates 5000 plants/ha at 1 × 2 m spacing. The plant canopy is managed judiciously with regular topping and hedging. Plants are topped 2 months after planting, i.e., in the month of October, for emergence of new shoots below the cut end. After appearance of new shoots, 50% of the shoots are pruned again in December-January for further induction of new shoots. Growth is initiated, flower differentiates, and well spread plant canopy is attained by the end of May. Heading back of entire shoots is repeated every year in September, May and January for dwarf tree canopy and better fruiting. An average yield of 12.5 t/ha is obtained after first year which reaches up to 55 t/ha after 3 years of planting (Mishra, 2013). The system is high inputs demanding for sustainability.

3. Citrus: Among the citrus crops, efforts have been made for high-density orcharding of orange, mandarin and grapefruit with increasing density. In India citrus (sweet orange, mandarin and grapefruit) is generally planted with a density of 250 to 350 plants/ha depending upon the variety, rootstock, soil fertility and the prevailing agro climatic conditions. In Haryana, a plant density of 375 plants/ha in Kinnow mandarin on Jathi Khatti rootstock was reported to be quite successful (Chundawat and Arora, 1981). In a study at IARI, New Delhi, Troyer Citrange and Karna Khatta rootstocks were reported to have the potentiality to increase its density with 1.8 × 1.8 m and 3 × 3 m spacing, respectively, accommodating 3000 and 1088 plants/ha.

In absence of a dwarfing rootstock, the practical approach would be to plant at optimum densities for early production and to thin the orchard (tree thinning) just prior to crowding. A care should be taken so that planting distance remains workable before and after thinning. Most of the progressive growers, plant their citrus orchards at 2.74 × 2.74 m with the intention of thinning later to make it 5.49 × 5.49 m when crowding occurs.

Plant size in HDP can also be controlled by removal of terminal portion (50% of its total growth) of branches, destroying apical dominance and stimulating lateral bud growth resulting in the bushy and compact center. Multiple shoots were allowed to grow just above the bud union on these plants to keep the terminal growth is check, provided adequate fruiting area and help easy management. Such Kinnow plants start giving profitable return in three years on Troyer Citrange and after four years on Karna Khatta rootstock with earlier fruit maturity, i.e., by the end of November and mid-December, respectively.

4. Banana: It is the most important food ingredient in terms of their gross value next to rice, wheat and milk products. Plant density had pronounced effect on crop duration. In wider plant density, more area of leaf surface is exposed to light, causing increased metabolism of plants, which leads to early physiological maturity and flowering, while duration of crop could be extended in closer spacing due to poor interception of light. Several factors are responsible for productivity of banana. Among these, cultivars, light interception, soil fertility and climatic conditions and soil moisture may play vital role.

Under normal spacing of 2.1 × 2.1 m (2267 plants/ha), banana cvs. Robusta and Dwarf Cavendish yield 50–60 t/ha. Nowadays, HDP in banana has been standardized for increasing the land, water and fertilizer use efficiency, to obtain maximum profitability with increased productivity. Naik (1963) suggested a spacing of 1.5 to 2.1 m for Dwarf Cavendish in Maharashtra. HDP at 1.8 × 3.6 m spacing with 3 suckers per pit with 4500 plants/ha for Robusta and Grand Naine cultivar, producing yield of 80–90 t/ha.

Mustaffa and Pandey (2010) recommended alternatively paired row planting of 5200 plants/ha at a spacing of 1.2 × 1.2 × 2.0 m for Dwarf Cavendish, 3800 plants/ha at a spacing of 1.5 × 1.5 × 2.0 m for tall varieties. This technology is highly suitable for micro-irrigation system, thereby a saving 30–40% in water, 25–30% in fertilizer and 40–50% increase in yield. With closer spacing in cultivar Martman (Rasthali), there was no reduction in bunch weight, number of hands and fingers per bunch (Bhan and Majumdar,

1961). Closer spacing of 1.2 × 1.2 m to 1.5 × 1.5 in Basrai banana brought a reduction in bunch weight and finger number per bunch compared to those in spacing of 1.7 × 1.7 m and 1.8 × 1.8 m. In cultivar Lacatan also, closer spacing of 1.2 × 1.2 m and 1.3 × 1.3 m accommodating 6044 and 5917 plants/ha, respectively, significantly reduced bunch weight and number of fingers per bunch in comparison to 4444 plants/ha when spaced at 1.8 × 1.8 m (Chundawat et al., 1982), whereas fruit yield in Nendran was increased from 43.95 to 72.12 t/ha when plant population was increased from 4500 to 7000 plants/ha (Anon, 1978). Patil et al. (1978) also found an increase in yield from 32.41 to 79.93 t/ha, when spacing was reduced from 2.0 × 2.0 m to 1.2 × 1.2 m.

5. Pineapple: It is predominantly grown in Kerala, Karnataka, West Bengal, Bihar, and North-Eastern regions. Solar radiation requirement for proper vegetative growth and fruit production is directly related with optimum plant density and planting system. Double row, three row, and four row systems have been tried for this crop to maximize solar energy harvesting.

Under well-designed plant orientation by optimizing population, high yield of acceptable size and quality fruits is achieved. In general adoption of 53,000 to 63,000 plants/ha has been recommended. Momin (1975) observed that double row system of planting was more suitable giving better performance with regards to growth, yield and quality of fruits under Assam conditions, whereas, in Kerala, most viable method is two-row bed with the spacing of 25 × 60 × 105 cm (Balakrishnan et al., 1977). In pineapple, yield largely depends upon average fruit weight and fruitage (percentage flowering). Significant reduction in fruitage was noticed when inter-plant and inter-row spacing were reduced to raise the density from 63,492 to 1,00,000 plants/ha (Anon, 1977), whereas, Chadha et al. (1973) recorded higher flowering percentage on increasing planting density to 63,000 plants. Das Biswas et al. (1987) observed an increase in fruit yield up to a density of 72,945 plants/ha and reduced thereafter. They also reported that 63,000 to 72,000 plant density is optimum under North Bengal conditions.

6. Papaya: Normally most of the papaya varieties are planted at a distance of 2.5 × 3.0 m or 2.5 × 2.5 m (accommodating 1333 or 1600 plants/ha) but with the development of Pusa Nanha variety, now-a-days it is possible to grow papaya commercially under HDP concept by planting at a distance of 1.25 × 1.25 m, accommodating 6400 plants/ha (Ram, 1989). Such orchards give 3–4 times more yield (60–65 t/ha) as compared to the 15–20 t/ha yields in traditional systems.

7. Litchi: Young litchi trees are trained to develop a strong trunk and a frame of scaffold branches well distributed around the tree which are sufficiently strong to support heavy load of fruits without limb breakage. Among the various training methods, modified leader has been reported superior.

Traditionally planted litchi plants (10 × 10 m) produced higher yield on per plant basis (82.30 kg/ tree or 82.30 qt /ha) but planting in double hedgerow (5 × 5 × 10 m) accommodating 222 plants/ha, result maximum yield per hectare (156.73 qt/ha) with reduced fruit cracking (Ray et al., 2008).

Pruning plays an important role in controlling growth and regulation of flowering and fruiting. Above ground, level pruning is mainly done for controlling tree shape and size. In new plantation up to 15 cm height, only one or two branches should be allowed. Pruning in old trees plays an important role in regulating and controlling growth, flowering and fruiting. It has been observed that pruning of bearing shoot 40 cm below the panicle significantly increased the fruit yield, fruit weight, and pulp recovery. Soil health management through orchard floor management, may also be a very effective means of controlling root development.

8. Apple: Earlier HDP is achieved by manipulation of pruning and fertilizer dose but nowadays for control in plant size different rootstock and spur type cultivars are available. A number of training systems such as spindle bush, dwarf pyramid, espalier, cordon, and polmate were initially recommended for intensive orchard in the worlds, but in India spindle bush, dwarf pyramid and cordon are found more suitable as training system to produce dwarf tree. Sharma (1989) observed that when spindle bushes on M7 rootstock and modified central leader on MM 106 rootstock, plants trained on spindle bushes had high tree volume, higher fruit set, high yield efficiency, firm fruit with higher TSS and anthocyanine pigment contents.

13.4 CONCLUSION

During the last four decades, the importance of HDP for higher productivity of fruit crops have been realized and now it has become one of the most successful tools of the Hi-Tech horticulture ensuring efficient use of land, water, nutrients and solar radiation with higher production per unit area. In maximizing the tree density along with efficient light distribution, some factors like canopy, nutrient, water, disease and pest management, etc., should be considered for better results. In the absence of research results, tentative decisions can be taken after judging the pattern of plantation and their results under field conditions. The results of plants per unit area should be optimum and no space should be allowed to go to waste. Many farmers in India have tried higher than recommended densities and achieved encouraging results.

KEYWORDS

- **dwarf cultivars**
- **high-density planting**
- **meadow orcharding**
- **pruning time**
- **rootstocks**
- **tropical fruits**

REFERENCES

Anonymous, (1977). *Annual Report* (p. 192). IIHR, Bangalore.

Anonymous, (1978). *Annual Report* (p. 77). Banana Research Station, Kannara, Trichur.

Balakrishnan, S., Veeraraghavan, P. G., & Natarajan, M., (1977). Influence of planting suckers in level ground and in trenches on fruit yield of pineapple. *Agri. Res. J., 15*, 190–191.

Bhan, K. C., & Majumdar, P. K., (1961). Spacing trials on banana in West Bengal. *Indian J. Agric. Sci., 31*, 149–155.

Bose, T. K., Mitra, S. K., Farooqi, A. A., & Sadhu, M. K., (1999). *Tropical Horticulture* (p. 29). Naya Prokash, Calcutta.

Chadha, K. L., Milenta, K. R., & Shikhamani, S. D., (1973). Effect of planting density on growth, yield and quality in Kew pineapple (*Ananas comosus* (L.) Merr.). *Indian J. Hort., 30*, 461–466.

Chundawat, B. S., & Arora, R. K., (1981). *Spacing Trial in Kinnow Mandarin* (p. 27). National Symposium on tropical, subtropical fruit crops, Bangalore (Abstr.).

Chundawat, B. S., Dave, S. K., & Patel, N. L., (1982). High density plantation in relation to yield and quality in Basrai Banana. *South Indian Hort., 30*, 175–177.

Das Biswas, S., Mitra, S. K., & Bose, T. K., (1987). Response of Kew pineapple to plant densities and calcium carbide. *Sci. Cult., 53*, 120–121.

Mishra, D., (2013). Improving orchard productivity through rejuvenation and canopy management. In: Singh, V. K., & Ravishankar, H., (eds.), *Recent Advances in Subtropical Fruit Crop Production* (pp. 35–39). PFDC, Central Institute for Subtropical Horticulture, Lucknow.

Momin, N. W., (1975). Studies on the effect of different planting patterns and plant population densities on growth, development and yield of Kew pineapple (*Ananas comosus* (L.) Merr.). *M.Sc. (Agri.) Thesis*. A. A. U., Jorhat.

Mustaffa, M. M., & Pandey, V., (2010). Research and developments in banana: A flag bearer of golden revolution. In: "*National Symposium on Conservation Horticulture" Organized by GBPUAT*. Pantnagar, Uttarakhand in collaboration with "Indian Society of Horticultural Research and Development, Uttarakhand" held on March 21–23, 2010 at Dehradun (Uttarakhand).

Naik, K. C., (1963). *Banana (In) South Indian Fruits and Their Culture* (p. 207). P. Varadachary and Co., Madras.

Pandey, S. N., & Majumdar, P. K., (1988). *Research Reports* (p. 34). Fruit research workshop, Subtropical and Temperate fruits, Pusa (Bihar).

Patil, S. K., Patil, D. R., & Amin, H. D., (1978). Studies on manurial investigations on banana varieties basrai and Harichhal. In: *Research Report and Project Proposal on Banana, Pineapple and Papaya Fruit Research Workshop Held at Univ. of Agric. Sciences, Bangalore*.

Ram, M., (1983). Pusa Nanha-a dwarfing papaya for closure spacing. *Indian Horticulture, 28*, 17–20.

Ram, S., & Sirohi, S. C., (1988). Studies on high density orcharding in mango cv. Dashehari. *Acta Hort., 231*, 339–344.

Ray, P. K., Kumar, R., & Rajan, R., (2008). Studies on high density planting in litchi. *National Seminar on "Production, Processing, Marketing and Export of Litchi for Economic Prosperity" Held from 8–11 June*. Mushari.

Sharma, D. D., (1989). Effect of different rootstock and training system on growth, cropping, nutrient status and water relations of apple tree. *PhD Thesis*. Dr. Y. S. Parmar University of Horticulture and Forestry, Solan.

Sharma, R. M., Singh, R. R., Ahmad, M., & Tripathi, V. K., (1998). Meadow Orchards: An intensive and mechanized system for apple production. *Indian Farmers Digest, 31*(12), 22–23.

CHAPTER 14

Enriching Vermicompost Using P-Solubilizing and N-Fixing Biofertilizers and Eco-Friendly Low-Cost Minerals

ARUN KUMAR JHA

Department of Soil Science and Agricultural Chemistry, Bihar Agricultural University, Sabour, Bhagalpur, Bihar, India, E-mail: jhaak_ss@rediffmail.com

ABSTRACT

Solid waste management has become an issue of increasing global concern as urban populations continue to rise; rapid economic development and urbanization have led to an increase in the generation of solid waste. In the majority of Indian cities, only 70–80% of generated Municipal Solid Waste (MSW) is collected and the remainder 20–30% remains unattended on streets causing infuriating situations. Numerous human health problems of the citizens are also associated with poor management of MSW. Recycling of MSW is the practical solution of the problem. Poor nutrient content in recycled product and higher recommended doses (RD) of these organic sources of plant nutrients for successful crop production have drawn attention of agricultural scientists, thinkers, planners, administrators, and politicians to enrich the vermicompost produced by using MSW. Enrichment of vermicompost using low cost minerals and bio-inoculants may offer a practical solution of the problem. An effort has been made to review the current status and future panorama of enriched vermicomposting in the present text.

14.1 INTRODUCTION

Global food as well as nutritional security along with environmental protection has become primary requirement of the world at present era. It has been accepted globally that soil fertility exhaustion due to imbalanced nutrient mining has degraded the soil. For farming systems to remain productive and to be sustainable in the long term, it is necessary to replenish the reserves of nutrients which are removed or lost from the soil especially by using the eco-friendly source of plant nutrients (Aveyard, 1988; Wani and Lee, 1992; Wani et al., 1995).

Domastic and farm waste contains huge quantity of plant nutrients (Anonymous, 2006; Bhiday, 1994). It is estimated that annually about 2,008 million tons of excretions is obtained in India out of which annual outturn of wet dung from cow and buffalo is 1,003 million tons (Panda and Hota, 2007). Beside this huge amount of excretions, 358 million tons compostable plant residues are also grown every year. Safe disposal of municipal solid waste (MSW) has also become a challenge for politicians, administrators, scientists and thinkers. Presently, the volume of waste generated from urban centers of the world is around 1,300 million tonnes per year (World Bank, 2012). However, the existing urban MSW production rate in India is 1, 09,598 tonnes per day (Hoornweg and Bhada-Tada, 2012). The Central Pollution Control Board (CPCB) has reported that 35,401 tonnes MSW is being generated every day from 59 major cities of the country during 2004–2005. However, survey report of the Central Institute of Plastic Engineering and Technology (CIPET) interferes that the MSW generation rate increased by 42% only within the five years and reached to 50,592 tonnes per day during 2010–11. These animal excretions, plant residues and municipal waste are required to be converted from waste to wealth very urgently especially for the sake of healthy environment, food security and nutritional security. But, in the contrary, majority of waste and residues are used for landfill due to lack of proper recycling.

There are numerous organisms (micro and macro) in nature having ability to convert waste into wealth. Earthworms are important biological organisms performing this activity very effectively. The earthworm population was found to be 8–10 times in uncultivated area in comparison to that counted in cultivated land (Anonymous, 2006). It indicates that worm's population is closely related to land degradation and owing to the reason, it may be used as a sensitive indicator of soil degradation. In the present era of agriculture, rate of land degradation is very high and due to the reason amalgamation of vermicompost in degraded soils has become very essential for building up its health. In the present text, effort has been made to suggest practical solutions of safe disposal of waste and production of enriched vermicompost.

14.2 THE EARTH WORM AND ITS ROLE IN AGRICULTURE

The earthworm is regarded as "farmer's friend" and "natural plowman." Aristotle described earthworm as an "intestine of the earth." However, Darwin stated that no other creature has contributed for building of earth as earthworm. Earthworms have potential to convert soil from unproductive to productive one. Earthworms eat unrecompensed plant residues, waste, and soil and convert them into valuable wealth containing useful organic acids, hormones and plant nutrients essentially required for soil health building and sustainable production of the crops.

But, the population of earthworms has decreased drastically after adoption of green revolution measures. Now, compost worms are being reared for vermicompost production, because it is quite easier than that to increase the population of earthworm in soil. Agricultural scientists and thinkers have realized that this is very high time to adapt vermiculture at large scale to combat the ill effects of exhaustive agriculture adopted in the past. Entre Pinoys (2010) has defined vermiculture as a science dealing with the breeding and raising compost worms. Earthworm has potential to reduce C:N ratio of waste and residues (Gandhi et al., 1997; Thompson and Nogales, 1999). Important role of compost worms in agriculture can be briefed as under:

- Self-employment to rural youth;
- Pollution free environment;
- Reduction in dependency of agriculture on chemical fertilizers;
- Improvement in soil health;
- Sustainable crop production;
- Strong foundation for organic farming.

14.3 WHAT IS VERMICOMPOSTING?

Conversion of waste and residues into a humus-like substance through the activities of compost worms is termed as vermicomposting. It is greatly different from composting (Gandhi et al., 1997), because, composting is a thermophilic; however, vermicomposting is mesophilic process. Earthworms are specialized to live in decaying matter and can degrade it into fine particulate materials high in available nutrients and considerable potential to improve soil activities (Sridevi et al., 2016). Thus, earthworm as a potential biological tool should be much better understand to make organic farming and sustainable development with the use of selected species of earthworms (Kale, 1998). Conclusively, vermicomposting is that process which is capable of transforming garbage into black gold (Anonymous, 2006). Beside nutritional and environmental importance of vermicomposting, it also offers employment to rural youth, which is very important for agriculture prime developing country like India.

14.4 IMPORTANCE OF VERMICOMPOSTING

Economically viable, environmentally sustainable and socially acceptable agricultural technologies are need of the time. Vermicomposting is that technology which combines all these intrinsic worth together (Karmengham and Rajasekhar, 2012). Vermicompost is superior to compost because it favors colonization of beneficial micro-organisms (Edwards, 2004). *Rhizobium*, *Azotobacter*, *Nitrobacter*, Actinomycetes and phosphate solubilizing micro-organisms proliferates nicely in vermicompost (Karmengham and Rajasekhar, 2012). Vermicompost improves resistance in plants against insect pest and diseases due to higher population of actinomycetes in it (Singh, 2009; Kumar and Shweta, 2011). Higher population of bacteria, fungi and actinomycetes has been found in the gut of earthworms and due to the reason; cast of the worms contains sufficient number of beneficial micro-organisms (Prakash and Karmegam, 2010). Use of bio-inoculants for enrichment of vermicompost improves quality of the produce (Kaushik et al., 2008). Interaction of earthworms and micro-organisms results very many beneficial enzymes and hormones which augments growth, and yield of the crops (Karmengam and Rajsekhar, 2012; Jayashree et al., 2011).

Low-grade rock phosphate is an eco-friendly mineral of but having no commercial and agricultural value until or unless it is mixed with phosphate solubilizing micro-organisms. Phosphorus content of vermicompost can be increased by mixing of rock phosphate and phosphate solubilizing micro-organisms into substrate. As more than 40% of Indian soil has become deficient in sulfur availability, so just to achieve optimum production from S-deficient soils, sulfur nutrition has become essential. Thus, enrichment of vermicompost with sulfur-containing minerals, viz. Gypsum or phosphogypsum has become important to increase production and productivity of pulses, oilseeds, and cereals in sulfur deficient soils.

14.5 INFLUENCE OF VERMICOMPOST ON SOIL HEALTH

Vermicompost supplies all essential nutrients to the soil. It improves physical, chemical and biological properties of the soil (Maheswarappa et al., 1999; Mitchell and Edwards, 1997). Incorporation of vermicompost into soils endows with congenial environment for growth of native micro-organisms. Vermicompost diminishes nutrient loss and improves use efficiency of chemical fertilizers. Vermicompost has potential to improve structure, texture, aeration and water holding capacity and diminishes soil erosion. This material supplies vitamins, enzymes and hormones viz. auxins and gibberellins to soil-plant system (Marinari et al., 2000). Sodic soil can also be ameliorated through the application of vermicompost (Sinha and Sinha, 2008). Biological nitrogen fixation also improves in vermicompost treated soils (Bhadauria and Ramakrishnan, 1996)

14.6 IMPORTANCE OF ENRICHED VERMICOMPOSTING IN MODERN AGRICULTURE

It is obvious from numerous experimental findings that there is a scope for quality improvement in vermicompost to regard it as an organic fertilizer (Padmavathiamma et al., 2008). Composting of organic wastes and rock phosphate mixture causes dissolution of insoluble phosphates and resulting in increasing the availability of P for plants (Kumar and Narula, 1999). It had also been observed by Kumar and Narula (1999) that some N_2-fixing bacteria also increase P availability through production of organic acids. *Pseudomonas, Azotobacter*, *Burkholderia*, *Agrobacterium*, and *Erwinia* have also showed their P-solubilizing potential during composting (Scervino et al., 2010; Eivazi and Tabatabai, 1977; Busato et al., 2012; Mohammady et al., 2010). Introduction of P-solublizing bacteria in substrate used for vermicomposting and rock phosphate mixture was found to be more effective to solubilize phosphorus than that introduced in soil-rock phosphate mixture (Kaushik and Garg, 2004).

It has been reported by Beauchamp et al. (2006) and Hendrikson (1990) that the population N_2-fixing bacteria used for conversion of organic wastes into vermicompost had not been reduced by earthworms. In contrary, population of N_2-fifing bacteria in final product got increased (Fischer et al., 1997). Introduction of N_2-fixing micro-organism in substrate used for vermicomposting improves N status of product.

In the light of above facts, enrichment of vermicompost is the need of time to reduce the RD of vermicompost for various crops and to make it popular among farming community.

14.7 VERMICOMPOSTING MATERIALS

The material required for vermicompost production should be bio-degradable and these are as under:

- Agricultural waste, e.g., crop residue, vegetable waste, sugarcane trash, etc.
- Biogas slurry.
- Excretions obtained from livestock and poultry birds.
- Waste from agro-industries.
- Forest litters.
- Biodegradable portion of urban and industrial waste.
- Biodegradable portion of hotel refuge.
- Aquatic and terrestrial weeds.
- Plant products viz. saw dust, pulp, etc.

Crop residues are the remnants of crop plant left after the harvest of crops. Only about one-third of the total crop residue generated in India is available for utilization in agricultural production. Only, biologically degradable and decomposable plant residues are commonly used for vermiculture. The availability of agricultural waste in India has been presented in Table 14.1. Every part of agricultural waste and agro-industry waste is not utilizable for vermicompost production. The utilizable waste of agricultural and agro-industry waste has been presented in Table 14.2. The advantages and disadvantages of common food of vermicompost have been presented in Table 14.3.

TABLE 14.1 Availability of Agricultural Waste in India

Agricultural Waste	Approximate Quantity (Million Tons/Year)
Bagasse and molasses of sugarcane	38.1
Crop residues	196.8
Fiber crop residues	3.3
Food grain residues	14.4
Fruit and vegetable residues	1.2
Oilseeds residues	34.4
Planting residues	17.4
Powder of flour mills	2.0
Pulse residues	5.8
Rice residues	18.0

TABLE 14.2 List of Wastes Tested and Found Suitable for Vermicompost Production

Source of Waste Generation	Utilizable Waste for Vermicomposting
	Agricultural Waste
Agricultural fields	Stubbles, weeds, husk, and straw.
Plantations and gardens	Stems, leaf matter, fruit rinds, stubbles and grass clippings.
Animal waste	Dung, urine and biogas slurry.
Urban solid waste	Kitchen waste from households, restaurants, biodegradable waste from market yards and places of worship and sludge from sewage treatment plants.
Mushroom production unit	Waste paddy or wheat straw used for mushroom production.
	Agro-Industry Waste
Food processing units	Peels, rinds and unused pulp of fruits and vegetable
Vegetable oil refineries	Press mud and seed husk.
Sugar factories	Press mud, fine bagasse, and boiler ash.
Breweries and distilleries	Spent wash, barley waste, and cast sludge.
Seed processing unit	Core of seeds, paper, and seeds after expiry date.
Aromatic oil extraction unit	Stems, leaves and flowers after extraction of oil.
Coir industry	Coir pith
Tissue culture units	Paper, agar and wasted plantlets.

Besides, agro-industry wastes like rice husk, coir pith, press mud, cotton lint, jute stick, and tea waste are also considered very good substrate for vermicomposting because, these are biodegradable and contain 0.5–1.55 N, 0.5–2.5% P_2O_5 and 0.5 to 3.0% K_2O. Oil cakes can also be used as a substrate for vermicompost production because of high NPK contents in it (1.5–5% N, 1.0–1.8% P_2O_5 and 1.0–1.8% K_2O), nutritional value of vermicompost produced by using these materials will also be high (Gupta, 2008).

Cattle dung is the main base of vermicomposting and is used in several combinations of agricultural waste including shed dropping, gram bran, kitchen waste, rice polish, semi crushed leaves, sludge, vegetable waste, wheat bran, weeds, etc. It has been reported by Gupta (2008) that the ideal ratio of cattle dungs: gram bran: wheat bran: vegetable waste for vermicompost production is 10:1:1:1. But, the generalized ratio of cattle dung and plant residues is 3:2. If there is a scarcity of cattle dung than plant residues or decomposable municipal waste should be wet by 2% aqua slurry of cattle dung. The annual excretions from livestock and human beings in India have been presented in Table 14.4. Panda and Hota (2007) calculated the NPK supplying potentials of excretions available in India and reported that 6.37 million tons N, 1.98 million tons, P_2O_5 and 2.67 million tons K_2O can be

TABLE 14.3 Advantages and Disadvantages of Common Food of Vermicompost

Food	Advantages	Disadvantages
Cattle manure	Good nutrition, natural food.	Weed seeds make pre-composting necessary
Poultry manure	High N content results in good nutrition and a high-value product.	High protein levels can be dangerous to worms, so must be used in small quantities, major adaptation required for worms not used to this feedstock.
Sheep/Goat manure	Good nutrition	Requires pre-composting (weed seeds), small particle size can lead to packing, necessitating extra bulking material
Hog manure	Good nutrition, produces excellent vermicompost	Usually in liquid form, therefore must be dewatered or used with large quantities of highly absorbent bedding
Rabbit manure	High N content, good nutrition, contains very good mix of vitamins and minerals; ideal earth-worm feed	Must be leached prior to its use because of high urine content, produces overheat if quantity is too large, availability is usually not good.
Fresh food scraps (e.g., peels, other food prep waste, leftovers, commercial food processing wastes)	Excellent nutrition, good moisture content, possibility of revenues from waste tipping fees	Extremely variable (depending on source); high N can result in overheating, meat and high-fat wastes can create anaerobic conditions and odors, and attract pests, can be included only after pre-composting
Pre-composted food wastes	Good nutrition, partial decomposition makes digestion by worms easier and faster, can include meat and other greasy wastes, less tendency to overheat.	Nutrition less than with fresh food wastes.
Biosolids (human waste)	Excellent nutrition and excellent product; can be activated or non-activated sludge, septic sludge, possibility of waste management revenues	Heavy metal and/or chemical contaminations (if from municipal sources); odor during application to beds (worms control fairly quickly), possibility of pathogen survival if process not complete
Seaweed	Good nutrition, results in excellent product, high in micronutrients and beneficial microbes	Salt must be rinsed off, as it is detrimental to worms; availability varies from region to region.

TABLE 14.3 *(Continued)*

Food	Advantages	Disadvantages
Legume hays	Higher N content makes these good feed as well as reasonable bedding.	Moisture levels not as high as other feeds, requires more input and monitoring
Corrugated cardboard (including waxed)	Excellent nutrition (due to high-protein glue used to hold layers together); worms like this material; possible revenue source from WM fees	Must be shredded (waxed variety) and/or soaked (non-waxed) prior to feeding
Fish, poultry offal; blood wastes; animal mortalities	High N content provides good nutrition; opportunity to turn problematic wastes into high-quality product	Must be pre-composted until past thermophillic stage

supplied to the soil through recycling of excretions (Table 14.5). However, the annual manorial potential of only bovine excreta is 2.822, 1.069, and 1.819 million tons N, P_2O_5 and K_2O, respectively. Residue and waste of onion can also be used as a substrate for vermicomposting (Sobana et al., 2016). Dry olive cake has been found good substrate for vermicomposting (Nogales et al., 1999). Maize straw can be used successfully for production of vermicompost (Manna et al., 1997). Zajonc and Sidor (1990) got success to produce vermicompost by using mixture of cotton waste and cattle dung in the ratio of 1:5. To feed scarps to worms, scarps should be chopped into small pieces and then used for vermicomposting after keeping it in a container for few days to increase the microbial population.

TABLE 14.4 Annual Excretion of Dung and Urine of Livestock and Human Beings in India

Animal Type	Daily Excretion		Annual Excretion		Total Excretion (Million Ton)	Percentage (%)
	Dung (Kg)	Urine (Liter)	Dung (Kg)	Urine (Liter)		
Cow and Buffalo	11.597	7.623	1002.587	658.901	1,661.488	82.71%
Sheep and Goat	0.300	0.200	12.228	7.918	20.146	1.00%
Pigs	2.000	2.000	4.596	3.990	5.586	0.43%
Poultry	0.068	-	3.395	-	3.395	0.20%
Other livestock	5.000	3.300	6.024	4.095	10.119	0.50%
Human beings	0.133	1.200	30.380	274.100	304.480	15.16%
Total			**1,059.210**	**949.004**	**2,008.214**	

14.8 WHAT SHOULD NOT BE FED TO COMPOST WORMS?

It is very important to note that metallic foils, plastic, chemicals, oils, patricides, soaps and paints are not fed to the worms. Worms are not allowed to eat sour fruits and their products/waste, onion and garlic cloves, extremely hot food, heavily spiced food, oleanders, poisonous plants, meat, chicken, dairy foods, dog and cat manure and any other acidic foods (Gupta, 2008).

14.9 MATERIALS USED FOR ENRICHMENT OF VERMICOMPOST

Microbial inoculants are mainly used for enrichment of vermicompst (Alikhani et al., 2016). Bihar Agricultural University (BAU), Sabour, Bhagalpur, Bihar has explored the possibility of vermicompost enrichment

TABLE 14.5 Manorial Potential of Livestock and Human Excretions

Animal	Annual Excretion (m. t.)		Chemical Composition (%)						NPK Potential (Million Tons)		
	Dung	Urine	N		P_2O_5		K_2O				
			Dung	Urine	Dung	Urine	Dung	Urine	N	P_2O_5	K_2O
Cow	744.57	480.15	0.15	0.20	0.10	0.01	0.05	0.20	2.03	0.79	1.33
Buffalo	258.02	178.75	0.15	0.20	0.10	0.01	0.05	0.20	0.75	.28	0.49
Sheep and Goat	12.23	7.92	0.65	1.70	0.50	0.02	0.03	0.25	0.21	0.06	0.02
Pigs	4.60	3.99	0.60	0.40	0.50	0.10	0.20	0.50	0.04	0.03	0.03
Poultry	3.40	-	0.80	-	0.60	-	0.30	-	0.03	0.02	0.01
Other livestock	6.02	4.10	0.50	1.20	0.30	-	0.30	1.00	0.08	0.02	0.07
Human beings	30.38	274.10	1.60	1.00	1.20	0.15	0.55	0.20	3.23	0.78	0.72
Total									**6.37**	**1.98**	**2.67**

through rock phosphate, phosphogypsum, *Azotobacter*, phosphate solubilizing bacteria and *Azolla* to augment nutritional status and to increase the population dynamics of beneficial microorganismss in vermicompost. These in richment materials are mixed at the time of pre-digestion @ rock phosphate–5.05 (wt/wt.), phosphogypsum–2.5% (wt./wt.), *Azotobacter*–50–100 g/t, phosphate solubilising bacteria–50–100 g/t and *Azolla*–up to 25% (Wt./Wt.).

In one study conducted at BAU, Sabour, Bhagalpur, Bihar, India, by the author, it was found that *Azolla* is an excellent enrichment material for vermicomposting and breeding of the worms. 25% replacement of cattle dung by *Azolla* increased *Eisenia fetida* population by 247% in 90 days. However, population increased only by 142.5% when only cattle dung was used as a substrate (Figure 14.1).

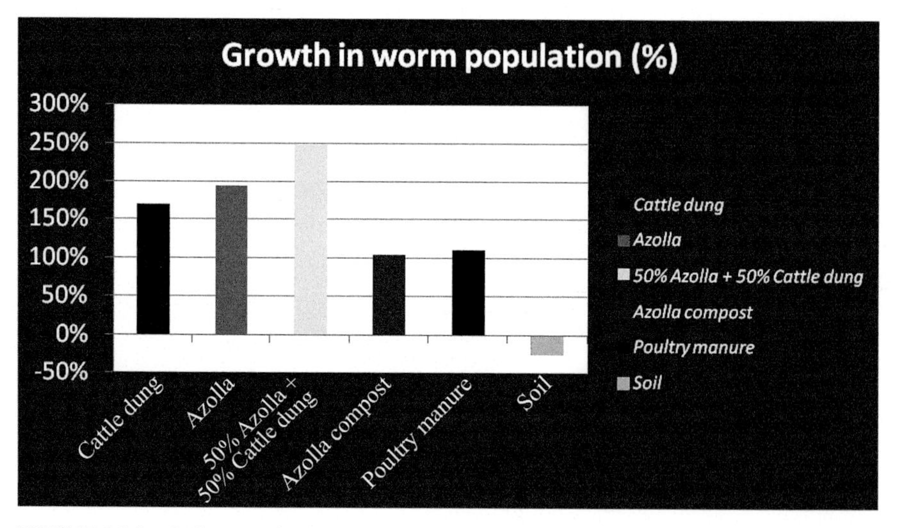

FIGURE 14.1 Influence of substrate on population of *Eisenia fetida* after 120 days of rearing.

14.10 DESCRIPTION OF WORMS USED FOR ENRICHED VERMICOMPOST PRODUCTION

Out of 3, 000 spices of earthworms identified so far, only a small number of species is useful for vermicompost production. Those species of earth worms which are used for production of vermicompost are called as "Compost Worm." *Eisenia foetida*, *Eudrilus eugeniae* and *Perionyx excavates* are used extensively in India for vermicomposting purposes. The first two are exotic and last one is indigenous. These species are most suitable because these are proliferic breeders with high multiplication rate, have short life cycles with less mortality and are voracious feeders which give out high quality vermicasts. They are easy to handle and survive very well throughout the year under varying weather conditions and are easily available.

On the basis of adaptability, compost worm can be divided into (1) Peregrine and (2) Endemic. Peregrine has wide range of adaptability and due to the reason; these are found in larger part of the world. *Eudrilus eugenae* and *Eisenia foetida* are important members of this group. However, endemic compost worms are found only in those areas where climatic variation is very less. However, On the basis of food habit, earth worms can be divided into (1) Phytophagus, and (2) Geophagus. Phytophagus worms eat organic matter; however, geophagus worms eat soil. Considering the ecological strategy, worms can be divided into (1) Epigeic, (2) Endogeic, and (3) Anecic. Population of epigeic worms are restricted to the upper horizon/(s) of the soil having sufficient amount of decomposed or un-decomposed organic matter. However, both, endogeic and anecic worms reside at comparatively lower horizons of soils and due to the reason it is also called "Soil Living Worms."

Eisenia foetida is commonly used in India in all part of the country, probably due to its high degree of resistance against climatic and managerial variation during vermicompost production. Due to its red color, it is also known as "Red Worm." Generally, it is 3–15 cm long and 0.3–0.5 cm thick worm and its weight ranging from 0.4 to 0.6 g. *Eisenia foetida* matures in 50–55 days and produces one cocoon in every three days.

14.11 AQUATIC AND ENVIRONMENTAL REQUIREMENT OF WORMS

1. **Water:** It is the basic requirement of worms. 85% of body weight of earthworm is constituted of water. Enormous volume of water from the worm's body is lost in the form of urine. Earthworms do not have any protective body cover and they have to keep the body surface moist as the body wall serves as respiratory organ. Due to the reason, they are releasing mucus constantly through the dorsal pores to keep the body wet. It is essential to maintain 60% moisture in the medium. Excess moisture or water stagnation creates anaerobic conditions in the medium and thus deters the growth of earthworms and also quality of the compost.
2. **Temperature:** The temperature of the earthworm feed should be in the range of 20°C to 35°C. Temperature greater than 45°C results into

desiccation of the body and moisture stress. Very low temperature is also deleterious to earthworms. Temperature below 0°C brings to a standstill the activities of earthworms.

3. **Requirement of Light:** Earthworm prefers dark to remain active. Light especially ultra-violet radiation causes injury to earthworms. Excessive exposure to light may prove fatal for earthworms. It is only why, vermicompost is produced under shade.
4. **pH:** For effective multiplication of earthworms, pH of the feeding material should be at neutral level, i.e., 7.0. Earthworm population is severely affected if the pH of the feed material is less than 4 and greater than 9. Optimal pH range of feed mixture suitable for the utmost activities of earthworm is 6.0–8.5. At both the extremes, rate of vermicompost production falls down. Thus, it is essential for efficient vermicompost production to maintain pH of the partially digested substrate used for vermicomposting. To maintain an ideal pH of the composting material, it is essential to use green matter along with the dry biomass and regulate the moisture in the medium (Joshi, 2005).

14.12 DIVERSITY IN VERMICOMPOSTING TECHNIQUES

Suitability of the technique depends on the purpose of vermicomposting. Small-scale vermicomposting is done either in the thermacol boxes or under straw-thatched shade or in orchard to meet out their personal demand of farming or kitchen gardening. Sometimes, rural youth also adopt small scale vermicompost for the sake of their self-employment. However, vermicompost is produced at large scale either for entrepreneurial purposes or for environmental protection through recycling of MSW. Enriched vermicompost production technique is being implemented by various institutions and agro-industries to solve the fertility-related problems in soils and to promote the application of vermicompost in agriculture by reducing its RD for various crops.

14.13 METHODS OF VERMICOMPOSTING

There are numerous methods of vermicomposting. Out of them, some important methods are as under:

1. Vermicompost production in pits below the ground.
2. Vermicompost production in heaps above the ground.

3. Vermicompost production in tanks above the ground.
4. Open method of vermicomposting in orchard.
5. In-situ Method of vermicomposting.

14.13.1 VERMICOMPOST PRODUCTION IN PIT

In this method, vermicompost is produced in one meter deep pits. This method of vermicomposting is facing lot of constraints especially in the area receiving sufficient rainfall. Probably due to the reason, this method is not popular among farming community.

14.13.2 VERMICOMPOST PRODUCTION IN HEAPS ABOVE THE GROUND

This method of vermicomposting is farmer's friendly. In comparison to the other methods, higher amount of vermicompost can be produced in equal span of the time. Heap method of vermicomposting is also suitable for large scale production of vermicompost because of easy operation of machine for the purpose. Bob cut loader is generally used for heaping and turning purposes to reduce the dependency of vermicomposting on manpower.

Heap of pre-digested material is prepared on sloppy floor. Width and height of the heap should be maintained 6 m and 3 m respectively, however, length depends on the space available in shade. Sufficient volume of water is sprinkled over the heap for 5–6 days to make the material cool. Compost worm are introduced in cool material @ 1000–2000/ton of substrate. Heap is covered with old gunny bags or paddy straw and water is sprinkled daily to maintain moisture level more than 60%. After three week, pre-digested material converts into tea like structure, called vermi-cast. After the formation of vermi-cast, irrigation is stopped for 3–4 days and cast is collected from upper surface of the bed. Vermi-cast is sieved and packed in an inner laminated bags to maintain the moisture between 20 and 25%.

14.13.3 VERMICOMPOST PRODUCTION IN TANKS ABOVE THE GROUND

In this method, tanks are constructed under shade. Dimension of pit as recommended by IRISAT is 15'LX5'WX3'H. However, Bihar government is popularizing the tank of 10'Lx 3'WX 2'6"H dimension by providing subsidy on this structure. The basic concept behind reduction of width and height of tank in Bihar model in comparison to ICRISAT model is to make it convenient for manual operation. Some farmers and entrepreneurs of Bihar have hybridized heap and tank method of vermicomposting by reducing height of the tank to 6"–12 inch."

Tank is filled with different materials in six layers. Description of the layers from bottom to top is as under:

1st layer – 7.5 cm thick layer of brick stones and sand followed by its sufficient moistening.
2nd layer – 7.5 cm thick layer of dry plant residue.
3rd layer – 5.0 cm thick layer of decomposed compost.
4th layer – Introduction of worms @1000–2000/ton pre-digested material.
5th layer – 45 to 60 cm thick layer of pre-digested waste.
6th layer – Mulching with old gunny bags.

Increase in number of tanks depends upon the availability of raw material. Composting worms may also be applied @ 1000–2000/t substrate on the upper portion of the bed. Moisture level of the substrate is maintained by sprinkling the water at regular interval. When upper portion of the material is converts into tea like structure (vermi-cast) then sprinkling of water is stopped for 2–3 days and cast is collected. Collected cast is sieved at 20–25% moisture and packed in an inner laminated bags to maintain moisture level of the product.

14.13.4 OPEN METHOD OF VERMICOMPOSTING

Though, first method of vermicompost production discussed above also comes under open method of vermicomposting. But, heap method of vermicomposting in old mango orchard is more successful than pit method. In BAU, Sabour, Bhagalpur, Bihar heaping of substrate in old mango orchard for vermicomposting has been practiced successfully. In this method, soil between the rows of mango is slightly raised and black polythene sheet is spread over the raised land. Pre-digested Waste and residue mixture is heaped over the polythene sheet and irrigated for 4–5 days. Worms are introduced into the heap @ 1 kg/ton substrate after becoming the substrates cool. During rainy season, heaps are covered with low height polythene parabola made with the help of bamboo logs.

14.13.5 IN-SITU METHOD OF VERMICOMPOSTING

This method is generally practiced in horticultural gardens. To provide nutrition to the fruit crops, at least 15 cm thick layer of waste/waste-residue mixture is spread in one-meter radius around the tree, 30 cm apart from the main trunk. Worms are applied @ 1000/tree. Waste is mulched with paddy straw after introduction of worms and light irrigation is allowed periodically till the conversion of waste into vermicompost. But, In-situ Method of vermicomposting has not been found to be better than vermicompost application with production point of view.

Out of various methods discussed above heap methods of vermicomposting has been found better than rest of the methods. Both, vermicompost production rate and worm multiplication rate were found to be higher in heap method of vermicomposting (Sunitha et al., 1997).

14.14 NUTRIENT CONTENT OF ENRICHED VERMICOMPOST

The level of nutrients in enriched vermicompost depends upon the nature of substrate, species of worm and enrichment materials used for vermicomposting. Due to the reason, nutrient content of vermicompost produced in the same unit varies from batch to batch. Nutritional status of enriched vermicompost has been found higher than normal vermicompost. Nutrient content of vermicompost and enriched vermicompost (enriched with rock phosphate, phosphogypsum, *Azotobacter*, *Azolla* and phosphate solubilizing bacteria) prepared from same substrates at BAU, Sabour, Bhagalpur has been presented in Table 14.6. It is pertinent to mention that Sulfur content and microbial population dynamics were determined only once.

Quality of vermicompost in India is being controlled under the guidelines mentioned in fertilizer control order (FCO). This order dictates that vermicompost satisfying following quality related guidelines can be permitted for marketing:

1. Moisture percent (by weight): 15–25.
2. Color: Dark brown to black.
3. Odor: Absence of foul odor.
4. Particle size: 90% material must pass through 4 mm sieve.
5. Bulk density (g/cm^3): 0.7–0.9.
6. Total organic carbon (% by weight): 18.0.
7. Total Nitrogen (%): 1.0.

8. Total P_2O_5 (%): 0.8.
9. Total K_2O (%): 0.8.
10. Maximum heaves metal content.
    Cadmium (ppm): 5.0.
    Chromium (ppm): 50.0.
    Nickel (ppm): 50.00.
    Lead (ppm): 100.0.

TABLE 14.6 Nutrient Content and Microbial Count in Vermicompost Produced at BAU, Sabour, Bhagalpur During 2013

Plant Nutrient/ Microbe	Content in Vermicompost	Content in Enriched Vermicompost
Nitrogen (%)	1.2–1.5	1.7–2.1
Phosphorus (%)	0.68–1.48	1.5–3.2
Potassium (%)	0.36–0.72	0.89–1.12
Sulfur (%)	0.80	1.26
Zinc (ppm)	56–100	81–113
Iron (ppm)	2,500–3,000	3,396–4,279
Copper (ppm)	25–50	50–52
Manganese (ppm)	250–429	360–466
Azotobacter (c.f.u./g dry compost)	-	2.4×10^9
P.S.B. (c.f.u./g dry compost)	-	5.2×10^9
Rhizobium (c.f.u. /g dry compost)	-	6.5×10^9

14.15 RECOMMENDED DOSE (RD) OF ENRICHED VERMICOMPOST FOR VARIOUS CROPS

Soil test values are the scientific basis to manures and fertilizers recommendation to achieve the sustainable yields of the crops. Manurial recommendation for same crop and equally fertile soil may differ in different agro-climatic zones. A little work has been done regarding recommendation of enriched vermicompost for various crops in different agro-climatic zones, probably due to late introduction of the technology. Impact of enrichment techniques on nutritional status of vermicompost is also one of the constraints behind unavailability of data regarding its recommendations for various crops. Thus, there is a need of further investigation to search out the best enrichment technique and to ascertain its dose for various soil-crop systems in different agro-climatic zones.

There is a great variation in the recommendations of vermicompost proposed by researchers. Recommendation of central research institute for dry land agriculture, Hyderabad is being presented in Table 14.7. ICAR Research Complex for NEH Region, Umiam, Meghalaya has generalized the recommendation of vermicompost, 5–6 t/ha for field crops, 3–5 kg/plant for fruit plants and 100–200 g/pot for plants grown in pots. However, recommendation of vermicompost for field crops and fruit plants, as prescribed by ICRISAT are 2–3 t/ha and 5–10 Kg/tree respectively. In vegetable crops, vermicompost is applied @ 1.0 t/ha in nursery to grow the healthy seedlings, whereas, at the time of transplanting, vermicompost should be applied @ 400–500 g/plant. Application of equal dose of vermicompost after 45 days of plant growth increases production of vegetable crops.

TABLE 14.7 Recommended Dose of Vermicompost for Various Crops

Name of Crop/(s)	Recommended Dose of Vermicompost (t/ha)
Rice	2.5 ton/hectare
Sugar cane	3.75 ton/hectare
Cotton	2.5 ton/hectare
Chili	2.5 ton/hectare
Groundnut	6.25 ton/hectare
Sunflower	3.75 ton/hectare
Maize	2.5 ton/hectare
Turmeric	2.5 ton/hectare
Grape	2.5 ton/hectare
Citrus, Pomegranate, Ber, and Guava	2.0 Kg/tree
Mango and Coconut	At planting – 2.0 Kg/tree; 1–5 Year old plant – 5.0 Kg/tree; 6–9 year old tree – 10 Kg/tree; ≥ 10 year old tree – 20 Kg/tree
Onion, Garlic, tomato, Potato, Ladies finger, Brinjal, Cabbage, and Cauliflower	2.5–3.75 t/ha
Teak, Red sandalwood and other forest trees	3 Kg/tree

14.16 MANIPULATION OF GROWTH AND YIELDS OF THE CROPS BY VERMICOMPOST APPLICATION

Vermicompost application has potential to increase yield of wheat (Desai et al., 1999), sorghum (Patil and Sheelavantar, 2000), sunflower (Devi and

Agarwal, 1998; Devi et al., 1998), pea (Reddy et al., 1998), and cowpea (Karmegam et al., 1999; Karmegam and Daniel, 2000). It has been reported by Vadiraj et al. (1998) that yields of coriander in vermicompost and fertilizer treated plots were at par. Integrated application of chemical fertilizers and vermicompost was to be found beneficial to increase flower yield in ornamental plants (Nethra et al., 1999).

14.16.1 PESTS AND PARASITES

Worms sometimes may also act as an agent for the spread of parasite, acting as reservoirs or intermediate host for many parasites and pathogens (Gupta, 2008). A large number of protozoanes, nematodes, rotifers, flat worms, mites, and dipteran larvae are found in the tissue and body fluids of compost worms. Among dipterans, a cluster fly parasitizes lumbricides. The parasitic flies like Onesia suvalipine and mite like Histostoma murchieae are parasitic on cocoons. Leeches are known to attack vermeries. Nematodes are found emerging from the cocoons of Endrilus eugeniae which affected their viability. Incidence of parasitism increased with the age of host. There are records of different species of parasitic nematodes from the coleomic cavity of the earthworms (Gupta, 2008). Cilliate and sporozoan parasites have also been isolated from the body of fluids and tissues of earthworms. The important pest and management related problems of vermicomposting and their solutions are as under:

14.16.2 EPILOGUE

Green revolution converted India from subsistence to surplus in food grain production. But, non-judicious use of agrochemicals for achieving this status has declined the population of beneficial flora and fauna in soil. Probably due to the avoidance of organic sources of plant nutrients and uncontrolled mining of essential plant nutrients from soil in green revolution era, a big challenge of agricultural sustainability is being faced by agricultural scientists, thinkers, planners, and administrators. The vital solution of the problem is large scale adoption of waste and residues recycling and use of recycled material with bio-inoculants to attain sustainability in agriculture. The earthworm is said "Farmer's Friend" and its casting is a good tool to retrieve soil health. Enrichment of vermicompost may reduce the RD of vermicompost to various crops and as a consequence this material will become

Problem	Cause	Solution
Bad smell from vermibed/tank	Overfeeding/Presence of non-compostable materials/ Exposed food scraps/over moistening of bed.	Stop feeding for two weeks; Remove non-compostable materials; Bury food completely; Mix in dry vermibed; Fluff bedding; Drill hole in tank/bed
Bin attracts flies	Food scraps exposed/Rotten food/Too much food especially citrus	Burry food completely; Avoid putting rotten food in bin; stop over feeding to worms
Drying of worms	Over moistening or over-drying of substrate/Extreme temperature/ insufficient aeration/Scarcity of food	Maintain optimum temperature and moisture through management; Provide food to the worms.
Worms crawling away	Conditions of the tank/bed is not good/high temperature in bed and cooler environment outside especially in summer/ More nutritious and desired food is available in nearby area	Improve conditions of bin/tank; Manage congenial environment for worms in bed/tank.
Mold forming	Too acidic condition prevails	Reduce acidity of substrate by mixing less acidic materials.
Drying of beds	Too much ventilation/sprinkling of water on beds in Insufficient amount	Reduce ventilation; sprinkling of water in proper amount
Water stagnation on bottom of tank	Watery scarps/Improper drainage facility/Over moistening	Improve drainage facility; Avoid over moistening of bed.

popular among farming community. Besides, this technology offers self-employment to rural youth, healthy and sustainable environment and food security to the ever-increasing human population. Adopting the enrichment technology of vermicomposting, vermicompost can be converted into more popular and beneficial agro-input. Thus there is a need of intensive scientific researches and wide-scale adaptation of enriched vermicomposting.

KEYWORDS

- **Azolla**
- **azotobacter**
- **bioinoculant**
- **plant nutrient**
- **pollution control**
- **soil health**
- **vermicompost**

REFERENCES

Alikhani, H. A., Hemati, A., Rashtbari, M., & Tiegs, S. D., (2016). Enriching vermicompost using P-solubilizing and N-fixing bacteria under different temperature conditions. *Communications in Soil Science and Plant Analysis*. doi: 10.1080/00103624.2016.1206913.

Anonymous, (2001). *Vermicomposting Technology for Waste Management and Agriculture: An Executive Summary*. PO Box 2334, Grants Pass, OR 97528, USA: Vermi Co. http://www.vermico.com/summary.htm (Accessed on 22 November 2019).

Anonymous, (2006). Vermicomposting: Recycling wastes into valuable organic fertilizer. *SAT e Journal, 2*(1), 1–16.

Anonymous, (2016). *Organic Farming*. TNAU portal, http://oacc.info.

Aveyard, J., (1988). Land degradation: Changing attitudes – why? New South Wales. *Journal of Soil Conservation, 44*, 46–51.

Beauchamp, C. J., Levesque, G., Prevost, D., & Chalifour, F. P., (2006). Isolation of free living dinitrogen-fixing bacteria and their activity in compost containing deinking sludge. *Bioresource Technology, 97*, 1002–1011.

Bhadauria, T., & Ramakrishnan, P. S., (1996). Role of earthworms in nitrogen cycle during the cropping phase of shifting agriculture (jhum) in northeast India. *Biology and Fertility of Soils, 22*, 350–354.

Bhiday, M. R., (1994). Earthworms in agriculture. *Indian Farming, 43*(12), 31–34.

Busato, J. G., Lima, L. S., Aguiar, N. O., Canellas, L. P., & Olivares, F. L., (2012). Changes in labile phosphorus forms during maturation of vermicompost enriched with phosphorus-solubilizing and diazotrophic bacteria. *Bioresource Technology, 110*, 390–395.

Desai, V. R., Sabale, R. N., & Raundal, P. V., (1999). Integrated nitrogen management in wheat-coriander cropping system. *Journal of Maharashtra Agricultural Universities, 24*(3), 273–275.

Devi, D., & Agarwal, S. K., (1998). Performance of sunflower hybrids as influenced by organic manure and fertilizer. *Journal of Oilseeds Research, 15*(2), 272–279.

Devi, D., Agarwal, S. K., & Dayal, D., (1998). Response of sunflower [*Helianthus annuus* (L.)] to organic manures and fertilizers. *Indian Journal of Agronomy, 43*(3), 469–473.

Edwards, C. A., (2004). *Earthworm Ecology* (2nd edn., pp. 256–292). CRC Press, Boca Raton, FL, USA.

Eivazi, F., & Tabatabai, M. A., (1977). Phosphatases in soils. *Soil Biology and Biochemistry, 9*, 167–172.

Entre, P., (2010). Retrieved from: http://www.mixph.com/2006/12/vermiculture-the-management-of-worms.html (Accessed on 22 November 2019).

Fischer, K., Hahn, D., Honerlage, W., & Zeyer, J., (1997). Effect of passage through the gut of earthworm *Lumbricus terrestris* L. on *Bacollus megaterium* studied by whole cell hybridization. *Soil Biology and Biochemistry, 29*, 1149–1152.

Gandhi, M., Sangwan, V., Kapoor, K. K., & Dilbaghi, N., (1997). Composting of household wastes with and without earthworms. *Environment and Ecology, 15*(2), 432–434.

Gupta, P. K., (2008). *Vermicomposting* (2nd edn., pp. 33–140). Agrobios India, Jodhpur.

Hendrikson, N. B., (1990). Leaf litter selection by detritivore and geophagous earthworms. *Biology and Fertility of Soils, 10*, 17–21.

Hoornweg, D., & Bhada-Tada, P., (2012). *What a Waste: A Global Review of Solid Waste Management*. No. 15, World Bank.

ICRISAT and APRLP, (2003). *Vermicomposting: Conversion of Organic Wastes into Valuable Manure* (p. 4). Andhra Pradesh, India: ICRISAT and APRLP.

IPCC, (2007). Climate change 2007: The physical science basis. In: *Contribution of Working Group I to the Fourth Assessment Report of the Intergovernmental Panel on Climate Cambridge University Press*. Cambridge, UK and New York, NY, USA.

Jayashree, S. J., Rathinamala, & Lakshmanaperumalsamy, P., (2011). Determination of heavy metal removal efficiency *Chrysopogon zizanioides* (Vetiver) using textile wastewater contaminated soil. *J. Environ. Sci. Technol., 4*, 543–551.

Kale, R. D., (1998). *Earthworm: Cinderella of Organic Farming* (pp. 1–88). Prism Publisher, Bangalore, India.

Karmegam, N., & Daniel, T., (2000). Effect of biodigested slurry and vermicompost on the growth and yield of cowpea [*Vigna unguiculata* (L.)]. *Environment and Ecology, 18*(2), 367–370.

Karmegam, N., Alagermalai, K., & Daniel, T., (1999). Effect of vermicompost on the growth and yield of greengram (*Phaseolus aureus* Rob.). *Tropical Agriculture, 76*(2), 143–146.

Karmengam, E., & Rajsekhar, K., (2012). Enrichment of biogass slurry vermicompost with *chroococcum* and *Bacillus megaterium*. *J. of Env. Sci. and Tech., 5*(2), 91–108.

Kaushik, P., & Garg, V. K., (2004). Dynamics of biological and chemical parameters during vermicomposting of solid textiler mill sludges mixed with cow dung and agricultural residues. *Bioresource Technology, 4*, 203–209.

Kaushik, P., Yadav, Y. K., Dilbaghi, N., & Garg, V. K., (2008). Enrichment of vermicomposts prepared from cow dung spiked solid textile mill sludge using nitrogen fixing and phosphate solubilizing bacteria. *Environmentalist, 28*, 283–287.

Kumar, R., & Shweta, (2011). Enrichment of wood waste decomposition by microbial inoculam prior to vermicomposting. *Bioresour. Technol., 102*, 1475–1481.

Maheswarappa, H. P., Nanjappa, H. V., & Hegde, M. R., (1999). Influence of organic manures on yield of arrowroot, soil physico-chemical and biological properties when grown as intercrop in coconut garden. *Annals of Agricultural Research, 20*(3), 318–323.

Manna, M. C., Singh, M., Kundu, S., Tripathi, A. K., & Takkar, P. N., (1997). Growth and reproduction of the vermicomposting earthworm *Perionyx excavatus* as influenced by food materials. *Biology and Fertility of Soils, 24*(1), 129–132.

Marinari, S., Masciandaro, G., Ceccanti, B., & Grego, S., (2000). Influence of organic and mineral fertilizers on soil biological and physical properties. *Bioresource Technology, 72*(1), 9–17.

Mitchell, A., & Edwards, C. A., (1997). The production of vermicompost using *Eisenia fetida* from cattle manure. *Soil Biology and Biochemistry, 29*, 3–4.

Mohammady Aria, M., Lakzian, A., Haghnia, G. H., Berenji, A. R., Besharati, H., & Fotovat, A., (2010). Effect of Thiobacillus, sulfur, and vermicompost on the watersoluble phosphorus of hard rock phosphate. *Bioresource Technology, 101*, 551–554.

Nethra, N. N., Jayaprasad, K. V., & Kale, R. D., (1999). China aster [*Callistephus chinensis* (L)] cultivation using vermicompost as organic amendment. *Crop Research, Hisar, 17*(2), 209–215.

Nogales, R., Melgar, R., Guerrero, A., Lozada, G., Beniteze, E., Thompson, R., Gomez, M., & Garvin, M. H., (1999). Growth and reproduction of *Eisenia andrei* in dry olive cake mixed with other organic wastes. *Pedobiologia, 43*(6), 744–752.

Padmavathiamma, P. K., Li, L. Y., & Kumari, U. R., (2008). An experimental study of vermibiowaste composting for agricultural soil improvement. *Bioresource Technology, 99*, 1672–1681.

Panda, H., & Hota, D., (2007). *Livestock and Human Wastes: Characteristics and Value: Biofertilizers and Organic Farming* (pp. 17–24). Gene-Tech Books, New Delhi.

Prakash, M., & Karmegam, N., (2010). Vermistabilization of press mud using *Perioneyx ceylanensis. Mich. Bioresour. Technol., 101*, 8464–8468.

Premono, E. M., Moawad, M. A., & Vlek, P. L. G., (1996). Effect of phosphate-solubilizing Pseudomonas putida on the growth of maize and its survival in the rhizosphere. *Indonesian Journal of Crop Science, 11*, 13–23.

Priti Joshi, P., (2005). '*Vermicomposting and Vermiwash Technologies'-Manual on Bio Inoculants*. Mahatma Gandhi Institute of Rural Industrialization, Wardha, Maharastra.

Reddy, R., Reddy, M. A. N., Reddy, Y. T. N., Reddy, N. S., Anjanappa, N., & Reddy, R., (1998). Effect of organic and inorganic sources of NPK on growth and yield of pea [*Pisum sativum*(L)]. *Legume Research, 21*(1), 57–60.

Scervino, J. M., Mesa, M. P., Mónica, I. D., Recchi, M., Moreno, N. S., & Godeas, A., (2010). Soil fungal isolates produce different organic acid patterns involved in phosphate salts solubilization. *Biology and Fertility of Soils, 46*, 755–763.

Singh, K., (2009). Microbial and nutritional analysis of vermicompost, aerobic and anaerobic compost. *Report of 40 CP Honors Project for Partial Fulfillment of Master in Environmental Engineering*. Griffith University, Brisbane, Australia.

Sinha, R. K., & Sinha, R., (2008). Role of plants, animals, and microbes in environmental management. *Environmental Biotechnology* (p. 315). Aavishkar Publishers, India, ISBN: 978-81-7910-229-9.

Sobana, K., Agnes, M. S., & Jegadeesan, M., (2016). *JECT, 5*(1), 193–200.

Sridevi, S., Prabu, M., & Tamilselvi, N. G., (2016). Bioconversion of water hyacinth into enriched vermicompost and its effect on growth and yield of peanut. *Int. J. Curr. Microbiol. App. Sci., 5*(9), 675–668.

Sunitha, N. D., Giraddi, R. S., Kulkarni, K. A., & Lingappa, S., (1997). Evaluation methods of vermicomposting under open field conditions. *Karnataka Journal of Agricultural Sciences, 10*(4), 987–990.

Tara Crescent, (2003). *Vermicomposting: Development Alternatives (DA) Sustainable Livelihoods*. http://www.dainet.org/livelihoods/default.htm (Accessed on 22 November 2019).

Thompson, R. B., & Nogales, R., (1999). Nitrogen and carbon mineralization in soil of vermicomposted and unprocessed dry olive cake ('Orujo seco') produced from two stage centrifugation for olive oil extraction. *Journal of Environmental Science and Health, Part B, Pesticides, Food Contaminants and Agricultural Wastes, 34*(5), 917–928.

Wani, S. P., & Leem, K. K., (1992). Biofertilizers role in upland crops production. In: Tandon, H. L. S., (ed.), *Fertilizers, Organic Manures, Recyclable Wastes and Biofertilisers* (pp. 91–112). New Delhi, India: Fertilizer Development and Consultation Organisation.

Wani, S. P., (2002). Improving the livelihoods: New partnerships for win-win solutions for natural resource management. *Paper Submitted in the 2nd International Agronomy Congress Held at New Delhi, India during 26–30 November 2002*.

Wani, S. P., Rupela, O. P., & Lee, K. K., (1995). Sustainable agriculture in the semi-arid tropics through biological nitrogen fixation in grain legumes. *Plant and Soil, 174*, 29–49.

World Bank, (2012). *What a Waste: A Global Review of Solid Waste Management*. Urban development series knowledge papers.

Zajonc, I., & Sidor, V., (1990). Use of some wastes for vermicompost preparation and their influence on growth and reproduction of the earthworm *Eisenia fetida*. *Pol'nohospodars-tvo (CSFR), 36*(8), 742–752.

CHAPTER 15

Protected Horticulture in India: An Overview

PARAMVEER SINGH, AJAY BHARDWAJ, and RANDHIR KUMAR

Department of Horticulture (Olericulture and Floriculture), Bihar Agricultural University, Sabour, Bhagalpur, Bihar, India

ABSTRACT

Horticulture sector is the most lucrative venture providing vast employment opportunities and increased earnings to the farmers in India and at global level. India is at second place in the production of various horticultural crops in the world. Protected cultivation is the modern approach to cultivate horticultural crops in a manner that it can provide high quality produce with increased quantity. Greenhouse/poly house/net houses is most commanding technology under varying climate for round the year and off time vegetable and flower production. Beside production, this technology can be used for virus free seedlings and hybrid seed production. This technology needs very vigilant planning, awareness and information about time-line of production and moreover, harvest time to coincide with high market prices, choice of varieties adopted for the off season environments. Greenhouse is an artificial structure works on the phenomenon which is known as greenhouse effect. The area coverage reported under protected cultivation in India is 110,000 hectares and global level is 275,000 hectares. Maharashtra and Gujarat covers 5,730.23 hectares and 4,720.72 hectares area under protected cultivation, respectively. Protected technology has in general two very basic components, first one is infrastructure and second one is production technology of crops. Infrastructure involves various engineering aspects of protected structure development. The second component production technology of crops involves scientific studies to develop the varieties suited for protected cultivation, choosing the type of crops and standardizing the production protocols. Higher intensity of cropping and intensive management necessitates higher labour requirements for these structures. The per unit labour requirement of greenhouse cultivated area (10 men/ha) is more than field cultivation (1 man/ha). Government of India is also providing 50% subsidy on total expenditure with a maximum cut off limit up to 4000 m^2 per beneficiary for adoption and installing greenhouses under National Horticulture Mission. For Indian farmers, this technology can help in making lucrative returns from various high value crops and will impart enough calibres to compete at International level.

15.1 INTRODUCTION

The horticulture sector has emerged as a powerhouse for overall agricultural advance in India (Mittal, 2007; Abou-Hadid and Ayman, 2005). It comprises a range of crops covering fruits, vegetables, flowers, ornamentals, medicinal and aromatic, tubers, spices and plantation crops (GoI, 2001). Presently it is the most lucrative venture providing vast employment opportunities and increased earnings to the farmers. For pushing agricultural growth rates beyond the targeted 4% level, this sector has tremendous potentialities (Bahadur, 2010). Considering these potentials, the government of India has also started giving emphasis on the development of the horticultural sector and more focus towards its important branch covering value addition. Since the 75 years plan development initiatives of this sector were given greater emphasis (Singh et al., 2004; Chand, 1996). These efforts resulted in the second-ranking of India at global level in the production field of vegetables and fruits.

India reserves its uniqueness for its agro-climatic diversity which exists from acute temperate to tremendous tropical areas and helps to grow all kinds of crops in their favorable environment. When we talk about horticultural crops, we have diverse crops that are in cultivation from acute hot to highly cool climate conditions. As per the statistics of horticultural crops, in 2014–15, the production was 280,986 thousand MT out of an area of 23,410 thousand ha. Out of this, the major contribution was from vegetables and flowers with the production of 166,566 and 2143 thousand MT with an area of 9417 and 249 thousand ha, respectively (NHB, 2015). As India is at second place in the production of various horticultural crops in the world, but while we talk about the pioneer in all crops, India is very far behind others. Among the fruits, the country receives its largest production of mango, banana, coconut, cashew, papaya, and pomegranate, and simultaneously, exporting the largest quantity of spices. India also holds the second rank in vegetable production globally, however, it is much behind if we have to provide balanced to every Indian. This condition is prevalent because of the adoption of traditional techniques and practices for cultivation under open fields which thereby results in lower yields and productivity. There is also a lack of full proof crop management technologies against biotic and abiotic stresses. The biotic and abiotic pressure not only decreases the productivity of crops but are accountable for reduced quality produce too specifically during adverse climatic conditions. Many of the farmers spray large amount of different chemicals to mitigate biotic stresses, which not alone escalates the net cost in cultivation but also enhances left over toxicity in the fresh produce, which is regarded as unsafe for health. To overcome, there are different ways. Those can be bringing more area under cultivation, use of hybrid seeds and use of well-proven agro-techniques, etc.

An additional way could be the cultivation of crops in protected environment. Population pressure, climate shift, squeezing land holdings, vulnerability of natural resources and ever-increasing demand for quality food are forcing us to make a shift towards protected cultivation. In the current scenario, the area of protected horticulture is 40,000 ha. The majority of this area is lying in north India. It can be estimated from this data that the full area is still not underutilization for protected horticulture, despite its importance in the creation of massive self-employments and direct correlation with the growth of national economy. Producing vegetable and flowers in protected conditions can easily increase the productivity by 3–5 times when compared with open field cultivation (Sabir and Singh, 2013). This technology is very potent for urban and peri-urban localities of India which can transform into fast-growing future markets for fresh produce supply.

Protected cultivation is the modern approach to mainly cultivate horticultural crops in a manner that can provide high-quality products with increased quantity. This technology has spreaded very fast in the last few decades at global level. It is precisely envisaged for high production, water, and land conservation and environment protection (Jensen, 2002). It involves growing of horticultural crops in a protected environment, where different plant growth-promoting aspects like temperature, relative humidity, light incidence, land, water, nutrition, etc. are maneuvered to attain high productivity in order to have a continuous availability of produce even in off-season also. Adoption of protected cultivation technology can add to the income of the farmers with additional remuneration for quality produce. A total of 115 countries at global level are enrolled in commercial protected cultivation (Sabir and Singh, 2013). The total acreage under protected cultivation is nearly about 623302 hectares while the protected vegetable cultivation area is 402981 hectares. Of this area, the 95,000-hectares area comes under soil-less/hydroponic culture systems (Hickman, 2011).

In India, protected cultivation technology has its roots since the early nineties in a commercial manner. The current area under protected cultivation in India is about 25,000 hectares while protected vegetable production covers 2000-hectare area. The constraints of decreasing land holdings, rapid urbanization, declining crop production, declining biodiversity and ever-increasing population, demand for food, especially vegetables are increased many times and thus to overcome this, protected technology has given a new aspect for producing more in per unit area (Sabir and Singh, 2013).

Nowadays, major exhaustive protected farming systems at the global level are owned by Dutch people who are fetching lofty outputs by employing most modern technologies (Goncharova, 2004). North Indian conditions are well suited for vegetable production because of fertile land, these tracts also face wide mercury range starting from 0° to 48°C during the year and don't allow round the year vegetable cultivation in open environment (Singh et al., 2011). During rainy seasons the crops become susceptible to various biotic agents. It has also been prevalent that high hill areas experience extreme cold –5° to –30°C temperatures and consequently creates problems to grow vegetables during winter months. Thus, protected horticulture, encompassing polyhouses, shade net houses, poly-tunnels, poly-mulches, etc. protects these crops from unexpected divergence in weather and controls the environment in protected structures and thereby reduce the effects of high or low temperature regimes as well as various biotic and abiotic factors (Negi et al., 2013).

Greenhouse/poly house/net houses are most commanding technology under varying climate for round the year and off-time vegetable and flower production. Beside production, this technology can be used for virus-free seedlings and hybrid seed production. The protected cultivation need has widened since last 10 years because this technology have helped in reducing various causes such as weed pressure, increasing moisture conservation, reduction of insect pests, higher production and efficient use of nutrients (Negi et al., 2013). Protected vegetable and flower cultivation are proven very successful in changing climatic conditions but also for stabilizing the unbalanced market prices prevailing almost every year in India. High-value horticultural crop cultivation is a marvelous and very remunerative venture in India. Under the new era of FDI (Foreign Direct Investment) in retail, the protected cultivation models have high prospects for increasing income of the growers who opt for quality and off-season vegetable and flower production (Singh et al., 2012). This technology is being profitably used for producing high value vegetable crops like big fruited tomatoes, cherry tomatoes, colored capsicums, parthenocarpic cucumbers and virus free seedlings (Singh et al., 2010). Though it's various advantages, this technology needs very vigilant planning, awareness, and information about the timeline of production and moreover, harvest time to coincide with high market prices, choice of varieties adopted for the offseason environments.

All types of protected structures may not suit the demand of all groups of farmers, because of different environments and geographical factors. They also have more initial costs in fabrication, running costs, etc. (Singh et al., 2012). Some of the low-cost technologies like low-cost polyhouses, low-pressure drip irrigation system and low-cost nursery raising technology are suited in different areas of the country and provide ample scope of agricultural advancement in the near future.

15.2 PROTECTED CULTIVATION

It is defined as a cropping practice in which the climate neighboring the plant is restricted to cater to the need for proper growth and development of the plant (Mishra et al., 2010). Types of protected cultivation technologies are available according to the widespread climatic conditions. Among them, polyhouse is a really useful technology for the round-the-year production of vegetable and flower crops. It is also known as Controlled Environment Agriculture, which is extremely productive, and protective to the environment (Jensen, 2002).

15.3 PRINCIPLE OF GREENHOUSE CULTIVATION

Greenhouse is an artificial structure having transparent covering material like polythene sheet, glass, fiber sheet, that transmits the sun energy inside it. This sun energy is absorbed by the crops and the objects and thereafter transmittance of light energy in the form of long-wavelength, which does not secrete out of non-transparent cladding material. At the end, this light energy gets trapped inside the structure and increases the surrounding temperature. This phenomenon is known as the greenhouse effect. This temperature rise leads to the fast-growing of vegetable and flower crops in winter months. During the summer period, air temperature in polyhouse is brought down by providing cooling instruments. A commercial greenhouse facilitates to produce desired crops without a gap in a year by maintaining inside temperature, relative humidity, carbon dioxide, photoperiod, temperature and plant nutrients that are needed for plants. Controlled climate and soil situation are given a chance for crops to articulate their yield potentials (Singh et al., 2011). Cladding materials used in greenhouses acts as choosy radiation filter. The solar radiations pass through it and trap the thermal energy inside the greenhouse, which is emitted by the objects that are kept inside.

15.4 WHY PROTECTED CULTIVATION?

Cultivation of vegetable and flower crops in open field conditions face many problems like high or low temperatures and humidity levels, excessive solar radiation, heavy rainfall, thunderstorms (Max et al., 2009), high biotic and abiotic stress pressure (Sringarm et al., 2013; Nguyen et al., 2009). Comparatively protected cultivation is the most appropriate approach for vegetable and flower cultivation. Off time cultivation of crops is possible in protected structures which otherwise is not possible in open fields. Besides, from protection to adverse climatic condition, the produce under protected cultivation is of high quality in aspects of shape, size, and colors (Sringarm et al., 2013). The surrounding climate inside polyhouse can be maneuvered easily utilizing various instruments. Many insects have need of UV light for their visualization; the UV opaque covering materials restrict the entry of the insects. Consequently, there is less use of harmful insecticides.

15.4.1 ADVANTAGES OF PROTECTED CULTIVATION

Protected vegetable and flower production judicially uses water and chemicals compared to open field conditions. The comparative benefits are:

1. Ensures the production of any plant at any place and throughout the year.
2. Overcoming adverse weather for vegetables and flowers production by opting specific systems.
3. Multi-cropping can be practiced on the same area.
4. Off time production is possible.
5. Quality and healthy seedlings production.
6. More crop productivity per unit of area with judicious use of resources.
7. High quality and clean products.
8. Cultivation is possible in remote areas.

9. Vertical cultivation using hydroponics, aeroponics techniques.
10. Seed production of expensive vegetables and flowers is possible.
11. Pesticide residue free produces having good nutrition.
12. Early raising of nurseries and protection of valuable germplasm.
13. Easy crop management.
14. Can be used for maintaining stock plants and cultivating grafted plantlets and micro propagated plants.

15.5 CONSTRAINTS OF PROTECTED/GREENHOUSE CULTIVATION

1. High initial cost investment required for construction and management of greenhouses.
2. A very important consideration in greenhouse cultivation is crop management. The various cultural operations in a greenhouse are intensive in nature and modified at times.
3. Pollination is restricted inside polyhouse for cross pollinated vegetables.
4. High cost involvement and unavailability of construction material for polyhouse at every place.
5. Lack of suitable instruments (tools) and machinery for use in protected cultivation.
6. Quality planting material at reasonable price.
7. Lack of organized marketing of flowers grown under cover has also become a problem as the middlemen are exploiting the produces by quoting low prices despite the better quality of the produce.
8. Lack of adequate cold storage facilities.

15.6 STATUS OF PROTECTED CULTIVATION

Environmentally controlled production of plants is in existence since Roman era. The Roman Emperor in 42 to 37 B.C. used to eat cucumbers daily produced through methods which are produced through protected systems. The growers were planting the cucumbers in handcarts and bringing them in the sun and later bringing them under protection in the homes in the night to save them from various stress elements. By the 16th century, European explorers created greenhouses to grow tropical plants in the temperate climates and the technology progressed from the Italian botanical gardens to wood made structures and then to structures of cast iron, and finally lead to glasshouses. After the arrival of plastics during the 2nd World War, a new segment in protected technology in the form of polyhouse formed. At present nearly 90% of the new greenhouses are being constructed by utilizing ultra-violet (UV) stabilized polythene sheets as the glazing material. Protected cultivation is being commercially practiced in 55 countries of the world for growing crops and this technology is replicating at a fast rate. China ranks first in the use of protected technology. The global scenario of protected cultivation is depicted in Table 15.1.

TABLE 15.1 Total Area Covered Under Protected Structures at Global Level

S. No.	Country	Area
1.	China	2,760,000 ha
2.	Korea	57,444 ha
3.	Spain	52,170 ha
4.	Japan	49,049 ha
5.	Turkey	33,515 ha
6.	Italy	26,500 ha
7.	Mexico	11,759 ha
8.	Netherlands	10,370 ha
9.	France	9,620 ha
10.	United States	8,425 ha

India's protected program came into existence by the advent of the Indo-Israel project which started at IARI, New Delhi in 1998. After the completion of this project, the Israeli experts went back in 2003, IARI, New Delhi itself started maintaining this facility and named it as Center for Protected Cultivation Technology (CPCT). Since then this center has refined many protected cultivation technologies. The area coverage reported under protected cultivation in India is 110 thousand hectares and the global level is 275,000 hectares (Mishra et al., 2010). Since then area coverage has consistently prolonged in India by 10%, the states have also come up with area expansion under protected cultivation are Maharashtra, Gujarat, Haryana, Punjab, Tamil Nadu, and West Bengal. Maharashtra and Gujarat cover 5,730.23 hectares and 4,720.72 hectares area under protected cultivation, respectively. In the European continent, Spain is the leader in protected cultivation covering 51,000 ha area at world level.

15.7 COMPONENTS OF PROTECTED HORTICULTURE

Protected technology has in general two very basic components, which are as follows:

1. Infrastructure; and
2. Production technology of crops.

Infrastructure involves various engineering aspects of protected structure development, including the type of cladding material to be used, supporting structural frames, fertigation and irrigation facilities, working tools, temperature controlling instruments, humidity controlling equipments and ventilation related accessories, etc. These inputs take care of various growth and development factors for a successful crop production like temperature, light, water, nutrients, etc. All these climatic factors can be managed easily with various engineering or electrical tools. The latest in this category is automated systems comprising various sensors. In order to regulate air ventilation which causes an increase in temperature, growth of pathogens inside the protected structures, installation of ventilation devices or through manual adjustment, ventilation can be regulated. For trapping light energy in terms of photosynthetic active radiation and photosynthesis, the choice of cladding material is very important. The availability of this cladding material, its quality, and its cost also play an important role in its selection. Another important component in the infrastructure category is irrigation and fertigation unit. In all these infrastructural aspects continuous upgradation is coming up with the latest discoveries and scientific inputs.

The second component which is production technology of crops is also an important factor because it needs scientific studies to develop the varieties suited for protected cultivation, choosing the type of crops which can give good production and standardizing the production protocols for harnessing bumper yield whether it involves various intercultural operations like training and pruning which are totally different which are adopted in field farming. Various pest and disease problems, weed infestations, nutrition requirements, pollination, harvesting practices, and abiotic stresses are not similar as in open field conditions. These practices are very specific. This field needs the utmost care and planning for scientific studies and in our country, it is still in much-neglected state.

15.8 DESIGN AND ORIENTATION OF PROTECTED STRUCTURES

Various protected structures for cultivation are polyhouse, low tunnel polyhouse, zero energy polyhouse, glasshouse, insect-proof net house, etc. Based on shape also protected structure are classified into lean to type, even span type, multi-span, slope type, etc. Based on utility, they may be temperature controlled and humidity controlled types. Based on construction type they may be of wooden/bamboo, pipe or truss framed and GI pipe framed, etc. Based on cladding materials they may be of plastic-film based, fiberglass-based and glass-based, etc. (Montero et al., 2005). Good agricultural practices require good ventilation and light transmission. In terms of the roof slope, computer simulations show that during the winter, increasing the roof slope from 11 to 45° can increase daily light transmission by nearly 10%. With regard to greenhouse orientation, there are two main factors that have to be balanced before choosing the best solution: light transmission and ventilation. For receiving high light transmission and to have good ventilation, polyhouse should be made in East-West orientation. But for getting uniform light the polyhouses should be built in North-South orientation in order to prevent the shadow of gutter and ridge in the daytime with the movement of the sun.

Higher intensity of cropping and intensive management necessitates higher labor requirements for these structures. The per-unit labor requirement of greenhouse cultivated area (10 men/ha) is more than field cultivation (1 man/ha). Thus, protected cultivation is like a factory-like approach with assured input-output relationships. As greenhouse cultivation is capital intensive, heavy financial investments are necessary in the beginning to construct and furnish the greenhouse with adequate environmental control devices. Depending upon the prevalent weather conditions and type of the crop, the initial investment could be from Rs. 350 to 3500 per square meter of floor area. It again depends upon construction type and material used. Covering materials made of plastic film are also divided as acrylic, polycarbonate, fiberglass-reinforced polyester, polyethylene film and polyvinyl chloride films, etc. (Montero et al., 2005). Plastic coverings have many benefits over glass coverings mainly in terms of cost. Plastics have more adaptation in various greenhouses designing because of its resistance against breakage, less weight and easy to install property.

15.9 TYPES OF STRUCTURES

15.9.1 LOW-COST STRUCTURE/GREENHOUSE

Low-cost types are made up of 700 gauge polythene sheet and come under the category of the zero-energy chamber and framed on bamboos/wooden poles with the help of ropes and nails. The size of these structures is dependent on space availability and purpose such as preventing the crops from rains. These structures are entirely dependent on the light energy from sun. In this type of structures mercury level are 6–10° higher than field conditions. If an UV stabilized sheet is being used as cladding material, there may be higher day temperature but low night temperature. The radiation which passes in inside is 30–40% lesser than the radiation falling on soils outside (Chakraborty and Sethi, 2015).

15.9.2 MEDIUM COST STRUCTURES

Generally they are quonset shaped polyhouses framed with GI pipes of 15 mm bore size and comes up with slight higher cost than low cost structures. They have UV stabilized polythene sheet of 800 gauge thickness with single layer. There is a provision of fan and pad system also for proper ventilation and dehumidification which are thermostatically controlled. They have a life span of about 20 years for frames and 2–3 years for glazing material (Chakraborty and Sethi, 2015).

15.9.3 HIGH COST STRUCTURE/GREENHOUSE

In these structures, the frames are built from galvanized iron/aluminum material with dome shape. Automated systems are installed which control humidity, temperature, CO_2 and light as per the need of plants and growers with various automated instruments. Floors and some side part are built with concrete. They have long durability and are about 5 times costlier than other structures. They require a qualified workforce, proper maintenance, care, and precautions while operating.

15.9.4 SHADE-NET HOUSES

Shade-net house is made of materials GI pipes, angle iron, wood or bamboo. It is a covered structure with plastics nets that are made of 100% polyethylene thread with specialized UV treatment having different shade percentages. It provides a partially controlled environment by reducing light intensity and heat during day time. Hence, year-round seasonal and off-time production is possible under these structures. Shade-nets come up in different shade factors, i.e., 15%, 35%, 40%, 50%, 75%, and 90%. To create optimum conditions for healthy crop growth, selection of the correct shade factor is an important criteria to enhance plant's productivity. Shade-net structures are generally recommended for raising horticultural crops in high rainfall areas. Cladding material is used for roofing. Sides are covered with wire mesh of various sizes. These structures show usefulness in North-Eastern Hilly areas (Nair and Barche, 2014).

15.10 SITE SELECTION FOR PROTECTED STRUCTURES

Protection is needed against high radiation from the sun in protected structures. Strong winds, heavy rain, and hail-storm are other important criteria's while selecting a site. The soil type, profile of the soil and location should be of high standards. Structures should be at a safe distance from industrial areas. Leveled land and high light-receiving areas are also prerequisites. If they are to be built on a slope area, there must be a provision of surface runoff. Proper drainage is also necessary. Continuous and enough water supplies with electrical source are also required near selected site.

15.10.1 IN USE PROTECTED STRUCTURES IN INDIA

1. Polyhouses with natural or modified ventilation.
2. Insect proof net houses.
3. Insect proof net cages.
4. Polyhouses with fan and pad system.
5. Walk-in tunnels (mostly used in dry temperate areas of Himachal Pradesh).
6. Solar greenhouses or Leh design greenhouses.
7. Poly low tunnels.
8. Underground trenches (mostly used in Leh and Ladakh area).

15.10.2 CAUSES OF GREENHOUSE DAMAGE

1. The profile used for frame, trusses and other places in protected structure is too light which is deformed by strong winds.
2. Structure not bearing the load of cladding material.
3. Poly film split due to sharp edges.
4. Not sufficiently secured foundation against uplifting forces.
5. Ventilation openings have a start point for damage of polyfilm.

15.11 PROVISION OF SUBSIDY BY THE GOVERNMENT

Government of India is providing 50% subsidy (Table 15.2) on total expenditure with a maximum cut off limit up to 4000 m^2 per beneficiary for adoption and installing greenhouses under National Horticulture Mission (NHM) (Nair and Barche, 2014).

TABLE 15.2 Subsidy by Government for Installing Greenhouses in India

Items	Pattern of Assistance (Rs/m^2) (Limited to 4000 m^2 per Beneficiary)		
	Tubular	Wooden	Bamboo
Greenhouse (Fan and Pad cooling System)	1465	-	-
Naturally ventilated polyhouse	935	515	375

15.12 CHOICE OF CROPS FOR PROTECTED HORTICULTURE

The size and economics of crop production determines the selection of crop to be grown inside polyhouse. That is why high value horticultural crops are more popular in the greenhouses. Colored capsicums, parthenocarpic cucumbers, cherry tomatoes, big fruited tomatoes, lettuce, spinach, beans, roses, gerbera, carnation, orchids, anthurium, chrysanthemum, strawberry, pot plants, etc. are preferred for protected cultivation in India and many countries. In addition, nursery seedlings, grafted planting material and hardening of tissue culture plants are other important commercial aspects of greenhouses technology.

15.13 PLANT CONTAINERS IN PROTECTED HORTICULTURE

The crop duration inside greenhouse is an important factor for making protected technology profitable. Short duration crops are preferred generally for protected cultivation. To minimize the pathogen load available in the soil with continuous cropping, containers are healthy choice.

15.13.1 BENEFITS OF USING PLANT CONTAINERS

1. Uniform and vigorous plant stand.
2. Provide quick and vigorous plant growth with less or without transplanting shock.

3. Easy maintenance of hygiene in greenhouse.
4. Easy to handle and grading.
5. Easy to shift for distant transportation.
6. Provide better drainage and aeration to the plants.
7. Easy monitoring of chemical characteristics of growing media.
8. Easy supply of required plant nutrition is possible.

15.14 IRRIGATION AND NUTRITION SYSTEMS FOR PROTECTED STRUCTURES

The entire growth and development of plants are dependent on water and nutrients. The plants with small root systems require high amount of photosynthates to be produced for getting handsome quantity of flowers per unit of area employing handy foliar systems. Hence regular and steady water and nutrient supply is must for steady production inside green-house system and the best option is to have drip irrigation cum fertigation system.

15.15 SYSTEMS FOR WATERING AND NUTRIENT SUPPLY

Micro-irrigation system can be used as a perfect choice for irrigating plants inside protected structure or greenhouse. The major challenge is to prevent the dispersal of water droplets on the leaves or flowers which otherwise create problems in the form of diseases and scorching from sunlight. To cope up with this challenge drip irrigation is best alternative inside protected structures. In this system of drip irrigation the injection of water is through nozzles which come out with high pressure with the help of a fitted motor. Drip irrigation system equipments are pressure pump for creating 2.8 kg/cm^2 pressure, Sand or silica or screen water filters and drip lines with mainline, sublines, dripper and emitters. Various types of drippers include Labyrinth, Turbo, Pressure compensating (with a membrane of silicon for steady flow rate), Button (easy to clean, good for pots and orchards) and pot drippers (long tube cones). The water discharge rate varies with the type of dripper used, few examples are given below:

1. 16 mm type for 2.8 kg/cm^2 pressure discharges and flow rate of 2.65 liters water per hour.

2. 15 mm type for 1 kg/cm^2 pressure discharges and flow rate of 1–4 liters water per hour.

15.15.1 TYPES OF FILTERS

Depending upon the available water source and its quality, different kinds of filters are used, which are:

1. Gravel filter: Water of open canals and reservoirs having organic impurities and algae, etc. are filtered through these types of filters with the help of beads of basalt or quartz.
2. Hydrocyclone: Used for filtration of river water having sand particles.
3. Disc filters: Used to filter fine particles mixed in water.
4. Screen filters: Made up of stainless steel filter having 120 mesh (0.13 mm) size. Generally used in second stage filtration.

15.16 SYSTEM OF FERTIGATION

An automatic mixing and distribution system is installed in the fertigation unit, which consists of pump and a mixture supplying device. The fertilizers are mixed individually in various tanks in a required ratio and then transported to the plants with the help of installed drippers.

15.17 POLYHOUSE HEATING

During the winter season, the temperature falls below the actual requirements of the plants, in that case, we need polyhouse heating mechanisms. Normally, sunlight is enough to maintain the mercury levels but in few cases according to the need of crop we need to heat the polyhouse for maintaining the optimum temperature for proper growth and development. Some methods generally employed for heating the polyhouse are:

1. Provision of underground heating tunnel in the protected structure.
2. Northern wall of polyhouse should be covered with jute cloth.
3. Covering of whole polyhouse with jute cloth in night.
4. Installation of solar panel driven devices in polyhouse.
5. Use of heating blowers.

15.18 POLYHOUSE COOLING

Cooling is required when mercury rises above 40°C inside the protected structure at noon in summer months. Side by side the relative humidity can also be maintained by following various measures which are:

1. Proper ventilation practices such as opening the side curtains in order to allow fresh air during the day time.
2. Exchanging of inside air with fresh air by the provision of fans.
3. Fitting of cooling system on Eastern or Western wall not only keeps down temperature but also maintains proper humidity in polyhouse.
4. Misting and fogging provisions for controlling the relative humidity and temperature.

15.19 CLADDING MATERIAL FOR PROTECTED STRUCTURES

Polythene is very good alternate option available at cheaper rates. Different forms of cladding materials are coming up now a day's which varies on their strength (heavy duty, unbreakable, light weight), UV stabilization and transparency. Among these new materials few are fiber glasses and polycarbonate panels. In humid and the areas receiving generally high temperature, plastics are mostly preferred against fiberglass because of their economical feasibility. Another forms which creates micro-ecosystem inside protected structures for proper growth conditions for plants in plastic cladding materials are LDPE (low density polyethylene)/LLDPE (linear low density polyethylene) which stay in original form up to 3–4 years.

15.20 ADVANTAGES OF GROWING ORNAMENTAL PLANTS UNDER GREENHOUSE IN INDIAN CONTEXT

- Abundant sunshine is available in the whole year especially in winters.
- The average radiation received at Nairobi is 462 cal/cm^2/day, which is at 1800 m AMSL and regarded as the best center for production of quality cut flowers, which is at par with radiation received at Bangalore (450 cal/cm^2/day at 1000 m AMSL).
- Ideal temperature.
- Shorter production cycle.

- Good production during the main international events when demand for flowers is high in International market.
- Diverse agro-ecology which helps in growing of any flower in India.

15.21 PROTECTED CULTIVATION AT BIHAR AGRICULTURAL UNIVERSITY (BAU), SABOUR

Bihar Agricultural University (BAU) started major initiative in establishment of infrastructure for hi-tech horticulture in the main campus at Sabour during the year 2011. Protected cultivation of vegetables and flowers in Bhagalpur district is a recent trend in hi-tech horticulture for production of high value crops, i.e., vegetables like tomato, colored capsicum, parthenocarpic cucumber, cherry tomato and flowers like Anthurium, Orchids, Carnation, Gerbera, Chrysanthemum, Dutch Roses, etc.

15.22 OBJECTIVES OF THE UNIT

1. To conduct research on different aspects of hi-tech horticulture.
2. To standardize technology for production of protected cultivation of flowers and vegetables.
3. To teach students about hi-tech horticulture activities.
4. To demonstrate technology to the farmers.
5. To develop human resource in this field of protected cultivation.

Infrastructure Facilities of the Unit

Sl. No	Infrastructure Facility	No. of Units	Total Area
1.	Naturally ventilated poly house	07	5140 m^2
2.	Shade net house cum polyhouse	03	1420 m^2

15.23 RESEARCH ACCOMPLISHMENTS

High value vegetable and flower crops and their varieties were evaluated in protected cultivation/polyhouse. The findings in brief are given hereunder.

- Cultivars Mini Angel, Infinity, RS 03602833, Aviva and KPCH-1 of parthenocarpic cucumber were evaluated in polyhouse. Cultivar RS 03602833 performed well in terms of fruit length (18.13 cm), fruit diameter (4.80 cm) and yield per vine (3.63 kg).

- In Orchid, cultivar D. Sonia produced maximum number of flower sticks and gave more earnings.
- In Anthurium, the cultivar Xavia followed by Angel performed better with respect to growth, quality and yield of spikes.
- In gerbera, 10 varieties viz. Laura, Szantal, Delfin, Newada, Olympia, Kormoran, Partrizia, Rock, Feliks, Samuraj were evaluated. Among all of these varieties patrizia produced maximum flowers per plant with longest stalk length (Singh, 2017).
- Various breeding advanced lines of parthenocarpic cucumber have been developed utilizing male flower induction with the use of silver thiosulfate solution @300 ppm.

Concluded RKVY Project:

1. Protected cultivation of vegetables and flowers in Bihar.

Ongoing Research Projects:

1. Selection of suitable vegetable and flower cultivars for protected cultivation in Bihar.
2. Development of parthenocarpic gynoecious lines in cucumber (*Cucumis sativus* L.) for protected cultivation.

15.24 CONCLUSION

Protected cultivation technology is a recent and a fast emerging technology. Seeing the vagaries of climate shift, it will turn a life-saving strategy owing to its various advantages. Popularizing this technology and doing need-based research is the need of the hour in order to fetch maximum returns in the near future utilizing available know-how and proper upkeep methods. For Indian farmers, this technology can help in making lucrative returns from various high-value crops and will impart enough calibers to compete at the International level. Pictorial representation of protected cultivation at BAU, Sabour is given in Figure 15.1.

(a) Polyhouse complex.

(b) Practical classes in polyhouse complex.

(c) Orchids in bloom.

(d) Dendrobium cultivar of orchid-Sonia.

(e) Anthurium var. Xavia.

(f) Farmers training on protected cultivation.

FIGURE 15.1 *(Continued)*

(g) Chrysanthemum in full bloom.

(h) Parthenocarpic cucumber at flowering.

(i) Bed preparation and transplanting.

(j) Ready for sale seedless cucumber.

(k) Carnation cultivar-Luna.

(l) Carnation cultivar-Red King.

FIGURE 15.1 *(Continued)*

(m) Rose cultivar- Sultan.

(n) Capsicum variety - Buffalo.

(o) Capsicum crop in polyhouse.

(p) Capsicum cultivar-Nikita.

(q) Tomato crop at fruiting stage.

(r) Profuse fruiting of tomato.

FIGURE 15.1 Photo gallery.

KEYWORDS

- **Center for Protected Cultivation Technology**
- **cultivation**
- **fertigation**
- **greenhouse**
- **polyhouse**
- **ultraviolet**

REFERENCES

Abou-Hadid, & Ayman, F., (2005). *High Value Products for Small Holder Markets in West Asia and North Africa: Trends, Opportunities and Research Priorities*. The Global Forum on Agricultural Research, No. 191, FAO, Vialedelle Termedi Caracalla, Roma (Italy).

Bahadur, S., (2010). *Horticulture-Key to India's Agriculture Growth.* Available at: http://www.commodityonline.com/news/horticulture-key-to-indias-agriculture-growth-34627–3–34628.html (Accessed on 25 November 2019).

Chakraborty, H., & Sethi, L. N., (2015). Prospects of protected cultivation of vegetable crops in North Eastern hilly region. *International Journal of Basic and Applied Biology, 2*(5), 284–289.

Chand, R., (1996). Ecological and economic impact of horticultural development in the Himalayas: Evidence from Himachal Pradesh. *Economic and Political Weekly, 31*(26), A93–A99.

GoI, (2001). *Report of the Working Group on Horticulture Development for the Tenth Five Year Plan.* TFYP Working Group: No. 14/2001, Planning Commission, Government of India.

Goncharova, N. A., Vander, V. J. A. A. M., & Verstegen, (2004). Changes in horticulture sector in the Netherlands. *Acta Hort, 655*, 319–331.

Hickman, G. W., (2011). *A Review of Current Data on International Production of Vegetables in Greenhouses* (p. 73). w.w.w.cuestaroble.com.

Jensen, M. H., (2002). Controlled environment agriculture in deserts tropics and temperate regions– A world review. *Acta Hort, 578*, 19–25.

Max, J. F. J., Horst, W. J., Mutwiwa, U. N., & Tantau, H. J., (2009). Effects of greenhouse cooling method on growth, fruit yield and quality of tomato (*Solanum lycopersicum* L.) in a tropical climate. *Sci. Hortic., 122*(2), 179–186.

Mishra, G. P., Singh, N., Kumar, H., & Singh, S. B., (2010). Protected cultivation for food and nutritional security at Ladakh. *Defense Science Journal, 61*(2), 219–225.

Mittal, S., (2007). *Can Horticulture be a Success Story for India?* Working Paper No. 197, New Delhi: Indian Council for Research on International Economic Relations.

Montero, J. I., Munoz, P., Anton, A., & Iglesias, N., (2005). Computational fluid dynamic modeling of night-time energy fluxes in unheated greenhouses. *Acta Horticulturae, 691*(1), 403–409.

Nair, R., & Barche, S., (2014). Protected cultivation of vegetables-present status and future prospects in India. *Indian Journal of Applied Research, 4*(6), 245–247.

Negi, V. S., Maikhuri, R. K., Rawat, L. S., & Parshwan, D., (2013). Protected cultivation as an option of livelihood in mountain region of central Himalaya India. *International Journal of Sustainable Development and World Ecology, 20*(5), 416–425.

Nguyen, T. H. N., Borgemeister, C., Max, J., & Poehling, H. M., (2009). Manipulation of ultraviolet light affects immigration behavior of *Ceratothripoidesclaratris* (Thysanoptera: Thripidae). *J. Econ. Entomol., 102*(4), 1559–1566.

NHM, (2013). *Report of the Joint Inspection Team on its Visit to Karnataka During 3rd to 12th January, 2013 to Review the Progress Under the National Horticulture Mission.* Ministry of Agriculture, Department of Agriculture & Cooperation, National Horticulture Mission, New Delhi.

Sabir, N., & Singh, B., (2013). Protected cultivation of vegetables in global arena: A review. *Indian Journal of Agricultural Sciences, 83*(2), 123–135.

Singh, B., Singh, A. K., & Tomar, B. S., (2010). In peri-urban protected cultivation technology to bring prosperity. *Indian Horticulture, 55*(4), 31–32.

Singh, B., Singh, A., & Kumar, M., (2012). In urban areas: Protected cultivation of vegetables. Phal Phul (Fruits and Flower), *33*(3), 3–6.

Singh, H. P., Prem, N., Dutta, O. P., & Sudha, M., (2004). *State of the Indian Farmer* (Vol. 11). New Delhi: Academic Foundation.

Singh, P., Bhardwaj, A., Kumar, R., & Singh, D., (2017). Evaluation of gerbera varieties for yield and quality under protected environment conditions in Bihar. *Int. J. Curr. Microbiol. App. Sci., 6*(9), 112–116.

Singh, W., Mushtaq, A., & Dar, (2011). Protected cultivation of tomato, capsicum and cucumber under Kashmir valley conditions. *Asian Journal of Science and Technology, 1*(4), 56–61.

Sringarm, K., Johannes, F. J., Max, S. S., Wolfram, S., Kumpiro, S., & Müller, J., (2013). *Protected Cultivation of Tomato to Enhance Plant Productivity and Reduce Pesticide Use.* Conference on International Research on Food Security, Natural Resource Management and Rural Development organized by the University of Hohenheim Tropentag Stuttgart, Germany.

CHAPTER 16

Use of Fly Ash in Agriculture

SANKAR CH. PAUL

Department of Soil Science and Agricultural Chemistry,
Bihar Agricultural College, Bihar Agricultural University, Sabour,
Bhagalpur, Bihar, India, E-mail: scpaul64@rediffmail.com

ABSTRACT

Fly ash is a coal combustion byproduct at thermal power plant has been considered as hazardous solid waste material. Its characteristic depends on several factors like nature of parent coal, combustion processes, nature of emission control devices, methods of handling and storage. It is mainly composed of Si, Al, Fe and Ca oxides with an average of 95 to 99%. Remaining parts are other macro and micronutrients except nitrogen and carbon. Therefore, it may be utilized as an important soil amendment or as a source of cost effective essential plant nutrients material that can improve various physicochemical and biological characteristics of soil. It minimizes the use of chemical fertilizers, soil amendments and quantity of irrigation water for crop production. Its use increases Si, P, K, Ca, B and other nutrients uptake in various crop plants. Several field studies reported that many crops successfully can be grown in waste land soil by using fly ash and edible parts of the crops are having within the safe limit of toxic elements along with good consumer acceptability by satisfying standards of food quality. Therefore, fly ash has immense potential in improving soil fertility and crop productivity in agriculture.

16.1 INTRODUCTION

Fly ash is a by-product of coal combustion at thermal power plants has been considered as challenging hazardous solid waste over the globe. Indian coal belongs to sub-bituminous, bituminous and lignite quality. Combustion of coals like bituminous, sub-bituminous and lignite for production of electricity in thermal plants generates this solid fly ash. Particles fly ashes are very small and get easily suspended in the atmosphere. Frequent exposure to ash causes irritation of eyes, nose, throat skin, and respiratory tract. Hence, careful assessment of immense quantity of fly ash needs for safe removal and utilization from thermal power generating plants. Therefore, fly ash management would remain a gigantic alarm of the coming years. It is not fully burnt and consists of mineral constituents in the coal. It requires large dimension of land, water, and energy for deposit. The physicochemical characteristics of fly ash differ with the nature of parent coal, the process of combustion, nature of emission control devices, methods of handling and storage. As a result, ash formed by burning of different coals has different compositions. Utilization of high ash containing (30–50%) coal in thermal plants leads to indiscriminate dumping of this hazardous industrial solid waste over the years (Basu et al., 2009). The coal ash has been recognized as a Green List waste under the Organization for Economic Cooperation and Development (OECD) rather a waste material. However, this industrial byproduct has been utilized inappropriately in many countries rather ignored like a waste material. About 112 million tons of dry coal fly ash produced annually in India from coal-burning for electric power generation. So, coal ash management becomes a serious environmental and human problem for India and it needs a mission mode approach. A large number of appropriate technologies and confidence-building programs have been developed across the country for the removal and utilization prospects of fly ash. At present, fly ash is being used in dike construction, landfilling and in other construction industries. It is used for the manufacturing of bricks, concrete blocks, ceramics, landfill application, road construction, insulating bricks, agriculture, recovery of metals and dam construction in several countries. Many investigations under field and pot culture on the potentiality of fly ash as an amendment in agricultural applications have been conducted by various research agencies, institutes at different places over the world. Here, utilization of coal fly ash in crop production and soil health is highlighted with the aim of serving opening up the usage of ash and reducing the hazardous effect of disposal on environment and economy.

It consists of Si, Al, Fe, Ca oxides with an average of 95–99%. Remaining part of is consists of macronutrients like P, K, Ca, Mg, S, and micronutrients like Cu, Fe, Zn, Mo, Mn, B, etc. except nitrogen and carbon.

The significant amount of coal fly ash is used in agriculture and development of wasteland that receives a potential to increase in yield of crops. Large amount of fly ash can be used for regeneration of wasteland soil/forestry plantations. Fly ash may be utilized as important soil amendments in the cultivation of Jatropa which is being cultivated on large scale in the country for production of bio-diesel.

With incorporation of fly ash as soil ameliorant or as a source cost effective essential plant nutrients with fertilizers and decomposed organic substances relating to the soil health and yield of crops, shows the followings:

- It facilitates to improve soil texture and decreases bulk density of sandy soil.
- It enhances soil permeability status and can conserve soil moisture content.
- It improves amount of pore space and water holding capacity of light textured soil, thereby improves aeration of soil.
- It optimizes soil reaction status.
- It decreases crust formation in soil.
- It improves soil fertility status and crop yield.
- It adds macronutrients and micronutrients in soil.
- It acts as a liming material for the reclamation of acidic soils and also can perform as the alternate of gypsum for reclamation of saline alkali soil.
- It lowers pest incidence.
- Crops produced on soil incorporated with fly ash are safely consumable for human beings and groundwater quality remains safe.

Considering these indispensable findings, fly ash is considered for agricultural crops, forestry crops, management of problem soils and/or wasteland soil management at 100 to 200 tons per hectare on one-time basis coupled with optimal dose of manures and fertilizers. The repeat application of ash in a same piece of land for long time would have considerable residual effect on succeeding crops.

Fly ash was applied successfully in many agricultural projects in several countries, like Australia, Germany, India, Japan, South Africa, UK, and USA. As a result, it minimizes use of fertilizer, soil amendment, and quantity of irrigation water. Combination of coal ash and organic substance produces significant effect in soil towards more fertile and helps to grow plant more easefully. Study has also revealed that combined application of fly ash and organic additives can be highly effective. Earthworm composted fly ash not only increase the crop production but also reduces the application of chemical fertilizers thereby minimizes the cost of cultivation. It can be used as pesticide in agriculture because of its fineness.

The fly ash application increases the uptake of Si, P, and K by the rice plants, but did not result in an excessive uptake of heavy metals in the submerged paddy soil. It is healthy substitute to other inorganic soil amendments to enhance the important nutrient balance in rice soils. Fly ash in combination with gypsum reclaims saline-alkali soils, resulting in saving of gypsum (50–75%). Various workers reported that low dose (2–4%) of fly ash has shown an enhancement of average plant height, root length, yield and biomass of wheat (*Triticum vulgare*), pea (*Pisum sativum*), gram (*Cicer arientinum*) and mustard (*Brassica junceca*), etc. Many crops were successfully grown in wasteland soil treated with fly ash, such as, paddy, wheat, maize, sunflower, tomato, potato, sugarcane, cabbage, onion, pea, carrot, okra, berseem, medicinal. Edible parts were assessed to verify quality, consumer acceptability, uptake of toxic heavy metals and it was found to be the uptake of toxic elements within the safe limits and satisfy the food quality standards. Many field experiments and pot culture studies on the potentiality of fly ash use as an amendment in agricultural applications have been performed by various agencies, research institutes all over the world. Therefore, the introduction of fly ash in agriculture is highlighted with the aim of helping opening up the usage of fly ash and reducing the environmental and economic impacts of disposal.

16.2 FLY ASH UTILIZATION AND DISPOSAL

Numerous technologies have been developed for beneficial use of fly ash and its safe management through research projects funded by Fly Ash Unit under the Ministry of Science and Technology, Government of India, since 1994. Propagation of these technologies would facilitate and encourage the fly ash utilization in the country by creating 'Self-sustaining technology demonstration centers' (CEA, 2015–16). The use of fly ash has increased from 6.64 million tons in 1996–97 (Tiwari et al., 2016) to the extent of 102.54 million tons in 2014–15. The highest quantity of fly ash utilization was achieved to 62.6% in the year 2009–10 and it was about 58.48% in the year 2011–12, about 61.37% in the year 2012–13, 57.63% in the year 2013–14 and 55.69% in 2014–15. During the 1st half of 2015–16, fly ash utilization of was 56.04% which is behind the fixed target (CEA, 2015–16).

Fly ash utilization data received from Thermal Power Stations/Power Utilities for the 1st half of the year 2015–16 has been compiled to determine the modes in which fly ash was utilized and the amount utilized in each mode. During the 1st half of 2015–16, the use of fly ash was maximum to the level of 41.97% of total fly ash utilized in the Cement sector, followed by 12.85% in bricks and tiles making, 11.21% in low lying area reclamation, 10.91% in mine filling, 7.67% in ash dyke raising, 4.87% in roads and embankments, 2.15% in agriculture, 1% in concrete and 7.32% in others etc.

16.3 CLASSIFICATION OF FLY ASH

Type of coal is one important factor that has the greatest effect on the characteristics of fly ash. The rank wise coals were arranges in order of carbon contents is: Peat < Lignite < sub-bituminous coal < bituminous coal < anthracite. Indian coal is of mostly sub-bituminous type followed by the rank of bituminous and lignite. The ash content in Indian coal ranges from 35% to 50%.

Fly ash has been categorized into two classes, class-F, and class-C based on the content of silica, alumina and iron oxide in it. Burning of older and harder anthracite, bituminous coal typically produces class F type fly ash. But younger lignite and sub-bituminous coals produces class C type ash which usually contains a significant amount of calcium hydroxide (CaO), also known as lime (Upadhyay and Kamal, 2007). Wang and Wu (2006) grouped fly ashes into two classes, i.e., class F that developed from anthracite, bituminous and sub-bituminous types of coals containing less than 7.0% calcium oxide, and class C formed by burning of lignite coal containing a considerable amount of liming material, up to 30%.

16.4 EFFECTS ON AGRICULTURE

According to Basu et al. (2009), fly ash has immense potentiality in growing crops because of its effectiveness in modification of entire soil characteristics for better crop performance. The considerable amount of essential plant nutrient elements like P, K, Na, Zn, Ca, Mg, Fe, Cu, Mn, B, etc. in fly ash enhances growth and yield of numerous of agricultural crops. But compared with other sectors, very limited quantity of the fly ash is used in agriculture. Going through various important research articles, researchers also expected that substitute of lime with fly ash in agriculture can lessen net CO_2 release in acid soil, thus reducing global warming. They reported further that fly ash has low bulk density (1.01–1.43 g cm^{-3}), hydraulic conductivity and specific gravity (1.6–3.1 g cm^{-3}) (Roy et al., 1981; Tolle et al., 1982; Mattigod et al., 1990).

Bhattacharya and Chattapadhyaya (2004) studied the possibility of augmenting the nitrogen status in the mixtures of fly ash and organic substances by implementing vermicompost technology. The used different ratios of fly ash and cow dung viz. fly ash alone, cow dung alone and fly ash + cow dung at 1:1, 1:3 and 3:1 ratios which were nurture with and without epigeic earthworm (*Eisenia foetida*) for 50 days. Among the three combinations, highest availability of N was recorded in 1:1 mixture of vermicomposted fly ash and cow dung.

Vimal Kumar et al. (2005) studied on fly ash utilization in varying agroclimatic conditions along with different soil-crop combinations and reported significant increase in the production of edible plant biomass without undesirable consequence on soil health and crop quality since the reasons including occurrence of negligible toxic heavy metal elements and radioactive elements in fly ash. They also inked that fly ash application to agricultural field supplements the crop produce with calcium and iron which are excellent for development of human beings with respect to nutritional point of view.

Yavarzadeh and Shamsadini (2012) reported on wheat crop that the fly ash application @ 100 t ha^{-1} along with optimum dose of NPK was better over optimum dose of NPK fertilizer alone and important soil properties also altered in the similar way. It was noticed that fly ash is beneficial for soil and crop when fly ash applied on soil more than 10% by weight (Adriano et al., 1980; Matsi and Keramidas, 1999; Scotti et al., 1999; Asokan et al., 2005). According to Adriano et al. (1980), wheat grain yield decreased at fly ash treatment @ 300 t ha^{-1} as compared to the normal farmers practice and also it was undesirable for soil health.

Aggarwal et al. (2009) studied on the effect of fly ash and nitrogen on germination, growth parameters and yield of wheat as well as sorghum crops and reported that germination percentage and early growth was adversely affected in wheat but increasing amount of fly ash did not cause any harmful effect in sorghum. But grain yield of both was slightly increased at higher levels of nitrogen and ash, i.e., 20 t ha^{-1} fly ash + 120 kg ha^{-1} N (wheat) or 40 kg ha^{-1} N (sorghum).

Many researchers (Garg et al., 2005; Page et al., 1979; Grewal et al., 2001; Hill and lamp, 1980; Martens, 1971; Basu et al., 2006; Sridhar et al., 2006; Thetwar et al., 2007) demonstrated that fly ash improved crop yield of wheat, alfalfa, barley, Bermuda grass, sabai grass, mungbean, white clover, etc. and improved the physicochemical characteristics of soil. Furr et al. (1978) demonstrated that field corn, alfalfa, sorghum, millet, carrots, onion, beans, cabbage, potatoes, and tomatoes grew well on slightly acidic soil, treated with unweathered fly ash @ 125 MT ha^{-1} and these crops tested higher concentration of As,

Se, Mg and B. Higher rate seed germination and root length of lettuce was found at application of weathered coal ash @ 5% (Lau and Wang, 2001).

Fly ash has immense potential in improving production of agricultural crops and fertility status of soil. The Indian fly ash is generally alkaline in nature and having macro and micronutrients in it and thus, improves soil quality. In fact, fly ash composed of all essential plant nutrient elements as present in soil except nitrogen and organic carbon. Central Power Research Institute, Bengaluru has developed some hollow and porous type globules from fly ash. If these globules are placed around crop, it absorbs lots of the moisture and retains it for longer period of time in the soil by resisting evaporation loss. This application facilitates to expand the gap between two irrigation cycles. It may also be considered as an insecticide and if it is incorporated with organic waste, it markedly supplements the efficiency of chemical fertilizers. The fly ash application in forestry crops and agriculture production is suitable because of its encouraging physicochemical properties including appreciable amounts of essential plant nutrient elements (Page et al., 1979). Coal fly ash has wonderful influence on soil physicochemical and biological properties and soil processes which have good effect on plants growth and development (Ukwattage et al., 2013).

Pandey et al. (2009) found different translocation system and accumulation of hazardous metals into edible parts of *Cajanus cajan* when grown on fly ash amended soil. Amendment with fly ash at ratios ranged between 25 and 100% increased soil pH from 3.47% to 26.39%, particle density from 3.98% to 26.14%, porosity of soil from 37.50% to 147.92% and maximum water holding capacity 163.16% to 318.42% in comparison to control. This amendment also decreased bulk density from 8.94 to 48.89% in the fly as amended soil as compared to non-ash amended soil.

Arivazhagan (2004) conducted research project on fly ash application on the cultivation of rice, wheat, maize, mustard, ragi, red gram, potato sugarcane, and banana crops and found to be increase in yield of all crops at fly ash level 50 ton per hectare over control.

Singh et al. (2012) found to be decrease in the content NH_4^+ -N, NO_3^--N, total N, organic carbon, organic matter, available phosphorus, and cation exchange capacity of soil in the post-harvest of rice that had been amended with fly ash (0–20%). Reduced content of NH_4^+-N and NO_3^-N at different rates of fly ash application was also reported by Singh and Agrawal (2010). Lee et al. (2006) reported increased content soil pH and increased the availability of phosphorus, silicon, and other mineralogical components, in a paddy soil that was amended with fly ash. They concluded that fly ash can improve the nutritional balance in soils of the paddy field (Lee et al., 2006).

Addition of fly ash stimulates activity of soil enzymes viz., urease, phosphatases, and dehydrogenase, etc. (Pati and Sahu, 2004). Amending fly ash in soil adds many essential elements (e.g., P, K, Ca, Mg, Fe, Zn, Cu, Mn, and B), which may alter the physico-chemical and biological characteristics of soil (Yeledhalli et al., 2007). Fly ash utilization in soil has become popular worldwide in the past few decades (Singh and Agrawal, 2008). More recently, researchers have investigated on the impact of fly ash on soil health and quality, especially on soil-microbial interactions and transformation of soil nutrients (Sarkar et al., 2012).

Generally, the addition of fly ash declines bulk density of soil, thus reducing soil porosity and increases water-holding capacity (Page et al., 1979; Pandey and Singh, 2010). Fly ash amended with sandy and silty soil increased water holding capacity by 8% as it does not able to retain water well (Chang et al., 1977). Singh and Agrawal (2010) observed when fly ash was used to modify soil, the level of soil available nutrients like potassium, calcium, magnesium, sodium, iron, etc. improved significantly. But Aitken and Bell (1985) reported the restricted use of fly ash in agriculture due to its high boron content. However, this problem with higher boron content in fly ash can be overcome if it is properly weathered. The mobility of calcium, magnesium and hydroxide ion in soil increased due to its of liming characteristics, which sequentially boosts bacterial growth (Surridge et al., 2009). However, high levels of toxic elements that may be deposited to the soil through addition of fly ash (Page et al., 1979) can hamper normal microbial metabolic processes (Pandey and Singh, 2010).

Padhy et al. (2016) studied paddy cultivation with varying levels of fly ash *viz.* 0.5, 1.0, 2.0, 4.0, 8.0, 10.0 kg m^{-2} and reported to increase in different growth parameters and yield of paddy grain up to 8.0 kg fly ash per meter square area. Pigment contents of leaf and enzymes activity found to be enhanced by fly ash application, and attained maximum at 10.0 kg fly ash per meter square area. Maximum protein content was found in rice seeds at 4.0 kg fly ash per meter square area. Important soil properties *viz.* soil pH, electrical conductivity, cation exchange capacity, water holding capacity, silt and clay percent and content of organic carbon improved with fly ash application. Combined application of fly ash and N_2-fixing cyanobacteria resulted further significant increase in most soil characteristics, different plant growth parameters and yield attributes. Combination of fly ash at 4.0 kg per meter square area and cyanobacteria. 1000 rice grain weight was maximum at fly ash level 4.0 kg m^{-2} combined with cyanobacteria addition. Soils and plants accumulate of phosphorus, potassium, iron and several micronutrients *viz.* Mn, Ni, Zn, Cu, Co and toxic elements like Pb, Cd, Cr, etc. as a result of fly ash addition, but the content Na remained almost same in soil and grain. Addition of cyanobacteria showed an ameliorating action on the content of toxic metals in soil and various plant parts. Cyanobacteria supplementation of 1.0 kg m^{-2} with fly ash 4.0 kg m^{-2} may be the best combination, since soil would be recharged with essential nutrients and toxic chemicals relative to lesser content, and cyanobacteria would cause minimizing toxic chemical loads in soil and plant parts with N_2-fixation.

Tripathi et al. (2009) conducted a field experiment on the yield and nutrition of wheat, maize, and eggplant in a sequence to see the effect of fly ash and reported that yield of grain and straw of first crop (wheat) was increased by 29.4% and 26.6% respectively over control. The residual effect of fly ash was also observed with yield increase of maize grain by 33.1% – (2nd crop) and eggplant by 18.4% (3rd crop). Thus, it clearly exhibits that appropriate amount of fly ash as soil amendment can help to enhance and sustain agricultural crop productivity. Likewise, Singh and Agrawal (2010) evaluated the effect of fly ash on growth various growth parameters of three important leguminous cultivars, mungbean and recorded significantly positive response of all growth parameters at 10% fly ash treatment for each cultivar tested.

Sinha and Gupta (2005) studied on the influence of various levels of fly ash at 10, 25, 50, 75 and 100%, (w/w) on growth of *Sesbania cannabina*. After 90 days, they recorded length of shoot and biomass yield increased up to the 25% fly ash level as compared to untreated soil. Other hands, Singh et al. (2008) studied the effects of fly ash (0%, 5%, 10%, 15%, and 20%) on *Beta vulgaris* and showed higher dose of fly ash (15 and 20%) caused considerable reductions in growth and biomass yield. Tsadilas et al. (2009) reported on a field experiment with an Alfisol incorporated with two fly ash rates (5.5 t ha^{-1} and 11 t ha^{-1}) and wheat (*Triticum vulgare*) crop that fly ash application increased the grain and biomass yield of wheat and the yield was increasing with the application rates.

A laboratory experiment on column leaching demonstrated that relatively small amount of fly ash addition to contaminated waste soils significantly reduced the toxic elements content in the effluent (Ciccu et al., 2003). Similarly, Dermatas and Meng (2003) found that fly ash addition to metal-contaminated soils efficiently reduced metals leaching well below their safe regulatory limits. Thus, use of fly ash appears to offer important metal immobilization potential.

Ipshita and Tarar (2014) worked on fly ash based pesticide in cotton together with soil application of ash and manure and stated that fly ash can be used effectively as biopesticide and enhanced the various soil physico-chemical characteristics. Fly ash based pesticide formulation augmented the productivity of cotton. Cotton boll weight and cotton yield was significantly increased by fly ash application at different rates. It can be effectively utilized as a low cost input and economically viable biopesticide by poor farmers.

Thakare et al. (2013) studied on growth and yield of Chili (*Capsicum annum*) using fly ash at 0, 2, 5, 10, 15, 20 and 25% levels. Plant growth was regularly monitored every month from the date of sowing. It was found that soil blended with fly ash at 5% to 10% concentration improved soil physical properties as well as different growth parameters and yield of Chili. A possible option for safe disposal of fly ash without serious harmful effects may be managed by using in agricultural field which may curtail the cost of fertilizer inputs and encourage the farmers' economy if incorporated in proper combination.

Buddhe et al. (2014) carried out field experiments on rice using magnetized fly ash (*Biosil*) as soil conditioner together with recommended dose (RD) of fertilizer, maintaining RD and vermicompost. An interesting result was observed on rice plant growth, crop productivity and post-harvest soil fertility at very low fly ash concentrations (150 to 900 kg ha^{-1}) which indicating the enhanced inherent soil-enriching power of fly ash. *Biosil* at 900 kg ha^{-1} was found to be the most favorable dose for rice crop growth and yield attributes as well as soil characteristics. The progress for rice crop growth was in the order of *Biosil* + RDF > VC > RDF and for soil characteristics was *Biosil* + RDF > RDF > VC. Hence, they suggested that fly ash application in an integrated manner would be the most advantageous for attaining maximum benefits from *Biosil*.

Patel et al. (2016) conducted an experiment on the effect of integrated application of organic and inorganic inputs together with fly ash on heavy metal accumulation in acid soils of Northern Hills zone (Chhattisgarh) and they observed that the maximum yield of grain and straw was documented by using of 75% RD + 60 t fly ash ha^{-1} + 5 t FYM ha^{-1}, which was not significantly different with 75% RD + 40 t fly ash ha^{-1} + 5 t FYM ha^{-1} and 75% RD + 20 t fly ash ha^{-1} + 5 t FYM ha^{-1} and control recorded lowest yield. Grain and straw yield increased with increased dose of fly ash combined with and without FYM. The content of DTPA extractable available heavy metals were significantly increased in increasing levels of fly ash treatments as compared to untreated control. Fly ash application at 20, 40, 60 t ha^{-1} integrated with chemical fertilizer and FYM significantly increased in heavy toxic metals load as compared to RD. The highest content of available Co,

Cr, Ni, Pb were observed in application of 75% RD + 60 t fly ash ha^{-1} + 5 t FYM ha^{-1}, while control showed lowest. Thus, integration of fly ash, chemical fertilizer, FYM in acid soil was useful in yield increasing of rice grain as compared to chemical fertilizer use only and heavy metal load in soil gradually increased with gradual increase in the levels of fly ash in due course.

16.5 RESEARCH UNDER BIHAR AGRICULTURAL UNIVERSITY (BAU)

1. One externally funded ad-hoc research project entitled: "*Potential use of fly ash as a source of nutrient for sustainable agriculture*" was started in 2007 with the following objectives:
   - i. To work out physicochemical and biological characters of fly ash.
   - ii. Impact of fly ash on physical, chemical and biological characters of soils of Bhagalpur.
   - iii. To evaluate fly ash as a supplier of plant nutrients, soil conditioner, and amendment.
   - iv. To work out the fertilizer value of fly ash customized by mixture with compost.
   - v. To evaluate the continuous application of fly ash with manure on yield of rice-wheat crops and soil properties.

   This research project carried out in the department of Soil Science and was funded by National Thermal Power Corporation (NTPC), Kahalgaon.

2. Another university-funded research project was run in the same department since 2013 entitled: "*Impact of Fly Ash on Physico-chemical Properties of Light Textured Soils of Koshi Region of Bihar in Relation to Crop Productivity*" and the experiment was conducted at flood-affected sand deposited waste field at Supaul district of Bihar. Under this, the following experiments have been carried out:

Experiment No. 1: An incubation study to observe the behavior of changes of fine sandy waste soil under the combined application of fly ash and cow dung manure.

The study was carried out with doses of fly ash treatments *viz.* 0.5, 1, 5, 10, 25, 50% by weight basis in sandy soil and fly ash- soil mixture (total 10 kg per pot) along with only fly ash (control 1) and only sandy soil (control 2) under two (2) different cow dung manure doses, i.e., 50 t ha^{-1} and 100 t ha^{-1} for six month. Soil sampling was made on a monthly basis. From data, it is opined that fly ash can play a significant role to modify uncultivable sandy soil to a favorable condition for plant growth by altering its key characteristics.

Experiment No. 2: To grow awareness and interest to farmer about fly ash use in agriculture

Farmers hesitated to apply fly ash at high quantity to the soil for growing crops as they know that ash has a burning action to the plant root and leaf. So, a fly ash introduction trial has been taken for growing awareness to the farmers about fly ash use in agriculture at the farmers' field at small scale.

Fly ash was applied in forestry plantation directly (Young plant, *Acacia*) in sandy soil at Basantpur block of Supaul with fly ash treatments of 0, 2.5, 5, 10 kg per plant with vermicompost (5 kg) and without vermicompost.

Results showed that plants were healthy having no bad effect on root systems and above ground parts. Consequently, farmers showed their interest on its further utilization in agricultural crops. It has been seen that plant growth better at higher dose of fly ash.

Experiment No. 3 Impact of fly ash on sand deposited waste soil in relation to growing of Elephant foot yam.

A field study was carried out on elephant foot yam at flood-affected area of Basantpur block of Supaul district of Bihar by using fly ash at five different levels combined with three levels of vermicompost in 2014 in randomized block design. Same experiment was repeated in 2015 at the same place. Fly ash levels were 0, 5, 10, 20, 30 tons/ha along with 0, 1.0, 2.0 ton vermicompost per hectare. It resulted that yield of elephant foot yam increased with increased levels of fly ash. Edible part of elephant foot yam was tested for heavy toxic metals results below detection level for all the elements by the instrument. Later, it has been suggested to use fly ash in non-edible crops in research council meeting. It was also suggested to apply fly ash to forestry crop plant only. Therefore, experiment was taken on Forestry plantation which is described below:

Experiment No. 4 Study on role of fly ash on growth of poplar tree plantation (Forestry plantation) along with metal content in leaf

Poplar tree saplings were planted in sand deposited waste soil with various levels fly ash. 3| × 3| × 3| sized pits were made and fly ash was incorporated in the pits by v/v basis. Fly ash levels were 0, 25, 50, 75% per plant with replicated thrice. No other external fertilizers and composts were applied, but water was applied to the pit to provide initial establishment of sapling. Initial soil sample was collected and initial data regarding saplings size like height from ground level, girth at ground level and girth at 1 m height were recorded. In conclusion, nutrient contained in the fly ash might be massive source of nutrient for growing plants in light-textured soil along with the condition of soil also gets modified by altering its physicochemical properties.

Experiment No.5 Effect of fly ash on Tal land (Heavy textured soil)

An experiment was conducted to observe the problems connected with tal soil which is heavy in texture. Due to occurrence of high percentage of clay particle soil become very hard when dry. Considering this problem, the experiment was formulated in tal soil with different fly ash doses. It was observed that, fly ash application to heavy textured soil resulted in upgrading of soil physicochemical characteristics like bulk density, water holding capacity and soil texture.

16.6 CONCLUSION

Fly ash may be a good source of essential plant nutrients elements and may keep the soil healthy. It may be used as soil conditioner. However, there is required to generate in depth knowledge for its scientific management and the best use for crop production. Keeping in view, this study appears to be very practical in research. Fly ash has immense potential in agriculture for its effectiveness in changing of soil health/quality and crop growth. On the other hand, since there is possibility to damage the atmosphere and human health directly or indirectly, long term conformity research programme is necessary to the best use of fly ash prior to planning agriculture.

KEYWORDS

- **calcium hydroxide**
- **fly ash**
- **National Thermal Power Corporation**
- **organization for economic cooperation and development**
- **recommended dose**
- **thermal power plants**

REFERENCES

Adriano, D. C., Page, A. L., Elseewi, A. A., Chang, A. C., & Strughan, I., (1980). Utilization and disposal of fly ash and other coal residue in terrestrial ecosystems, review, *J. Eviron. Quality*, *9*, 333–334.

Aggarwal, A., Singh, G. R., & Yadav, B. R., (2009). Utilization of fly ash for crop production: Effect on the growth of wheat and sorghum crops and soil properties, *Journal of Soil Physics*, *9*, 20–23.

Agrawal, G. K., Sarkar, A., Righetti, P. G., Pedreschi, R., & Carpentier, S., et al., (2013). A decade of plant proteomics and mass spectrometry: Translation of technical advancements to food security and safety issues. *Mass Spectrom. Rev*., *32*(5), 335–365.

Aitken, R. L., & Bell, L. C., (1985). Plant uptake and phytotoxicity of boron in Australian fly ashes. *Plant Soil*, *84*, 245–257.

Arivazhagan, K., (2004). *Project Report Submitted to NTPC-Simhadri, Dadri, Talcher, Thermal and Vindhyachal Entitled: Showcase Project on Ash Utilization in Agriculture in and Around the Thermal Power Station Areas*.

Asokan, P., Saxena, M., Yadav, B., & Verma, M., (2005). Wheat crop productivity of water logged soil influenced by properties of coal combustion residues, *Ecol. Env. Cons*., *12*(1), 25–39.

Basu, M., Mahapatra, S. C., & Bhadoria, P. B. S., (2006). Exploiting fly ash as soil ameliorant to improve productivity of Sabaigrass (*Eulaliopsis binata* (Retz) C. E.) under acid lateritic soil of India. *Asian J. Plant Sci*., *5*(6), 1027–1030.

Basu, M., Pande, M., Bhadoria, P. B. S., & Mahapatra, S. C., (2009). Potential fly ash utilization in agriculture: A global review. *Progress in Natural Science, 19*, 1173–1186.

Bhattacharya, S. S., & Chattapadhyaya, G. N., (2004). Transformation of nitrogen during vermicomposting of fly ash. *Waste Manage Res*., *22*(6), pp. 488–491.

Buddhe, S. T., Thakre, M., & Chaudhari, P. R., (2014). Improvement in rice crop productivity and soil fertility in field trial with magnetized fly ash soil conditioner. *Annals of Applied Bio-Sciences, 1*, A28–A39.

CEA, (2015–2016). *Report on Fly Ash Generation at Coal/Lignite Based Thermal Power Stations and Its Utilization in The Country For The Year 2015–2016*. Central Electricity Authority, New Delhi.

Chang, A. C., Lund, L. J., Page, A. L., & Warneke, J. E., (1977). Physical properties of fly ash amended soils. *J. Environ. Qual*., *6*, 267–270.

Ciccu, R., Ghiani, M., Serci, A., Fadda, S., Peretti, R., & Zucca, A., (2003). Heavy metal immobilization in the mining-contaminated soils using various industrial wastes. *Miner. Eng*., *16*, 187–192.

Dermatas, D., & Meng, X., (2003). Utilization of fly ash for stabilization/solidification of heavy metal contaminated soils. *Eng. Geol*., *70*, 377–394.

Furr, A. K., Parkinson, T. F., Gutenmann, W. H., et al., (1978). Elemental content of vegetables, grains and forages field grown on fly ash amended soils. *J. Agric. Food Chem*., *26*, 357–359.

Garg, R. N., Pathak, H., Das, D. K., et al., (2005). Use of fly ash and biogas slurry for improving wheat yield and physical properties of soil, *Environ. Monit. Assess*., *107*, 1–9.

Grewal, K. S., Yadav, P. S., & Mehta, S. C., et al., (2001). Direct and residual effect of fly ash application to soil on crop yield and soil properties. *Crop Res*., *21*, 60–65.

Hill, M. J., & Lamp, C. A., (1980). Use of pulverized fuel ash from Victorian brown coal as a source of nutrients for pasture species. *Aust. J. Exp. Agric. Anim. Husb*., *20*, 377–384.

Inam, A., & Sahay, S., (2015). Effect of a thermal power plant waste fly ash on leguminous and non-leguminous leafy vegetables in extracting maximum benefits from P and K fertilization. *Pollution, 1*(3), 297–304.

Ipshita, S., & Tarar, J. L., (2014). Use of fly ash as bio-pesticide for cotton plant. *International Journal of Current Research in Life Sciences, 3*(12), 087–090.

Lau, S. S. S., & Wong, J. W. C., (2001). Toxicity evaluation of weathered coal fly ash amended manure compost. *Water Air Soil Pollut., 128*, 243–254.

Lee, H., Ha, H. S., Lee, C. S., Lee, Y. B., & Kim, P. J., (2006). Fly ash effect on improving soil properties and rice productivity in Korean paddy soil. *Bioresour Technol., 97*, 1490–1497.

Martens, D. C., (1971). Availability of plant nutrients in fly ash. *Compos. Sci., 12*, pp. 15–19.

Matsi, T., & Keramidas, V. Z., (1999). Fly ash application on two acid soils and its effect on soil salinity, pH, B, P and on rye grass growth and composition. *Environ. Pollut., 104*, pp. 107–112.

Mattigod, S. V., Dhanpat, R., Eary, L. E., et al., (1990). Geochemical factors controlling the mobilization of inorganic constituents from fossil fuel combustion residues: I. review of the major elements, *J. Environ Qual., 19*, 188–201.

Padhy, R. N., Nayak, N., Dashmohini, R. R., Rath, S., & Sahu, R. K., (2016). Growth, metabolism and yield of rice cultivated in soils amended with fly ash and cyanobacteria and metal loads in plant parts. *Rice Science, 23*(1), 22–32.

Page, A. L., Elseewi, A. A., & Straughan, I. R., (1079). Physical and chemical properties of fly ash from coal-fired power plants with special reference to environmental impacts. *Residue Rev., 71*, 83–120.

Pandey, V. C., & Singh, N., (2010). Impact of fly ash incorporation in soil systems. *Agric. Ecosyst. Environ., 136*, 16–27.

Pandey, V. C., Abhilash, P. C., Upadhyay, R. N., & Tewari, D. D., (2009). Application of fly ash on the growth performance and translocation of toxic heavy metals within *Cajanus cajan* L.: Implication for safe utilization of fly ash for agricultural production. *J. Hazard Mater, 166*, 255–259.

Patel, K., Singh, R. N., & Tedia, K., (2016). Integrated effect on inorganic and organic fertilizers with fly ash on heavy metal accumulation on acid soils of Northern Hills zone (Chhattisgarh). *Proc. of International Conference on Natural Resource Management: Ecological Perspective at SKUAST-Jammu* (p. 730).

Pati, S. S., & Sahu, S. K., (2004). CO_2 evaluation and enzyme activities (dehydrogenase, protease and amylase) of fly ash amended soil in presence and absence of earthworms (Under laboratory condition). *GeoDerma, 118*, 289–301.

Roy, W. R., Thiery, R. G., Schuller, R. M., et al., (1981). Coal fly ash: a review of the literature and proposed classification system with emphasis on environmental impacts. *Environmental Geology Notes*, 96. Champaign, IL: Illinois State Geological Survey.

Sarkar, A., Rakwal, R., Agrawal, S. B., Shibato, J., Ogawa, Y., Yoshida, Y., Agrawal, G. K., & Agrawal, M., (2010). Investigating the impact of elevated levels of ozone on tropical wheat using integrated phenotypical, physiological, biochemical and proteomics approaches. *J. Proteome Res., 9*(9), 4565–4584.

Schneider, T., & Riedel, K., (2010). Environmental proteomics: Analysis of structure and function of microbial communities. *Proteomics, 10*, 785–798.

Scotti, A., Silva, S., & Botteschi. G., (1999). Effect of fly ash on the availability of Zn, Cu, Ni and Cd to chicory. *Agriculture Ecosystem and Environment, 72*, 159–163.

Singh, A., & Agrawal, S. B., (2010). Response of mungbean cultivars to fly ash: Growth and yield. *Ecotoxicol. Environ. Saf., 73*, 1950–1958.

Singh, A., Sarkar, A., & Agrawal, S. B., (2012). Assessing the potential impact of fly ash amendments on Indian paddy field with special emphasis on growth, yield, and grain quality of three rice cultivars. *Environ. Monit. Assess, 184*, 4799–4814.

Singh, A., Sharma, R. K., & Agrawal, S. B., (2008). Effects of fly ash incorporation on heavy metal accumulation, growth and yield responses of Beta vulgaris plants. *Bioresour. Technol., 99*, 7200–7207.

Singh, R. P., & Agrawal, M., (2008). Potential benefits and risks of land application of sewage sludge. *Waste Manage, 28*, 347–358.

Sinha, S., & Gupta, A. K., (2005). Translocation of metals from fly ash amended soil in the plant of *Seabania cannabina* L. Ritz: Effect on antioxidants. *Chemosphere, 61*, 1204–1214.

Sridhar, R., Duraisamy, A., Kannan, L., et al., (2006). Utilization of thermal power station waste (fly ash) for betterment of plant growth and productivity. *Pollut. Res., 25*(1), 119–122.

Surridge, A. K. J., Merwe, A., & Kruger, R., (2009). Preliminary microbial studies on the impact of plants and South African fly ash on amelioration of crude oil polluted soils. *In World of Coal Ash (WOCA) Conference*.

Thakare, P. B., Mahendra, C. D., & Pokale, W. K., (2013). Growth of chili plants (*Capsicum Annuum*) in fly ash blended soil. *African Journal of Basic and Applied Sciences, 5*(5), 237–241.

Thetwar, L. K., Desmukh, N. C., Jangde, A. K., et al., (2007). Studies on the effects of fly ash and plant hormones on soil metabolic activities. *Asian J. Chem., 19*(5), pp. 3515–3518.

Tiwari, M. K., Bajpai, Dr. S., & Dewangan, Dr. U. K., (2016). Fly ash utilization: A brief review in Indian context. *International Research Journal of Engineering and Technology, 03*(04), 949–956.

Tolle, D. A., Arthur, M. F., & Pomeroy, S. E., (1982). *Fly Ash Use for Agriculture and Land Reclamation: A Critical Literature Review and Identification of Additional Research Needs*. RP-1224–1225. Columbus, Ohio: Battelle Columbus Laboratories.

Tripathi, R. C., Masto, R. E., & Ram, L. C., (2009). Bulk use of pond ash for cultivation of wheat-maize egg plant crops in sequence on a fallow land. *Resour. Conserv. Recyc., 54*, 134–139.

Tsadilas, C. D., Shaheen, S. M., & Samaras, V., (2009). Influence of coal fly ash application individually and mixing with sewage sludge on wheat growth and soil chemical properties under field conditions. In: *15th International Symposium on Environmental Pollution and its Impact on Life in the Mediterranean Region* (p. 145). Bari, Italy.

Ukwattage, N. L., Ranjith, P. G., & Bouazza, M., (2013). The use of coal combustion fly ash as a soil amendment in agricultural lands (with comments on its potential to improve food security and sequester carbon). *Fuel, 109*, 400–408.

Upadhyay, A., & Kamal, M., (2007). *Characterization and Utilization of Fly Ash, B.Tech Thesis*. National Institute of Technology, Rourkela, Odisha.

Vimal, K., Gulab, S., & Rajendra, R., (2005). *Fly Ash: A Material for Another Green Revolution*. Fly Ash India, New Delhi.

Wang, S., & Wu, H., (2006). Environmental-benign utilization of fly ash as low-cost adsorbents. *J. Hazard. Matter., 136*, 482–501.

Yavarzadeh, M. R., & Shamsadini, H., (2012). Safe environment by using fly ash and vermicompost on wheat. *International Conference on Transport, Environment and Civil Engineering (ICTECE 2012)*.

Yeledhalli, N. A., Prakash, S. S., Gurumurthy, S. B., & Ravi, M. V., (2007). Coal fly ash as modifier of physicochemical and biological properties of soil. *Karnataka J. Agric. Sci., 20*(3), 531–534.

CHAPTER 17

Natural Resource Management and Land Use Planning

C. D. CHOUDHARY, B. K. VIMAL, SHWETA SHAMBHAVI, and RAJKISHORE KUMAR

Department of Soil Science and Agricultural Chemistry, Bihar Agricultural University, Sabour, Bhagalpur, Bihar, India, E-mail: binodvimal099@gmail.com (B. K. Vimal)

ABSTRACT

At present quo, the impact of green revolution will be more visible become discernible in agriculture with incorporation of modern tools of Remote Sensing and GIS for characterization of natural resource and land use planning (LUP). The physiographic, climate, rainfall and potential water (surplus/deficit) for agricultural demarcated into eight agro-ecological regions with imposed some limitations. Data collected from 160 meteorological stations across the country and imply the concept of moisture adequacy index (MAI), and dominant soil group's proposed 29 agro-ecological zones adopted by FAO/UNESCO. Micro-morphology and landform of alluvial settings as happened in Indo-Gangetic Plains (IGP) for Bihar. Hitherto, diverse types of geomorphic features *viz., Tal, Chour*, and *Kewal* (Heavy-clay *vertisols,* medium-clay *chromosols*) and regular encounters of flood threat as well, and drain off the fertile soil. Southern portion of Ganga, there are vast stretches of backwater (one lakh ha) known as Tal area located in the districts of Patna, Nalanda, Lakhisarai, Munger, and Bhagalpur having limited option for *rabi* season. Whereas, vast stretches of land occupying (1.8 lakh ha) Diara land. Soils of eastern parts contain large volumes of fresh to weakly altered plagioclase and smectitic types of mineralogical composition. The major soils *Entisols, Inceptisols, Alfisols, Mollisols, Aridisols,* and *Vertisols* located in IGP and geological information's are sketched under the digital geological environment. The technology for the management of salt-affected soils developed mainly due to dominance of magnesium, needs refinement especially observed in *Aridisols.*

Agro-climatic Zone-I, II, and III comprise Alluvial Plains North of River Ganges and Chhotanagpur Plateau Regions formed natural levee. Zone-I has rich alluvial soils and a good scope of groundwater exploitation with vagarious fruit crops offers in this region *viz.,* mango, litchi, and banana. In light of field crops (rice, wheat, and maize), oilseed (rapeseed and mustard), spices-turmeric, ginger, dhania, Chilies, garlic, methi, sonfand mangaraila, tubers (sweet potato), tobacco, sugarcane, etc. taken as major high-value crops predominate in these areas. Whereas, Zone-II (North-East Alluvial Plain) comprises highly flood-prone areas affected mostly by Kosi and Mahananda rivers. The soils having low in nitrogen, low to medium in available phosphorus and potash. Whereas, zinc and boron observed low status and toxicity of manganese has been reported in these some patches. Heavy leaching results in soil acidification resulting in micro-nutrient deficiency and nonsetting of seeds in cereals and pulses was prevalent in these regions. Rice, jute, maize, summer pulses (moong), summer millets and sugarcane are important crops. There is a tremendous scope for exploiting waterlogged area by growing winter (Boro) rice. Maize is very high performing crop of this zone like Zone-I. In N-E portion of the zone, pineapple is an important crop and in the southern portion banana is predominant crops. This zone alike Zone-I, there is good scope for promotion of fisheries and honey production.

17.1 INTRODUCTION

The soils are valuable natural resources which are directly or indirectly associated with agricultural economics. Generally, clay soils appear near low land ecology of river basins and locally known as *Tal, Chour* and *Kewal* in Bihar, India. Tillage problem, tree less ecology and single cropping system are the geospatial features that may be observed in heavy clay soils. Soil types include a heavy-clay vertisol, both black and grey types, as well as medium-clay chromosols. In this context, soil survey towards agricultural LUP is an important part for the sustainability of agriculture practices. It provides adequate information in terms of landform; natural vegetation as well as characteristics of soils that can be utilized for land resources management and development (Manchanda et al., 2002).

17.1.1 BRIEF HISTORY OF SOIL SURVEY AND L.U.P. SCHEME, SABOUR

Prior to the implementation of the First Five Year Plan, information of soils of Bihar was meager especially on soil mapping towards their genesis, characterization, classification and suitable land-use plan. In order to bridge the gap, a high powered committee was constituted and on recommendation, the Soil Survey Scheme under the Department of Agriculture, Govt. of Bihar was created and established in 1954 with headquarter at Sabour with the following objectives to accomplish:

1. To identify and determine the important characteristics of soils of the state;
2. To classify into a defined system and other classification units;
3. To establish and delineate boundaries of different soils on map;
4. To correlate and predict the adaptability of soils to various crops, fodders and trees; their behavior and productivity under different management systems and the yields of adapted crops under desired sets of management practices;
5. To find out soil classification system most suited to the conditions and needs of the state.

During the period of First Five Year Plan to Third Five Year Plan, reconnaissance soil survey of Saharsa district was initiated and later on gradually started the survey work for the entire undivided Bihar including the Jharkhand and completed by 1965–66 and the laboratory analysis were accomplished accordingly in later dates. In this process, Topo sheets (Scale-1:63, 360) of different locations were used (Tables 17.4 and 17.5)

During the reconnaissance soil survey, nearly 17,000 soil profiles were exposed, their morphological and other visual features were studied and some relevant soil characteristics were determined in situ. General features like slope, topography, relief, vegetations, existing crops and cropping pattern, drainage conditions and water table were recorded. Altogether 80,000 soil samples from 17,000 profiles were collected and analyzed in laboratories subsequently.

Over 500 soil series were identified for the whole undivided Bihar based on the first-hand district wise exhaustic correlation with the properties were recorded. All these soils were further grouped into 24 Broad Soil Associations and mapped (Broad Soil Association Map of Bihar (Scale 1:10,00,000) was formed with well defined cartographic procedures (Figure 17.1). based on different pedological and associated features and called as in the second phase, Gov. of Bihar has initiated to enlarge the extensive program of irrigation potentials in the state and different irrigation commands, i.e., Gandak, Kosi, Mahananda, Sone, Kiul-Barua, Swarnrekha River commands were constituted and transferred the administrative control of Soil Survey Scheme, Sabour to Rajendra Agricultural University, Pusa and its man powers were engaged in different irrigation command agencies for conducting 'Detailed Soil Survey Work for the preparation of land use plans. There were altogether about 6 lakh hectares of land were covered and more than 100 detailed survey reports were published and submitted to the respective user agencies. During the third phase, Soil Survey and LUP Scheme (R.A.U. Pusa) entered into an agreement with the National Bureau of Soil Survey and LUP, Nagpur (ICAR) in the year 1986 which marked the beginning of phase-III. Under this agreement, our Scientists have been trained to use False Color Composite imageries of the satellite for preparing soil inventories and land use plans of Bihar. As a result, the Soil Resource Atlas was published in May, 1992. All these soil resource data may be computerized which might form the basis for future scientific investigations and LUP of the soils of Bihar.

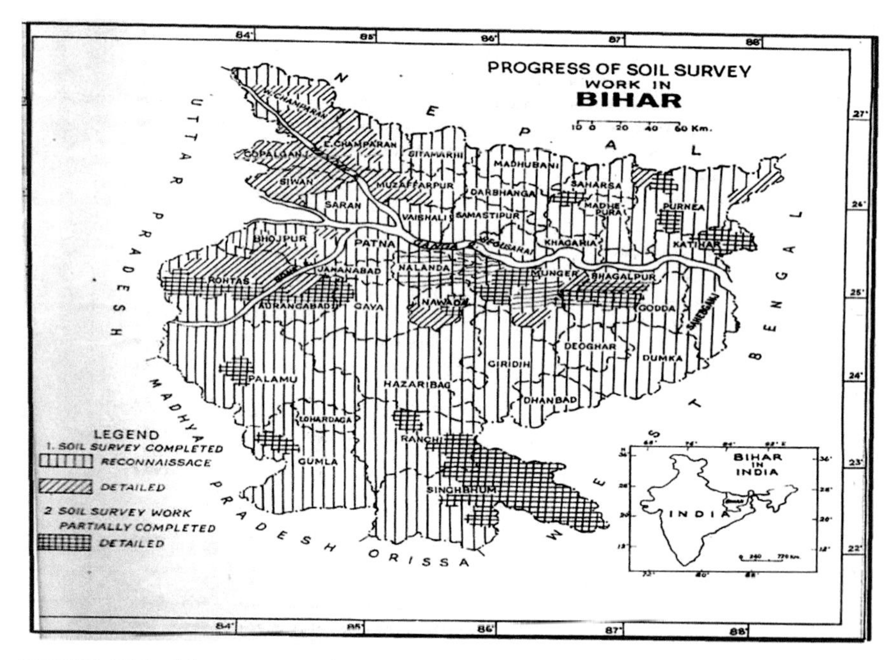

FIGURE 17.1 Map represents the progress of soil survey work.

Keeping the importance of digital Soil Surveying, the advanced cartographic systems were introduced by RAU, Pusa and well developed Remote Sensing and Image processing laboratory was established in 2010. Three Research Projects on preparing Soil Inventory with Soil Health Management for Khagaria, Bhagalpur, Munger, Jamui, Sheikhpura and Nalanda districts were covered successfully under the remote sensing and GIS techniques.

17.2 GEOGRAPHICAL SET-UP OF UNDIVIDED BIHAR

The State of Undivided Bihar, covering an area of 173,866 sq. km. and stretching between 21° 58' and 27° 23' North latitude and 83° 20' to 88° 00' east longitudes, is divided into three physiographic regions. They are (i) alluvial plains north of Ganges River receiving sediments from rivers originating in the Himalayas; (ii) alluvial plains south of the Ganges River receiving sediments from rivers originating in Chhotanagpur plateau; and (iii) Chhotanagpur plateau. The parent materials, topography, age, and vegetation of these physiographic region differ appreciably, there being also noted differences in climate especially in the moisture regime. These differences have resulted in the formation of many soil types in the state which differ significantly between different regions as well as within the region itself. Later on the undivided Bihar was divided into the southern most part known as Jharkhand and the northern part was called Bihar.

Jharkhand state lies between 21° 58'2" to 25° 8'32" North latitude and 83° 19'05" to 87° 55'03" East longitude covering an area of nearly 7.97 m.ha. and accounted for nearly 2.4% of total geographical area (TGA) of the country. It is bounded on the east by West Bengal, on the west by Chhattisgarh, on the north by Bihar and on the south by Orissa. Summary of geographical distribution of Land is given in Table 17.1.

Granite and Gneissic plateau surface predominantly cover the red soils patches which act as parent material in predominant areas in Jharkhand state (Tables 17.2 and 17.3). These soils are frequently observed in a catenary sequence and demarcated in upland, shallow to medium depth, well to excessively drained, prone to erosion, low water holding capacity, high permeability, low base exchange capacity (BEC) having low fertility status (Figure 17.2). In light of soil texture, heavier texture moving downwards of soil profile, reddish yellow to yellow and yellowish grey color is the typical characteristics of these vulnerable zones. With the incorporation with advance technology of remote sensing and GIS for red soils patches, model developed (Vimal et al., 2016) for spectral response to demarcate the

TABLE 17.1 Distribution of Geographical Land in Different Units of Old Bihar

		Area (%)	Area in (Lakh ha)
1.	Total geographical area		79.71
2.	Total cultivable land		38.00
3.	Net area sown	28.08%	25.75
4.	Current fallow land	11.12%	8.87
5.	Other fallow land	08.46%	6.75
6.	Forest	29.20%	23.28
7.	Barren land	7.20%	5.74
8.	Non-agricultural Uses	8.60%	6.86
9.	Pasture and other grazing land	2.48%	1.97
10.	Cultivable wasteland	3.44%	2.74
11.	Irrigated land	12.73%	3.007
12.	Cropping Intensity	116%	

TABLE 17.2 Distribution of Different Soil Order Predominating in Jharkhand State

District	Alfisols	Inceptisols	Entisols	Miscellaneous	Vertisols
Bokaro	62	21.4	12.9	4.0	
Deogarh	50	42	06.3	1.7	
Dumka	44	44	10	2.3	
Pakur	80	17	00	2.0	0.6
Palamu	54	20	22	4.5	
Ranchi	71	17	10	2.0	
East Singhbhum	71	13	13	2.5	
West Singhbhum	31	33	35	1.0	
Simdega	27	38	34	1.0	
Saraikela	54	27	16	2.3	
Sahebganj	41	37	10	12	
Chatra	52	13	34	1.0	
Dhanbad	70	8	18	4.5	
Garhwa	54	15	30	1.0	
Giridih	63	18	17	1.1	
Godda	43	39	16	2.0	
Gumla	43	40	16	0.7	
Hazaribagh	72	8	18	2.2	
Jamtara	51	39	6.7	3.3	
Koderma	41	35	23	1.1	
Latehar	64	7.2	28	1.0	
Lohardagga	53	25	21	1.0	

red soil patches in Bihar soils and it will be validate for all over the country and they were validate the soil pH (Vinay et al., 2013) having acidic (5.0 to 5.6) in reaction by digital image processing (DIP). *Lithic Haplustalfs*, *Rhodustalfs* are typical characteristics up to soil series level for upland soils and *Haplustalfs* and *Typic paleustalfs* for medium sloped soils in hilly region of Jharkhand state. Soils with grey color indicated neutral in reaction along with high clay content and high fertility status for lowland patches (Don Soils).

TABLE 17.3 Rocks and Minerals

Sr. No	Districts	Rocks and Minerals
1.	Santhal Pargana Rajmahal Trap	Dolerit Basalt, Granite, Quarzite, Amphibolite, Dolerite, Pegmatite, Marble Schist, Mica and Hornblende Schist.
2.	Hazaribagh	Mica And Granite Gneiss, Hornblende Gneiss And Calcschist Pegmatite, Gabro, Granite and Gondwana Rocks.
3.	Singhbhum	Garnet, Granite-Mica-Schist, Phyllite, Dalma Lava, Quartzite, Arkasani Granophyr, Titanium, Copper, Manganese and Dolerite.
4.	Dhanbad	Granite, Granite Gneiss, Mica Schist, Quartzite, Gondwana Rock, Coal and Sand Etc.
5.	Palamau	Mica And Hornblende Schist, Granite Geniss, Calc-Schist, Granite, Gabrovindhyan Rocks, Sand Stone, Limestone, etc.
6.	Ranchi	Granite, Granite Gneiss, Hornblende and Biotit Gneiss, Mica Schist, Quartzite, Amphibolites and Pegmatite.

17.2.1 GEOLOGICAL INFORMATION

The present state of Bihar is endowed with best soils (Terai, recent and old alluvium) along with subtropical climate, consisting of moderate rainfall, hot and desiccating summer and frost-free cold winter. It is located between 240 17’ 10” N to 270 31’ 15” N Latitude and 830 19’ 15 E to 880 17’ 40” E Longitude. Geographical area of Bihar state (94,154 lakh hectares), having total human population of 10.41 crores as per 2011 census of which male was 54,278,157 and female 49,821,295. The population density is 880 persons per square kilometer out of which 89.56% are rural population and 10.44% are urban. On account of major population being rural, agriculture constitutes the main livelihood and was 79.2% as per 1981 estimates. The annual growth rate of population is 2.84% (2001) as against 2.34% per annum (1991).

As per the present quo, Bihar stands the largest producer of vegetables after the second-largest producer of fruits. About 80% of the state's population, employed in agriculture, highest in agricultural production as compared to the national average. Among the fruit crops, Litchi, guava, mango, and pineapple are the major crops, whereas, brinjal, lady's finger, cauliflower, cabbage, rice, wheat, maize, and sugarcane are the main agricultural produce contributing by Bihar state and play a pivotal role in the Indian economy. Bihar having diversity of climate, plenty of rainfall and good heterogeneity towards soils texture and structure, provide ideal habitat for agriculture. It encounters flood threat as well, and drains off the fertile soil. Hence, our major challenge to conserve properly and prepare proper LUP for diversity of the crops. Southern part of state faces droughts almost every year affecting production of crops mainly paddy crops and due to severe flood situation, farmers face distress and unable to deploy his strength in agriculture.

17.3 SOILS, AGRO-CLIMATIC CONDITIONS, RAINFALL, AND LAND USE PATTERNS IN BIHAR

17.3.1 SOILS OF BIHAR

Due to differences in physiographic, lithological impact, parent materials, climate, topography, and age have resulted in the formation of different kind of soil which varies from sandy to very heavy clay in texture, red to dark grey in color, recent alluvium to old alluvium and Tal land in their makeup and highly fertile to practically unfertile wasteland.

Based on the Soil Survey data developed by the Scientists of Soil Survey and LUP Scheme, Sabour, the following information's have been compiled and mapped with advanced cartographic techniques (Figure 17.2).

17.3.2 GEOLOGICAL INFORMATION

Micro morphology and landform studies particularly in alluvial settings as happened in IGP for Bihar. Geomorphic surfaces influence the pedogenesis factor for Bihar plains *accompying* middle part of IGP with regular insisting the intense mineral weathering. Soils of eastern parts contain large volumes of fresh to weakly altered plagioclase and smectitic clay types of mineralogical composition. *Vertic* types of features with clay pedo-features in southern parts of Bihar remarked illuviation of fine clay which governs the pedogenic process. Expansive types of characteristics as seen in the old alluvium which become a part of Tal lands.

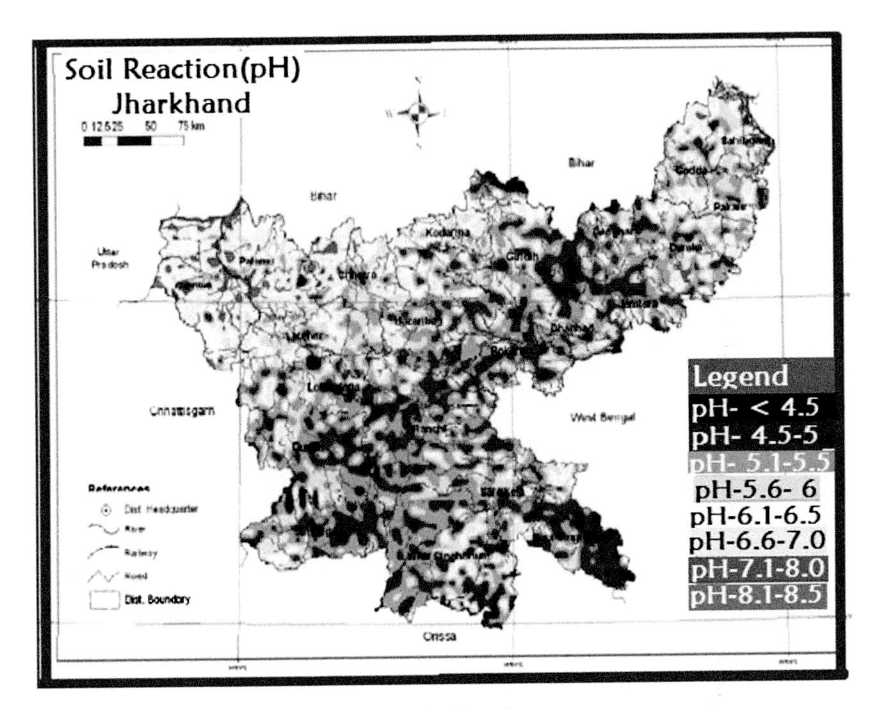

FIGURE 17.2 Soil reaction (pH) map of Jharkhand.

The major soil-forming processes as evidenced from the field (as seen in the Soil profile) as well as the laboratory records suggested about the information's of calcification, leaching, *lessivage, salinization* are the *pedogenetic* features, *alkalinization* and *gleization* govern by alkali soils and *homogenization*, and *argilliturbation* are typical characterizes of Tal land with vertices types of character accompanying major portion in soils of Bihar. *Entisols, Inceptisols, Alfisols, Mollisols, Aridisols,* and *Vertisols* located in IGP (Shankarnarayana and Sarma, 1982; Bhattacharyya et al., 1993; Ray et al., 2006) and geological information's for Bihar soils developed are sketched under the digital geological environment (Figure 17.3). Indian sub-continent exhibits a diversity of landscapes and climatic conditions demarcated by soils and vegetation and show a significant relationship between climate, soils, and vegetation. While preparing the land use plan or a suitable cropping pattern keep keen interest in mind, the combined effect of climate, soil, land forms, topography and vegetation of the state of Bihar (Tables 17.4 and 17.5).

FIGURE 17.3 Mineral map of Bihar.

TABLE 17.4 Different Land Use Plan in Bihar

Total Geographical Area		Area (in lakh ha.) 93.60
1.	Forests	6.16
2.	Barren and uncultivable land	4.37
3.	land put to non-agricultural uses	16.38
4.	Culturable wasteland	0.46
5.	Pasture and grazing land	0.18
6.	Miscellaneous trees and groves	2.31
7.	Other fallow	1.39
8.	Current fallow	5.67
9.	Net area sown	56.68
10.	Total cropped area	79.95
11.	Area sown more than once	23.27
12.	Cropping intensity	1.41

In the year 1954, Carter divided into six climatic regions, varies from arid to per humid demarcated to Indian condition. According to the Thornthwaite system (climatic classification), Murthy and Pandey (1978) correlate (physiography, climate, rainfall, and potential water surplus/deficit) agricultural regions demarcated into eight agro-ecological regions. Though, this system of classification imposed some limitations. Data collected from 160 meteorological stations across the country and imply the concept of MAI, and dominant soil group's proposed 29 agro-ecological zones (Subramaniam, 1983) with the possible 36 combinations pattern adopted by FAO/UNESCO. The delineated of 40 soil-climatic zones based on major soil types and moisture index (Krishnaft, 1988).

TABLE 17.5 Area, Production and Productivity for Major and Minor Crops in Bihar

Crops	Area (Lakh ha.)	Production (Lakh Tones)	Productivity (q/ha)	
			Bihar	India
Rice	35.96	57.11	15.88	19.9
Wheat	20.81	48.54	22.03	26.2
Maize	6.39	13.90	21.75	16.6
Barley	0.28	0.38	13.57	18.6
Ragi	0.24	0.25	15.83	14.0
Other millets	0.11	0.05	22.73	-
Tur	0.43	0.53	12.33	7.6
Gram	0.79	0.79	10.00	7.8
Lentil	1.79	0.70	9.97	-
Khesari	1.61	1.60	9.94	-
Peas	0.25	0.18	7.15	-
Moong	1.81	1.77	9.77	-
Other pulses	1.52	0.39	7.31	4.7
Rape and Mustard	0.89	0.70	7.87	8.7
Linseed	0.44	0.25	5.68	3.4
Til	0.04	0.02	5.00	3.1
Castor	-	-	-	-
Others	0.20	0.18	8.92	-
Sugarcane	0.94	39.89	422.20	719.9
Potato	1.44	13.78	84.98	-
Onion	0.18	1.91	107.32	-
Chilies	0.05	0.06	10.57	-
Other spices	0.10	0.10	10.28	-
Jute	1.49	10.95	13.26	19.5
Mesta	0.25	1.78	12.22	10.6
Tobacco	0.17	0.18	10.13	-

The irrigation potentiality has been created for 26.17 lakh ha in Bihar. The consumption of this irrigation potential is only 16.36 lakh hectare.

Sehgal et al. (1987) prepared a bio-climatic map of northwest India, with reference to dry month (more than moisture deficit of 40% in a month or more PET). The fifteen broad agro-climatic zones in the pattern of physiography and climate demarcated by the Planning Commission as part of the mid-term appraisal of the planning targets of the VII Plan (1985–1990). If we review the whole situation in the State of Bihar, it appears to be more applicable which requires the immediate attention of the planners and the Scientists together. Factually the knowledge about natural resources is very poor and scattered. Therefore, it is high time that an earnest effort is made to organize the knowledge and put them in an order that can be gainfully utilized.

Earlier Bihar State has been divided into six Agro-climatic situations, crop adaptability and cropping pattern (NARP, 1982):

1. North-west alluvial plains;
2. North-east alluvial plains;
3. South-Bihar alluvial plains;
4. Central and north-eastern plateau;
5. Western plateau;
6. South-eastern plateau.

Later on, NBSS, Nagpur recognized five Agro-Ecological Zones in Bihar. They are:

1. Hot sub-humid (North Bihar Plains – Moist).
2. Hot sub-humid (South Bihar Plains – Dry).
3. Hot sub-humid (South-west high land – Dry).
4. Hot sub-humid (Eastern Chotanagpur Plateau – Dry).
5. Hot humid (Mahananda Plain – Moist).

However, such divisions of agro-climatic zones and agro-ecological zones serve the purpose of planning at a broader scale of the national level. When we consider such divisions on State level, we find that this requires further subdivisions for the preparation of the Land Use Plan and its execution at district levels.

17.3.3 GROWING PERIOD

The concept of the growing period is essential to AEZ, and provides seasonal land resource appraisal. In tropical areas exist too dry during part of the year for crop growth to ensure crop stands without irrigation, whereas, temperate climatic regimes in winter season exert by cold temperatures. The growing period offers moisture and temperature conditions are suitable for crop production. Therefore, the Bihar state has been broadly delineated into the following three Agro-climatic zones based on rainfall, temperature, terrain, and soil characteristics.

- Zone I – North West alluvial plains.
- Zone II – North East alluvial plains.
- Zone III – South Bihar alluvial plains (Which are further subdivided into zone III A and III B?)

17.4 GENERAL FEATURES OF AGRO-CLIMATIC ZONE-I, II AND III

17.4.1 ALLUVIAL PLAINS NORTH OF RIVER GANGES

West and East Champaran, Gopalganj, Siwan, Saran, Sitamarhi, Muzaffarpur, Vaishali, Madhubani, Darbhanga, Begusarai and Samastipur comprises with an area of 32,665 km^2. This zone comprises of 18.77% of State area and 31–51% of State population. Average population density is 673 person/km^2.

The parent materials for the soils of this region have been deposited by the following rivers and their tributaries, i.e., (i) Kosi and Mahananda in the east, (ii) the Adhwara river system of the central North, (iii) Gandak, Burhi Gandak and Sarju in the West, and (iv) the Ganges in the south all originating in the Himalayas. As in the steep highlands of the Himalayas the physical disintegration of rocks and minerals predominance over chemical weathering, as well as due to frequent high floods in these rivers, the sediments deposited by them are richer in coarser fragments like sands and silts over clay. The Kosi, Mahananda and Adhwara system of rivers on one hand and the Gandak, BurhiGandak and Sarju on the other show tendencies to deposit sodium salts and calcium carbonate in their sediments respectively. This is directly related to the mineralogy of their catchment area in the Himalayas.

The rainfall in this region shows a gradual decrease both in the South and Western directions, for example, the North-Eastern part of Purnea and Kishanganj districts receive a rainfall of more than 2000 mm which gradually decreases to 1500 mm in the Northern part of Champaran. In the south, it shows a gradual decrease from 1500 mm in the southern part of Purnea to 1000 mm in the Western part of Saran district.

17.4.2 ALLUVIAL PLAINS SOUTH OF RIVER GANGA

The parent materials for soils of this region, in the most part except a strip in the North along river Ganga, have been deposited as sediments brought by rivers originating in the uplands of Chhotanagpur plateau where chemical decomposition of rocks and minerals predominate over their physical disintegration. Thus the parent material (alluvium) of this region, enrich with finer fragments like clays and poorer in silt as compared to the sediments of the north. Moreover, the sediments become heavier as we move away from the plateau or from the mainstream of the rivers.

The rainfall in this region shows a gradual decrease in the Western direction. It decreases from about 1,500 mm in the alluvial North-Western part of Santhal Parganas which adjoins Bhagalpur to about 875 mm in the Western part of Shahabad district.

17.4.3 CHHOTANAGPUR PLATEAU REGIONS

The parent materials of the soils of Chhotanagpur plateau have been derived on the various kinds of rocks found in this region. The rocks of Chhotanagpur plateau are of various kinds and ages.

Rocks found in major parts of the districts of Ranchi, Hazaribagh, Western part of Santhal Parganas, Southern part of Palamau, Northern part of Singhbhum, South Bhagalpur, South Munger, and South Gaya are various kinds of Archean gneiss and schists. Of these, the rocks occurring in the Northern parts of Hazaribagh and continuous areas in Munger and Gaya as well as around the township of Mandu in Hazaribagh are highly micaceous. The Gondwana and Archean country rocks of eastern Santhal Parganas are covered by trap-rocks known as Rajmahal traps. Rocks of Dhanbad, South-Eastern part of Hazaribagh and some parts of Palamau districts are coal-bearing Gondwana and Dharmars. The rocks of Adhowra hills of Shahabad district are Vindhyan sandstone, shales, and limestones. The country rocks of the central part of Palamau and Singhbhum districts are highly intersected with numerous dikes and sills of basic and ultrabasic rocks. The southern part of Singhbhum district has rich iron ore rocks of the Dharwar system. High-level laterites are found as cappings on the Naterhat plateau regions of Ranchi, Palamau districts and also in the Rajmahal hills of Santhal Parganas. Low-level laterites are found both over the ground or buried under the soil in the south-eastern part of Singhbhum district (Figure 17.4).

The average height of the plateau being 1,500 m above M.S.L. the average mean annual temperature is lower than that of the alluvial regions. The rainfall is highest in the East and South 1500 to 2000 mm which gradually decreases in the West and North. A part of the Palamau district in the North West of the plateau and its adjoining areas in the Gaya district are subjected to frequent droughts though rainfall is about 875 to 1000 mm only.

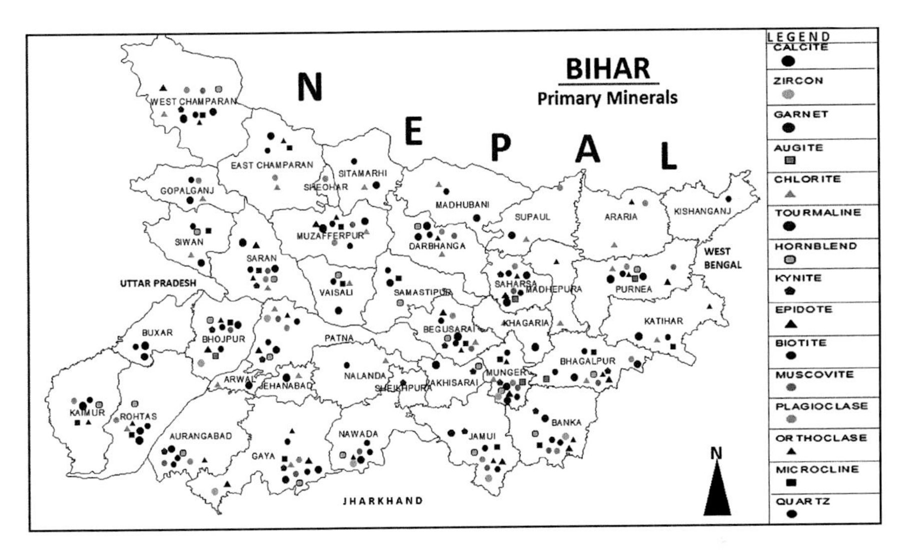

FIGURE 17.4 Available primary mineral in rocks.

17.4.4 CLIMATE

The climate is characterized by having tropical humid to sub-humid type. The rainfall in the zone decreases from North to South as a result of which district of Vaishali in the South and southwest area of Siwan receive on an average only 1000 mm rainfall per annum. The highest rainfall received by West Champaran is 1548 mm. Southwest monsoon rains usually start in the 2nd week of June and continue up to Ist week of Oct, Maximum rainfall occurs in July and August. This zone offers alluvial plains are slopy towards the southeast direction with very low gradient as evident by the direction is which rivers flow except the north-west areas of West Champaran. However, the Rivers move eastward direction with the natural levee, finally drain into the Ganga. As a result, there are vast waterlogged areas in the districts of Saran, Vaishali, and Samastipur. These areas are near flatness of the landscape, gets flooded during rains.

17.4.5 RIVER BASINS IN BIHAR

LUP's and cropping systems in Bihar are badly affected by these river basins. The entire north Bihar is flood-prone. Crops, as well as populations, suffered badly regularly. Soil erosion and sand deposition in cultivated lands always change the situations (Figure 17.7 and Table 17.6).

There are three major river basins in north Bihar.

1. **Gandak Basin:** It comprises of the basins of Gandak, BurhiGandak and Bagmati river valley. This valley covers the districts of Saran, East and West Champaran, Gopalganj, Siwan, Muzaffarpur, Vaishali, Samastipur, Khagaria, and Darbhanga.
2. **Adhwara Basin:** It comprises of the basins of the Bagmati, Kamla Balan and BurhiGandak and covers the major part of Darbhanga, Madhubani, Samastipur and Begusarai districts.
3. **Kosi and Mahananda Basins:** It comprises the basins of Kosi and Mahananda which covers part of Darbhanga, the whole of Saharsa, supaul, Purnia Kishanganj, and Katihar districts. The Kosi and Ganga have been marked by numerous changes in its course during historical time. This has far-reaching effect on the geomorphological and geo-hydrological status of the terrain. The effects are noticeable in the flood plains and waterlogged regions found in the tract.

TABLE 17.6 Basin Wise Flood Prone and Protected Area of Bihar

Sl. No.	Basin	Catchment Area (km²)	Length of River	Flood Prone Area (km²)	Protected Area (km²)	Embankment Constructed (km)
1.	Ganga	19322	445	12920	4300	537.81
2.	Kosi	11410	260	10150	9300	797.90
3.	BurhiGandak	9601	320	8210	4010	656
4.	Kiul-Harohar	17225	—	6340	—	7.00
5.	Punpun	9026	235	6130	260	40.60
6.	Mahananda	6150	376	5150	1210	51.69
7.	Sone	15820	202	3700	210	51.69
8.	Bagmati	6500	394	4440	3170	313.73
9.	Kamla Balan	4488	120	3700	2810	155.50

17.5 SOILS UNDER DRAINAGE SYSTEM

The area south of the Ganga River can be divided into two distinct geographical units, the alluvial plains alongside of the Ganga River and hard rock terrain mostly hills or sub-plateau to its south and east. It includes the districts of Rohtas, Bhojpur, Kaimur, Patna, Gaya, Nalanda, Nawada, Aurangabad, Munger, Jamui, Bhagalpur and Banka. The plains of South Bihar are widest towards the west and middle but gradually taper towards east. The south Gangetic plains are stable and not subjected to floods except in certain localized areas. The soils of South Bihar plains have shaped by the sediments of Ganges and rivers like Karmnasa, Sone, Sakri, Panchane, Kiul, Harohar, Badua, Chandan, Falgu, Barnar and many small rivers originating from Jharkhand (Table 17.8). The sub-plateau and plain regions of South Bihar are drained in the river Ganga through karmnasa, Sone, Punpun, Kiul, Falgu, Sakari and other small rivers.

Nowadays, remote sensing and GIS are known as space-based technology for natural resource management and agricultural LUP. Many organizations are interested to adopt this technology for enhancement their business or research work and lots of funds are being invested on these fields. Remote sensing and GIS (Geographic Information Systems) technology provide important tool for the assessment and monitoring of the natural resources whereas optical remote sensing makes use images of the earth's surface by detecting the solar radiation reflected from targets on the ground by visible, near-infrared and short-wave infrared sensors. In continuation of this technology, Hyper-spectral, thermal and microwave remote sensing have also been incorporated. Different materials *viz.* soil, water and vegetation of earth surface, reflect and absorb at different wavelengths and they are categorized as targets by differentiated by spectral reflectance signatures towards remotely sensed images (Table 17.7).

TABLE 17.7 Source Wise Irrigated Area in Bihar

Source	Gross Irrigated Area (Lakh ha)	Net Irrigated Area (Lakh ha)
Canals	12.93	10.11
Tanks	1.46	1.17
Tube wells	24.33	18.34
Well's	0.25	0.23
Other sources	6.54	5.35
Total	45.50	35.20

TABLE 17.8 Geological Information of Different Districts of Bihar and Jharkhand

Sl. No.	Districts	Rocks from Which the Soil Formed
1.	Bhagalpur	Mica, Horblende schist, Quartzite, Amphibolite.
2.	Munger, Patna, Rajgir, Kharagpur and Sheikhpura	Quartzite, Conglomerate, Jasper Quartzite, Slate, Phyllite, Mica-Schist, Intrusive granite and converted gneiss.
3.	Gaya	Mica, Hornblende schist, Granite, Hornblende gneiss, Quartzite Amphibolite, Syenite, Dolerite and Pegmatite.
4.	Shahabad	Vindhyan Rocks, Sand Stone, Slate, etc.
5.	Santhal Pargana, Rajmahal Trap	Dolerite Basalt, Granite, Quarzite, Amphibolite, Dolerite, Pegmatite, Marble Schist, Mica, and Hornblende Schist.
6.	Hazaribagh	Mica and Granite gneiss, Hornblende gneiss and Calc-schist, Pegmatite, Gabro, Granite and Gondwana rocks.
7.	Singhbhum	Garnet, Kyanite-Mica-Schist, Phyllite, Dalma lava, Quartzite, Arkasani Granophyr, Titanium, Copper, Manganese and Dolerite.
8.	Dhanbad	Granite, Granite gneiss, Mica schist, Quartzite, Gondwana rocks, i.e., Coal and sand etc.
9.	Palamau	Mica and Hornblende schist, Granite gneiss, Calc-schist, Granite, Gabro Vindhyan rocks, i.e., Sand stone, limestone etc.
10.	Ranchi	Granite, Granite gneiss, Hornblende and Biotite gneiss, Mica schist, Quartzite, Amphibolite and Pegmatite.

17.6 SOIL EROSION MAP

In north Bihar, recurrence floods boost up unstable crop production exerting the area about 17 lakh ha. Vast area (1.6 lakh ha) *accompanying* low lands with 1 to 5-meters depth of water which makes agricultural management difficult, the result being poor harvest. Despite large surface and groundwater resource is low lands with 1 to 5-meters depth of water which makes agricultural management difficult, the result being poor harvest.

The State of Bihar suffered from erosion of soil by water. The problem of markable wind erosion is being felt in West and East Champaran, Araria and Kisanganj districts of North Bihar. Much larger area of the state is under the influence of water erosion due to high intensity of rainfall, undulating land topography, over exploitation of forest area and great pressure of human and animal on the agricultural land. The river Ganga is the main drainage channel flows from west to east and distinctly passes through different soil types and climatic conditions. The erosion of soil by water is more pronounced in the southern part of the state away from the river Ganga comprising sub-plateau regions of state. The districts covered are Munger, Jamui, Banka, Kaimur, Rohtas, Aurangabad, Gaya and Nawada. 5.43 lakh ha. land is under severe soil erosion in these districts. North Bihar experiences almost every year the worst effect of soil erosion in the form of flood. To minimize the flood, soil conservation measures such as afforestation and plantations, etc., are required to be done at a large scale.

17.7 RIVER BASIN MAP OF BIHAR

Uplands in general, to the extent of 3.2 lakh ha are high in P- fixation capacity. Both area and intensity of soil salinity is on increase leading to low crop productivity and lowered nutrient efficiency. There is a need to bridge the gap by adopting suitable reclamation measures and farming technology for such large areas. Widespread deficiency of zinc is prevalent. Crops have been found to differ in their response to Zn-application. Similarly, a deficiency of B at places has been observed. Soils are becoming poor in organic matter.

Despite large surface and groundwater resources, the gross cropped area brought under irrigation is only a little over 27% in north Bihar. The exploitation of water resources is also low and in-efficient. Conjunctive use of surface and groundwater could have helped in controlling drainage problems, has been so far neglected. Drainage has not received adequate attention due to obsessions with the creation of new irrigation potential. Little research priority has been attached to these potent aspects.

- High water table in Gandak Command area: Priority needs to be attached in order to save the land going under waterlogging and salinity hazards.
- The progress of groundwater exploitation is seriously hampered due to inadequate and erratic supply of electric power.

17.7.1 SOIL HEALTH STATUS OF AGRO-ECOLOGICAL ZONE-II AND III

Healthy soil offers healthy sound environment with profitable foundation, productive and endeavor with agricultural systems. Understanding the formation of soil process and mechanism for support plant growth and regulate environmental quality remarked by management practices, convey crop and soil management system, improves and maintains soil health. At present quo, status of soil health of 25 districts of Bihar was conducted. There were about 20,000 soil samples from different physiographic levels collected, processed and analyzed in the laboratory. Organic carbon, Av. phosphate, and Av. potash were determined and estimated in terms of Nutrient Index for all the districts superlatively and these data were mounted on the GIS maps (Figure 17.7). Based on these studies, recently soil site suitability model has developed by Jayanti et al. (2015) by considering the soil health indicator for Katihar district and validate the model in representative five selected soil series by pedological study. Now a day, deficiency of micronutrients is burning issue as per concern of human health. Hence for keen interest of different workers, macronutrients and micronutrient mapping (Choudhary et al., 2016a, 2016b) and have been done for Tal land areas in Patna district for sustained the soil health.

17.8 SOIL ASSOCIATION MAP OF OLD BIHAR

The planning of systematic farming system should be laid down by keeping in mind about the responsible crops.

Suitable cropping and finally the farming systems for specific situations viz. flood prone, water-logged and saline soils need to be developed perfected and extended. Animal and other allied components need to be incorporated.

- High weed population: weed-control problem is intensifying.

17.8.1 ALLIED SECTORS

- Complete absence of suitable agro-processing units and lack of suitable storage facilities within the socio-economic preview of the farmers.
- Lack of agro-industries particularly for fruit and vegetable preservation. Family-based systems and techniques need to be developed.
- Agro-forestry and social forestry programs have not got proper attention.

17.8.2 INFRA-STRUCTURE

- Limited supply of improved quality seed of grain and horticultural crops.
- Problem of proper communication, marketing facilities, and suitable modern amenities in villages.

- Very low use of improved implements/machines.

The zone has rich alluvial soils and a good scope of groundwater exploitation. It is rich in fruits like mango, litchi, and banana. The important field crops are rice, wheat, maize, oilseed (rapeseed and mustard), spices-turmeric, ginger, dhania, Chilies, garlic, methi, sonf and mangaraila, tubers (sweet potato), tobacco, sugarcane, etc.

- Sugarcane industry is fairly well developed. There is a scope for its modernization. The productivity of sugarcane could easily be increased by about 50% in the next decade provided adequate help and support (rural roads, irrigation, required inputs and electricity) to the farmers are provided.
- Fruit and vegetable processing industries need strengthening. The causes of sickness of one at Hajipur may be looked into and required measures for research and development could be strengthened.
- The area can export spices. The spices curing/processing and packaging industry can be developed. Tobacco is an important cash crop. Tobacco processing industry has a scope.
- Maize and sweet potato-based industry has immense scope. The value of maize may be increased if starch, corn oil and other industries are established.

17.9 ZONE-II (NORTH-EAST ALLUVIAL PLAIN)

North-East Alluvial Plains comprises the district of Saharsa, Madhepura, Purnea, Araria, Katihar, Naugachhia sub-division of Bhagalpur and Khagaria. The zone is fairly densely populated (530 persons/km^2). It has an area of 11.69% of the state and shares 13.9% population.

17.9.1 SOIL MANAGEMENT

- Highly flood-prone areas affected mostly by Kosi and Mahananda rivers which are responsible for silt depositions in cultivated fields creating major problems for sustainable crop productions.
- Sandy to loamy soils, very highly permeable and strongly acidic non-calcareous soils of light texture especially in the North-Eastern part of Purnea.

- In respect of primary nutrients, soils very poor in nitrogen, very poor to medium in available phosphorus and potash in very light-textured soils. Whereas, deficiency of zinc and boron was observed in this region. But, toxicity of manganese has been reported in these areas.
- Heavy leaching of the soil has increased soil acidity and caused nutritional problems resulting in mico-nutrient deficiency and perhaps in nonsetting of seeds in cereals and pulses. Hence, delineation of micro-nutrient deficiencies and their amelioration is necessary.

17.9.2 WATER MANAGEMENT

- Recurrence of floods in large areas and consequently waterlogged areas to the extent of 1.6 lakh ha.
- Poor irrigation facility- Irrigation is available to the tune of only 21% of cropped area.
- Lack of suitable irrigation and water management practices for light textured soils.

17.9.3 FARMING SYSTEM

- Lack of suitable farming system for different geomorphologic situations viz.
    i. Flooded area between the embankments.
    ii. Sandy uplands having little irrigation facilities.
- Lack of information on causes for poor seed setting in gram and wheat and evolving suitable varieties and agro-technique for the crops of this region.
- Lack of work on varietal improvement of plantation and fruit crops and suitable agro-technique.
- No sufficient program on fish culture and agroforestry, etc.
- Non-availability of the relevant packages of practices for crops of these regions and lack of information on the post-harvest technology.
- Hot humid climate of this zone causes heavy insect pest damage. Considerable damage is caused to Jute crop due to root-rot. Nature, extent, and damage due to insects and pests and development of control measures need to be undertaken.

Rice, jute, maize, summer pulses (moong), summer millets and sugarcane are important crops. There is a tremendous scope for exploiting the water-logged area by growing winter (Boro) rice. Maize is very high performing crop of this zone like Zone-I. In N-E portion of the zone pine-apple is an important crop and in the southern portion banana. Problems facing these two fruit crops need to be looked into. Side by side some fruit preservation/canning industry may be developed. The zone has along with Zone-I good scope for fisheries development honey productions.

17.10 ZONE-III (SOUTH BIHAR ALLUVIAL PLAINS)

Since this zone comprises of the entire South Bihar Plains, which vary a great deal in soil, climate, irrigation facilities and even nature of crops and cropping. The zone has been divided into: III (A) comprising the districts of Bhagalpur, Banka, Jamui, Seikhpura, Lakhisarai and Munger and III (B) with districts of Patna, Bhojpur, Rohtas, Aurangabad, Gaya, Jahanabad, Arwal, Nawadah, and Nalanda. The density of population is about 465 persons/km^2. It has 25.75% of the state's area and supports 29.8% of the population.

17.10.1 GEO-MORPHOLOGY

Through 72.7% of the cultivated area is under irrigation, yet the substantial area in Aurangabad, Rohtas, and Kaimur districts are drought-prone. In the south of Ganga, there is vast stretches of backwater (one lakh ha) known as Tal area located in the districts of Patna, Nalanda, Lakhisarai, Munger and Bhagalpur which remain inundated in the Kharif season facilitating only one crop during the rabi season. Vast stretches of land under Diara areas are (1.8 lakh ha). The technology for the management of salt-affected soils developed mainly due to the dominance of Magnesium needs refinement. Erosion map and drainage map was laid out by SS and LUP, BAC, Sabour (Figures 17.5–17.8).

17.10.2 WATER MANAGEMENT

- Heavy to very heavy textured soils having pressure phases or slicken sides below causing them to have poor to very poor internal drainage.
- Lack of adequate irrigation facilities for Tal and Diara areas.

- Low-water-use-efficiency in Sone command areas. There is a need to train the farmers for the efficient use of irrigation water to check in rise of water table.

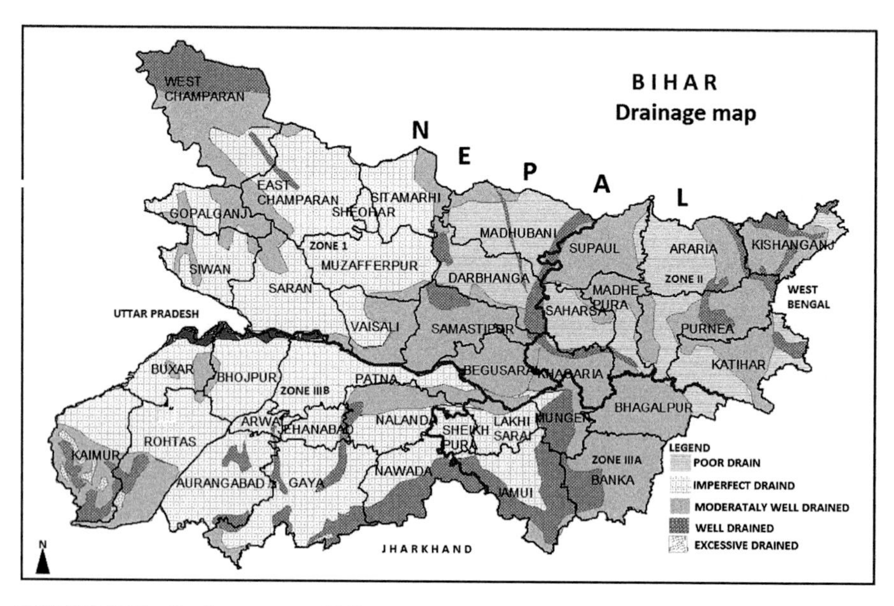

FIGURE 17.5 Drainage map of Bihar.

FIGURE 17.6 Bihar erosion map.

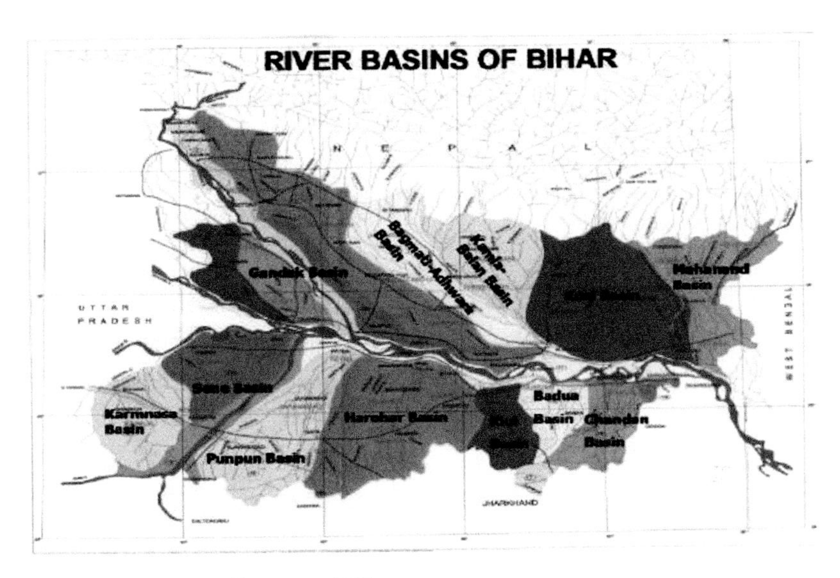

FIGURE 17.7 Soil fertility map of Bihar.

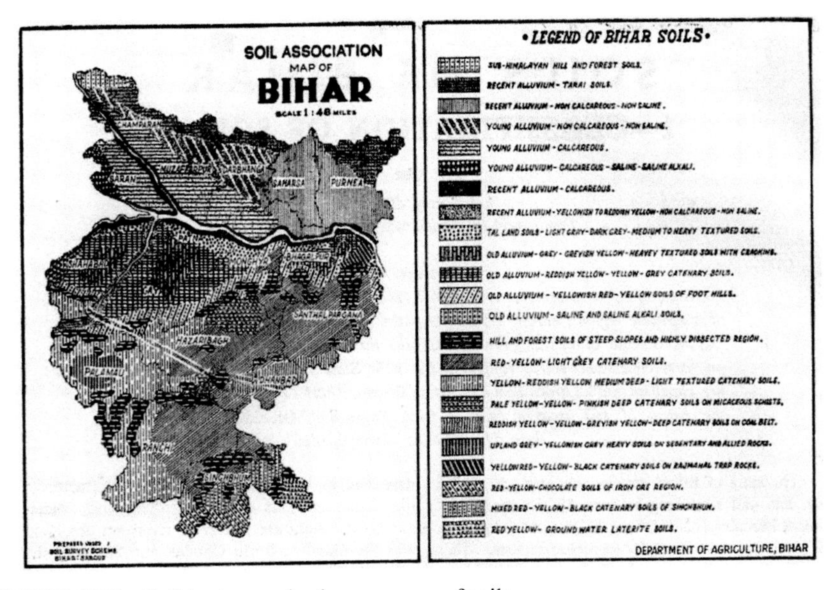

FIGURE 17.8 Soil texture and calcareousness of soils.

17.10.3 FARMING SYSTEMS

- Low and uncertain productivity due to floods monsoon (Kharif) season and due to soil moisture stress and drought during pre and post-flood period (Winter and Summer months). Mono-cropping, thus, is a common practice in such areas.

- Suitable soil management technology and cropping patterns for the rapidly hardening heavy textured soils is yet to be identified.
- Research on rainfed agriculture for upland conditions of this zone requires special attention for development of this zone IIIA.
- Non-availability of high yielding varieties of pulses crops suitable for Tal areas.
- Pisciculture programs should also be undertaken in this zone as an enterprise.

17.10.4 ALLIED SECTOR/INFRASTRUCTURE

- Food processing industries should be developed.
- Storage and marketing facilities especially in Tal and Diara areas require special attention.

17.11 CROP PLANNINGS

With anticipated better soil and water resource management, new innovation in agricultural production systems, improved infrastructure and input supply facilities and accelerated pace of information dissemination, both productivity and total production from a unit land area will increase appreciably. The impact of green revolution will be more widely visible in such areas, which so far have remained neglected.

1. **Tal Areas:** This area will remain to be efficient zone for pulses, viz. gram and lentil as pure and mixed with wheat during rabi. In irrigated Tal areas, rabi maize or wheat can also be grown. Moisture conservations and timely cultural practices are the very important management tools.
2. **Diara Lands:** Early maize-wheat/gram/linseed/forage crops/potato/sweet potato/fallow/vegetable particularly parwal, tomato, green pea and other cucurbits in rainfed condition. Early maize: Rabi maize/wheat/rai/potato in irrigated areas.

In the above background of anticipated improvement in various production components, following cropping systems have been proposed, which need to be popularized and adopted situation-wise (Table 17.9).

TABLE 17.9 Proposed Cropping Systems of Different Agro-Ecological Zones in Bihar

Agro-Ecological Zones	Proposed Cropping Systems Based on Soils			Plantation Crops
Zone-I (North-Western Gangetic Plains)				
District covered- East and West	Kharif	Rabi	Summer	Litchi, Mango, Guava,
Champaran, Sheohar, Sitamarhi,	Early Paddy	Maize	----	Forestry Plantations
Darbhanga, Siwan, Madhubani,	Maize	Potato	Moong	should be given importance
Saran Muzaffarpur, Vaishali	Maize/Early	Mustard	Moong	particularly in foothills
Gopalpur. Samastipur, Begusarai	Paddy	Wheat	Moong	as well on the bank
	G M	Sugarcane	Sugarcane	of Canals, ridges and
	Groundnut	Wheat	Moong	roadsides.
	Paddy	Sugarcane	Sugarcane	
	Maize	Potato	Onion	
	Maize	Sweet Potato	Moong/Berseem	
Agro-Ecological Zone-II (North-Eastern Plains)				
	Paddy	Wheat	Jute	Litchi, Mango, Guava,
	Jute	Mustard	Maize	Forestry plants in
	Early Paddy	Potato	Groundnut	northern parts should
District Covered- Supaul,	Banana	Vegetables	Vegetables/Chilies,	be promoted. Makhana,
Madhepura, Saharsa, Purnia,	Jute	Maize	Pulses	Singhara should
Katihar, Kishanganj, Araria,	Paddy	Potato	Onion/Ananas	also be promoted in
Khagaria	Jute	Black Gram	Vegetables	waterlogged areas.
	Rice	Wheat	Rice	

TABLE 17.9 *(Continued)*

Agro-Ecological Zone-III A and B: South Gangetic Plains and Plateau Regions				
District Covered	**Maize**	**Arhar**	**Vegetables**	**Plantation Crops**
Bhagalpur, Banka, Munger,	Paddy	Maize/Onion	Paddy	Mango, Litchi, Guava,
Jamui, Lakhisarai, Sheikhpura,	Paddy	Wheat	Vegetables	Banana, Caster,
Nawada, Gaya, Aurangabad,	Paddy	Potato	Onion	Shisham, Shal, and other
Kaimur, Rohtas, Buxar, Bhojpur,	Paddy	Sugarcane	Sugarcane	forestry crops.
Jehanabad, Patna, Nalanda	Paddy	Vegetables	Maize	
	Paddy	Potato	Moong	
	Early Paddy	Vegetables	Vegetables	
	Paddy	Tori	Moong	

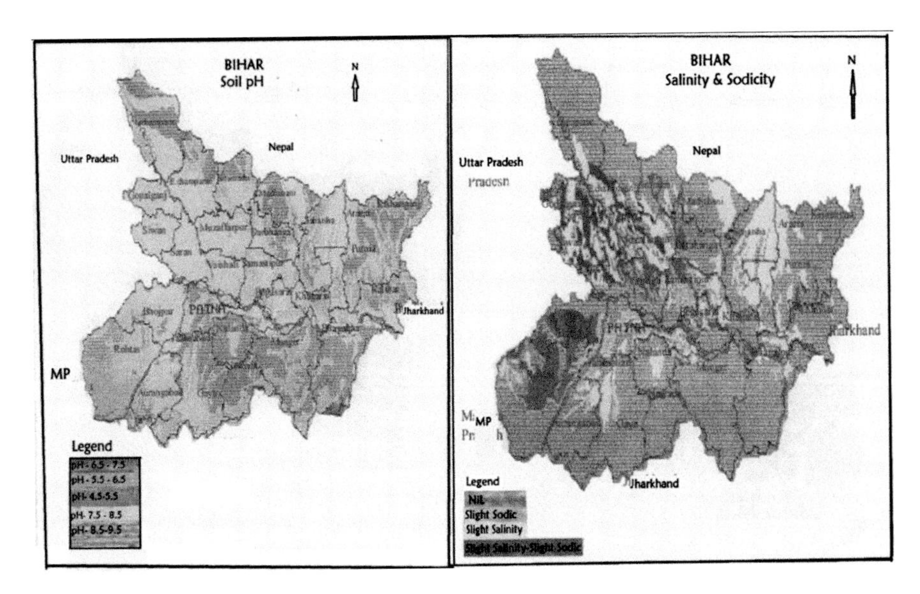

FIGURE 17.9

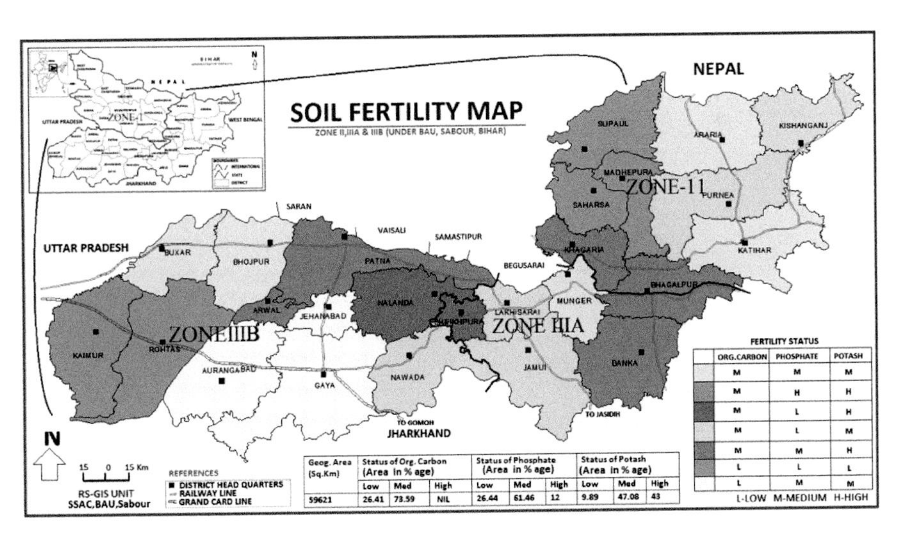

Geog. Area (Sq.Km)	Status of Org. Carbon (Area in % age)			Status of Phosphate (Area in % age)			Status of Potash (Area in % age)		
	Low	Med	High	Low	Med	High	Low	Med	High
59621	26.41	73.59	NIL	26.44	61.46	12	9.89	47.08	43

	ORG.CARBON	PHOSPHATE	POTASH
	M	M	M
	M	H	H
	M	L	H
	M	L	M
	M	M	H
	L	L	L
	L	M	M

FIGURE 17.10

KEYWORDS

- **base exchange capacity**
- **Chour land**
- **Daira land**
- **land use planning**
- **remote sensing and GIS**
- **Tal land**

REFERENCES

Bhattacharyya, T., Pal, D. K., & Deshpande, S. B., (1993). Genesis and transformation of minerals in the formation of red (Alfisols) and black (Inceptisols and Vertisols) soils on Deccan basalt in Western Ghats, India. *J. Soil Sci., 90*, 263–276.

Binod, K. V., Rajkishore, K., Choudhary, C. D., Sunil, K., Rakesh, K., Singh, Y. K., & Ragni, K., (2016). Spectral signature of red soil patches and their acidity: A case study of Banka district, Bihar, India. *Journal of Applied and Natural Science, 8*(2), 874–878.

Choudhary, C. D., Singh, Y. K., Ragni, K., Shewta, S., Kumar, J., & Rajkishore, K., (2016a). Soil mapping of some selected soil series and micronutrient status of Tal land areas in Patna districts using remote sensing and GIS. *Progressive Research: An International Journal, 11*(Special II), 793–795.

Choudhary, C. D., Singh, Y. K., Shewta, S., Ragni, K., & Rajkishore, K., (2016b). Soil characterization and macronutrient status of Tal land areas in Patna districts using remote sensing and GIS. *Eco and Env. Cons., 22,* S357–S362.

Jayanti, P., Ghanshyam, C. D., Choudhary, B. K., Vimal, S. K., & Rajkishore, K., (2015). *Eco and Env. Cons., 21*, S103–S108.

Krishnan, A., (1988). Delineation of soil climatic zones of India and its practical application in agriculture. *Fertilizer News, 4*, 11–19.

Manchanda, M. L., Kudrat, M., & Tiwari, A. K., (2002). Soil survey and mapping using remote sensing. *Tropical Ecology, 43*, 61–74.

Murthy, R. S., & Pandey, S., (1978). Delineation of agro-ecological regions of India. In: *Proc. Commission V, 11th Congress of the Int. Soc. Soil Sci.* Edmonton, Canada.

Ray, S. K., Bhattacharyya, T., Chandran, P., Sahoo1, A. K., Sarkarl, D., Durge, S. L., Raja, P., Maurya, U. K., & Pal, D. K., (2006). On the formation of cracking clay soils (Vertisols) in West Bengal. *Clay Research, 25*(2), 141–152.

Sehgal, J. L., (1987). Soil site suitability evaluation for cotton. *Agropedology, 1*, 49–63.

Shankarnarayana, H. S., & Sarma, V. A. K. (1982). Soils of India. Benchmark Soils of India, 41–70.

Subramaniam, A. R., (1983). Agro-ecological zones of India. *Archives for Metrology and Bioclimatology Series, 32*, 329–333.

Vinay, K., Vimal, B. K., Rakesh, K., Rakesh, K., & Mukesh, K., (2013). Determination of Soil pH using digital image processing technique. *Journal of Applied and Natural Science, 6*(1), 14–18.

PART III
Crop Protection

CHAPTER 18

Plant Disease Management Through Application of Nanoscience

ABHIJEET GHATAK

Department of Plant Pathology, Bihar Agricultural University, Sabour, Bhagalpur, Bihar, India, E-mail: ghatak11@gmail.com

ABSTRACT

Over application of fungicides and other pesticides in the agricultural field lead to cause an environmental hazard. The introduction of biocontrol agents and different cultural methods in the crop fields has not been rendered promising results in disease management. Therefore, the utmost need for another option in managing plant diseases may provide a satisfactory solution. The advent of nanoscience in the field of agriculture is playing a vivid role in supplying nutrition and managing diseases. This chapter discusses the different activities of nanoscience conducted at BAU Sabour for uplifting and unveiling the scope of nanoparticles in controlling plant diseases. Apart from various attempts, the primary efforts were made to manage soil-borne pathogens using biodegradable nanochitosan which is generating encouraging results for sustainable or long-term effect under soil ecology system. This chapter covers a few preliminary works on the management of fungal phytopathosystem using nanoparticles of different categories.

18.1 INTRODUCTION

Photosynthetic plants are the major source of carbon to the non-photosynthetic organisms. It is well known that over thirty thousand plant species are invaded by a large number of harmful organisms. The phytopathogens and insects share a huge volume of this detrimental group (Pieterse and Dicke, 2007). These harmful creatures influence the plant growth and deteriorate the production (yield) drastically. Therefore, managing plant diseases and insect infestation is a challenging task, which is directly associated with the improvement of human welfare.

There are various approaches can be applied to cope-up against the biotic problems in different phytopathosystems. In nature, plants are being exposed to a wide array of microorganisms and insects. All such organisms, get in exposure to plants, are not injurious. A large fraction of these minute creatures are beneficial to plants and to the ecosystem. A well-balanced interaction of plants with beneficial and detrimental microorganisms and insects for aboveground and belowground environments retains the ecology sustainable (Pineda et al., 2010). Therefore, a sustainable approach is believed paramount among the series of plant disease management approaches. Sustainable management is a full-proof tool to enhance plant growth by achieving the increased crop yield or by reducing the quantum of loss.

In the modern era of commercial agriculture, to get a quick remedy, the growers rely more on chemical pesticides rather to follow a sustainable approach. The chemical pesticides not only imply damage to the ecology but also its effects could be observed to the next generation. Apart from these problems of chemical pesticides the resistance development is again an issue with agro-chemicals. This imparts to develop a novel strategy for plant disease management. In the era of 'smart management technology,' the application of nanotechnology in plant disease management has a tremendous scope. Nanotechnology supports the precise utilization of individual atoms or molecules for their application to physical, chemical, and biological systems, and sometimes, involving action by forming a complex (by integration) system. A particle size is considered nanometer when it ranges between 1 and 100 nm. Chemically, the nanoparticles should have alterable physical properties, target specificity, rigidity with structure (non-changing ability), and delayed particle aggregation. The surface property of a nanoparticle may also include the photoemission, increased electrical and heat conductivity along with advanced surface catalytic activity (Garg et al., 2008; Shrestha et al., 2007).

Exploitation of nanoparticles for management of plant diseases is a 'bottom-up' approach where the foremost reaction participates into the action is reduction/oxidation. This could relate to the microbial enzymes, with reducing or antioxidant properties, in general responsible for reduction of metal compounds into their respective nanoparticles (Mohanpuria et al., 2007). In the applied part, researchers are attempting several ways to unveil the fungal inhibitory capability of different nanoparticles. So far, effect of silver nanoparticles has drawn the attention of researchers for inhibiting the phytopathogenic fungal growth under controlled condition (Min et al., 2009; Kim et al., 2009, 2012). However, only a few workers presented the significant effect of silver nanoparticles under in-planta interaction (Jo et al., 2009; Lamsal et al., 2011). Similarly, Fateixa et al. (2009) observed the inhibitory effect of silver nanoparticles on postharvest pathogen *Aspergillus niger*. Various authors also explained the mechanism of silver nanoparticles using electron microscope. Ouda (2014) found the antifungal property of silver nanoparticles on different cell wall components of two fungal pathogens, *Alternaria alternata* and *Botrytis cinerea*. The author observed deterioration of sugar, protein, n-acetyl glucosamine and lipid content in both of the fungi under treatments of silver nanoparticle. Application of nanoparticles for disease management is rather more developed in veterinary science and in the medical science. Many works have been presented for the animal bacterial inhibition by application of nanoparticles (Feng et al., 2000; Pal et al., 2007; Salem et al., 2015; Smekalova et al., 2016).

Nearly 60 countries have launched many-a-few national programs in nanotechnology supported by research and development grants. The Indian programs on nanotechnology are also shouldered by the Ministry of Science and Technology, Government of India. The program is launched under 'Mission on Nano Science and Technology' (nanomission.gov.in). Only a few centers in India are focusing on this discipline of research; work on nanoparticles is devoted to dig-out the significant opportunities for high proficient chemicals with the less hazardous trait (Li et al., 2007). Achievements of research on nanoscience in India can be estimated with a fact of publication; India ranked 6th in 2009, whereas previously, its rank verified with 17th rank in 2000 (Beumer and Bhattacharya, 2013). However, India's contribution in the field of patenting is very meager. Indian institutes granted a less than 3.5% patenting (based on nanotechnological research) filed in the Indian Patent Office.

Contribution of nanotechnological research is largely supported by 'research institutes,' which was assessed with merit and number of scientific publication (Parvathamma and Gobbur, 2011). Being a state agricultural university (SAU), Bihar Agricultural University (BAU) has taken a vital step by establishing the nanoscience and nanotechnology unit in July, 2015. The team at BAU Sabour is working to achieve the goal of phytopathogen management using application of nanoparticles. At preliminary stage, we evaluated different laboratory synthesized nanoparticles to various phytopathogenic fungi. Biological characterization of different phytopathogenic fungi to a series of nanoparticles is under process. This chapter will also furnish a few works on fungal phytopathosystem management conducted at BAU Sabour.

18.2 APPLICATION OF NANOTECHNOLOGY IN AGRICULTURE: AN OVERVIEW

Nanotechnology has great scope in agriculture, although it has not been fully explored for agricultural systems. The technology is exploiting for many facets of agricultural system (Figure 18.1).

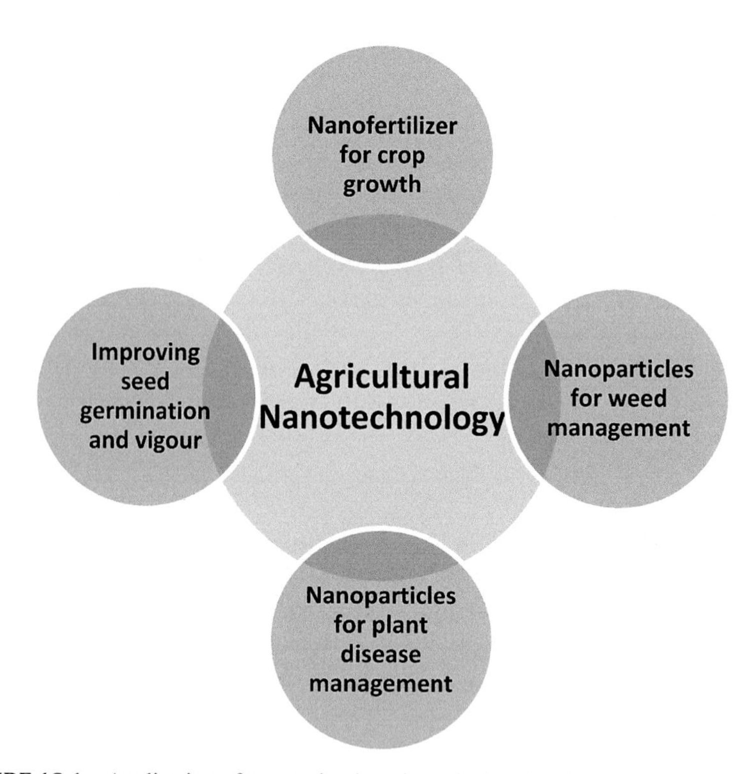

FIGURE 18.1 Application of nanotechnology in agricultural system.

Improvement in seed germination and seedling vigor is observed in nanoparticle treated seeds. Germination rate along with root length and aboveground dry biomass were found greater for monocot and dicot wetland plants with exposure to silver nanoparticle (Yin et al., 2012). Similarly, Maity et al. (2016) observed that metal nanoparticles also influenced the germination and yield of fodder crops, oat (*Avena sativa*) and berseem (*Trifolium alexandrinum*). Nanoparticle treatments resulted in better seedling emergence rate, tiller number and seed yield. However, Lin and Xing (2007) found phytotoxic effect of nanoparticles supported with reduced germination and root growth.

Fertilizers occupy for pivotal function in agricultural production nearly 35–40% of the productivity. It is well noted that approximately 40–70% N, 80–90% P, and 50–90% K from the fertilizer are lost, and do not reach to its targeted site due to various reasons like leaching of chemicals, drift, evaporation, hydrolysis by soil moisture, and photolytic and microbial degradation in the soil (Ombodi and Saigusa, 2000). This ultimately resulted in economic and environmental loss in terms of pollution. Scientists are now paying attention to mitigate the hazardous impact from residual fertilizers. Therefore, application of nanoparticles in the form of fertilizer is getting hike in agricultural field. The slow-release ability from such compound is an additional feature that sustain for whole of the cropping season (Solanki et al., 2015). Nanostructured fertilizers have the ability to increase the nutrient use efficiency. The mechanism of such fertilizers is delivery at targeted site with slow and controlled release of nutrient compounds.

Nanotechnological knowledge is also applying for the management of weeds. However, not much work has been observed under this category of crop management. Lim et al. (2012a) successfully developed a nano-emulsion formulation incorporating with herbicide glyphosate, which ultimately resulted in well-dispersed nanoparticles. The developed optimum entrapped glyphosate with the nano-emulsion improved the physicochemical characteristics of the nanoformulation indicating such nano-emulsion system could ameliorate in the improvement of herbicide by reorganizing the bioactivity and bioavailability. Significantly lower surface tension can be observed in the herbicide developed through nano-emulsion than a commercial formulation (Lim et al., 2012b). The narrow-leaved weed (*Eleusineindica*) managed at a lower dose of nano-emulsion of herbicide glyphosate (Roundup®; ED_{50} was 0.40 kg a.e./ha) compared to the commercial formulation of this herbicide that took a higher dose (ED_{50}0.48 kg a.e./ha) when applied as such. Therefore, the nano-emulsion herbicide ensures the increased penetration and uptake of herbicide glyphosate.

Sustainability remains the central idea of modern agriculture, particularly when the world is desperate for a 'second green revolution.' Under this circumstance there is need of a significant and effective strategy ensuring fruitful results in plant disease management. The present scenario of agricultural systems discourages the non-judicious use of chemical pesticides. At the same time the growers are eagerly looking for the replacement of such hazardous chemicals. Nanotechnology ensures the reduction in rate of chemical (pesticide) amount with enhanced time of its active phase (Nair et al., 2010; Ghormade et al., 2011). Among the metal nanoparticles used as pesticides; silver has drawn special attention of researchers. Several workers have found great inhibitory property of silver nanoparticle to various phytopathogenic fungi (Kasprowicz et al., 2010; Min et al., 2009; Jo et al., 2009; Park et al., 2006).

18.3 GLIMPSES OF NANOPARTICLE APPLICATION FOR DIFFERENT FUNGAL PATHOSYSTEMS

18.3.1 PHYTOPATHOGENIC FUNGI TO METALLIC NANOPARTICLES

Various phytopathogenic fungi were evaluated for its response to metallic nanoparticles (Figure 18.2). Two foliar fungi and a root-infecting fungus were tested for this study. The foliar fungi included *Bipolaris sorokiniana* causing spot blotch of wheat and *Alternaria* spp. causing blight of mustard; similarly, *Fusarium oxysporum* causing lentil wilt disease was involved in this study. Mycelia growth was assessed on different concentrations of copper and silver nanoparticles. Fungal growth was assessed in nanoparticle amended medium and comparison was made with the growth assessed on non-amended medium. Our result strongly supported the previously discussed results (Min et al., 2009; Kim et al., 2009, 2012); greater mycelia growth inhibition obtained for silver nanoparticle for all the tested fungi. The similar result obtained in a pot experiment for collar rot disease with different formulations of copper and silver nanoparticles. No diseased seedling was detected for the treatment of silver nanoparticles; however, 8–17% infected seedlings were observed in copper nanoparticle treatment. This experiment was conducted under artificially created heavy disease pressure; the control pots found with nearly 80% infected seedlings. We got a clue from the series of nanoparticle-amendment experiments that deep-colored spores are more susceptible to metal nanoparticle. *F. oxysporum* with white to grayish mycelium was least affected with either nanoparticles used. Our results further invite deep investigation on impact of nanoparticle on melanin synthesis in phytopathogenic fungi.

18.3.2 PHYTOPATHOGENIC FUNGI TO BIODEGRADING NANOPARTICLES

The sclerotia-producing phytopathogenic fungus, *Sclerotium rolfsii* causing collar rot of chickpea, was evaluated for sclerotia germination

(Figure 18.3). Harvested sclerotia were dipped in the colloidal suspension of chitosan-based nanoformulation for the presented period. The assessment was made on day 6 when no further germination was detected. Even a 4-h dipping in the nanochitosan colloid significantly reduced sclerotial germination. One day (24-h) dipping resulted in less than 40% sclerotial germination. Complete inhibition in germination of sclerotia was recognized for a dipping period of one and one-third day (32-h). No variation in germination of sclerotia was detected for 0-h dipping in nanochitosan colloid and in water. Min et al. (2009) also found effectiveness of colloidal formulation with silver nanoparticle in reducing the sclerotial germination of different sclerotia-producing fungi. Similarly, Mendes et al. (2014) also observed the great inhibition property of silver colloid nanoparticles to the soybean seed fungus (*Phomopsis* sp.). One of the pot experiments also revealed with significant reduction in chickpea collar rot disease by using the biodegrading nanoparticle of chitosan origin. Considerable disease appeared in control pots but the nanochitosan rendered 69–74% disease inhibition. These results indicate for great potentiality of chitosan-based (biodegrading) nanoformulation for application of soil-borne phytopathogenic fungus, *S. rolfsii*. This primary information could be useful to develop a novel fungicide for soil-borne disease management.

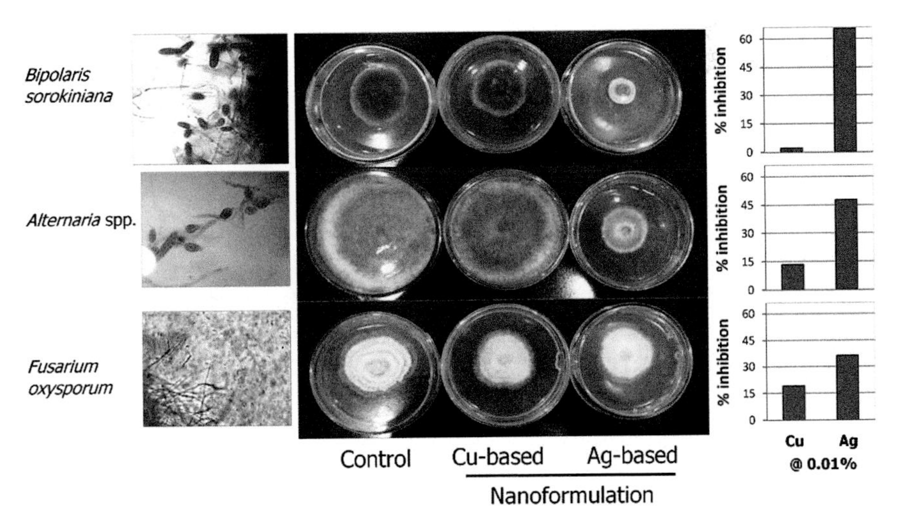

FIGURE 18.2 Response of three phytopathogenic fungi to copper and silver nanoparticles.

18.3.3 EXPLOITING NANOCLAY AS FUNGICIDE

Application of nanoclay was tested for lentil collar rot disease (*Sclerotium rolfsii*). Plants (7-days-old) were grown in pot-mix and sclerotia of the fungus were inoculated near the collar zone of the plant. Scheduling of nanoclay application was tested for 3 hours before, and 3, 6, and 24 hours after inoculation. Within a treatment 280–325 plants were assessed for disease development. A trend of increased disease percent was observed with delaying the nanoclay application. Approximately, 50% low disease incidence was recorded in nanoclay application after 24 hours of inoculation compared to inoculated control without nanoclay application. The result strongly suggests the more early nanoclay application the less disease appearance. However, this result should be reinvestigated for other phytopathosystems.

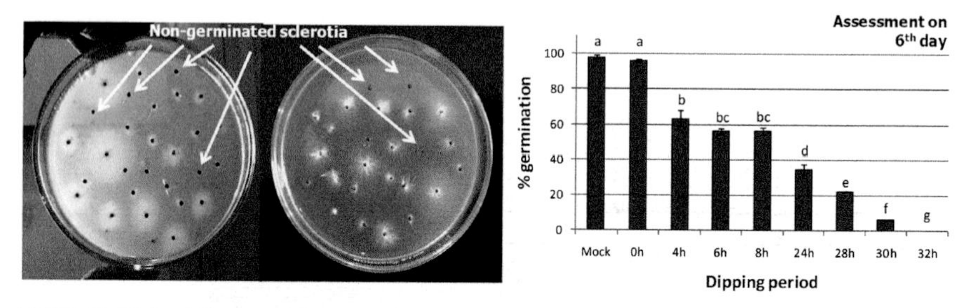

FIGURE 18.3 Dipping effect of sclerotia of *Sclerotium rolfsii* in nanochitosan colloid.

18.4 CONCLUSION

The advent of new technologies in modern agriculture definitely increased production. However, the progress involving plant disease management in a parallel direction is now advocating to hunt some other options or to find a sustainable way. In this connection, the reduction of harmful chemicals is a mandatory target although the growers do not rely on the significant impact of biopesticides, because this method requires a comparably long time to perform. Therefore, the other way as a replacement of chemical pesticide would be introducing the nanoparticle for plant disease management. The nanochemicals used in a meager quantity, therefore, reducing chemical load on the ecosystem, and they are very much effective in a persistent way to act. Many authors found the considerable inhibitory impact of silver nanoparticles on several phytopathogenic fungi. Our study also supports this fact. Additionally, the study discussed in this chapter recommends exploring for chitosan-based nanoformulation because of its biodegrading nature.

KEYWORDS

- **disease management**
- **fungal pathosystems**
- **nanoclay**
- **nanoparticles**
- **phytopathogenic fungi**

REFERENCES

Beumer, K., & Bhattacharya, S., (2013). Emerging technologies in India: Developments, debates and silences about nanotechnology. *Science and Public Policy*, 1–16.

Fateixa, S., Neves, M. C., Almeida, A., Oliveira, J., & Trindade, T., (2009). Anti-fungal activity of SiO_2/Ag_2S nanocomposites against *Aspergillus niger*. *Colloids and Surfaces B: Biointerfaces, 74*, 304–308.

Feng, Q. L., Wu, J., Chen, G. Q., Cui, F. Z., Kim, T. M., & Kim, J. O., (2000). A mechanistic study of the antibacterial effect of silver ions on *Escherichia coli* and *Staphylococcus aureus*. *J. Biomed. Mater. Res., 52*, 662–668.

Garg, J., Poudel, B., & Chiesa, M., (2008). Enhanced thermal conductivity and viscosity of copper nanoparticles in ethyleneglycol nanofluid. *J. Appl. Phys., 103*, 074301.

Ghormade, V., Deshpande, M. V., & Paknikar, K. M., (2011). Perspectives for nanobiotechnology enabled protection and nutrition of plants. *Biotechnol. Adv., 29*, 792–803.

Jo, Y. K., Kim, B. H., & Jung, G., (2009). Antifungal activity of silver ions and nanoparticles on phytopathogenic fungi. *Plant Dis., 93*, 1037–1043.

Kasprowicz, M. J., Kozioł, M., & Gorczyca, A., (2010). The effect of silver nanoparticles on phytopathogenic spores of *Fusarium culmorum*. *Can J. Microbiol., 56*(3), 247–253.

Kim, S. W., Jung, J. H., Lamsal, K., Kim, Y. S., Min, J. S., & Lee, Y. S., (2012). Antifungal effects of silver nanoparticles (AgNPs) against various plant pathogenic fungi. *Mycobiology, 40*(1), 53–58.

Kim, S. W., Kim, K. S., Lamsal, K., Kim, Y. J., Kim, S. B., Jung, M., Sim, S. J., Kim, H. A., Chang, S. J., Kim, J. K., & Lee, Y. S., (2009). An *in vitro* study of the antifungal effect of silver nanoparticles on oak wilt pathogen *Raffaelea* sp. *Microbiol. Biotechnol., 19*(8), 760–764.

Lamsal, K., Kim, S. W., Jung, J. H., Kim, Y. S., Kim, K. S., & Lee, Y. S., (2011). Application of silver nanoparticles for the control of *Colletotrichum* species *in vitro* and pepper anthracnose disease in field. *Mycobiology, 39*(3), 194–199.

Li, Z. Z., Chen, J. F., Liu, F., Liu, A. Q., Wang, Q., et al., (2007). Study of UV-shielding properties of novel porous hollow silica nanoparticle carriers for avermectin. *Pest Manag. Sci., 63*, 241–246.

Lim, C. J., Basri, M., Omar, D., Rahman, M. B. A., Salleh, A. B., & Rahman, R. N. Z. A., (2012a). Physicochemical characterization and formation of glyphosate-laden nano-emulsion for herbicide formulation. *Industrial Crops and Products, 36*, 607–613.

Lim, C. J., Basri, M., Omar, D., Rahman, M. B. A., Salleh, A. B., Rahman, R. N. Z. A., & Selamat, A., (2012b). Green nano-emulsion intervention for water-soluble glyphosate isopropylamine(IPA) formulations in controlling *Eleusineindica* (*E. indica*). *Pesticide Biochemistry and Physiology, 102*, 19–29.

Lin, D. H., & Xing, B. S., (2007). Phytotoxicity of nanoparticles: Inhibition of seed germination and root growth. *Environmental Pollution, 150*, 243–250.

Maity, A., Natarajan, N., Vijay, D., Srinivasan, R., Pastor, M., & Malaviya, D. R., (2016). Influence of metal nanoparticles (NPs) on germination and yield of oat (*Avena sativa*) and Berseem (*Trifolium alexandrinum*). *Proc. Natl. Acad. Sci., India, Sect. B Biol. Sci.* doi: 10.1007/s40011–016–0796-x.

Mendes, J. E., Abrunhosa, L., Teixeira, J. A., De Camargo, E. R., De Souza, C. P., & Pessoa, J. D. C., (2014). Antifungal activity of silver colloidal nanoparticles against phytopathogenic fungus (*Phomopsis* sp.) in soybean seeds. *International Journal of Biological, Veterinary, Agricultural and Food Engineering, 8*(9), 928–933.

Min, J. S., Kim, K. S., Kim, S. W., Jung, J. H., Lamsal, K., Kim, S. B., Jung, M., & Lee, Y. S., (2009). Effects of colloidal silver nanoparticles on sclerotium forming phytopathogenic fungi. *Plant Pathol. J., 25*, 376–380.

Mohanpuria, P., Rana, K. N., & Yadav, S. K., (2008). Biosynthesis of nanoparticles: Technological concepts and future applications. *Journal of Nanoparticle Research, 10*, 507–517.

Nair, R., Varghese, S. H., Nair, B. G., Maekawa, T., Yoshida, Y., & Kumar, D. S., (2010). Nanoparticulate material delivery to plants. *Plant Sci., 179*, 154–163.

Ombodi, A., & Saigusa, M., (2000). Broadcast application versus band application of polyolefin-coated fertilizer on green peppers grown on and isol. *J. Plant Nutr., 23*, 1485–1493.

Pal, S., Tak, Y. K., & Song, J. M., (2007). Does the antibacterial activity of silver nanoparticles depend on the shape of the nanoparticle? A study of the gram-negative bacterium *Escherichia coli*. *Appl. Environ. Microbiol., 73*(6), 1712–1720.

Park, H. J., Kim, S. H., Kim, S. J., & Choi, S. H., (2006). A new composition of nanosized silica-silver for control of various plant diseases. *Plant Pathol. J., 22*, 295–302.

Parvathamma, N., & Gobbur, D. S., (2011). Scientific productivity and Impact of Indian polymer scientists (1992–2006): An informetric study. *COLLNET Journal of Scientometrics and Information Management, 5*(2), 205–215.

Pieterse, C. M., & Dicke, M., (2007). Plant interactions with microbes and insects: From molecular mechanisms to ecology. *Trends Plant Sci., 12*, 564–569.

Pineda, A., Zheng, S. J., Van Loon, J. J. A., Pieterse, C. M. J., & Dicke, M., (2010). Helping plants to deal with insects: The role of beneficial soil-borne microbes. *Trends Plant Sci., 15*, 507–514.

Salem, W., Leitner, D. R., Zingl, F. G., Schratter, G., Prassl, R., Goessler, W., Reidl, J., & Schild, S., (2015). *Antibacterial Activity of Silver and Zinc Nanoparticles Against Vibrio Cholerae and Enterotoxic Escherichia coli.*

Shrestha, S., Yeung, C. M. Y., Nunnerley, C., & Tsang, S. C., (2007). Comparison of morphology and electrical conductivity of various thin films containing nano-crystalline praseodymium oxide particles, Sens. *Actuators A: Phys., 136*, 191–198.

Smekalova, M., Aragon, V., Panacek, A., Prucek, R., Zboril, R., & Kvitek, L., (2016). Enhanced antibacterial effect of antibiotics in combination with silver nanoparticles against animal pathogens. *The Veterinary J., 209*, 174–179.

Solanki, P., Bhargava, A., Chhipa, H., Jain, N., & Panwar, J., (2015). Nano-fertilizers and their smart delivery system. In: Rai, M., Ribeiro, C., Mattoso, L., & Duran, N., (eds.), *Nanotechnologies in Food and Agriculture* (pp. 81–101). Springer.

Yin, L., Colman, B. P., McGill, B. M., Wright, J. P., & Bernhardt, E. S., (2012). Effects of silver nanoparticle exposure on germination and early growth of eleven wetland plants. *PLoS One, 7*(10), e47674. doi: 10.1371/journal.pone.0047674.

CHAPTER 19

Status of Insect Pests of Cereals in India and Their Management

ANIL, TARAK NATH GOSWAMI, and SANJAY KUMAR SHARMA

Department of Entomology, Bihar Agricultural University, Sabour, Bhagalpur, Bihar, India.
E-mail: aniljakhad@gmail.com (Anil)

ABSTRACT

Cereals occupy the important place in area, production and diet composition across the world. These crops suffer due to ravages of insect pests which are considered as an important limiting factor in their production. Global warming, aberrant weather, changing cropping patterns and adoption of technologies alter the insect abundance, distribution and pest associated losses over the time and space. This chapter focuses on the status of insect pests of important cereals crops like, rice, wheat, maize, sorghum and pearl millet in India. Besides, the emerging insect pests of cereals in India and work done by the university on insect pests management in cereals are also highlighted. Management practices of important insect pests of cereal crops are also summarized in the chapter.

19.1 INTRODUCTION

Globally, the cereals constitute a major proportion of the human diet and India is the second-largest producer of rice, wheat, and other cereals. During 2013–14, India produced 106.54, 95.85, 24.26 and 5.39 million tonnes of rice, wheat, maize and sorghum over the acreages of 43.95, 31.19, 9.43 and 5.82 million hectares, respectively (Anonymous, 2014a). These crops suffer due to the ravages of various insect pests besides other biotic stresses. Additionally, global warming, aberrant weather, and changing cropping patterns are playing important roles in emerging insect pest problems. The insect pests associated crop losses in India are around 15.7% which account for the annual losses of US$ 36 billion (Dhaliwal et al., 2015). After the green revolution in the country, the pest associated losses in cereal crops are 21.3% which increased by 15.9% compared with the losses during the pre-green revolution period. The crop-wise details of such a shift in pest-associated losses are presented in Table 19.1. The pest associated losses in rice, wheat, maize, and both sorghum and millets are 25.0, 5.0, 25.0 and 30.0%, respectively. This indicates the respective increase in such losses by 15.0, 2.0, 20.0 and 26.5% after the green revolution.

TABLE 19.1 Changes in Pest Associated Losses (%) in Cereals During Pre- and Post-Green Revolution in India (Adapted from Dhaliwal et al., 2007)

Crop	Pre-Green Revolution (the Early 1960s)	Post-Green Revolution (the Early 2000s)	Changes in Pest Associated Losses
Rice	10.0	25.0	+ 15.0
Wheat	3.0	5.0	+ 2.0
Maize	5.0	25.0	+ 20.0
Sorghum and millets	3.5	30.0	+ 26.5
Average	5.4	21.3	+ 15.9

19.2 INSECT PESTS OF CEREALS

The summarized status of insect pests of cereals including rice, wheat, maize, sorghum, and pearl millet in India is mentioned hereunder.

19.2.1 RICE

More than 100 species of insects and mites are associated with rice and only 20 of these are considered economically important in India. The details of important insect and mite pests are given in Table 19.2. Insects namely, yellow stem borer, brown planthopper (BPH), white-backed planthopper, leaf folder, gundhi bug, and gall midge are having national significance. Except for these insects, few other insects and mites are having regional significance and their distribution is limited to specific areas within the country. For instance, the infestation of caseworm is observed in low-lying and waterlogged areas of eastern India. Similarly, the white grubs infesting

TABLE 19.2 Insect and Mite Pests of Rice in India and Their Distribution (Prakash et al., 2014)

Insect Pest	Scientific Name	Order	Family	Distribution
Yellow stem borer	*Scirpophaga incertulas* (Walker)	Lepidoptera	Crambidae	Throughout the country
Brown planthopper	*Nilaparvata lugens* (Stal.)	Hemiptera	Delphacidae	
White-backed plant hopper	*Sogatella furcifera* (Horvath)	Hemiptera	Delphacidae	
Leaf folder	*Cnaphalocrosis medinalis* (Guenee)	Lepidoptera	Crambidae	
Gundhi bug	*Leptocorisa acuta* (Thunberg)	Hemiptera	Coreidae	
Gall midge	*Orseolia oryzae* (Wood-Mason)	Diptera	Cecidomyiidae	
Termite	*Odontotermes obesus* (Rambur)	Isoptera	Termitidae	Rainfed upland areas irrigated rice-wheat system
Swarming caterpillar	*Spodoptera mauritia* (Boisduval)	Lepidoptera	Noctuidae	Low lying areas in Bihar, Odisha, West Bengal, Assam, Jharkhand, Chhattisgarh, and Punjab
Green leafhopper	*Nephotettix nigropictus* (Stal.), *N. virescens* (Distant)	Hemiptera	Cicadellidae	Bihar, West Bengal, Assam, Orissa, Madhya Pradesh, Andhra Pradesh, and Tamil Nadu
Rice hispa	*Dicladispa armigera (*Oliver)	Coleoptera	Chrysomelidae	Bihar, West Bengal, Assam, Odisha, Meghalaya, Mizoram, Tripura, Punjab, Himachal Pradesh, Uttar Pradesh, and Uttarakhand
Climbing cutworm	*Mythimna separata* (Walker)	Lepidoptera	Noctuidae	Coastal areas, Haryana, Punjab, and Uttar Pradesh
Caseworm	*Nymphula depunctalis* Guenee	Lepidoptera	Pyralidae	Low lying and waterlogged areas in eastern India
Thrips	*Stenchaetothrips biformis* (Bagnall)	Thysanoptera	Thripidae	Upland areas in Odisha, Andhra Pradesh, Madhya Pradesh, Punjab, Haryana, Assam, and Tamil Nadu
Mealybug	*Brevennia rehi* (Lindinger)	Hemiptera	Pseudococcidae	Upland areas in Uttar Pradesh, Bihar, West Bengal, Odisha, Madhya Pradesh, Tamil Nadu, Kerala, Pondi-cherry and Karnataka

TABLE 19.2 *(Continued)*

Insect Pest	Scientific Name	Order	Family	Distribution
Panicle mite	*Steneotarsonemus spinki* Smiley	Acari	Tarsonemidae	Andhra Pradesh, Odisha, West Bengal, Gujarat, and western Uttar Pradesh
Leaf mite	*Oligonychus oryzae* (Hirst)	Acari	Tetranychidae	Eastern India and Andhra Pradesh
Root weevil	*Echinocnemus oryzae* (Marshall)	Coleoptera	Curculionidae	Haryana, Punjab and Tamil Nadu
White grub	*Holotrichia* spp.	Coleoptera	Melolonthidae	Hilly areas
Black bug	*Scotinophara coarctata* (Fabricius)	Hemiptera	Pentatomidae	Andhra Pradesh, Tamil Nadu, and Kerala
Blue beetle	*Leptisma pygmaea* Baly	Coleoptera	Chrysomelidae	Kerala, Maharashtra and Tamil Nadu

rice are limited to hilly areas. In Bihar, the insects like swarming caterpillar, green leafhopper, rice hispa, and mealybug are important regional insect pests of rice.

19.2.2 WHEAT

This crop is relatively less attacked by the insect pests and around 8 insect pests are considered economically important in wheat (Table 19.3). Insects like termite, wheat aphid, armyworm, American pod borer, pink stem borer, and shoot fly are of national significance. However, wheat thrips and ghujia weevil are limited to the few areas. For instance, the infestation of ghujia weevil in wheat is found in Uttar Pradesh.

TABLE 19.3 Insect Pests of Wheat in India and Their National/Regional Importance (Satyagopal et al., 2014)

Insect Pest	Scientific Name	Order	Family	Importance
Termite	*Odontotermes obesus* (Rambur), *Microtermes obesi* Holmgren	Isoptera	Termitidae	National
Wheat aphid	*Sitobion avenae* (F.), *S. miscanthi* (Takahashi)	Hemiptera	Aphididae	
Armyworm	*Mythimna separata* (Walker)	Lepidoptera	Noctuidae	
American pod borer	*Helicoverpa armigera* (Hübner)	Lepidoptera	Noctuidae	
Pink stem borer	*Sesamia inferens* (Walker)	Lepidoptera	Noctuidae	
Shoot fly	*Atherigona naqvii* Steyskal, *A. oryzae* Mall	Diptera	Muscidae	
Wheat thrips	*Anaphothrips favicinctus, Haplothrips tritici* (Kurdjumov)	Thysanoptera	Thripidae	Regional
Ghujhia weevil	*Tanymecus indicus* Faust	Coleoptera	Curculionidae	

19.2.3 MAIZE

More than 130 insect species are associated with maize and 12 of which are considered economically important insect pests of maize in India. The details of these insect pests are given in Table 19.4. Maize stem borer, pink stem borer, and shoot fly are considered important pests at the national level. Other insects like white grub, cutworm, hairy caterpillar, aphid, armyworm, pyrilla, thrips, termite and chafer beetle are limited to specific areas of the country.

TABLE 19.4 Insect Pests of Maize in India and Their National/Regional Importance (Kumar et al., 2014)

Insect Pest	Scientific Name	Order	Family	Importance
Maize stem borer	*Chilo partellus* (Swinhoe)	Lepidoptera	Crambidae	National
Pink stem borer	*Sesamia inferens* (Walker)	Lepiodptera	Noctuidae	
Shoot fly	*Atherigona* spp.	Diptera	Muscidae	
White grub	*Holotrichia consanguinea* Blanchard	Coleoptera	Scarabaeidae	Regional
Cutworm	*Agrotis ipsilon* (Hufnagal) *A. segetum* (Denis and Schiffermuller)	Lepidoptera	Noctuidae	
Hairy caterpillar	*Amsacta albistriga* Walker	Lepidoptera	Arctiidae	
Aphid	*Rhopalosiphum maidis* (Fitch)	Hemiptera	Aphididae	
Army worm	*Mythimna separata* (Walker)	Lepidoptera	Noctuidae	
Pyrilla	*Pyrilla purpusilla* Walker	Hemiptera	Lophopidae	
Thrips	*Anaphothrips sudanensis* Trybom	Thysanoptera	Thripidae	
Termite	*Microtermes obesi* Holmgren	Isoptera	Termitidae	
Chafer beetle	*Chiloloba acuta* (Wiedmann)	Coleoptera	Scarabaeidae	

19.2.4 SORGHUM

Out of 150 insect pests, only 20 insect and mite species are considered economically important pests of sorghum in India (Table 19.5). Shoot fly, stem borer, midge, white grub are distributed throughout the country and therefore, these pests are of national significance. Insects like armyworm, cutworm, grasshopper, pyrilla, earhead caterpillars, shoot bug, earhead bug and aphid are limited to few specific areas. Similarly, the distribution of spider mite is limited to specific areas like Karnataka, Punjab, Maharashtra and Tamil Nadu.

TABLE 19.5 Important Insect and Mite Pests of Sorghum in India and Their Distribution (Anonymous, 2014b)

Insect Pest	Scientific Name	Order	Family	Distribution
Shoot fly	*Atherigona soccata* (Rondani)	Diptera	Muscidae	Throughout the country
Stem borer	*Chilo partellus* (Swinhoe)	Lepidoptera	Crambidae	
Midge	*Stenodiplosis sorghicola* (Coquillett)	Diptera	Cecidomyiidae	
White grub	*Holotrichia consanguinea* Blanchard	Coleoptera	Scarabaeidiae	
Armyworm	*Mythimna separata* (Walker)	Lepidoptera	Noctuidae	Karnataka, Maharashtra
Cutworm	*Agrotis ipsilon* (Hufnagal)	Lepidoptera	Noctuidae	—
Grasshopper	*Hieroglyphus* spp.	Orthoptera	Acrididae	Karnataka, Rajasthan, Tamil Nadu, Madhya Pradesh, Uttar Pradesh, Haryana, Delhi
Pyrilla	*Pyrilla perpusilla* (Linnaeus)	Hemiptera	Lophopidae	Maharashtra, Madhya Pradesh
Earhead caterpillars	*Helicoverpa armigera* (Hübner), *Eublemma* spp.	Lepidoptera	Noctuidae	Gujarat, Andhra Pradesh, Tamil Nadu, Haryana, Madhya Pradesh, Delhi, Maharashtra
	Cryptoblabes spp.	Lepidoptera	Pyralidae	
Shoot bug	*Peregrinus maidis* (Ashmead)	Hemiptera	Delphacidae	Andhra Pradesh, Madhya Pradesh, Karnataka, South India, Tamil Nadu, Bihar, Maharashtra
Earhead bug	*Calocoris angustatus*	Hemiptera	Miridae	Karnataka, Andhra Pradesh, Tamil Nadu, Uttar Pradesh
Aphid	*Rhopalosiphum maidis* (Fitch)	Hemiptera	Aphididae	Andhra Pradesh, Delhi, Gujarat, Maharashtra
Spider mite	*Oligonychus indicus* Hirst	Acari	Tetranychidae	Karnataka, Punjab, Maharashtra and Tamil Nadu

19.2.5 PEARL MILLET

Out of various insect pests associated with pearl millet, around 12 insect species are considered important and details of those are presented in Table 19.6. White grub and cutworm are considered important insect pests at the national level. Other insects like shoot fly, grasshopper, white ants, grey weevil, stem borer, earhead bug, hairy caterpillar, earhead worm, blister beetle, and chaffer beetle are limited to few areas within the country.

TABLE 19.6 Important Insect Pests of Pearl Millet in India and Their National/Regional Importance (Anonymous, 2014c)

Insect Pest	Scientific Name	Order	Family	Importance
White grub	*Holotrichia consanguinea* Blanchard	Coleoptera	Scarabaeidiae	National
Cutworm	*Agrotis ipsilon* (Hufnagal)	Lepidoptera	Noctuidae	
Shoot fly	*Atherigona soccata* (Rondani)	Diptera	Muscidae	Regional
Grasshopper	*Hieroglyphus* spp.	Orthoptera	Acrididae	
White ants	*Chrotogonu* sp.	Isoptera	Termitidae	
Grey weevil	*Myllocerus* sp.	Coleoptera	Curculionidae	
Stem borer	*Chilo partellus* (Swinhoe)	Lepidoptera	Crambidae	
Earhead bug	*Calocoris angustatus* Lethierry	Hemiptera	Miridae	
Hairy caterpillar	*Spilosoma obliqua* Walker	Lepidoptera	Arctiidae	
Earhead worm	*Cryptoblabes gnidiella* Milliere	Lepidoptera	Pyralidae	
Blister beetle	*Mylabris pustulata* Thun.	Coleoptera	Meloidae	
Chaffer beetle	*Rhizotrogus majalis* (Razoumowsky)	Coleoptera	Scarabaeidae	

19.3 EMERGING INSECT PEST OF CEREALS

The crop plants are damaged by more than 10,000 species of insect and the level of infestation has been changing due to various factors like global warming, aberrant weather conditions, changing cropping patterns, adoption of technologies and modification in farming practices. Additionally, various pesticides associated problems like pesticide resistance, resurgence, and environmental contamination are also accountable for the problem of emerging insect pests in specific regions. Various insects that have become or expected to become serious pests on cereal crops are summarized in Table 19.7. Insects like BPH, green leafhopper, and gall midge have become the serious pests of rice. The outbreaks of BPH occurred in various states like Karnataka, Andhra Pradesh, Madhya Pradesh, Odisha, and Tamil Nadu during the 1970s and 1980s. New biotypes of gall midge have been reported since the cultivation of gall midge resistance varieties in the country. The problem of aphids in wheat has recently emerged and now it has gained national significance. Additionally, the shoot fly has emerged as an important and regular pest of the late sown wheat crop since the adoption of semi-dwarf varieties. The infestation of pink stem borer is increased in wheat, maize, and sorghum. The problem of maize stem borer and midge has become more prominent in sorghum. Additionally, several polyphagous insects like termites, white grubs, hairy caterpillars, etc., are also becoming prominent in specific areas.

TABLE 19.7 Insect Pests That Have Become or Are Expected to Become Serious Pests on Cereal Crops in India (Sharma et al.,1999, 2005; Sharma, 2010; Dhaliwal et al., 2010)

Insect Pest	Scientific Name	Crop(s)
Brown plant hopper	*Nilaparvata lugens* (Stal.)	Rice
Green leafhopper	*Nephotettix* spp.	Rice
Gall midge	*Orseolia oryzae* (Wood-Mason)	Rice
Wheat aphid	*Sitibion miscanthi* (Takahashi)	Wheat
Shoot fly	*Atherigona naqvii* Steyskal	Wheat
Pink stem borer	*Sesamia inferens* (Walk.)	Maize, sorghum, wheat
Aphid	*Rhopalosiphum maidis* (Fitch)	Maize
Midge	*Stenodiplosis sorghicola* (Coquillett)	Sorghum
Maize stem borer	*Chilo partellus* (Swinhoe)	Sorghum
Polyphagous insects (termite, white grubs, etc.)	Several species	Many crops

19.4 WORK DONE IN THE UNIVERSITY

The Bihar Agricultural University (BAU), Sabour has conducted work on the management of insect pest of cereals in the form of projects and student's thesis. Kumari et al. (2015) conducted work on the management of BPH infesting rice, stem borer infesting maize and for the development of integrated pest management (IPM) modules for pest complex in rice and maize. The work on biological control of insect pests was initiated and Goswami et al. (2015) conducted the quantitative estimation of spider fauna in rice ecosystem from zone IIIA in Bihar. Similarly, Yadav et al. (2016) studied the species composition of spider-fauna in paddy ecosystem throughout the cropping period. The effect of chemical and non-chemical pest management modules on predatory spiders in paddy-ecosystem was studied by Goswami et al. (2016). Keeping in view the higher infestation of insects in aromatic rice, Patel (2016) studied the population dynamics of yellow stem borer on aromatic rice and evaluated newer chemical for its management. Besides, the university has also initiated research work on ecological pest management of rice.

19.5 MANAGEMENT OF IMPORTANT INSECT PESTS OF CEREALS

The management practices including cultural, physical, biological, chemical, etc., for important insect pests of cereals are discussed.

19.5.1 YELLOW STEM BORER, SCIRPOPHAGA INCERTULAS

- Removal and destruction of rice stubbles at the time of first plowing after harvesting the crop decrease the carry-over of the hibernating population if insect to the next crop.
- Since the eggs of stem borer are laid near the tip of the leaf, clipping off the tips of seedlings before transplanting can reduce the carry-over to the main field.
- Seedlings root dip treatment for 12 or 14 hours before transplanting with chlorpyriphos (0.02%) can provide protection up to 30 days against stem borer.
- Conserve and augment the natural enemies like spiders, dragonflies, and damselflies.
- Install pheromone trap @ 5 traps ha^{-1} for monitoring and @ 20 traps ha^{-1} for mass trapping of the male. Besides, light traps @ 1 trap ha^{-1} can also be used for mass trapping of the adult male and females.
- Release egg parasitoid, *Trichogramma japonicum* @ 50,000 ha^{-1} from 30 days after transplanting at weekly intervals.
- If the field shows 2 egg masses/m^2 or >10% dead heart or > 5% white ear:
    - Apply granular insecticides like cartap hydrochloride 4G @ 20–25 kg or fipronil 0.3GR @ 15 kg or phorate 10G @10 kg ha^{-1} in irrigated conditions.

- Spray cartap hydrochloride 50SP @ 600 g or triazophos 40EC @ 875 ml or monocrotophos 36SL @ 1100 ml ha^{-1} in non-irrigated conditions.

19.5.2 BROWN PLANTHOPPER, *NILAPARVATA LUGENS*

- Closer spacing of 15×10 cm creates a favorable microclimate in the field for the rapid development of hopper population. Hence, a spacing of 20×15 cm should be followed.
- Alternate drying and wetting the field during peak infestation of the hopper and draining out of standing water from the field 2–3 times check the population to a large extent.
- Alleys (30 cm wide) after every 3 meters of rice planting provide proper aeration to the crop which ultimately restricts the multiplication of pest and it is also helpful in the insecticidal spray.
- Grow resistant varieties like IR-8, Yijaya, IR-20, Mudgo, TKM-6, Parijat, Shakti, Triveni, Rajendra masuri-1, Pant dhan-2, Raj Shree, Bhudev, Suraksha, Kunti, Shaktiman, etc.
- High doses of nitrogenous fertilizers should be avoided as it favors the fast multiplication of planthoppers.
- Release *Cyrtorhinus lividipennis* @ 50–75 eggs/m^2 at 10 days intervals.
- If the population of BPH is >5–10 insects $hill^{-1}$:
    - Spray imidacloprid 17.8SL @125 ml or thiamethoxam 25 WG @ 100 g or buprofezin 25SC @ 400 ml or monocrotophos 36 SL @ 1500 ml ha^{-1}.
    - Repeat the application of insecticide if hopper population persists beyond a week after first application and while spraying the nozzle should be directed at the basal portion of the plants.
    - The same group of chemical should not be repeated.

19.5.3 GALL MIDGE, *ORSEOLIA ORYZAE*

- Plough the ratoons of an earlier crop to reduce the infestation.
- Clean grassy weeds and wild rice from surrounding areas as they serve as alternate host.
- Careful timing of planting can avoid damage; once the crop passes the tillering stage, the plant is not suitable for infestation.
- A considerable build-up of midge population on grasses near the rice crop can be avoided by removing the grasses.

- Seedling root dip in 0.02% chlorpyriphos for 12 hours before transplanting protects the crop for 25–30 days.
- Release larval parasitoid like *Platygaster oryzae*, if possible.
- If the damage is more than 5%:
  - Spray the crop with chlorpyriphos 20 EC @ 625 ml or quinalphos 25EC @ 500 ml ha^{-1}.
  - Alternatively, apply cartap hydrochloride 4G @ 25 kg or fipronil 0.3G @ 15 kg or phorate @ 12.5 kg ha^{-1} in the standing water in the field.
  - The same chemical should not be used repeatedly.

19.5.4 WHEAT APHID, *SITOBION* SPP.

- Grow 4 rows of maize/sorghum/pearl millet around the field as a barrier crop.
- Properly monitor the population of aphids through yellow pan.
- If the population is >5 aphids/earhead:
  - Spray thiamethoxam 25WG @ 50 g or quinalphos 25EC @ 250 ml or dimethoate 30 EC @ 375 ml ha^{-1}.

19.5.5 PINK STEM BORER, *SESAMIA INFERENS*

- Installation of pheromone traps for monitoring is useful.
- Release of *Trichogramma japonicum* @ 50,000/ha at weekly intervals.
- Spray triazophos 40EC @ 875 ml or monocrotophos 36SL @ 1.4 liters or chlorpyriphos 20EC @ 2.5 liters ha^{-1}.

19.5.6 MAIZE STEM BORER, *CHILO PARTELLUS*

- Destruction of stubble, weeds and other alternate hosts.
- Installation of pheromone traps is useful in monitoring.
- Removal and destruction of the dead heart are useful.
- Release *Trichogramma chilonis* @ 1,00,000 per ha at weekly intervals.
- Spray the crop after 2–3 weeks of sowing as soon as borer injury to the leaves is noticed with fenvalerate 20EC @ 100 ml or cypermethrin 10EC @ 100 ml or deltamethrin 2.8EC @ 200 ml ha^{-1}.
- Alternatively, apply carbofuran 3G or phorate 10G @ 0.5–1.5 kg in the whorls of the plants.

19.5.7 SHOOT FLY, ATHERIGONA SOCCATA

- The crop sown from early June to the second week of July normally escapes the attack of shoot fly.
- Grow resistant varieties like CHS-7, CHS-8, Indian Sorghum types IS-5566, 5285 and 5613.
- Install fish meal traps @ 12 traps ha^{-1} for mass trapping of shoot flies.
- Apply carbofuran 3G @ 12.5 kg or phorate 10G @ 10 kg ha^{-1} in furrows before sowing.
- Alternatively, spray malathion 50EC @ 1.25 liters ha^{-1}.

19.5.8 MIDGE, STENODIPLOSIS SORGHICOLA

- Both early and late maturing varieties should not be grown in the same area as it would provide the pest with a continuous supply of flowers. Further, the varieties having the same flowering and maturity time would reduce midge damage considerably.
- Since the damaged and aborted seeds are the main sources of carryover of the pest from one season to another, collect and burn the panicle and post-harvest trashes.
- Spray malathion 50EC @ 1.0 liter ha^{-1}. Only earhead should be treated at 90% panicle emergence followed by the second spray after 4–5 days.

19.6 CONCLUSION

The cereal crops have experienced changing trends of insect infestations since the advent of green revolution in the country. Few of the minor pests have emerged as important pests and caused significant yield losses at regional and national levels. Various factors like global warming, aberrant weather, changing cropping patterns and adoption of technologies alter the insect abundance, distribution, and pest associated losses. Work-related to survey and surveillance of insect pests across the country should be regularly conducted for getting proper information on changing trends of insect pests and associated losses in different crops including cereals. Additionally, the management practices for emerging insect pests of crops are needed to be devised or modified accordingly well in advance.

KEYWORDS

- **brown planthopper**
- **cereals**
- ***Chilo partellus***
- **emerging pests**
- **management practices**
- ***Scirpophaga incertulas***

REFERENCES

Anonymous, (2014a). *Agricultural Statistics at a Glance*. Directorate of economics and statistics, Ministry of Agriculture, Government of India.

Anonymous, (2014b). *AESA Based IPM-Sorghum* (p. 64). Department of agriculture and cooperation, Ministry of Agriculture, Government of India.

Anonymous, (2014c). *AESA Based IPM-Pearl Millet* (p. 51). Department of agriculture and cooperation, Ministry of Agriculture, Government of India.

Dhaliwal, G. S., Dhawan, A. K., & Singh, R., (2007). Biodiversity and ecological agriculture: Issues and perspectives. *Indian Journal of Ecology, 34*(2), 100–109.

Dhaliwal, G. S., Jindal, V., & Mohindru, B., (2015). Crop losses due to insect pests: Global and Indian scenario. *Indian Journal of Entomology, 77*(2), 165–168.

Goswami, T. N., Kumari, K., Anil, & Kole, B., (2015). Quantitative estimation of spider fauna in rice ecosystem of Zone IIIA in Bihar. *Environment and Ecology, 33*(2), 783–785.

Goswami, T. N., Yadav, M., Anil, & Ray, S. N., (2016). Effect of chemical *vis-à-vis* non-chemical pest management on predatory spiders in paddy-ecosystem. *Journal of Applied and Natural Science, 8*(1), 437–440.

Kumari, K., Bhagat, A. P., & Panwar, G. S., (2015). Development of integrated pest management in key crops of Bihar. *Final Report (RPF III)*. Directorate of Research, Bihar Agricultural University, Sabour (Bhagalpur) 813210, India.

Patel, V. K., (2016). Population dynamics and management of yellow stem borer, *Scirpophaga incertulas* (Walker) in aromatic rice. *M.Sc. Thesis* (p. 41). Department of Entomology, Bihar Agricultural University, Sabour.

Prakash, A., Bentur, J. S., Prasad, M. S., Tanwar, R. K., Sharma, O. P., Bhagat, S., et al., (2014). *Integrated Pest Management for Rice*, p. 43.

Satyagopal, K., Sushil, S. N., Jeyakumar, P., Shankar, G., Sharma, O. P., Sain, S. K., et al., (2014). *AESA Based IPM Package for Wheat*, p. 58.

Sharma, H. C., (2010). Global warming and climate change: Impact on arthropod biodiversity, pest management, and food security. In: Thakur, R. K., Gupta, P. R., & Verma, A. K., (eds.), *Souvenir, National Symposium on Perspectives and Challenges of Integrated Pest Management for Sustainable Agriculture* (pp. 1–14). Nauni, Solan, Himachal Pradesh, India.

Sharma, H. C., Dhillon, M. K., Kibuka, J., & Mukuru, S. Z., (2005). Plant defense responses to sorghum spotted stem borer, *Chilo partellus* under irrigated and drought conditions. *International Sorghum and Millets Newsletter, 46*, 49–52.

Sharma, H. C., Mukuru, S. Z., Manyasa, E., & Were, J., (1999). Breakdown of resistance to sorghum midge, *Stenodipiosis sorghicola. Euphytica, 109*, 131–140.

Yadav, M., Goswami, T. N., Anil, & Ray, S. N., (2016). Species composition of spider-fauna in paddy ecosystem throughout the cropping period at Sabour, Bihar, India. *Ecology, Environment and Conservation, 22*(2), 719–722.

CHAPTER 20

Major Insect Pests of Vegetable Crops in Bihar and Their Management

TAMOGHNA SAHA,[1] NITHYA CHANDRAN,[2] and B. C. ANU[1]

[1]*Department of Entomology, Bihar Agricultural University, Sabour, Bhagalpur, Bihar, India, E-mail: tamoghnasaha1984@gmail.com (T. Saha)*

[2]*Division of Entomology, Indian Agricultural Research Institute, New Delhi–110012, India*

ABSTRACT

Vegetables are the major ingredients of the Indian diet. Although, numerous strands are present which demarcate it's effectively, mainly insects and diseases. Various insects and mites result in damage to vegetable crops at diverse periods of growth namely aphid, thrips, whitefly, leafhopper, two-spotted spider mite, pumpkin beetles, hadda or epilachna beetles, brinjal shoot and fruit borer, tomato fruit borer, tobacco caterpillar and fruit fly. An amalgamation of insect pest control program is of foremost and usually works well to suppress damage and maintain the insects population below the level it can cause any economic loss to yield, i.e., EIL. The precise way to recognize the insect damage is the beginning of a successful pest control program. The pest management program includes use of strategies that involve different mechanical and cultural practices, use of tolerant plant varieties, growing insect pest and disease-free healthy seedlings, performing appropriate field sanitation particularly confiscating the fallen leaves, debris, undertaking weekly field monitoring to observe whether insects are present, conservation of farmers friendly insect (natural enemies) and biological control agents, application of need-based insecticides and safeguarding of society approach for all the recommended strategies to maximize benefits.

20.1 INTRODUCTION

Among the countries after China, India stands second in vegetable production, about 1,62,897 thousand metric tons. The present area in India is around 9396 thousand ha under vegetable cultivation (NHB, 2015). Majority of Indian population are vegetarian, and a per capita consumption 135 g per day as against the endorsed 300 g per day. Thus is still very less than suggested diet level. In nearby expectations, there is a requirement of around 5–6 million tons of food may be feed our 1.3 billion population of Indian supposed by the year 2020 (Paroda, 1999). Vegetable export of India is going down due to grown domestic need and other limitations in crop production. The major restraining factor, comprise the extensive crop devastations due to rising of pest menace. In many situations, there is 100% yield failure for viral diseases vectored by insects (Shivalingswami et al., 2002). Among them Insect pests play a major role in vegetable crops and yield loss are given (Table 20.1).

TABLE 20.1 Percent Yield Losses Due to Major Insect Pests in Vegetable Crops in India

Crop/Pest	Yield Loss (%)	Crop/Pest	Yield Loss (%)
Tomato		**Cabbage**	
Fruit borer (*Helicoverpa armigera*)	24–73	Diamondback moth (*Plutella xylostella*)	17–99
Brinjal		Cabbage caterpillar (*Pieris brassicae*)	69
Fruit and shoot borer (*Leucinodes orbonalis*)	11–93	Cabbage leaf webber (*Crocidolomia binotalis*)	28–51
Chilies		Cabbage borer (*Hellulaundalis*)	30–58
Thrips (*Scirtothrips dorsalis*)	12–90	**Cucurbits**	
Mites (*Polyphagotarsonemus latus*)	34	Fruit fly (*Bactrocera cucurbitae*)	
Okra		Bitter gourd	60–80
Fruit borer (*H. armigera*)	22	Cucumber	20–39
Leafhopper (*Amrasca biguttula biguttula*)	54–66	Ivy gourd	63
Whitefly (*Bemisia tabaci*)	54	Musk melon	76–100
Shoot and fruit borer (*Earias vittella*)	23–54	Snake gourd	63
		Sponge gourd	50

Recently changes in the cropping pattern and climate, and initiation of various high yielding varieties/hybrids which are input-intensive, a change in insect status has been noticed. This has resulted in dramatic changes in pest scenario leading to minor pests assuming major status, like serpentine leaf miner *Liriomyza trifolii* (Burgess), become a challenging problem in tomato hybrid growing pockets. Vegetable growers are mainly relying on chemical pesticides to face the insect's problem. It accounts for 13–14% of total pesticides consumption, as against 2.6% of cropped area (Sardana, 2001). Because of indiscriminating exploitation of synthetic pesticides there has been serious ecological consequences such as effect on nontarget organisms, shattering of fauna natural enemy, remnants in edible commodities including mineral water that are packed and pure, eventually resistance to the synthetic pesticides, to which we totally depend. In the Indian farm management system, bio-intensive pest management (BIPM) is the latest tendency that is attracting more and more farmers to enhance their produce's income. This consequence is because of increasing awareness among the end-users and concerned to the deterioration of ecological conditions between the eco-campaigners. Hence, it must be kept in our minds that it is not possible 100% alternatives to chemical pesticides and they are must when situations are demanding. Thus pest management the BIPM practices for major vegetable crops evolved by scientific approaches are discussed (Dhandapani et al., 2003). The major objective of this chapter is improvement in vegetables production per unit area by solving insect pest's problems through management practices. Among all the strategies the chief tactics to deal with the insects is to know insects' life cycle, behaviors, their natural environment, and feeding habits, and to identify between the harmful insect pests and the one that are actually helping hand in biological control. Pest identification is the most important step because misidentification is a common cause of control failure and knowing the correct pest is the key to select control measures that work effectually.

20.2 RECENT CHANGES IN INSECT PESTS' SITUATION IN VEGETABLE CROPS

A number of the pests under vegetable crops are slowly achieving the major pest status in diverse provinces of the country due to alteration in the ecosystem and habitats (Table 20.2). *Helicoverpa armigera* in tomato, whitefly *Bemisia tabaci*, leaf miner *Liriomyza trifolii* on tomato and cucurbitaceous crops, fruit fly on fruits and vegetables, mealybugs on several horticultural crops, gall midge on brinjal, okra stem fly and bitter gourd leafhopper, red spider mite on okra, brinjal, cowpea, Indian bean, and nematodes on several vegetable crops are few examples (Rai et al., 2014). In current times, the pests that are minor one and had less occurrence like gall midge (*Asphondylia capparis*), slowly becoming a serious trouble in capsicum and brinjal and Chili in the states of Karnataka and Andhra Pradesh and in brinjal in Chhattisgarh. Diverse species of mealybugs in papaya and vegetables have strengthened their severity of incidence in different parts of India due to gradual changes in weather in the past one decade. A number of national and global pests like white grub, termite, gram pod borer and *Spodoptera litura* that are polyphagus are becoming a severe problem and increasing their host-horizon. This visualizes that there is a requirement to reassess the cropping system for reversion and values should be given to vegetables with sufficient pest management strategies. In this background, the pest management system was extended for a sustainable crop protection strategy against the background of escalating pesticide use and deleterious effect of residues in the environment (Rai et al., 2014).

TABLE 20.2 Shifting Insect Pests Scenario in Vegetable Crops in India

Insect Pest	Major Host	Currently Infesting Crops	References
Serpentine leaf miner, *Liriomyzatrifolii* (Burgress)	Tomato	Brinjal, Cowpea, French bean, Leafy vegetables, Cucurbits	Srinivasan et al., 1995
Mealybug, *Phenacoccus solenopsis* Tinsley	Cotton	Brinjal, Tomato, Chili, Okra, Cucumber, Pumpkin	Chaudhary, 2006; Halder et al., 2013
Hadda beetle, *Henosepilachna vigintioctopunctata* Fab.	Brinjal	Bitter gourd, Cowpea	Rajapaske et al., 2005
Fruitborer, *Helicoverpa armigera* (Hubner)	Gram, Cotton, Tomato, Cabbage	Peas, Chili, Brinjal, Okra	Puri et al., 2000
Cabbage butterfly, *Pierisbrassicae* Linn	Cabbage, Cauliflower, Mustard	Knol-Khol, Radish	Puri et al., 2000
Red spider mite, *Tetranychus urticae* Koch.	Okra, Brinjal	Cucumber, Cowpea, Indian bean	Mahto and Yadav, 2009

20.3 MAJOR INSECT PESTS PROBLEM IN VEGETABLE CROPS

20.3.1 A. SOLANACEOUS CROPS

Brinjal shoot and fruit borer; *Leucinodes orbonalis* (Crambidae: Lepidoptera).

20.3.1.1 HOST RANGE

L. orbonalis Guenee is monophagous pest, feeding mainly on brinjal (eggplant); however, several other Solanaceous plants are reported to be hosts of this insect. In globally brinjal cultivation, *L. orbonalis*als occur on diverse host plants. The majority are *Solanum melongena* Linnaeus (eggplant), *and Solanum tuberosum* Linnaeus (potato), but there are several minor host, like *Ipomoea batatas* Linnaeus (sweet potato), *Lycopersicon esculentum* Mill (tomato), *Pisumsativum* var. *arvense* Linnaeus (Austrian winter pea) *Solanum indicum* Linnaeus, *Solanum myriacanthum* Dunal, and wild host *Solanum gilo* Raddi (gilo), *Solanum nigrum* Linnaeus (black nightshade) (CABI, 2007). In addition, *Solanum anomalum* Thonn (Singh et al., 2009), and *Solanum macrocarpon* Linnaeus (Kumar and Sadashiva, 1996) are uncultivated hosts of *L. orbonalis.*

20.3.1.2 DISTRIBUTION

The pest is reported from areas of brinjal cultivation in Africa, in the south portion of Sahara and Southern-Eastern region of Asia, including China and the Philippines (CABI, 2007). This pest attains most vital and the first categories pest of Nepal, Thailand, Srilanka, India, Cambodia, Bangladesh, Philippines, Laos and Vietnam (AVRDC, 1994). The pest is serious in those regions having humid and warm climate (Srinivasan, 2009).

20.3.1.3 LIFE CYCLE

In Lepidoptera, *L. orbonalis* undergoes four stages of development: egg stage, larval stage, pupal stage, and adult. The longest growth period is larval period, afterwards pupal period and incubation period.

1. **Egg:** During the night egg-laying (oviposition) takes place and eggs are laid singly lying on the abaxial plane of the tender leaves, green

stems, flower buds, or fruits and number of eggs laid by *L. orbonalis* varies from 80 to 253 (Alpuerto, 1994), however, as per report it lays up to 260 (FAO, 2003). Mostly they laid eggs in the early morning hours either singly or in the groups on the lower surface of the leaves (CABI, 2007). The eggs are flattened, oval with 0.5 mm in diameter and the color of the egg is creamy-white but before hatching the color changes to red (Alam et al., 2006).

2. **Larva:** The larval stage completed 12–15 days during summer and 14–22 days in the winter season (Rahman et al., 2006). Five instars have to be crossed from larval to the pupal stage.
3. **Pupa:** The mature larvae emerge from the infested shoots and fruits and further gone for pupation in the dried shoots and leaves or in plant remains fallen on the ground within tough silken cocoons. There were confirmations of the existence of cocoons at soil depths of 1 to 3 cm (Alam et al., 2003). They pupate on the plane they finger first (FAO, 2003). The pupal period completes 6 to 17 days depending upon heat (Alam et al., 2003).
4. **Adult:** The adult moth is tiny white with 40-segmented antennae (Sexena, 1965) and having marks on their forewings of 20 to 22 mm widen. The full maturity gained by adult in 10 to 14 calendar days. Longevity of adults lasts 1.5 to 2.4 calendar days for males and 2.0 to 3.9 days for females. The oviposition period ranges from 1.4 to 2.9 days, respectively.

20.3.1.4 NATURE AND EXTENT OF DAMAGE

L. orbonalis attacks mostly in fruits/pods that are vegetative stage, flowering and fruiting stage, they even attack on growing points and inflorescence (CABI, 2007). The larval percent is higher on fruits followed by shoots, flowers, buds of flower and leaf midribs. Within an hour of hatching, *L. orbonalis* larva bores into the nearest young shoot, blossom, or fruit. Soon after boring into shoots or fruits, they block the entrance hole (feeding tunnel) with excreta (Alam et al., 2006). Larval feeding inside the fruit results in destruction of fruit tissue. In harsh cases, decaying was common (Neupane, 2001). Fruits damage particularly in autumn is extremely severe and the entire crop can be ruined (Atwal, 1976). The pest is active throughout the year at several places having modest climate but its action is badly affected by severe cold.

20.3.1.5 MANAGEMENT

1. **Resistant Varieties:** It is used as the farmer's first line of protection against insect pests and most safest and compatible technique with other management approaches (Lit, 2009). Numerous attempts have been taken in South Asia to develop varieties resistant to *L. orbonalis*, but past 40 years of efforts, development of commercial cultivars with considerable level of resistance has not been done. But, it requirements the further exercise in future due to its apparent effectiveness (Srinivasan, 2008).
2. **Sex Pheromones:** These are chief constituent of IPM programs and are mostly utilized to observe as well as mass-trap the male insects. The pest management approach based on sex pheromone for managing the *L. orbonalis* has minimized pesticide use and enhanced the performance of natural enemies in Indo-Gangetic plains (IGP) of South Asia (Srinivasan, 2012). Zhu et al. (1987) recognized (E)-11-hexadecenyl acetate as the chief chemical constituent of *L. orbonalis* sex pheromone in China.
3. **Cultural Methods:** It includes cutting and removing of infested twigs and branches that prevent the spread of *L. orbonalis*. Likewise, the periodic pruning of damaged shoot, their collection and flaming help in reduction in pest infestation. Pruning will not unfavorably distress the plant growth with yield (Talekar, 2002). It is mostly imperative in the initial stages of the crop development and it must be carried on until the final harvest. Additionally, timely destruction of pest damaged eggplant shoots as well as fruits at normal intervals, reduced the pest (Srinivasan, 2008). Weekly exclusion of infested fruits and shoots resulted in the utmost load of strong fruits and least incidence of infested fruits among the treatments (Duca et al., 2004).
4. **Bio-Pesticides and Other Bio-Control Agents:** Chatterjee and Mondal (2012) revealed that *Bacillus thuringiensis* (Bt) (Berliner) worked well in reducing the infestation and increasing yield. Bt formulation Biolep was most effective than chemical insecticide (Sharma, 2002) and efficacy increased in consecutive days of spray. Among the parasitoid, the egg parasitoid *Trichogramma chilonis* Ishii have been demonstrated to be most effective (Krishnamoorthy, 2012) and it was slightly more efficient than *Trichogramma evanescens* Westwood in parasitizing the egg in test tube and in confined conditions.

5. **Botanical Pesticides:** Muregesam and Murugesh, (2009) confirmed the efficiency of Nimbecidine. Neem cake extract @ 5% and *Calotropis gigantean* (Linnaeus) @ 5% were also quite effective in *Kharif* crop suppressing fruit damage by more than 50%. The oil produce of plant based products was found to be effective for *L. orbonalis* management. Both Rabi (59.91%) and Kharif (60.20%) neem oil @ 2% was the most effective treatment both followed by Nimbecidine @ 2 ml/l (57.42%) (Murugesam and Murugesh, 2009).
6. **Chemical Methods:**
   - Spraying of rynaxypyr (Coragen 20 SC) 0.006%, flubendiamide (Fame 480 SC) 0.01%, spinosad (Spintor 45 SC) 0.0135% and emamectin benzoate (Proclaim 5 WG) 0.0025% (Table 20.3) provided superior control of *Leucinodes orbonalis* Guenee on eggplant (Saha et al., 2014).
   - Spraying of spinosad 45 SC @ 0.5 ml/l at 60 DAT followed by indoxacarb 14.5 SC @ 0.5 ml/l at 75 DAT and followed by emamectin benzoate 25 WG @ 0.4 g/l at 90 DAT gave (Table 20.4) superior control of shoot and fruit borer (Saha et al., 2014)

20.3.2 EPILACHNA BEETLE OR HADDA BEETLE

- *Henosepilachna vigitioctopunctata* (Curculionidae: Coleoptera): *Henosepilachna vigitioctopunctata* Fab. and *Epilachna dodecastigma* (Wied.) (Curculionidae: Coleoptera) is a popular pest of brinjal and potato. But in recent years, its severe prevalence was reported in cowpea as plants feeder in various parts of the country particularly eastern U.P. and Bihar. In excess of 80%, cowpea leaves were damaged by the adults and grubs of this beetle during the summer season (Halder et al. 2011). Likewise, its severe prevalence was also recorded from bitter gourd, *Momordica charantia* in Jammu (Jamwal et al., 2013), Allahabad in U.P. (Maurice and Ramteke, 2012) wild bitter gourd, *Momordica dioica*in Maharashtra (Deshmukh et al., 2012) and Ivy gourd, *Coccinia grandis* Maurice and Ramteke, 2012) in Allahabad.

20.3.2.1 LIFE CYCLE

Adult lives for one month to more than two months. The yellow elliptical eggs are placed generally ventral surface of the leaves in batches of hundreds,

TABLE 20.3 Effect on Different Insecticides on Shoot and Fruit Infestation by *Leucinodes orbonalis*

Treatments (Concentration)	% Damage by Shoot and Fruit Borer				Mean No. of Holes/ Fruit	No. of Larvae Observed/ Plot	Mean Yield (q/ha) (Pooled Data Over Three Years)
	% Shoots Infestation (Pooled Data Over Three Years and Three Sprays)	% Reduction in Shoots Infestation Over Control	% Fruits Infestation (Pooled Data Over Three Years and Three Sprays)	% Reduction in Fruits Infestation Over Control			
Spinosad 45 SC (0.0135%)	7.60 (15.97)	51.99	16.48 (23.94)	57.36	0.75 (1.11)	3.73 (2.05)	317.88
Indoxacarb 15.5 SC (0.007%)	9.73 (18.17)	38.53	22.04 (27.99)	42.98	1.23 (1.30)	5.10 (2.36)	289.82
Emamectin benzoate 5 WG (0.0025%)	8.57 (17.02)	45.86	19.13 (25.92)	50.50	0.87 (1.17)	4.35 (2.18)	305.50
Rynaxypyr 20 SC (0.006%)	5.67 (13.73)	64.18	12.59 (20.78)	67.42	0.40 (0.94)	2.36 (1.65)	346.69
Flubendiamide 480 SC (0.01%)	6.73 (15.01)	57.49	14.35 (22.25)	62.87	0.56 (1.03)	3.17 (1.91)	326.13
Deltamethrin 1EC + Triazophos 35EC (0.036%)	10.80 (19.17)	31.78	26.62 (31.04)	31.12	1.87 (1.54)	7.13 (2.75)	281.33
Cypermethrin 20EC (0.04%)	10.65 (19.04)	32.72	23.86 (29.23)	38.27	2.20 (1.63)	8.40 (2.98)	274.99
Untreated check	15.83 (23.41)	-	38.65 (38.43)	-	3.73 (2.05)	18.37 (4.33)	185.50
S. Em. (±)	0.67	-	0.69	-	0.11	0.18	3.22
CD (p=0.05)	2.05	-	2.10	-	0.35	0.54	9.78

Figures in parentheses are arcsine √ p transformations,

TABLE 20.4 Effect of Different Pest Management Modules on the Infestation and Damage by *Leucinodes orbonalis* Guenee

Modules	Mean of Shoot Infestation (%)	Percent Reduction in Shoot Infestation Over Control (%)	Mean of Fruit Infestation (%)	Percent Reduction in Fruit Infestation Over Control (%)	Mean No. of Holes/ Fruit	Mean No. of Larvae/Fruit
M1	8.88 (17.32)	37.46	19.80 (26.41)	47.28	2.32 (8.58)	1.74 (7.52)
M2	7.28 (15.65)	48.73	18.94 (25.79)	49.57	1.84 (7.61)	1.38 (6.68)
M3	5.10 (13.04)	64.08	15.18 (22.92)	59.58	1.32 (6.52)	0.90 (5.40)
M4	14.20 (22.12)	-	37.56 (37.78)	-	4.96 (12.84)	4.34 (11.98)
S.Em (±)	0.40	-	0.68	-	0.67	0.50
CD (p=0.05)	1.22	-	2.11	-	2.07	1.53

(M3= Spraying of spinosad 45 SC @ 0.5 ml/l at 60 DAT followed by indoxacarb 14.5 SC @ 0.5 ml/l at 75 DAT and followed by emamectin benzoate 25 WG @ 0.4 g/l at 90 DAT)

glued to the surface of the leaves in a vertical position. The hatching acquires place in 3–4 days. The freshly hatched grub is yellowish and turns to cream yellow white, when full grown. The grub is broad in front and narrows posteriorly and is covered with spiny structures all over (http://agritech.tnau.ac.in/crop_protection/brinjal/ brinjal_4.html).

20.3.2.2 NATURE OF DAMAGE

The minute grubs on hatching start damaging by feeding on the fresh matter of the leaf surface leaving veins and veinlets. The grown up grubs become voracious feeders, found in batches. Both the grubs and adults confine their feeding activities generally to the lower surface of leaves. The infestation of the hadda beetle is more on the lower surface of leaves resulting in:

- Damaged leaves show a lace-like appearance as the greenish matter in connecting the veins is eaten away (skeletonization of leaves).
- Affected leaves, depending upon the area damaged, drying up and in severe infestation, presenting sickly appearance.

20.3.2.3 MANAGEMENT

- Maintenance of larval-pupal endoparasitoid, *Pediobius foveolatus* (Eulophidae: Hymenoptera) which results up to 36.6% suppression in beetle population within and around Varanasi, Uttar Pradesh (Halder et al., 2011).
- Islam et al. (2011) found that *Ricinus communis* leaf extract had maximum larvicidal toxicity (LC50 = 18.40%) besides significant reduction in both oviposition and egg-hatch, prolonged larval duration ($P<0.001$), and restrained pupae formation and emergence of adult.
- Black muscardine fungi *Metarhizium anisopliae*@ 5 g/l gave good management against epilachna beetle (Anonymous, 2014).
- Vishwakarma et al. (2011) was also found that, Epilachna beetle (74.91%) was significantly control by *Beauveria bassiana*, when used @ 3.0 g/l of water.
- In chemical pesticide malathion @ 1 kg a.i. /ha or cypermethrin @ 0.4 ml/l gave good management against this polyphagous pest (Rai et al., 2014).

20.3.3 SOLENOPSIS MEALY BUG

- *Phenacoccus solenopsis* (Pseudococcidae: Homoptera)
  This pest earlier known as a cotton pest but now is a recent menace to a number of cultivated crop plants. Amongst 84 host plants of about 28 families documented at the Central cotton growing zone of India up to 2009, sixty plant species as of twenty families belonged to weeds (Vennila et al., 2010). At present, they nourish on the host plants including oil seeds, pulses, cereals, vegetables, fruits, ornamental crops in addition to many weeds including *Parthenium*. In vegetable crops, they mostly attack the plants belonging tomalvaceae (okra), solanaceae (brinjal, tomato, chilly, potato), cucurbitaceae (cucumber, all melons, pointed gourd and gourds) (Halder et al., 2013).

20.3.3.1 NATURE OF DAMAGE

Moreover sucking the plant sap, they also secret the profuse quantities of honey dew which kept on the plants and make black sooty mold and thereby reducing the photosynthetic movement of the plants.

20.3.3.2 MANAGEMENT

- Exclusion of alternate hosts and unwanted weeds like *Parthenium, Vernonia* from the field helps in pest incidence reduction.
- Ants play major role in spread the mealy bug along with giving protection to them against natural enemies. So, destroying of ant's colonies during preparation of land is recommended.
- Removal and flaming the infested plants helps in reduction of pest from the crop field.
- Apply fish oil resin soap (FORS) @ 20 g/lit of water (Kumar et al., 2011) or entomopathogenic fungi *Verticillium lecanii* (2 × 108 cfu/ml) @ 5 g/lit of water provide better management (Halder et al., 2013).
- In severe infestation, apply Buprofezin 25% SC @ 1 ml/l or Acephate 75 SP @ 1 gm/l of water is suggested (Rai et al., 2014).
- In another study Arif et al., 2008 reported that profenofos is most effective against mealybug and gave maximum mortality within 24 hours.

20.3.4 TOMATO

Fruit Borer

- *Helicoverpa armigera* (Lepidoptera: Noctuidae)

20.3.4.1 HOST RANGE

The tomato fruit borer has an extensive array of host plants but among the vegetable crops, the pest prefers tomato and corn. Among the other vegetables that are mostly attacked by fruit borer are the solanaceous, cruciferous and also cucurbitaceous vegetables. Many common weeds act as larval hosts (Capinera, 2000).

20.3.4.2 LIFE CYCLE

Adult female lays oval and heavily ridged eggs singley on leaves, mostly just below the topmost flowers of the upper canopy. Hatched larvae like to bore small green fruits but sometimes if the fruits are not present they feed on buds, flowers or stems. Larvae either complete their development inside an individual fruit or shift to another fruit. Mature larvae are present in the soil and form a cell 2–4 inches deep and pupate inside. The egg to adult development completes about 30 days (Capinera, 2000).

1. **Eggs:** They are carved and whitish cream in color, laid singly.
2. **Larva:** The neonate larvae may nurture up to 7 mm long, with yellowish white to reddish brown in color. The dark spot on the body owing to the dark colored spiracles and tuberculation bases.
3. **Pupa:** Body brownish in color, pupation mostly in soil, leaf surface and crop rubble.
4. **Adult:** Females are pale with light brown yellow heavy moth, Male-Fair greenish moth with V formed speck.
5. **Forewing:** Emerald green to light brown with a dark brown round spot in the middle.
6. **Hind wing:** It is light smoky white with a broad black external margin (http://agritech.tnau.ac.in/crop_protection/tomato/tomato_1.html)

20.3.4.3 NATURE OF DAMAGE

Larvae generally bore into the fruit generally at or near calyx. Infested fruit are considered unmarketable and generally rot due to attack of secondary microorganisms (Capinera, 2000).

- Adolescent larvae feed on young foliage.
- Fully grown larvae bore into fruits and form circular holes.
- Penetrate only a fraction of its body inside fruit and feed on inner part of plant (http://agritech.tnau.ac.in/crop_protection/tomato/tomato_1.html).

20.3.4.4 MANAGEMENT

- **Cultural Method:**
    1. Deep plowing before transplanting helps to kill the pupae hiding in the soil.
    2. Grow pest-resistant varieties.
    3. Early transplanted crop escapes the damage of this pest.
    4. Intercropping of yellow marigold as a trap crop in a row after every 6–9 rows of tomato helps in checking this pest. Raising of the nursery should be 15 days prior to tomato nursery, so that both of these may be transplanted at the same time. There will high egg-laying by this pest on marigold as compared to tomato plants. The eggs and neonate larvae might be collected or the marigold plants have to be drenched with Bt insecticide to kill the larvae.
    5. Trimming of field bunds to destroy the existing rodent burrows.

- **Mechanical Control:**
    1. Pick up the caterpillars and kill them.
    2. Provide perching places for birds in the ground so that they may pick up the larvae from the crop.

- **Use of Traps:**
    1. Use pheromone and light traps to supervise and collect male moth population.

- **Bio-Logical Control:**
    1. Grow cowpea or pulses on the bunds to build up natural enemy fauna.

2. Release *Trichogramma chilonis* parasitoids twice after mid-March at 2–5 day interval @ 500,000 parasitized per hectare.
3. Spray bio-pesticides like Bt formulations viz., Biolep, Bioasp, Delfin, Dipel (@ 400 gm per acre) or NPV preparations @ 300 LE per acre.

- **Botanical Insecticides:**
  1. Spray nimbecidine @2.5 liter per hectare 12 days after release of *Trichogramma.*

- **Chemical Control:**
  1. Spray indoxacarb 14.5% SC @ 8 ml/10 L or flubendiamide 20 WG @ 5 g/10 l of water (http://agritech.tnau.ac.in/crop_protection/tomato/tomato_1.html)

20.3.5 SERPENTINE LEAF MINER

➢ *Liriomyza trifolii* (Agromyzidae: Lepidoptera)
This leaf miner, *Liriomyza trifolii* Burgess (Agromyzidae: Lepidoptera) which is a native of the Southern part of the United States of America and Central America has extended to other countryside's in the seventies alongside with Chrysanthemum flowers (Puri and Mote, 2004). It is suspected that this group has gone to Karnataka along with planting materials during 1990–91 (Viraktamath et al., 1993) and then disseminated numerous parts of the country like Maharashtra, Tamil Nadu, Andhra Pradesh, Gujarat, Delhi, Uttar Pradesh, West Bengal, Haryana, and Madhya Pradesh. Srinivasan et al. (1995) described 78 host plants covering fiber crops, pulses, vegetable crops, ornamentals and flowers, fodder crops, green manuring, and weeds comes under 16 families. In vegetables, three species of this insect reported and they recorded in infesting tomato, Cowpea, French bean, cluster bean, cucumber, summer squash, all kind of melons, etc. Extent of loss in French bean with 15–70%, cucumber and 35% in tomato has been so far reported by Krishna Kumar (1998) from Karnataka.

20.3.5.1 NATURE OF DAMAGE

Larvae mine into the foliages and in severe damage, leaves dry up.

- Drying and dropping of leaves.

20.3.5.2 MANAGEMENT

- It has been observed that higher dose of nitrogen favors the population build-up of this insect. Thereby sensible application of nitrogenous (N_2) fertilizer results in suppression in the pest infestation in prevalent areas.
- Periodically infested leaves should be collected, confiscated and wiped out, i.e., buried or burnt.
- The insect can also be managed by utilizing yellow sticky traps in the ground for monitoring the incidence (Lopez, 2010) and mass trapping of adults and utilizing the natural enemies. Recently, a new parasitoid *Neochrysocharis farmosa* (Eulophidae: Hymenoptera) has been notified from this leaf miner. During February-March, 37.5% was parasitized at the Varanasi region, Uttar Pradesh, India. In Hawai, Gyaana, and Senegal this insect was managed by inundative release of parasitoids *viz.*, *Hemiptarsenus, Chrysocaris* and *Chrysonomyia* sp. (Puri and Mote, 2004).
- In another study, it was also recorded that marigold (*Tagetes erecta*) act as an attractant for this leaf miner alongside it also attract huge amount of parasitoid *N. farmosa.*
- Application of Neem seed kernel seed extract (NSKE) 4% along with a sticker are observed to be effectious to suppress the leaf miner.
- In case of havoc infestation, foliar spray with imidacloprid 17.8 SL @ 0.3 ml/l of water prior to flowering at initial stages of crop development is effective (Nadagouda et al., 2010)
- Rai et al. (2014) reported that foliar spray with imidacloprid 17.8 SL @ 0.35 ml/l of water during the initial stage of the crops prior to flowering and application of dichlorvos 76EC @ 0.5 ml/l of water in severe damage throughout the reproductive stage of the crop is beneficial.

20.3.6 CHILI

1. **Chili thrips**
   - ➢ *Scirtothripsdorsalis* (Thripidae, Thysanoptera)

     Thrips are minute insects and having fringed wings, serious during dry periods of high temperature.

20.3.6.1 APPEARANCE

Adults are slender, yellowish-brown, having apically pointed wings about 1 mm in length. The female has long narrow wings with the fore margin fringed with long hairs. Nymphs are like adults in figure and straw yellow in color and are smaller in size.

20.3.6.2 LIFE CYCLE

Adult female lay minute dirty white eggs under the leaf tissues and young shoots. The fecundity rate is 30–40. Egg hatching was completed within 4–7 days; larval stage within 5–10 days; prepupal stage within 1–2 days; pupal stage was completed within 2–3 days, and Egg to adult stage of the Chili thrips is of about 15–35 days (http://agropedia.iitk.ac.in/ content/Chili-thrips).

20.3.6.3 NATURE OF DAMAGE

Both adult and lacerate the leaf and suck the oozing sap from young developing leaves. Affected leaves show upward curling along the margin and get crinkled and reduced in size. They also feed on floral parts and fruits. Fruit damage results in rough brown patches affecting their quality and drastically reducing the market price (NHM, 2012).

20.3.6.4 MANAGEMENT

- Intercropping with agathi *Sesbania grandiflora* to give shades that control the thrips infestation.
- Continuous cropping of Chili after a sorghum-more infestation to thrips.
- Avoid mixed cropping of Chili and onion—both the crops are highly infested by thrips.
- Shower of water over the seedlings to manage the population of thrips.
- Seedling was treated with imidacloprid 70% WS @ 12 g /kg of seed.
- Foliar spray of the following insecticide viz., Thiacloprid 21.7% SC @6.0 ml/10 lit.; Dimethoate 30% EC @1.0 ml/lit.; Oxydemeton-Methyl 25% EC @1.0 ml/lit (http://agropedia.iitk.ac.in/content/management-strategies-Chili-thrips-scirtothrips dorsalis).

20.3.7 CHILI MITE

- *Polyphagotarsonemus latus* (Tarsonemidae: Acarina)

20.3.7.1 APPEARANCE

Adults are larger in size, elliptical, broad and the body color is yellow and also has four pairs of legs. Nymphs are minute with white transparent color and a semi-transparent band on dorsal surface while inactive larva is stretched at both sides and sedentary (immovable).

20.3.7.2 LIFE CYCLE

Eggs are lying over the ventral side of the leaf. Oviposition period exists for 1–3 days. Total larvae period were completed within 1–2 days. Pupal period was completed within 24 h or a day. Adult life exists from 3–5 days. The total life cycle is finished within 6.5–10 days (http://agropedia.iitk.ac.in/content/Chili-mite).

20.3.7.3 NATURE OF DAMAGE

Mite affected leaves of the Chili plant turn towards underside and look like an inverted boat. As young and adult mites attack the tender leaves, almost all new leaves curled downward and lower surface appears silvery whereas, the upper surface of leaves look more dark green as compared to healthy ones. Further, the plant growth is checked and the internodes become shorter (Shankar and Raju, 2012).

20.3.7.4 MANAGEMENT

- **Cultural Practice:**
  - Planting or growing of tolerant cultivars like Guntur types.
  - Nutrient and water management also suppress the mite population.

- **Biological Control:**
  Conservation of potential predators like *Amblyseius ovalis* in Chili crop.

- **Chemical Management:**
  1. **Karnataka:** Foliar spray of vertimec 1.9 EC @ 0.2 ml/lit or fenazaquin 10 EC @ 2 ml/lit or dicofol 18.5 EC @ 2.5 ml/lit for controlling of mites.
  2. **Tamil Nadu and Spice Board:** Foliar application acaricide *viz.* wettable sulfur 50WP @ 5 g m/lit or dicofol 18.5 EC @ 2.5 ml/lit of water and monocrotophos spray should be avoided.
  3. **National Horticulture Board:** Spray Phasalone 3 ml/l (Serious infestation) or dicofol 5 ml/lit of water (http://agropedia.iitk.ac.in/content/Chili-mite).

20.4 *B. MALVACEOUS* CROPS

20.4.1 OKRA

Okra shoot and fruit Borer

> *Earias vitella, E. Insulana* (Noctuidae: Lepidoptera)
>
> Okra fruit borer are active throughout the year reaching peaks during March-May and August-October. Female lays green colored eggs with longitudinal ridges on buds, flowers, and fruits. Like brinjal shoot and fruit borer, the infestation comes first in shoots and after blossoming, feed exclusively on fruits (NHM, 2012).

20.4.1.1 APPEARANCE

1. **Earias Vittella:** Adult Moths are medium-sized (13–15 mm) with pale brownish white thorax. Hind wing are silvery-creamy white. While in forewings in *E. vittella* are light white with a thick wedge-shaped horizontal green band in the center. Larva is brownish-white with number of brown and creamy white markings. Eggs are round, single and light bluish-green color with longitudinal ridges.
2. **Earias Insulana:** Adult -Forewing identically silvery green. Larva-Brownish with dorsum showing a white standard longitudinal line. Pupa-Brown and boat formed.

20.4.1.2 LIFE CYCLE

Adult female laying eggs on flower buds, inflorescence, and fruits. Maximum eggs are lying on top of the apical bud. The incubation period is 4–5 days, the larval stage and pupal stage ranges from 15–21 and 7–8 days respectively. Pupation takes place in turn upside down boat-shaped glossy cocoons. The life cycle lasts about 30–40 days (http://uasr.agropedias.iitk.ac.in/sites/default/files/Okra%20Shoot%20and%20fruit%20borer%20pdf.pdf).

20.4.1.3 NATURE OF DAMAGE

During the initial stage of the crop, larvae bore into tender shoots and tunnel downward. The growing point is killed; drooping of shoots downwards and side shoots emerge. Later on, when fruiting bodies appear caterpillars bore in the buds and fruits. The damaged buds drop down and the fruits turn from the point of injury. The larva penetrates inside the fruit and feeds on the developing seeds. The damaged fruits are not suitable for consumption purposes. As a result heavy loss in seed manufactures (Sharma, 2014).

20.4.1.4 MANAGEMENT (SHARMA, 2014)

- **Cultural Control:**
  1. Removal of unwanted weeds.
  2. Avoid ratoon crop.
  3. Summer and early sown crop can escape from this pest.
  4. Use balanced fertilizers as high nitrogen attracts more insects.
  5. Develop resistant varieties.
  6. Avoid cultivation of or growing of okra in the proximity of cotton fields.
- **Mechanical Control:**
  1. Collect damaged shoots as well as fruits and burn or bury them deep in soil.
  2. Remove crop debris and weeds, which harbor this insect.
  3. Installation of Pheromone traps @ 5 traps per hectare.
- **Biological Control:** Conserve the parasitoids and predators by border or intercropping with sesame or cowpea or sorghum. If available the parasitoids or predators can be released in crop.
- **Chemical Control:**
  1. Spray 400–500 ml Malathion 50 EC or 400–500 g carbaryl 50WP in 250–300 liters water per acre.
  2. Crop grown exclusively for seed purpose may be protected from this pest by spraying 400 ml cypermethrin 10EC or 200 ml

decamethrin 2.8 EC in 250-liters water or 75 ml fenvalerate 20 EC per acre. Use only recommended insecticides with its proper dose and water.

3. Spray of emamectin benzoate 5% SG @ 4g/10 l of water.

20.4.2 WHITEFLY

- *Bemesia tabaci* (Aleyrodidae: Hemiptera)
  It acts as a vector transmitting yellow vein mosaic (YVM) disease and also polyphagous in nature. They are laying eggs on the ventral surface of leaves.

20.4.2.1 APPEARANCE

Adults are 1.0–1.5 mm long and the color of the body is yellowish and somewhat dusted with white waxy dust-like substances. Both the wings are whitish in color and have prominent long hind wings.

20.4.2.2 LIFE HISTORY

The adult females are laying eggs singly under the surface of the leaves, eggs laid on an average 80–110 egg. Eggs are stalked and hatch within 3–5 and 5–33 days during summer and winter. The nymphs suck cell sap and have three nymphal stages to form the pupae within 9–14 and 17–81 days during summer and winter. Afterwards, pupae convert into whiteflies within 2–8 days. Lifecycle lasts up to 14–100 days based on environmental conditions (http://www.krishisewa.com/articles/disease-management/233-okra-ipm.html).

20.4.2.3 NATURE OF DAMAGE

Both the adult and the nymph feed by sucking leaf sap. They excrete honeydew, which results in a sooty mold. At present many varieties and hybrids resistant to YVM are existing in market and hence, disease management is very easy.

- Chlorotic spots on the foliages which further coalesce appearance uneven yellowing of leaf tissue.

- Severe infestation leading to early or early defoliation.
- Sooty mold appearance.
- Vector of YVM virus.

20.4.2.4 MANAGEMENT

- Collection and exclusion of unwanted weed like *Abutilon indicum* (Thuthi), *Chrozophorerottlari* (Purapirakkai), *Solanum nigrum* (Manathakali/Milaguthakali, Chukkittikeerai) and *Hibiscus ficulensus* from the ground and nearby fields and maintenance of field cleanliness.
- Sowing should be done on time with suggested spacing, usually broader spacing and sensible use of proposed dose of fertilizers, mainly nitrogenous and water management or managing of irrigation water is necessary to stop the excessive vegetative development and insect establishment. Delay sowing might be avoided and the development of the crop must not be extended beyond its usual period.
- Filed sanitation is very important to get a pest-free crop. Alternate hosts like eggplant, tomato, sunflower, and tobacco may be stayed away from the main crop. In case there is an unavoidable cultivation, crop protection method should be taken to these crops also.
- Watching the movements of the adult whiteflies by installing yellow sticky and pan traps at 1 foot long above the crop canopy and also in situ counts.
- Foliar spray of plant products either singly or in combination with the suggested dose of insecticide (at 2 ml/l of water)
- The extract neem seed kernel 5% and neem oil at 5 ml/l of water.
- Fish oil rosin soap 25 kg at 1 kg in 40 lit of water.

20.4.2.5 CHEMICAL CONTROL

- Acetamiprid 20% SP 100 g/ha.
- Chlorpyrifos 20% EC 1.25 l/ha.
  Synthetic pyrethroids should be avoided in cotton to overcome the problem of whitefly.
- Cypermethrin and deltamethrin cause outbreak of whiteflies (http://agritech.tnau.ac.in/crop_protection/cotton/crop_prot_crop_insectpest%20_cotton_13.html)

20.5 C. CRUCIFEROUS CROPS

20.5.1 CABBAGE, CAULIFLOWER

Diamond Back Moths (DBM)

- *Plutella xylostella* L. (Plutellidae: Lepidoptera)
  P. xylostella, is considered one of the most serious insect of cruciferous vegetable crops globally. Major epidemics of *P. xylostella* are mostly in the grounds that are often sprayed with insecticides. Therefore, lack of useful natural enemies and establishment of insecticide resistance are supposed to be the key reasons of rising pest incidence of *P. xylostella* in many parts of the country. This insect causes 50 to 80% failure in marketable yield and resistance to many insecticides. This pest outbreak was recorded on cauliflower in Aligarh for the period of September to first fortnight of October 2006 (Uthamasamy et al., 2011).

20.5.1.1 HOST RANGE

It infests essential cruciferous crops viz., broccoli, cauliflower, cabbage, radish, Chinese cabbage, knol-khol, turnip, Brussels sprout, rabi mustard, collard, rape seed, saishin, patchouli, watercress and kale (Uthamasamy et al., 2011).

20.5.1.2 LIFE CYCLE

The bionomics of this insect has been observed in the laboratory and natural conditions in relation to ecological factors (Kalyanasundaram, 1995). Eggs are laying in clusters generally on the adaxial side of leaves close to veins and sometimes on both the sides. The eggs are tiny, whitish yellow and 0.5 mm in length and each female can lay 164 eggs in field conditions. Egg period lasts from two and six days. Newly emerged larvae are fair white with dull brown head as matured caterpillars are light green in color assessing 10 mm in long. Small hairs become noticeable on green grown-up larvae which wriggle aggressively on smallest stroke. The larvae nourish for different stage of time ranging between 14 and 21 days ahead of pupation. The larvae of first instars mine into leaves up to the first molting, after which they nourish exteriorly and generally attain four caterpillar instars. Male larvae can be differentiated from the female owing to the occurrence of white prominent gonads on the fifth abdominal segment of final instars. Pupation occurs near the midrib on the ventral side of the leaf in a thin apparent glossy cocoon loosely spun by the caterpillar. Pupal length is 6 mm with light brown in shade. Pupal period exists for four days in summer and five days in rainy season.

Moths are minute grayish and size of the wing is 14 mm when expanded. Male moth wings are folded externally and rising towards their tops appearing a row of three diamond formed yellow marks along the center of the back. The longevity of adults varies from 6–13 days, females survive shorter than males. Mating of adults occurs at evening on the similar day of appearance; mating exists for one to two hours and females were mating only once. Females are laying eggs after mating and oviposition maintain for 10 days with a summit on the first day of emergence (Uthamasamy et al., 2011).

20.5.1.3 NATURE OF DAMAGE

The newly hatched caterpillar feeds by scrapping the epidermis leaving behind typical white patches. Later instar caterpillars make holes in leaves and when the newly emerged seedlings are bothered, the growing tip is eaten away and the curd is not formed (Sharma, 2014).

- Young and immature larvae or caterpillar cause small yellow mines on leaves.
- Scrapping of epidermal leaf tissues producing typical whitish patches on leaves.
- Full-grown mature larvae create openings in the foliages and feeds on curd.

20.5.1.4 MANAGEMENT

- Exclusion and damage of all crop wreckages and stubbles after harvesting of crop.
- Use trap crop as mustard in 2:1 ratio (cabbage: mustard) to attract *P. xylostella* for laying eggs at least ten days before transplanting of major crop.
- Install Pheromone traps @12/ha.
- Larval parasitoid: *Diadegma semiclausm* @100,000/ha (Hills-below 25–27°C) *Cotesi aplutellae* (plains) at 20,000 release from twenty days after transplanting.

- Bt var kurstaki @ 2 g/lit.
- Extract of Neem seed kernel @ 5%.
- Cartap hydrochloride 0.5% at 10, 20, and 30 days after sowing (nursery) and primitive stage.
- Spraying mustard crop with DDVP76 WSC 0.076% to avoid dispersal of the caterpillar.
- Foliar application of chlorantraniliprole 18.5% SC or chlorfenapyr 10% SC or flubendiamide 20% WG @ 0.1 ml per litre of water (http:// agritech.tnau.ac.in /crop_protection/colecrop/cole_1.html).

20.5.2 TOBACCO CATERPILLAR

➢ *Spodoptera litura* (Lepidoptera: Noctuidae)

20.5.2.1 APPEARANCE

The length of the moths is about 22 mm and their body is pale brown in color as the forewings are darkish brown with white spots and hind wings are white with a brown margin.

20.5.2.2 LIFE-CYCLE

Eggs are lying in clusters about 300 eggs and their masse enclosed with buff-colored hairs acquired from the mother's body. Egg period is about 3–6 days. The caterpillar feed in groups in cluster for the initial few days and then disseminated to feed individually. They are going through 6 stages and are fully grown within 16–30 days. The mature caterpillar enters the soil where pupation happened. The pupal stage exists about 7–14 days and the moth, on emergence, live for 7–10 days. The life cycle is ended within 30–50 days and the insect finished eight generations in a year (Kumaranag et al., 2014).

20.5.2.3 NATURE OF DAMAGE

The caterpillar scrapping the greenish matter of the tender leaves and defoliate the plant. Mainly they are nocturnal and cause heavy damage, to the leaves and heads of cabbage and cauliflower (Kumaranag et al., 2014).

20.5.2.4 MANAGEMENT

- Proper sanitation in field and plowing helps to kill the pupae in the soil
- Castor maybe used as border and trap crop.
- Manually collection and damage of grown-up larvae.
- Light trap has to be installed @1/ha.
- Pheromone traps have to be installed (Pherodin SL) @ 15/ha to attract adult male moths.
- Foliar application of Sl NPV @ 1.5 × 1012 POBs/ha + 2.5 kg crude sugar + 0.1% teepol.
- Foliar application of chlorpyriphos 20 EC 2 lit/ha or DDVP 76 WSC 1 lit/ha.

20.5.3 CABBAGE APHID

- *Brevicorynebrassicae, Lipaphiserysimi, Myzuspersicae* (Hemiptera: Aphididae)

20.5.3.1 APPEARANCE

The body of nymphs is yellowish-green and adults suck plant sap and weakening plants. These are tiny (2–2.5 mm) fine, pear-shaped, wings are yellow or green or wingless insects (Kumaranag et al., 2014).

20.5.3.2 LIFE-CYCLE

The birth capacity of an adult female is up to 20–130 nymphs. Their development is very fast and complete fed within 7–10 days. In a year about 45 generations can take place. In autumn and spring, wings forms are emerged and spring and they move from one field to another and one locality to another (Kumaranag et al., 2014).

20.5.3.3 NATURE OF DAMAGE

Adults and nymph suck the plant sap and result in weakening of plants. Infested portions become faded and malformed. High humidity supports the fast growth of this pest. The aphids are mostly found on the ventral part of the leaves and terminal surface of the plant.

20.5.3.4 MANAGEMENT (KUMARANAG ET AL., 2014)

- In case of rape-seed mustard, early sowing should be taken.
- Install yellow sticky trap to check aphid population.
- Foliar application of dimethoate @ 2 ml and acetamiprid 20% SP @ 0.2 gm per liter of water.

20.6 *D. CUCURBITACEOUS* CROPS

20.6.1 BITTER GOURD, POINTED GOURD, CUCUMBER, MUSKMELON, ETC.

Fruit fly
Bactrocera cucurbitae (Tephritidae: Diptera)

20.6.1.1 APPEARANCE AND NATURE OF DAMAGE

Usually, the females are laying eggs in delicate, young fruit tissues by piercing them with the help of ovipositor. Immediately after hatching, the larvae feed into the soft tissue and form galleries. The fruit further decays or becomes deformed. Newly emerged larvae leave the dead necrotic region and shift to newly emerged tissue, where they usually introduce various pathogens and hasten fruit decomposition (Dhillon et al., 2005). The full-grown mature larva goes out from the fruit by creating one or two exit openings for pupation in the soil. The pupation occurs in the soil at a deepness of 0.5–15 cm. It also lays eggs inside the corolla of the cucurbits flower, tap root, leaf stalk, and stem. The larvae successfully grew up in these plant parts and feed inside (Weems et al., 2001). The fruits infested in the initial stages not succeed to grow properly and drop down from the plant. As the maggots spoils the fruits inside, it quite hard to manage this insect with insecticides. Hence, there is a requirement to use different alternative methods to control or formulate an integrated pest management (IPM) approach for successful management of this insect.

20.6.1.2 MANAGEMENT (SHARMA, 2014)

- **Cultural Control:**
  1. Grow one line of maize after every 10 meters of cucurbit crops to provide shelter for files during the dark period. Such rows should be sprayed with insecticides like quinalphos 25EC @ 1 ml/l to kill such flies.
  2. Frequent raking of soil helps to kill pupae in soil.

- **Mechanical Control:**
  1. Before start ripening, fruits should be harvested.
  2. Damaged fruits have to be collected and buried deep in soil (at least 3 feet).
  3. Use pheromone/lure traps (Cuelure) to collected and kill flies.
  4. Use vinegar + Sugar syrup in a pot or in a pumpkin as a bait to control fruit flies.

- **Chemical Control:**
  1. Use poison bait or/and bah spray, prepared as given below:
     - **Poison Bait:** Mix 20 ml malathion 50 EC and 500 g molasses or Gur/jiggery and 20 g yeast hydrolysate, Put small amount of poison bait at different places in field to attract flies for egg laying.
     - **Bait Spray:** Spray 400 ml malathion 50 EC mixed with 250 l water and 1 kg molasses/gur per acre. Repeat at 10 days period if required.

20.6.2 RED PUMPKIN BEETLE

- *Aulacophora foveicollis* (Chrysomelidae: Coleoptera)

  It is a regular and major insect of a broad range of cucurbitaceous crops, especially bottle gourd, sweet gourd, watermelon, bitter gourd and muskmelon. This insect is polyphagous in character (Doharey, 1983). Both adult and larva damage the crop and reason heavy loss to plantlets and tender foliage and blossoms (Rahaman and Prodhan, 2007; Rahaman et al., 2008). The pest is broadly dispersed in diverse parts of the globe, particularly in Africa, Asia, Australia, and south Europe. In India, it occurs all over the state but is more familiar in the north-western parts.

20.6.2.1 APPEARANCE AND NATURE OF DAMAGE

The dorsal portion of the mature beetle is deep orange, while the ventral portion is black (plain beetle). The beetle seems to be oblong about 5–8 mm in long and 3.5–3.75 mm in girth, with six legs. The posterior portion of the abdomen possesses delicate white hairs. They are laying cluster eggs of 8–9 in the damp soil close to the bottom of the host plant or dead leaves. From eggs to larvae it took 6–15 days. The eggs are extended and brown in shade. Caterpillar is creamy white in shade with brown color heads and is about 10–12 mm long. Beetles start laying eggs after about 7 days of emergence and complete five generations starting from March to October. Another beetle which is identified as the banded pumpkin beetle is also orange colored but has four very distinct large black markings on its back (one on every angle of the wing cover up). Creamy, yellow-colored larvae nourish on the roots, shoots and fruits touching the soil. The infested roots and underground stems might decay due to infection caused by the saprophytic fungi. Adult beetles are voracious feeder and feed the leaf lamina and making irregular holes on leaves with netlike appearance. The maximum infestation occurs during the cotyledon stage. The damaged plants might be shriveling, and resowing/planting may become essential under severe case of infestation. The young and immature fruits of the damaged plants may be dried up, whereas the fully grown fruits become inappropriate for human eating (Kamal et al., 2014).

They are strong fliers, very active in hot weather, and take quick flight when disturbed. Sometimes, damage becomes very severe if it is not taken measures in time. Losses due to infestation are quite visible, which can goes up to 35%–75% at the plantlet stage (Yamaguchi, 1983). In some cases, the damage due to this insect has been recorded up to 30%–100% in the ground (Khan et al., 2012). During spring, the grubs defoliate the seedlings of cucurbits to such an extent that the crop has to be sown repeatedly three to four times (Mahmood et al., 2005). It results in delay of the produce for marketing of different cucurbits and therefore reduction in grower's income.

20.6.2.2 MANAGEMENT

- **Cultural Method:**
  1. Since adult beetles hibernate among plant debris, clean cultivation helps in reducing its attack.
  2. Grow a few scattered cucurbit plants in January and collect beetles from them and kill them. Frequent raking of soil.

- **Biological Control:** Vishwakarma et al. (2011) observed that treatment with entomopathogenic fungi *Beauveria bassiana* resulted in maximum reduction.
  Khan and Wasim (2001) observed maximum repellency against beetles in treatment comprising of neem extracts mixed with benzene. The plant extract of *Parthenium* spp. was found to be very effective in managing the red pumpkin beetle (Ali et al., 2011).
- **Chemical Control:** During initial infestation, applications of carbaryl (0.1%) or malathion (0.5%) suppress the damage successfully (Hasan et al., 2011).

20.7 CONCLUSION

Vegetables insect pests can be successfully controlled by incorporating proper field selection, growing insect-free transplants, planting early, controlling weeds and diseases, using insect traps, monitoring fields regularly, timing and foliar application of insecticides correctly, and immediate damage of crop on completion of harvest to ensure success in pest management. Insecticides can control insect pests effectively when used judiciously. Implementing of IPM practices can enhance the environmental benefits, and improve the vigor of vegetables and the farm.

KEYWORDS

- **brinjal shoot and fruit borer**
- **chilli thrips**
- **diamond back moth**
- **fruit fly**
- **management**
- **vegetable crops**

REFERENCES

Alam, S. N., Hossain, M. I., Rouf, F. M. A., Jhala, R. C., Patel, M. G., Rath, L. K., et al., (2006). *Implementation and Promotion of an IPM Strategy for Control of Eggplant Fruit and Shoot Borer in South Asia* (p. 74). Technical Bulletin No. 36. AVRDC publication number06–672. AVRDC - The World Vegetable Center, Shanhua, Taiwan.

Alam, S. N., Rashid, M. A., Rouf, F. M. A., Jhala, R. C., Patel, J. R., Satpathy, S., et al., (2003). *Development of an Integrated Pest Management Strategy for Eggplant Fruit and Shoot Borer in South Asia* (p. 56). Technical Bulletin No. 28. AVRDC Publication No. 03–548. AVRDC – The WorldVegetable Center, Shanhua, Taiwan.

Ali, H., Ahmad, S., Hassan, G., Amin, A., & Naeem, M., (2011). Efficacy of different botanicals against red pumpkin beetle (*Aulacophora foveicollis*) in bitter gourd (*Momordica charantia* L.). *Journal of Weed Science Research, 17*(1), 65–71.

Alpuerto, A. B., (1994). Ecological studies and management of brinjal fruit and shoot borer, Leucinodes orbonalis Guenee. *Indian Journal of Agricultural Sciences, 52*(6), 391–395.

Arif, M. J., Hussain, S., Shahid, M. R., Abbas, G., & Gogi, M. D., (2008). *In vivo* and *in vitro* bioassay of potential insecticides against mealy bugs: A cotton worse catastrophe in Pakistan. In: *Proceedings of 2nd Congress on Insect Science-Pest Management in Global Context* (p. 130). Punjab Agricultural University, Ludhiana.

Atwal, A. S., (1976). *Agricultural Pests of India and Southeast Asia* (p. 529). Kalyani Publishers. New Delhi, India.

AVRDC, (1994). *Eggplant Entomology* (p. 88). Control of eggplant fruit and shoot borer. Progress Report. Asian Vegetable Research and Development Center, (AVRDC), Shanhua, Taiwan.

CABI, (2007). *Crop Protection Compendium*. CAB International. Available at: http://www.cabicompendium.org/cpc (Accessed on 25 November 2019).

Capinera, J. L., (2000). *Corn Earworm: Helicoverpa (=Heliothis) zea (Boddie) (Lepidoptera: Noctuidae) UF/IFAS EENY-145.* http://creatures.ifas.ufl.edu/veg/corn_earworm.htm (Accessed on 25 November 2019).

Chatterjee, M. L., & Mondal, S., (2012). *Sustainable Management of Key Lepidopteran Insect Pests of Vegetables*. ISHS Actahorticulturae. Available at: http://www.actahort.org/books/958/958_17.htm (Accessed on 25 November 2019).

Deshmukh, P. S., Chougale, A. K., Shahasane, S. S., Desai, S. S., & Gaikwad, S. G., (2012). Studies on biology of Haddabeetle, *Epilachna vigintioctopunctata* (Coleptera, Coccinillidae): A serious pest of wild better gourd, Momordicadioica. *Trends in Life Sciences, 1*(3), 46–48.

Dhandapani, N., Umeshchandra, R. S., & Murugan, M., (2003). Bio-intensive pest management (BIPM) in major vegetable crops: An Indian perspective. *Food, Agriculture and Environment, 1*(2), 333–339.

Dhillon, M. K., Singh, R., Naresh, J. S., & Sharma, H. C., (2005). The melon fruit fly, *Bactrocera cucurbitae*: A review of its biology and management. *Journal of Insect Science, 5*, 40.

Doharey, K. L., (1983). Bionomics of red pumpkin beetle, *Aulacophora foveicollis* (Lucas) on some fruits. *Indian Journal of Entomology, 45*, 406–413.

Duca, A. A., Arida, G. S., Punzal, B. S., & Rajatte, E. G., (2004). Management of eggplant fruit and shoot borer, *Leucinodes orbonalis* (Guenee): Evaluation of farmers' indigenous practices. *Philippine Entomologist, 18*(2), 172–173.

FAO, (2003). *Eggplant Integrated Pest Management an Ecological Guide* (p. 177). FAO inter-country programme for integrated pest management in vegetables in South and Southeast Asia. Bangkok, Thailand.

Halder, J., Rai, A. B., & Kodandaram, M. H., (2013). Compatibility of neem oil and different entomopathogens for the management of major vegetable sucking pests. *National Academy Science Letters, 36*(1), 19–25.

Halder, J., Rai, A. B., Kodandaram, M. H., & Dey, D., (2011). *Pediobius foveolatus* (Crowford): A promising bioagents against *Epilachna vigitionctopunctata*. Uttarakhand, In: *National Conference on Horti-Business Linking Farmers with Market* (p. 133). Held at Dehradun.

Hasan, M. K., Uddin, M. M., & Haque, M. M., (2011). Efficacy of malathion for controlling red pumpkin beetle, *Aulacophora foveicollis* in cucurbitaceous vegetables. *Progress Agriculture, 22*(1&2), 11–18.

Kalyanasundaram, M., (1995). "Bioecology and management of diamondback moth *Plutel-laxylostella* (Linnaeus) on cauliflower". *PhD Thesis* (p. 138). TNAU, Coimbatore.

Kamal, M. M., Uddin, M. M., Shajahan, M., Rahman, M. M., Alam, M. J., Islam, M. S., Rafii, M. Y., & Latif, M. A., (2014). Incidence and host preference of red pumpkin beetle, *Aulacophora foveicollis* (Lucas) on cucurbitaceous vegetables. *Life Science Journal, 11*(7), 459–466.

Khan, M. M. H., Alam, M. Z., Rahman, M. M., Miah, M. I., & Hossain, M. M., (2012). Influence of weather factors on the incidence and distribution of red pumpkin beetle infesting cucurbits. *Bangladesh Journal of Agricultural Research, 37*(2), 361–367.

Khan, S. M., & Wasim, M., (2001). Assessment of different plant extracts for their repellency against red pumpkin beetle (*Aulacophora foveicollis* Lucas.) on muskmelon (*Cucumismelo* L.) crop. *Journal of Biological Sciences, 1*(4), 198–200.

Krishnakumar, N. K., (1998). *Bio-Ecology and the Management of the Serpentine Leaf Miner, Liriomyza trifolii on Vegetable Crops*. Final report of ICAR AD-hoc scheme, IIHR, Bangalore.

Krishnamoorthy, A., (2012). Exploitation of egg parasitoids for control of potential pests in vegetable ecosystems in India. *Comunicata Scientiae, 3*(1), 1–15.

Kumar, N. K. K., & Sadashiva, A. T., (1996). Solanummacrocarpon: A wild species of brinjal resistant to brinjal shoot and fruit borer, *Leucinodes orbonalis* (Guen.). *Insect Environment, 2*(2), 41–42 (Abs.).

Kumar, R., Nitharwal, M., Chauhan, R., Pal, V., & Kranthi, K. R., (2011). Evaluation of ecofriendly control methods for the management of mealy bug, *Phenacoccus solenopsis* Tinsley in cotton. *Journal of Entomology*. doi: 10.3923/je.2011.

Kumaranag, K. M., Kedar, S. C., Thodsare, N. H., & Bawaskar, D. M., (2014). Insect pests of cruciferous vegetables and their management. *Popular Khety, 2*(1), 80–83.

Lit, M. C., (2009). Combined resistance of eggplant, *Solanum melongena* L., to leafhopper, Amrascabiguttula (Ishida) and the eggplant borer, *Leucinodes orbonalis* Guenee. *M.Sc. Thesis* (p. 191). University of Los Baños, Laguna, Philippines.

López, R., CaRmona, D., VinCini, A. M., Rubbianesil, G., & Caldiz, D., (2010). Population dynamics and damage caused by the Leafminer *Liriomyza huidobrensis* Blanchard (Diptera: Agromyzidae), on seven potato processing varieties grown in temperate environment. *Neotropical Entomology, 39*(1), 108–114.

Mahmood, T., Khokhar, K. M., Hussain, S. I., & Laghari, M. H., (2005). Host preference of red pumpkin beetle, Aulacophora (Raphidopalpa) foveicollis (Lucas) among cucurbit crops. *Sarhad Journal of Agriculture, 21*(3), 473–475.

Maurice, N. G., & Ramteke, P. W., (2012). Development of *Epilachna vigintioctopunctata* (Fabricius) (Coleoptera: Coccinellidae) on essential and alternative foods. *International Journal of Advanced Pharmaceutical and Biological Sciences, 2*(3), 199–204.

Murugesam, N., & Murugesh, T., (2009). Bioefficacy of some plant products against brinjal fruit borer, *Leucinodes orbonalis* Guenee (Lepidoptera: Pyrallidae). *Journal of Biopesticides*.

Nadagouda, S., Patil, B. V., Venkateshalu, & Sreenivas, A. G., (2010). Studies on development of resistance inserpentine leaf miner, *Liriomyzatrifolii* (Burgess) (Agromyziidae, Diptera) to insecticides. *Karnataka Journal of Agricultural Sciences, 23*(1), 56–58.

National Horticulture Mission (NHM), (2012). *Integrated Pest Management: Schedule for Vegetables*. http://midh.gov.intechnologyIPM-Schedule-for-vegetables.pdf (Accessed on 25 November 2019).

Paroda, R. S., (1999). *For a Food Secure Future*. The Hindu Survey of Indian Agriculture.

Patel, M. G., Jhala, R. C., Vaghela, N. M., & Chauhan, N. R., (2010). Bio-efficacy of buprofezin against mealy bug, *Phenacoccus solenopsis* Tinsley (Hemiptera: Pseudococcidae) an invasive pest of cotton. *Karnataka Journal of Agricultural Sciences, 23*(1), 14–18.

Puri, S. N., & Mote, U. N., (2004). Emerging pets problems in India and critical issues in their management: An overview. In: Subrahamaniyam, B., Ramamurthy, V. V., & Singh, V. S., (eds.), *Frontier Areas of Entomological Research. Proceedings on the National Symposium on Frontier Areas of Entomological Research* (pp. 13–24). Division of Entomology, IARI, New Delhi-110012, India.

Rahaman, M. A., & Prodhan, M. D. H., (2007). Effects of net barrier and synthetic pesticides on red pumpkin beetle and yield of cucumber. *International Journal of Sustainable Crop Production, 2*(3), 30–34.

Rahaman, M. A., Prodhan, M. D. H., & Maula, A. K. M., (2008). Effect of botanical and synthetic pesticides in controlling Epilachna beetle and the yield of bitter gourd. *International Journal of Sustainable Crop Production, 3*(5), 23–26.

Rahman, M. M., (2006). *Vegetable IPM in Bangladesh: Department of Entomology*. Bangabandhu Sheikh Mujibur Rahman Agricultural University Gazipur, Bangladesh. Available at: http://ipmworld.umn.edu/chapters/rahman.htm (Accessed on 25 November 2019).

Rai, A. B., Jaydeep, H., & Kodandaram, M. H., (2014). Emerging insect pest problems in vegetable crops and their management in India: An appraisal. *Pest Management in Horticultural Ecosystems, 20*(2), 113–122.

Rai, A. B., Loganathan, M., Halder, J., Venkataravanappa, V., & Naik, P. S., (2014b). Eco-friendly approaches for sustainable management of vegetable pests. *IIVR Technical Bulletin-53, Published by Indian Institute of Vegetable Research* (p. 104). Varanasi, Uttar Pradesh-221305.

Saha, T., Nithya, C., & Randhir, K., (2014). Evaluation of different pest management modules for the insect pest complex of brinjal during Rabi season in zone - III of Bihar. *The Bioscan, 9*(1), 393–397.

Saha, T., Nithya, C., & Ray, S. N., (2014). Field efficacy of newer insecticides against brinjal shoot and fruit borer, *Leucinodes orbonalis* Guenee (Lepidoptera: Pyralidae) in Bihar. *Pesticide Research Journal, 26*(1), 63–67.

Sardana, H. R., (2001). Integrated pest management in vegetables. In: *Training Manual-2, Training on IPM for Zonal Agricultural Research Stations*, pp. 105–118.

Saxena, P. N., (1965). The life history and biology of *Leucinodes orbonalis* Guen. *Journal of Zoological Society of India, 17*(1&2), 64–70.

Shankar, U., & Raju, S. V. S., (2012). Integrated pest management in vegetable eco-system. In: Abrol, D. P., & Uma, S., (eds.), *Ecologically Based Integrated Pest Management* (pp. 619–650). New India Publishing Agency, New Delhi (India).

Sharma, D. R., (2002). Bioefficacy of certain insecticide and biopesticides against major pest of brinjal under field condition. *M.Sc. Thesis* (p. 160). Indian Agriculture Research Institute, New Delhi, India.

Sharma, S. S., (2014). Latest trends in pest management in vegetable crops. In: Saini, R. K., Yadav, G. S., & Beena, K., (eds.), *Novel Approaches in Pest and Pesticide Management in Agro-Ecosystem* (pp. 75–89). CCSHAU Press, Hisar (India).

Shivalingswami, T. M., Satpathy, S., & Banergee, M. K., (2002). Estimation of crops losses due to insect pests in vegetables. In: Sarath, B. B., Varaprasad, K. S., Anitha, K., Prasada, R. R. D. V. J., Chakrabarthy, S. K., & Chandurkar, P. S., (eds.), *Resource Management in Plant Protection* (Vol. 1, pp. 24–31).

Singh, D. K., Singh, R., Datta, S. D., & Singh, S. K., (2009). Seasonal incidence and insecticidal management of shoot and fruit borer (*Leucinodes orbonalis* Guenee) in brinjal. *Annals of Horticulture, 2*(2), 187–190.

Srinivasan, K., Virakmath, C. A., Gupta, M., & Tewari, G. C., (1995). Geographical distribution, host range and parasitoids of serpentine leaf miner, *Liriomyzatrifolii* (Burgess) in South India. *Pest Management in Horticultural Ecosystem, 1*(2), 93–100.

Srinivasan, R., (2008). Integrated pest management for eggplant fruit and shoot borer (*Leucinodes orbonalis* Guenee) in south and Southeast Asia: Past, present and future. *Journal of Biopesticides, 1*(2), 105–112.

Srinivasan, R., (2009). *Insect and Mite Pests on Eggplant: A Field Guide for Identification and Management* (p. 64). AVRDC Publication No. 09–729. AVRDC - The World Vegetable Center, Shanhua, Taiwan.

Srinivasan, R., (2012). Integrating bio-pesticides in pest management strategies for tropical vegetable production. *Journal of Biopesticides, 5*(supplementary), 36–45.

Talekar, N. S., (2002). *Controlling Brinjal Fruit and Shoot Borer: A Simple, Safe and Economical Approach* (p. 4). International Cooperators' Guide, AVRDC Pub. #02–534. Asian Vegetable Research and Development Center, Shanhua, Taiwan.

Uthamasamy, S., Kannan, M., Senguttuvan, K., & Jayaprakash, S. A., (2016). Status, damage potential and management of diamond back moth, *Plutellaxylostella* (L.) in Tamil Nadu, India. In: *The 6th International Workshop on Management of the Diamondback Moth and Other Crucifer Insect Pests*. AVRDC - The World Vegetable Center.

Vennila, S., Ramamurthy, V. V., Deshmukh, A., Pinjarkar, D. B., Agarwal, M., Pagar, P. C., Prasad, Y. G., Prabhakar, M., Kranthi, K. R., & Bambawale, O. M., (2010). *A Treatise on Mealy Bugs of Central Indian Cotton Production System* (p. 39). Technical Bulletin No. 24, NCIP[M, Pusa Campus, New Delhi.

Virakmath, C. A., Tewari, G. C., Gupta, M., & Srinivasan, K., (1993). American serpentine leaf miner is a new threat to crops. *Indian Farming, 43*(2), 10–12.

Vishwakarma, R., Chand, P., & Ghatak, S. S., (2011). Potential plant extracts and entomopathogenic fungi against Red pumpkin beetle, *Raphidopalpa foveicollis* (Lucas). *Annals of Plant Protection Science, 19*(1), 84–87.

Vishwakarma, R., Prasad, P. H., Ghatak, S. S., & Mondal, S., (2011). Bio-efficacy of plant extracts and entomopathogenic fungi against Epilachna beetle, *Henosepilachna vigintiocto-punctata* (Fabricius) infesting bottle gourd. *Journal of Insect Science, 24*(1), 65–70.

Weems, H. V., Heppner, J. B., & Fasulo, T. R., (2001). *Melon Fly, Bactroceracucurbitae Coquillett (Insecta: Diptera: Tephritidae)*. Florida Department of Agriculture and Consumer Services, Division of Plant Industry, and Fasulo, T. R., University of Florida. University of Florida Publication EENY-199. https://edis.ifas.ufl.edu/in356 (Accessed on 25 November 2019).

Yamaguchi, M., (1983). *World Vegetables: Principles, Production and Nutritive Values* (p. 415). Springer, Netherlands.

Zhu, P. C., Kong, F. L., Yu, S. D., Yu, Y. Q., Jin, S. P., Hu, X. H., & Xu, J. W., (1987). Identification of the sex pheromone of eggplant borer *Leucinodes orbonalis* Guenèe (Lepidoptera: Pyralidae). *Zeitschriftfür Naturforschung, C (Biosciences), 42*(11&12), 1343–1344 (Abs.).

WEB ADDRESS

http://agritech.tnau.ac.in/crop_protection/colecrop/cole_1.html (Accessed on 25 November 2019).

http://agritech.tnau.ac.in/crop_protection/tomato/tomato_1.htm (Accessed on 25 November 2019).

http://agropedia.iitk.ac.in/content/chili-mite (Accessed on 25 November 2019).

http://agropedia.iitk.ac.in/content/chili-thrips (Accessed on 25 November 2019).

http://agropedia.iitk.ac.in/content/management-strategies-chili-thrips-scirtothripsdorsalis (Accessed on 25 November 2019).

http://uasr.agropedias.iitk.ac.in/sites/default/files/Okra%20Shoot%20and%20fruit%20 borer%20pdf.pdf (Accessed on 25 November 2019).

http://www.krishisewa.com/articles/disease-management/233-okra-ipm.html (Accessed on 25 November 2019).

http://www.krishisewa.com/articles/disease-management/233-okra-ipm.htm (Accessed on 25 November 2019).

CHAPTER 21

Emerging Viral Diseases of Vegetable Crops: An Outline and Sustainable Management

MOHAMMAD ANSAR, A. SRINIVASARAGHAVAN, MONIKA KARN, and ANIRUDDHA KUMAR AGNIHOTRI

Department of Plant Pathology, Bihar Agricultural University, Sabour, Bhagalpur, Bihar, India,
E-mail: ansar.pantversity@gmail.com (M. Ansar)

ABSTRACT

Vegetable production is challenged by various pest and diseases, among them viruses are emerging as major limiting factor for cultivation of vegetables crops across the world. A diverse range of viral population have been investigated which encircles solanaceous, leguminous, cucurbitaceous and cruciferous hosts. Insect-vectors are recognized as major virus dispersal mediator in vegetable hosts providing epidemiological edge to the plant viruses. On of probable cause of emergence of plant viruses in vegetable ecosystem may be attributed to host-shifting behavior of insect vectors. Whitefly transmitted Geminiviruses (Genus Begomovirus) are emerging as major viruses infecting cucurbitaceous crops in the recent past. Potyvirus persuade serious threat to vegetable cropping system which are mostly transmitted by aphids and few are by seed. Reports suggests that, more than 14 vegetable hosts are being suffered due to Potyvirus in Indian sub-continent. Cucumber mosaic virus having the wider host range (>1,000 plant species), besides the cucurbits it affects major vegetable like pepper, spinach, lettuce, tomato and bean. It has been considered as emerging potential virus in vegetable pathosystem. Off late thrips transmitted Tospoviruses are also emerging as a limiting factor for the vegetable cultivation in India. Among the five species, four members are known to be considerably affecting vegetable cultivation in India. *Peanut/*

Groundnut bud necrosis virus is known to infect wide host including solanaceous crops. Emergence of new viruses and their adaptability in various hosts is major challenge towards sustaining vegetable production to meet the demand of burgeoning population. Moreover, most of viruses are transmitted by insect vectors which makes it more challenging to understand their epidemiology. Long practicing of conventional approaches will lead to economic loss, therefore integrated and eco-friendly tactics are essential to develop sustainable management strategies. In order to control the insect vector, alternative application of various insecticides with integrated cultural operations are found effective. There is need to focus on accelerating the development of resistant cultivar/variety in the cropping complex to reduce the economic loss without harming the environment.

21.1 INTRODUCTION

There has been a considerable increase in the food production since the last decade across the globe. Keeping in view of burgeoning population and shrinking resources, it is important to maintain the rate of food production without compromising the nutritional standards. Food and nutritional security can be achieved by increasing production of vegetables which will help to resolve issue of food as yield of vegetable crops is 4 to 10 times more than food grains. Therefore, vegetables play a crucial role on food front as they are cheapest source of natural foods and can worthily complement the main cereals of the diet. Vegetables are prosperous and reasonably cheaper and rich sources of protein, vitamins, minerals and essential amino acids. They play a vital part in up keeping of the digestion process by counter balancing the acidic environment created during digestion of pretentious and fatty foods. The daily recommendation vegetables for a healthy diet include 75–125 g of green leafy vegetables, 85 g of other vegetables and 85 g of roots and tubers with other food. India is the second largest producer of vegetables in world with an annual production of 87.53 million tonnes from 5.86 million hectares instead of 14.4% production globally (FAO statistics, 2009). A range of vegetables are grown under field conditions in the diverse agro-climatic zones of the country facilitating it to grow almost fresh vegetables year-round to fulfill the increased demand for nutritional requirements in both rural and urban areas. Tomato, potato, chili pepper, okra, leafy vegetable, and onion are economically important and occupy almost 50% of the total area under production in India. Looking at the potentiality of yield and importance it contributes significantly to India's agricultural exports. Vegetable crops bare a wide range of potential pests, consequently, are more prone to various biotic and abiotic stresses. Among biotic stresses, repeated occurrences of fungal, bacterial and viral diseases are the major causes of reduced productivity. Viruses by inducing mosaics, curling, stunting and wilting in vegetable crops cause a significant loss in the quality and quantity of the produce. The viruses infecting plants are classified into 73 genera and 49 families. Though, data represents only the viruses infecting cultivated host which corresponded to only a small portion of the total plant species (Roossinck, 2011). Viruses infecting wild hosts have not given much emphasis, but exist studies showed tremendously such relations between wild hosts and their viruses do not interact to cause disease in the host plants. Plant viruses have developed different strategies for their transmission in various plant species. The involvement of vector is more prominent in the case of virus transmission due lack of mobility in Plants. The virus genus belonging to various families are interacting by various means like insect-vector which played a significant role in dissemination. Among virus groups, whitefly transmitted geminiviruses, aphid transmitted Potyviruses and thrips transmitted Tospoviruses covered a wide range of solanaceous, cucurbitaceous and leguminous hosts.

21.2 THE VIRUSES: A MIGHTY PATHOGEN OF VEGETABLE PATHOSYSTEM

21.2.1 WHITEFLY TRANSMITTED GEMINI VIRUSES

Begomoviruses are the leading and most important genus under the family *Geminiviridae* having single-stranded DNA as their genome and characterized by, incomplete icosahedral geminate particle structure. Begomoviruses (type species: *Bean golden mosaic virus*) are transmitted by whiteflies and encompass either a monopartite (DNA-A) associated with beta satellite or a bipartite (DNA-A and DNA-B) genome organization, infecting dicotyledonous plants (Gutierrez, 1999, 2000; Mansoor et al., 2003; Jeske, 2009). The DNA-A of bipartite and the single component of monopartite begomoviruses contain five open reading frames (ORFs), but occasionally it may be six, one (AV1) or two (AV1 and AV2) in the viral sense strand and four (AC1 to AC4) in the complementary sense. Both the DNA-A and DNA-B are approximately 2.7 kb in size. The DNA-B having two ORFs e.g., BV1 and BC1 in V-sense and C-sense, respectively. In India so far begomoviruses

TABLE 21.1 Molecular Studies of Important Begomoviruses Associated with Vegetable in India

Name of the Virus	Crop Infected	Mono-Bipartite	Association of Satellite(s)	Symptoms Produced	References
Bhendi yellow vein, mosaic virus	Bhendi/ Okra	Monopartite	Betasatellite	Vein clearing, yellowing Reduced size of leaves and fruits	Jose and Usha, 2003; Kulkarni, 1924
Okra enation leaf curl virus	Bhendi/ Okra	Monopartite	Alp/ha satellite	Vein clearing, yellowing	Chandra et al., 2013
Bitter gourd yellow, mosaic virus	Bittergourd		-	Leaf yellowing and mosaic	Raj et al., 2005a
Chili leaf curl virus	Chili	Monopartite	Betasatellite	Yellowing, leaf curling, stunting, and blistering, shortening of internodes	Shih et al., 2006; Chattopadhyay et al., 2008
Cucumis yellow, mosaic disease, associated virus	Cucumis	Bipartite (suspected)	–	Leaf yellowing and mosaic	Raj and Singh, 1996
Dolichos yellow, mosaic virus	French bean, cowpea, dolichos bean	Bipartite	-	Leaf yellowing	Varma and Malathi, 2003; Balaji et al., 2004; Girish et al., 2005
French Leaf curl virus	French bean,	Monopartite	Betasatellite	Severe leaf curling	Naimuddin et al., 2014
Chili Leaf curl virus	Chili	Monopartite	Betasatellite	Curling and chlorosis	Bhatt et al., 2016
Tomato leaf curl, New Delhi virus,	Wide range of vegetables	Bipartite	Rare betasatellite	Curling, mosaic, Puckering	Moriones et al., 2017; Pratap et al., 2011; Agnihotri et al., 2018
Tomato leaf curl, Gujarat virus,	Tomato, Beans	Bipartite	-	Curling, mosaic	Chakraborty, 2003; Naimuddin et al., 2014
Tomato leaf curl, Karnataka virus	Tomato		Betasatellite	Curling mosaic, stunting	Chatchawankanphanich et al., 1993
Tomato leaf curl Joydebpur virus	Tomato, chili, Egg plant	monopartite	Betasatellite	Curling, mosaic, stunting	Venkataravanappa et al., 2014; Shih et al., 2007; Tiwari et al., 2013; Ansar et al., 2018

have been documented relatively 16 different groups of crop. A comprehensive study has been explored on these viruses on their genetic diversity, functions of viral proteins, host interactions, virus-derived transgenic resistance, and linked alp/ha or betasatellites. In vegetable pathosystem, both mono and bipartite begomoviruses played a significant role by inducing severe mosaic, curling and stunted growth symptom in diverse families (Table 21.1).

21.2.1.1 CUCURBIT INFECTING BEGOMOVIRUSES

Whitefly transmitted viruses causing severe curling and mosaic has been reported form several parts of India, which was suspected to be caused by begomoviruses. The PCR assay has been conducted using a degenerate primer of whitefly transmitted geminivirus (Deng 541F and 540R) for their confirmation. The diverse symptom has been reported from mild to severe mosaic in different cucurbits. Major cucurbits like sponge gourd, bitter gourd, pumpkin, and ridge gourd are severely affected by mosaic along with the shortening of internodes, mottling, stunting, puckering of leaf lamina and fruit deformities (Figures 21.1 and 21.2).

FIGURE 21.1 Severe mosaic in cucurbits, (A) ridge gourd (B) pumpkin (C) sponge gourd (D) cucumber.

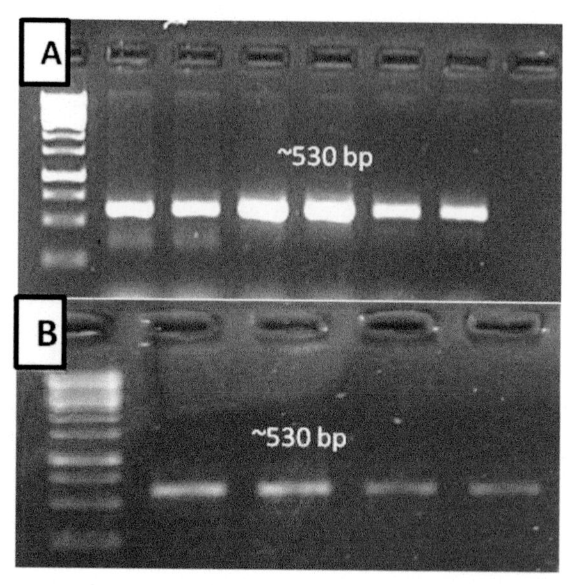

FIGURE 21.2 PCR mediated amplification CP gene of whitefly transmitted geminivirus using Deng 541F/540R primer (A) sponge gourd and bitter gourd (B) pumpkin and cucumber.

In summer and *Kharif* season cucurbits covered a major part among vegetables. Moreover, under protected cultivation several cucurbits are included with capsicum and tomato. Due to prevalence of tropical conditions and the survival of whitefly throughout the year, number of begomoviruses has been reported from cucurbits. The incidence of mosaic varied according to their host, in sponge guard 34%, ridge guard (47%), bitter guard (27.6%), pumpkin (16.3%) and cucumber (56.3%) observed at Vegetable farm in Bihar Agricultural University (BAU) (Figure 21.3). Cucumber crop severely affected by mosaic at earlier growth stage, resulting less and rudimentary fruit formation noticed.

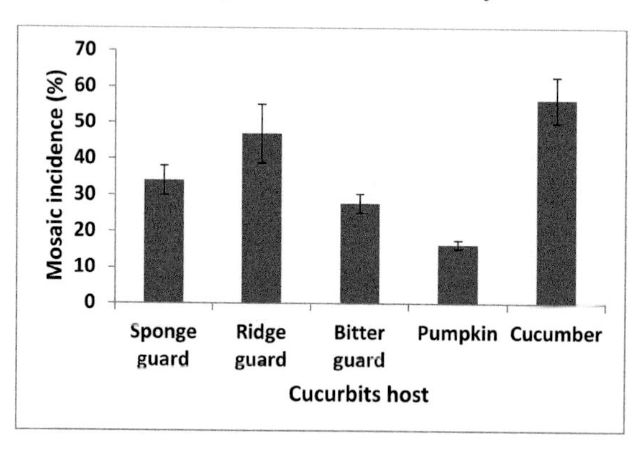

FIGURE 21.3 Disease incidence in cucurbit hosts caused by begomoviruses.

21.2.1.2 FRENCH BEAN MOSAIC

Rajma (French bean) crop is affected by several viruses, among them yellow mosaic is one of the emerging disease. The disease is observing since the last two years which affects whole crop. Prominent symptom includes severe mosaic, stunted growth and malformed buds. Infected plants showed several abnormalities like crinkled leaves, which failed to attain normal pod size. The DNA samples were tested with whitefly transmitted geminivirus specific primer which showed positive results. Subsequently, the sample were tested with four specific primers targeting CP gene of *Dolichos yellow mosaic virus*, *Horsegram yellow mosaic virus*, *Mungbean yellow mosaic India virus* and *Tomato leaf curl Gujrat virus*. Among them, *Mungbean yellow mosaic India virus* found positive in four out of six samples by producing ~800 bp band. The disease is highly influenced due to shifting of viruliferous white-flies from other leguminous hosts (Figure 21.4).

FIGURE 21.4 (A) Positive amplification of CP gene in PCR assay of *MYMIV* infected rajma plants, (B) Field view of *MYMIV* Infection in Rajma (C) severe mosaic.

21.2.2 POTYVIRUS: A SERIOUS THREAT OF VEGETABLE CROPPING SYSTEM

The genus Potyvirus (named as type member, Potato virus Y) is the largest genus of the family *Potyviridae*. It contains at least 200 definite and tentative species (Berger et al., 2005) which cause significant losses in agricultural, horticultural crops (Ward and Shukla, 1991). They affect a wide range of mono and dicotyledonous plant species and have been found in all parts of the world (Gibbs and Ohshima, 2010). Potyviruses are reported to transmit through mechanical and aphids in a nonpersistent manner (Gibbs et al., 2003; Poutaraud et al., 2004; Fauquet et al., 2005). Fewer in number they may also be transmitted through the seeds of their hosts (Johansen et al., 1994). Furthermore, they are transmitted through infected planting materials such as cuttings and tubers (Shukla et al., 1994). Even though Potyvirus distribution is worldwide, they are most common in tropical and sub-tropical countries (Shukla et al., 1998).

The virion size of potyviruses ranges from 700 to 900 nm containing a monopartite, single-stranded RNA genome (10 kb). Potyviruses are known for the induction of characteristic pinwheel or scroll-shaped inclusion bodies in the cytoplasm of the infected cells. Inclusion bodies are cylindrical, formed by a virus-encoded protein and can be considered as the most important phenotypic key factor for assigning viruses to the Potyvirus genus (Milne, 1988; Shukla et al., 1989). Many Potyviruses also induce cytoplasmic as well as nuclear inclusion bodies which are generally amorphous in nature. The virion RNA is infectious and provides both the genome and viral messenger RNA. The genomic RNA is encoded into polyproteins which are subsequently processed by the action of three viral-encoded proteinases into functional products. Approximately, more than 4000 potyvirus sequences available in database of gene bank.

The NCBI database suggests that there are over 55 potyviruses are reported from India and of which 40 are economically important. These potyviruses infect a range of economically important crops, including potato, brinjal, capsicum, papaya, cowpea, common bean, tuber crops, cucurbits (Table 21.2) which causes considerable economic losses (Mali and Kulthe, 1980; Bhat et al., 1999; Srinivasulu and Gopal, 2010; Babu et al., 2012). Chili causing dark green mottling and distortion symptoms on leaves in major growing areas of the eastern region of India. The disease distributed throughout the area and the mottle incidence ranged up to 75%. A severe mosaic disease of gherkin (*Cucumis anguria* L.) in south India was identified (Srinivasulu et al., 2010) which is an important cucurbitaceous vegetable crop grown in the southern states of India like Andhra Pradesh, Karnataka and Tamil Nadu for slicing and pickling. Pepper veinal mottle virus (PVMV) affecting *Capsicum annuum* L. in India (Nagaraju and Reddy, 1980) which severely hampered the chili production.

21.2.2.1 ZUCCHINI YELLOW MOSAIC VIRUS (ZYMV)

Zucchini yellow mosaic virus (ZYMV) is the member of genus Potyvirus has particles which are flexuous filaments with a modal length of 750–800 nm, containing a single-stranded positive-sense RNA. The virus also known as Muskmelon yellow stunt virus was first reported on zucchini in Italy (Lisa et al., 1981; Lecoq et al., 1981). Huge variability has been reported in ZYMV, for instance, 22 isolates have been grouped into three pathotypes according to their reaction on muskmelon line PI 414723 (Pitrat and Lecoq, 1984). *ZYMV* is geographically distributed in many areas of the world where cucurbits crops are grown. In India it was reported in Pune, Maharashtra (Verma et al., 2004) However, ZYMV has been recorded in several countries like Algeria, Egypt, Mauritius, Morocco (Brunt et al., 1990).

TABLE 21.2 Potyviruses Associated with Vegetable Crops in India

Sl. No.	Virus Species	Host
1	*Amaranthus mosaic virus*	Red Amaranth
2	*Bean common mosaic virus*	Bean, Tomato
3	*Blackeye cowpea mosaic virus*	Cowpea
4	*Chili veinal mottle virus*	Pepper
5	*Dasheen streak mosaic virus*	Taro, Elephant food yam
6	*Lettuce mosaic virus*	Lettuce
7	*Onion yellow dwarf virus*	Allium, Garlic
8	*Papaya ringspot virus*	Cucurbits
9	*Potato virus A*	Potato
10	*Potato virus Y*	Potato, Tomato
11	*Sweet potato feathery mottle virus*	Sweet potato
12	*Turnip mosaic virus*	Radish
13	*Yam mild mosaic virus*	Yam
14	*Zucchini yellow mosaic virus*	Gherkin

ZYMV causes severe symptoms inducing huge damage to marrow or zucchini squash (*Cucurbita maxima*), muskmelon (*Cucumis melo*), cucumber (*Cucumis sativus*) and watermelon (*Citrullus lanatus*). Further, symptoms on foliage include mosaic, yellowing, shoe-string, and stunting. Fruits are also deformed, twisted, and covered with protuberances.

1. **Host Range:** The host range of *ZYMV* is narrow which includes mainly cucurbit species.
2. **Transmission:** The virus is transmitted by different species of aphid in a non-persistent manner. *Myzus persicae*, *Aphis gossypii*, and *Macrosiphum euphorbia* are most common. The virus is seed and mechanically transmitted at a low rate (Greber et al., 1989).

21.2.2.2 PEPPER VEINAL MOTTLE VIRUS (PVMV)

PVMV belongs to genus Potyvirus consisting of 770 nm long and 12 nm wide particles. PVMV was first accepted as a distinct member of a virus group which was initially designated the *Potato virus Y* group. Though, afterward renamed the Potyvirus group (Harrison et al., 1971). The occurrence of PVMV mainly in African countries, it also affects *Capsicum annuum* L. crops in the Indian subcontinent (Nagaraju et al., 1980; Singh and Lal, 1988). Different strains of the virus experimentally transmissible to 35 species of Solanaceous family and other nine species belongs to five others families e.g., Aizoaceae, Amaranthaceae, Apocynaceae, Chenopodiaceae, and Rutaceae (Brunt et al., 1978; Igwegbe and Waterworth, 1982; Ladipo and Roberts, 1977; Prasada Rao, and Yaraguntaiah, 1979). The nature and severity of symptoms are dependent on type of host, virus strain, and environmental conditions. PVMV causes mottle, leaf distortion, veinal chlorosis and vein banding in pepper, and mosaic in tomato. Severe strains may cause leaf and stem necrosis in tomato.

1. **Host Range:** The primary hosts of PVMV are hot and sweet pepper, tomato and eggplant.
2. **Transmission:** As similar ZYMV, in nature, PVMV is also transmitted by non-persistently by several aphid species like *Aphis gossypii*, *A. crassivora*, *A. spiraecola*, *Myzus persicae* and *Toxoptera citridus*. Mechanically virus may transmit by artificial inoculation. Moreover, the virus is not seed-borne in nature.

21.2.3 ASSOCIATION OF CUCUMBER MOSAIC VIRUS (CMV) WITH DIVERSE VEGETABLE CROPS

In the family Bromoviridae, the Cucumber mosaic virus (CMV) is an important member which belongs to the genus *Cucumovirus.* Owing its wide host range (>1,000 plant species) and economic impact, it has been considered as key virus in vegetable pathosyatem. The virus genome is organized into three single-stranded messenger sense genomic RNAs (RNAs 1, 2, and 3). RNAs 1 and 2 codes for components of the replicase complex, and RNAs 2 also codes for the 2b protein which played the role in suppression of gene silencing. The 3a protein encoded by RNAs 3 required for virus movement and the coat protein (CP), which is expressed from subgenomic RNAs 4. The CP has provides the shape of the viral particles and movement, vectors transmission and symptom expression (Crescenzy, 1993 and Palukaitis et al.,

1992). An associated satellite RNA molecules with some CMV strains which are small, linear, noncoding and single-stranded RNA molecules that depend on the helper virus (CMV) for their replication, encapsidation, and transmission. However, no sequence similarity has been found with the helper virus genome. CMV strains can be divided into two subgroups, I, and II, the previously it divided into subgroups IA and IB based on their biological, serological and molecular properties. Asian strain considered in sub-group IB, whereas other individuals of sub-group I fall under subgroup IA. The nucleotide sequence identity between CMV sub-group II and I strains ranges from 69 to 77%, whereas > 90% identity within subgroup (Palukaitis et al., 2003).

CMV is mainly transmitted by different aphids species. It can also be spread mechanically by agricultural operations. However, the stability if CMV found low than other viruses like Tobacco mosaic virus resulting no easily transmission by human. CMV is quickly acquired by all instars of aphids, usually less than one minute of feeding (nonpersistent). However, the ability of transmission declines and lots of virus particles in few hours. Several aphid species (60–80) are involved in transmission of CMV and it also transmitted by seeds. CMV may overwinter in perennial weeds, flowers and often roots of crop plants (Zitter and Murphy, 2009). During the spring season, the virus emerges with plants and comes out with new leaves, where the aphids acquired carried to other healthy hosts.

21.2.3.1 IMPORTANT VEGETABLE CROPS TARGETED BY CMV

1. **Cucurbits:** Majority of cucurbits are susceptible to CMV, with varying range of symptom. Severe epinasty, downward twisting of the petiole and leaf surface reduction is frequently appears in early season of infection in summer squash. Plants infected at premature stage are severely stunted and leaves are malformed. Fruits are often unmarketable because of distinct roughness on the surface (zucchini plant and fruits). Infection of vining crops, like muskmelon, showing severely stunted growing tips, and even though fruit may not show symptoms but renders poor quality.
2. **Pepper:** Depending upon infection stage, foliar symptoms of pepper plants may vary. The initial flush of symptoms includes a chlorosis of newly emerged leaves that may occur basal portion and finally whole leaf. Subsequently, chlorotic mosaic developed on leaves of varied degrees of deformation including sunken interveinal lamina with protruding primary veins. These leaves also have a dull light

green looks as contrasting to the dark green, rather shiny leaves of healthy pepper plants.

3. **Spinach:** CMV infection in spinach is often referred as blight. The symptoms can vary depending upon the variety, plant age, temperature, and virus strain. Characteristic symptom includes leaf chlorosis, which can increases to cause severe blighting of the growing point and ultimate plant death. In addition mottling, leaves can shows narrowing, wrinkling with vein distortion and inward leaf roll appearance.
4. **Lettuce:** Symptoms of CMV infection in lettuce having leaf mottling, severe bumpiness and occasional necrosis within the leaf tissue. Usually, stunted growth observed if infection taking place at an early stage of plant growth.
5. **Tomato:** Infection of CMV in tomato plants at an early stage is yellow, bushy and considerably stunted. The leaves may show a mottle which resembles to tomato mosaic disease. The most characteristic symptom of CMV is filiformity or shoestring-like leaf blades. The symptoms may be temporary, bottom or newly developed top leaves showing severe symptoms, while the middle leaves appears normal. Severely affected plants produce few numbers of fruits, which are usually small, often mottled or necrotic, with delayed maturity.
6. **Bean:** Viral infection in beans consisted of leaf curl, green mottle, and blistering, roughness along the main veins involving only a few leaves. Foliar symptoms are most clear, moreover, pod infection and loss are maximum when plants are infected before bloom. Early infected plants may yield fewer pods because flower abortion and abnormal development. The pods are mostly coiled, mottled and reduced in size.

21.2.4 ROLE OF ORHOTOSPOVIRUSES IN VEGETABLE PATHOSYSTEM

The genus *Orthotospovirus* (*Tospovirus*) of the family *Tospoviridae* is a serious threat for vegetable cultivation globally (Pappu et al., 2009; Kunkalikar et al., 2011). The genus name resulting from its first member *Tomato spotted wilt virus* (TSWV) observed in Australia in 1915. Later it was recognized as thrips transmitted spotted wilt disease of tomato. The virus is identified in 85 different families having 1090 different plant species (Sherwood et al., 2000; Parrella et al., 2003). So far, the genus is known to have more than 20 different viruses from all over the world (Pappu et al., 2009). In India, symptoms similar to TSWV were first observed in Nilgiri hills on tomato cultivar 'Marglobe' during 1964 (Todd et al., 1975). Later, the disease was reported from several crops like peanut (Ghanekar et al., 1979), tomato (Prasada Rao et al., 1980), peas (Rao et al., 1984), cowpea, chili, egg plants and cluster bean (Krishnareddy and Verma, 1990). Later, based on serological relationships the virus was found to be different from TSWV and designated as *Peanut bud necrosis virus* (Reddy et al., 1992). In plant parasitic viruses, orthotospoviruses have distinct particle morphology, genomic arrangement, and expression strategies. The virus particles are pleomorphic and 80–120 nm in diameter with surface projections possess two glycoproteins. The genome includes three RNAs referred to as large (L), medium (M) and small (S). The L RNA is in negative-sense while the M and S RNAs are ambisense. The L RNA codes for the RNA-dependent RNA polymerase (RdRp) while the M RNA for the precursor of two glycoproteins (GN and GC) and a non-structural protein (NSm). The N protein and another non-structural protein (NSs) coded by the S RNA. NSm and NSs were shown to function as movement protein and silencing suppressor, respectively (Tsompana and Moyer, 2008). The NSm of TSWV was recently shown to act as an avirulence determinant during the interaction between TSWV and resistant pepper containing the *Tsw* gene. The three genomic RNAs are tightly linked with the N protein to form ribonucleoproteins (RNPs). The lipid envelope consisting of two virus-coded glycoproteins and a host-derived membrane enclosed these RNPs.

Tospoviruses are transmitted in circulative and propagative manner by several species of thrips in (Whitfield et al., 2005). Thrips are minute insects found in a variety of habitats across the world. Several species are known to transmit the virus among them *Frankliniella occidentalis* (western flower thrips), *F. fusca* (tobacco thrips), *Thrips tabaci* (onion thrips), and *T. palmi* (melon thrips) are important. The thrips-tospovirus relationship is very specific, only a few viruses are known to acquire and transmit by known thrips species. At least 10 species of thrips have been reported to transmit the virus (Pappu et al., 2009) and their mode of transmission is very unique e.g., only larval stages can acquire the virus and transmitted by adults.

21.2.4.1 GEOGRAPHICAL DISTRIBUTION AND ECONOMIC IMPORTANCE

Tospoviruses are emerging as a limiting factor for the vegetable production in India (Kunkalikar et al., 2011). Four out of five tospoviruses reported in India which are known to be seriously affecting vegetable cultivation (Table 21.1). *Peanut/Groundnut bud necrosis virus* (PBNV/GBNV) is known to infect wide host including solanaceous crop. The virus infecting major vegetables like tomato (*Lycopersicum eculentum*), chili (*Capsicum annuum*), potato (*Solanum tuberosum*), brinjal (*Solanum melongena*), pea (*Pisum sativum*),

Amaranthus (*Amarnthus sp.*) (Mondal et al., 2012; Sharma and Kulshrestha, 2014). Few viruses emerged earlier in India like *iris yellow spot virus* (IYSV) on onion (Ravi et al., 2006), garlic (*Allium sativum*) (Gawande et al., 2010), *Capsicum chlorosis virus* (CaCV) on tomato (Kunkalikar et al., 2007) and chili (Krishnareddy et al., 2008). Cucurbitaceous crops especially watermelon is being seriously affected by *Watermelon bud necrosis virus* (WBNV) (Jain et al., 1998). Among four tospoviruses infecting vegetables, GBNV and WBNV are distributed throughout the subcontinent. In recent reports, increasing natural infection of GBNV on new crops and its geographical expansion are alarming. The occurrence of GBNV and WBNV in an epidemics proportion in different parts of India is well documented. Tospovirus infections in chili ranges 90–100% have significant concern in Khammam and Warangal districts of Andhra Pradesh (Gopal et al., 2010). Besides groundnut and tomato, severe loss has been recorded in early grown potato crop due to stem necrosis disease caused by GBNV in northern India (Singh et al., 1997; Ansar et al., 2015). Cultivation of watermelon seriously affected due to an outbreak of WBNV in southern part of India, forcing farmers to crop shifting due to total crop loss (Singh and Krishnareddy, 1996) (Table 21.3).

TABLE 21.3 Important Tospoviruses Affecting Vegetable Crops in India

S. No.	Tospovirus	Acronym	Vegetable Host	Major Symptoms on Vegetable Crops
1.	Groundnut bud necrosis virus	PBNV/ GBNV	Tomato, Chili, Pea, Brinjal, Cucurbits, Potato	Chlorotic and necrotic ring spot chlorotic apical bud necrosis and Stem necrosis
2.	Irish yellow spot virus	IYSV	Onion, Garlic	Yellow stripes, spindle-shaped chlorotic lesions/rings
3.	Watermelon bud necrosis virus	WBNV	Cucurbits, Tomato	chlorotic mottling, yellow spots/ patches, mild crinkling, Bud necrosis, dieback vines
4.	Capsicum chlorosis virus	CaCV	Capsicum, *Amaranthus* sp. chili, watermelon	Chlorotic ring spot, necrotic ring spot and apical bud necrosis

21.3 SUSTAINABLE MANAGEMENT OF VIRAL DISEASES IN VEGETABLE CROP

In vegetable pathosystem, viruses play an important role which affects the productivity throughout the world. Adaptation of one or two methods is not sufficient to combat against the diseases. A sole dependency on insecticidal application promotes the development of resistance in insect vectors which affects the subsequent population. Execution of conventional practices leads to encourage the infestation of insect and other biological entities which have direct or indirect role in virus transmission. Moreover, a popular cultivar not always stable, a period of time the resistance may breakdown which leads to develop the epidemic/endemic. Therefore, a combined approach is helpful in formulation of sustainable control strategies.

21.3.1 INSECTICIDE APPLICATION

Vector management is one of the important and effective strategies of viral disease management. Systemic insecticides with different mode of action are available for the effective control of insect vectors. For the management of thrips transmitted tospoviruses, insecticides constitute an important tactic. An integrated approach of seed treatment with Imidacloprid followed by spraying control the thrips vector resulting effective management of WBNV disease in watermelon recorded (Kamanna et al., 2010). Application of high and frequent doses (400 g a.i.//ha at 3- or 5-day intervals) of Dimethoate resulted in reduction of bud necrosis in groundnut, whereas low rates with longer intervals application (100 g a.i./ /ha at 7- or 10-day intervals) showed higher incidence (Mandal et al., 2012). Application of systemic insecticides, Thiamethoxam 25WG (0.03%) reduces the population leafhoppers (2.83/3 leaves) and whiteflies (0.93/3 leaves) on okra whereas spray of imidacloprid 70WG (0.04%) suppress the population 3.49 and 1.30 per 3 leaves, respectively (Sa/ha et al., 2016). Foliar application reduced the population of whiteflies resulting lowest incidence of yellow vein mosaic (YVM) disease observed. Effect of six different insecticides including Cyazypyr, Flupyradifurone, Pyrafluquinazon and Sulfoxaflor on transmission of Tomato yellow leaf curl virus by the whiteflies biotype B to compare them with two established insecticides like Pymetrozine and Zeta-cypermethrin/bifenthrin combination. Percentage of virus infected tomato seedlings symptom tended to be lowest in treated with Flupyradifurone (Smith and Giurcanu, 2014).

21.3.2 CULTURAL OPERATIONS

Disease management through cultural practices includes the action undertaken by human beings in order to avoid and control the disease by influencing or interfering the cropping system. Cultural disease management might be the only a strategy which is viable in some cases. This may include reducing the rate of spread/transmission of an established disease or planting a crop at a site where neither virus and nor vector is present. Practices that lowering down the initial level of inoculums contain selection of appropriate planting materials, destruction of crop residues, elimination of viral affected plants that act as source of inoculums. For the removal of infected plants, early or fields level diagnosis play an important role e.g., *Potato virus X* and *A* can be diagnosed under field condition using immuno-dipsticks in potato (Ansar and Singh, 2016). This will be helpful to eliminate infected plants resulting formation of infected tubers and further dissemination may prevent. Selection of suitable planting materials or using certified seeds making assurance that disease is not carry forwarded. Continuous growing of a single variety/cultivar with genetic homogeneity and emergence of virulent strain of viruses will leads to rapid development of epidemics. Therefore, incorporation and rotation of various varieties may be the suitable option for management of disease. Adjustment in planting/sowing time is one of the safest way to escape the several viral disease, e.g., tomato leaf curl, potato stem necrosis and okra enation mosaic.

21.3.2.1 EFFECT OF PLANTING TIME ON VIRAL INCIDENCE

Modification in planting time/period plays a key role to escape the vector or viral infection in vegetable pathosystem. An experiment was conducted for controlling whitefly transmitted leaf curl disease in tomato and chili crops at BAU research field. The trial was designed to transplant the seedlings at different intervals. The firs transplanted (last week of October) crop expressed more disease in comparison to subsequent planting at 15 days interval. There was slight reduction of tomato leaf curl incidence in second planting (middle November). However, chili leaf curl was found low in early planting (last week of March), but disease was much increased in delayed planted crop with 43.7% in 2013–14 cropping season (Figure 21.5a and b).

21.3.2.2 BORDER/BARRIER CROP

Growing of border crops for management of vector-borne viral disease given the satisfactory results. Whitefly transmitted geminiviruses like tomato leaf curl virus (TLCV), cucurbit mosaic viruses and okra enation leaf curl virus can be minimized using two-row maize plantings. Moreover, the aphid vector can be trapped on maize plants which are responsible for transmission of several potyviruses (Figure 21.6).

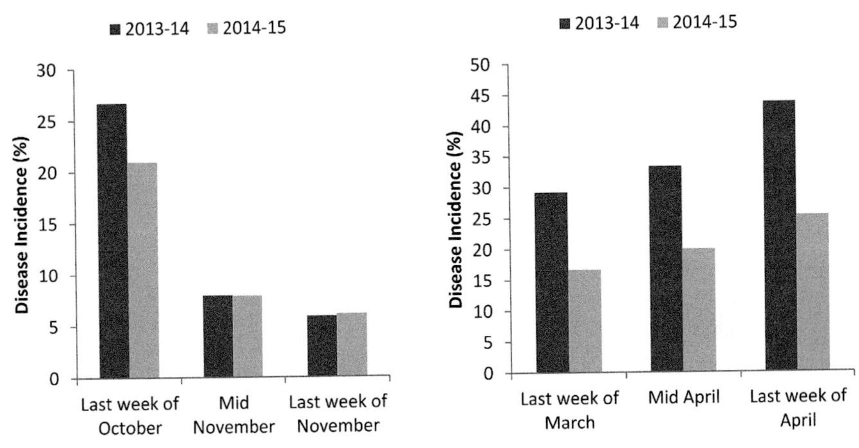

FIGURE 21.5 (a) Effect of staggered planting on Tomato leaf curl disease, (b) effect of staggered planting on chili leaf curl disease.

FIGURE 21.6 (a) Maize border in tomato crop to prevent the entry of insect vectors, (b) Severe infestation of aphids on maize leaf.

21.3.2.3 APPLICATION OF INSECTICIDES INTEGRATED WITH BORDER CROP

A field experiment for management of okra YVM disease, integrated application of insecticide along with border cropping with maize showed good results in order to minimize the disease. A treatment consisting of seed treatment and spray with imidacloprid and neem oil spray found effective to reduce the vector population (1.73/plant). However, sequential sprays of different insecticides reduced the whitefly population e in comparison to control. Treatment details along with whitefly and disease data presented in Table 21.4 (Ansar et al., 2014). Lowest incidence of mosaic (16.18%) was found in treatment having sequential application of various insecticides.

TABLE 21.4 Effect of Different Combination Insecticides Against Yellow Vein Mosaic of Okra

Treatments	Vector Population	Disease Incidence
T_0: Seed treatment with Goucho @ 0.5 g/l and sowing of border two rows with maize	4.73	34.37
T_1:T_0 + Spraying of Acephate @ 1.5 g/l + Neem oil @ 2.0 ml/l	4.10	21.78
T_2:T_0 + Spraying of Imidacloprid @ 0.5 ml/l + Neem oil @ 0.2 ml/l	1.73	15.47
T_3:T_0 + Spraying of Admire @ 2 g/15 l + Neem oil @ 2.0 ml/l	2.37	21.41
T_4:T_0 + Spraying of Hostathion @ 10 ml/l + Neem oil @ 2.0 ml/l	3.03	23.39
T_5:T_0 + First spraying of T_1, 2nd spraying of T_2, 3rd spraying of T_3, 4th spraying of T_4	1.37	16.18
T_6:T_0 + Neem oil spray @ 2.0 ml/l	4.83	27.38
T_0 + Water spray	7.20	38.01

Source: Trends in Biosciences 7(24): 4157–4160, 2014 (modified).

21.3.3 RESISTANCE BREEDING

Improvement of virus resistant cultivars has been a part of the plant breeder's tool since long time. Incorporation of resistant or tolerant cultivars in vegetable is one of the best options to minimize the losses due to viruses. Particularly at the stage, when there is growing public issues about the environmental pollution and residual effects on produce due to the arbitrary use of hazardous chemicals and emergence of new species and races. For the development of resistant cultivars and pre-breed lines, sources of resistance are the precondition and backbone of breeding program. Such sources may occur in the native cultivars, landraces, folk cultivars, semi-wild relatives and related species of the vegetable crops. For the resistant sources of viruses (Table 21.5) in different vegetable crops against major diseases and insect vector are useful for the vegetable breeders for the impending breeding program.

TABLE 21.5 Vegetable Varieties Resistant/Tolerant to Viruses and Insect-Vectors

Crop	Virus Disease	Resistance Source
Tomato	*Tomato leaf curl virus (TLCV)*	H-88-78-1, *L. hirsutum* f. Glabratum, *L. peruvianum* (LA 385), *L pimpinellifolium* and *L. hirsutum*, HS 101 *L. hirsutum* (LA 386, LA 1777, PI 390513, *L. glandulosum* (EC 68003) and *L. peruvianum*, PI 127830 and PI 127831, H-88-78-1, H-88-78-2, H-88-87
	Tomato yellow leaf curl	UPVTY 1, 3, 6, 9, 17, 53
	Orthotospoviruses	EC8630 and EC5888
Chili	*Pepper leaf curl Virus*	EC-497636, CM-334, IC-383072, IC-364063, BS-35, GKC-29, Pant C-1, Pusa SadaBahar, Tripura Collection, CO-309, NMCA-40008, BhutJolokia, Taiwan-2 and VR-339
	Thrips	NP46A, × 1068, × 743, × 1047,
		BG-4, × 226, × 230, × 233,
		EC119475B, Caleapin Red, Chamatkar,
Pumpkin	*Zucchini yellow mosaic virus, Watermelon Mosaic Virus*	C. lundelliana, C. martenezii Bohn and Whitaker, 1964, C. *ecuadorensis*, *C. faetidistima*, *C. martenezii*
Musk melon	*Cucumber green mottle mosaic virus*	VRM 5-10 (DVRM-1), VRM 29-1, VRM 31-1 (DVRM-2), VRM 42-4 and VRM 43-6*C. africanus*, *C. ficifolius*, *C. Anguria,* Chang Bougi, Hannah's Choice (Potyviruses) Phoot and Kachri (var. momordica)
Watermelon	*Watermelon mosaic*	PI 595203 (*Citrullus lanatus* var. *lanatus*), WM-1, WM-2, WM-3, WM-4,
	Watermelon bud necrosis virus	Durgapur selection, RHRWH-2 and EC-393243
Pumpkin	*Zucchini yellow mosaic virus, Watermelon Mosaic Virus*	C. lundelliana, C. martenezii C. *ecuadorensis*, *C. faetidistima*, *C. martenezii*
Cucurbita pepo	*Cucumber mosaic virus*	PI438699
	Tomato leaf curl New Delhi virus	*C. moschata*
Pea	*Pea Seed Borne mosaic virus*	PI 193586, PI 193835, X78123, X78126, X78127
	Pea Enation mosaic virus	OSU-547-29, OSU-559-6, OSU-546-3, OSU-584-16

TABLE 21.5 *(Continued)*

Crop	Virus Disease	Resistance Source
Cowpea	*Yellow mosaic virus*	IC 97767, IC 97829, IC 97787, IC 259084, IC 523658, IC 546883, IC 546884, Cowpea 263, NDCP 8, KLS 10, BC 244002, Arka Garima, Kashi Unnati and Kashi Kanchan
	Cowpea aphid borne virus	Purple Knuckle Hull-55, MNC-03-731C-21 and CNCx284-66E
	Leaf hoppers	TVu59, 123, 662, VITA 3
	Aphids	IT84S-2246-4, TVu 36, TVu 62, TVu 1889, TVu 2896
	Thrips	TVu 1509, IT84S-2246-4
French Bean	Bean common mosaic virus	Phaseolus coccineus, Morena, Monvisa, Niveo Oregon 54, SP6C, SP17B
Potato	*Groundnut bud necrosis virus*	CHIP-I, EX/A-680-16, J/95-221, JX-214, JX-24, JX-371, MP/97-621, MP/97-644, MS/92-2505, MS/95-117
Okra	*Yellow vein mosaic virus*	NIC-9303A, NIC-6308, NIC-3322, NIC-9408, NIC-3325, EC-329375, K-4409, *A. crinitus*, *A. angulosus*, *A. manihot*, *A. pungens*, and *A. tetraphyllus*
	Okra enation leaf curl virus	BRO-1

21.4 CONCLUSION

Vegetables are being the essential part that fulfills the nutritional demands of increasing population. These crops are severely hampered by various biotic factors, among them viruses playing a significant role. Since last decade diverse population of viruses have been explored, encompassing legumes, root crops and vegetables. Present investigation emphasized on the molecular interactions between viruses and their hosts. It is well clear that majority of viruses are transmitted by insect vectors which are making more challenging to the management task. Considering the emerging viruses belonging to the genus begomovirus, orthotospovirus, potyvirus and cucumovirus in vegetable pathosystem, possible efforts should be channeled to manage more effectively. Integrated approaches are essential to justify the sustainable management program. Preventive application of insecticides is routine approach to controlling insect-vectors. However, prolonged application single insecticide leads to develop resistance. An alternative application of different molecules with integrated cultural operations will helpful in managing vectors efficiently. Integration of resistant cultivar in the cropping system is the safest way to minimize the loss in an eco-friendly manner. For the development of resistant varieties against the viruses, sources of resistance are the prerequisite. Therefore, local cultivars, landraces, wild relatives and allied species of the vegetable should be undertaken in improving suitable cultivar.

KEYWORDS

- **capsicum chlorosis virus**
- **coat protein**
- **cucumber mosaic virus**
- **insecticides**
- **iris yellow spot virus**
- **open reading frames**

REFERENCES

Agnihotri, A. K., Mishra, S. P., Tripathi, R. C., Ansar, M., Srivastava, A., & Tripathi, I. P., (2018). First natural co-occurrence of tomato leaf curl New Delhi virus DNA-A and chili leaf curl beta satellite on tomato plants (*Solanum lycopersicum* L.) in India. *Journal of*

General Plant Pathology, 84, 414. https://doi.org/10.1007/s10327–018–0807–2 (Accessed on 25 November 2019).

Ansar, M., & Singh, R. P., (2016). Field diagnosis and temporal progress of and in tarai region of Uttarakhand. *Journal of Mycology Plant Pathology, 46*(4), 368–373.

Ansar, M., Akram, M., Agnihotri, A. K., Srinivasaraghavan, A., & Saha, T. N., (2018). Characterization of leaf curl virus in chili and overwintering role of nightshade in linkage between chili and tomato. *Journal of Plant Pathology*. doi: org/10.1007/s42161–018–0182-z.

Ansar, M., Akram, M., Singh, R. B., & Pundhir, V. S., (2015). Epidemiological studies of stem necrosis disease in potato caused by Groundnut bud necrosis virus. *Indian Phytopathology, 68*(3), 321–325.

Ansar, M., Saha, T., Sarkhel, S., & Bhagat, A. P., (2014). Epidemiology of okra yellow vein mosaic disease and its interaction with insecticide modules. *Trend in Biosciences, 7*(24), 4157–4160.

Babu, B., Hegde, V., Makeshkumar, T., & Jeeva, M. L., (2012). Rapid and sensitive detection of poty virus infecting tropical tuber crops using genus specific primers and probes. *African Journal of Biotechnology, 11*(5), 1023–1027.

Balaji, V., Vanitharani, R., Karthikeyan, A. S., Anbalagan, S., & Veluthambi, K., (2004). Infectivity analysis of two variable DNA B components of Mungbean yellow mosaic virus-Vigna in *Vigna mungo* and *Vigna radiata*. *Journal of Biosciences, 29*, 297–308.

Berger, P. H., Adams, M. J., Barnett, O. W., Brunt, A. A., Hammond, J., Hill, J. H., Jordan, R. L., Kashiwazaki, S., Rybicki, E., Spence, N., et al., (2005). Potyviridae. In: Fauquet, C. M., Mayo, M. A., Maniloff, J., Desselberger, U., & Ball, L. A., (eds.), *Virus Taxonomy* (pp. 819–841). VIIIth Report of the International Committee on Taxonomy of Viruses. London: Elsevier/Academic Press.

Bhat, A. I., Varma, A., Pappu, H. R., Rajamannar, M., Jain, R. K., & Praveen, S., (1999). Characterization of a potyvirus from eggplant (*Solanum melongena*) as a strain of potato virus Y by N-terminal serology and sequence relationships. *Plant Pathology, 48*, 648–654.

Bhatt, B. S., Chahwala, F. D., Rathod, S., & Singh, A. K., (2016). Identification and molecular characterization of a new recombinant begomovirus and associated betasatellite DNA infecting Capsicum annuum in India. *Archives of Virology, 161*(5), 1389–1394.

Brunt, A. A., Kenten, R. H., & Phillips, S., (1978). Symptomatologically distinct strains of pepper veinal mottle virus from four West African Solanaceous crops. *Annals of Applied Biology, 88*, 115–119.

Brunt, A. K., & Crabtree, G. A., (1990). *Viruses of Tropical Plants* (p. 707). CAB International, Wallingford, UK.

Chakraborty, S., Pandey, P. K., Banerjee, M. K., Kalloo, G., & Fauquet, C. M., (2003). Tomato leaf curl Gujarat virus, a new begomovirus species causing a severe leaf curl disease of tomato in Varanasi, India. *Phytopathology, 93*, 1485–1495.

Chandran, S. A., Packialakshmi, R. M., Subhalakshmi, K., Prakash, C., Poovannan, K., Nixon, P. A., Gopal, P., & Usha, R., (2013). First report of an alp/ha satellite associated with okra enation leaf curl virus. *Virus Genes, 46*, 585. doi: org/10.1007/s11262–013–0898-y.

Chatchawankanphanich, O., Chiang, B. T., Green, S. K., Singh, S. J., & Maxwell, D. P., (1993). Nucleotide sequence of a gemini virus associated with tomato leaf curl from India. *Plant Disease, 77*, 1168.

Chattopadhyay, B., Singh, A. K., Yadav, T., Fauquet, C. M., Sarin, N. B., & Chakraborty, S., (2008). Infectivity of the cloned components of a begomovirus: DNA beta complex causing chili leaf curl disease in India. *Archives of Virology, 153*, 533–539.

Crescenzy, A., (1993). Cucumber mosaic cucumovirus populations in Italy under natural epidemic conditions and after a satellite mediated protection test. *Plant Disease, 77*, 28–33.

FAO statistics, (2009). http://www.fao.org/faostat/en (Accessed on 25 November 2019).

Fauquet, C. M., Mayo, M. A., Maniloff, J., Desselberger, U., & Ball, L. A., (2005). Virus taxonomy. In: *Eighth Report of the International Committee on Taxonomy of Viruses*. San Diego (CA): Elsevier Academic Press.

Gawande, S. J., Khar, A., & Lawande, K. E., (2010). First report of Iris yellow spot virus on garlic in India. *Plant Disease, 94*, 1066.

Ghanekar, A. M., Reddy, D. V. R., Iizuki, N., Amin, P. W., & Gibbons, R. W., (1979). Bud necrosis of groundnut (*Arachis hypogaea*) in India caused by tomato spotted wilt virus. *Annals of Applied Biology, 93*, 173–179.

Gibbs, A. J., Mackenzie, A. M., & Gibbs, M. J., (2003). The "potyvirid primers" will probably provide phylogenetically informative DNA fragments from all species of potyviridae. *Journal of Virological Methods, 12*, 41–44.

Gibbs, A., & Ohshima, K., (2010). Potyviruses and the digital revolution. *Annual Review of Phytopathology, 48*, 205–223.

Girish, K. R., & Usha, R., (2005). Molecular characterization of two soybean-infecting begomoviruses from India and evidence for recombination among legume-infecting begomoviruses from South-East Asia. *Virus Research, 108*, 167–176.

Gopal, K., Muniyappa, V., & Jagadeeshwar, R., (2010). Management of peanut bud necrosis disease in groundnut by botanical pesticides. *Archives of Phytopathology and Plant Protection, 44*(13), 1233–1237.

Greber, R. S., Persley, D. M., & Herrington, M. E., (1989). Some characteristics of Australian isolates of zucchini yellow mosaic virus. *Australian Journal of Agricultural Research, 39*, 1085–1094.

Gutiérrez, C., (1999). Geminivirus DNA replication Cell. *Molecular Life Science, 56*, 313–329.

Harrison, B. D., Finch, J. T., Gibbs, A. J., Hollings, M., Shepherd, R. J., Valenta, V., & Wetter, C., (1971). Sixteen groups of plant viruses. *Virology, 45*, 356–363.

Igwegbe, E. C. K., & Waterworth, H. E., (1982). Properties and serology of the strain of Pepper veinal mottle virus isolated from eggplant (*Solanum melongena L.*) in Nigeria. *Phytopathologische Zeitschrift, 103*(1), 9–12.

Jain, R. K., Pappu, H. R., Pappu, S. S., Krishanareddy, M., & Vani, A., (1998). Watermelon bud necrosis tospovirus is a distinct virus species belonging to serogroup, I. V., *Archives of Virology, 143*, 1637–1644.

Jeske, H., (2009). Geminiviruses. *Curr. Top. Microbiol. Immunol., 331*, 185–226.

Johansen, E., Edwards, M. C., & Hampton, R. O., (1994). Seed transmission of viruses: Current perspectives. *Annual Review of Phytopathology, 32*, 363–386.

Jose, J., & Usha, R., (2003). Bhendi yellow vein mosaic disease in India caused by association of a DNA β satellite with a begomovirus. *Virology, 305*, 310–317.

Kamanna, B. C., Jadhav, S. N., & Shankarappa, T. H., (2010). Evaluation of insecticides against thrips vector for the management of watermelon bud necrosis virus (WBNV) disease. *Karnataka Journal of Agricultural Science, 23*, 172–173.

Krishnareddy, M., & Verma, A., (1990). Prevalence of tomato spotted wilt virus in Delhi. *Indian Phytopathology, 43*, 252.

Krishnareddy, M., Usha, R., Kumar, A., Madhavireddy, K., & Pappu, H. R., (2008). First report of Capsicum chlorosis virus (Genus Tospovirus) infecting chili pepper (*Capsicum annuum*) in India. *Plant Disease, 92*, 1469.

Kulkarni, G. S., (1924). Mosaic and other related diseases of crops in the Bombay Presidency. *Poona Agriculture College Magazine, 6*, 12.

Kunkalikar, S. R., Poojari, S., Arun, B. M., Rajagopalan, P. A., Chen, T. C., Yeh, S. D., Naidu, R. A., Zehr, U. B., & Ravi, K. S., (2011). Importance and genetic diversity of vegetable-infecting tospoviruses in India. *Phytopathathology, 101*, 367–376.

Kunkalikar, S. R., Sudarsana, P., Rajagopalan, P., Zehr, U. B., Naidu, R. A., & Ravi, K. S., (2007). First report of Capsicum chlorosis virus in tomato in India. *Plant Health Prog*. doi: 10.1094/PHP-2007–1204–01.

Ladipo, J. L., & Roberts, I. M., (1977). Pepper veinal mottle virus associated with a streak disease of tomato in Nigeria. *Annals of Applied Biology, 87*, 133–138.

Lal, S. B., & Singh, S., (1988). Identification of some virus diseases of vegetable crops in Afghanistan. *Plant Protection Bulletin (Faridabad), 36*(2), 83–89.

Lecoq, H., Pitrat, M., & Clement, M., (1981). Identification characterization d'un potyvirus provoquant la maladie du rabougrissement jaune du melon. *Agronomie, 1*, 827–834.

Lisa, V. G., Boccardo, G. D., Agostino, G., Dellavalle, & Aquilio, M. D., (1981). Characterization of a potyvirus that causes zucchini yellow mosaic. *Phytopathology, 71*, 667–672.

Mali, V. R., & Kulthe, K. S., (1980). A seed borne *potyvirus* causing mosaic of cowpea in India. *Plant Disease, 64*, 925–928.

Mandal, B., Jain, R. K., Krishnareddy, M., Krishna, K. N. K., Ravi, K. S., & Pappu, H. R., (2012). Emerging problems of tospoviruses (Bunyaviridae) and their management in the Indian subcontinent. *Plant Disease, 96*(4), 468–479.

Mansoor, S., Briddon, R. W., Zafar, Y., & Stanley, J., (2003). Geminivirus disease complexes: An emerging threat. *Trends Plant Science, 8*, 128−134.

Milne, R. G., (1988). Taxonomy of the rod-shaped filamentous viruses. In: Milne, R. G., (ed.), *The Plant Viruses: The Filamentous Plant Viruses* (Vol. 4, pp. 3–50). New York (NY): Plenum Press.

Moriones, E., Praveen, S., & Chakraborty, S., (2017). *Tomato Leaf Curl New Delhi Virus: An Emerging Virus Complex Threatening Vegetable and Fiber Crops Viruses, 9*(10), 264. doi: org/10.3390/v9100264.

Nagaraju, R., & Reddy, H. R., (1980). Occurrence and distribution of bell pepper viruses around Bangalore. *Current Research UAS Bangalore, 10*(9/10), 155, 156.

Naimuddin, K., Akram, M., & Agnihotri, A. K., (2014). Molecular evidence for the association of Tomato leaf curl Gujarat virus with a leaf curl disease of *Phaseolus vulgaris* L. *Journal of Phytopathology, 163*, 58–62. doi: org/10.1111/jph.12255.

Naimuddin, K., Akram, M., Pratap, A., & Yadav, P., (2013). Characterization of a new begomovirus and a beta satellite associated with the leaf curl disease of French bean in northern India. *Virus Genes, 46*, 120–127. doi: 10.1007/s11262–012–0832–8.

Palukaitis, P., & Garcia-Arenal, F., (2003). Cucumoviruses. *Advance Virus Research, 62*, 241–323. doi: 10.1016/S0065–3527(03)62005–1.

Palukaitis, P., Roossinck, M. J., Dietzgen, R. G., & Francki, R. I. B., (1992). Cucumber mosaic virus. *Advance Virus Research, 41*, 281–349.

Pappu, H. R., Jones, R. A. C., & Jain, R. K., (2009). Global status of tospovirus epidemics in diverse cropping systems: Successes achieved and challenges ahead. *Virus Research, 141*, 219–236.

Parrella, G., Gonalons, P., Gebre-Selassie, K., Vovlas, C., & Marchoux, G., (2003). An update of the host range of tomato spotted wilt virus. *Journal of Plant Pathology, 85*, 227–264.

Pitrat, M., & Lecoq, H., (1984). Inheritance of zucchini yellow mosaic virus resistance in *Cucumis melo* L. *Euphytica, 33*, 57–61.

Poutaraud, A., Desbiez, C., Lemaire, O., Lecoq, H., & Herrbach, E., (2004). Characterization of a new potyvirus species infecting meadow saffron (*Colchicum autumnale*). *Archives of Virology, 149*, 1267–1277.

Prasada, R. R. D. V. J., & Yaraguntaiah, R. C., (1979). The occurrence of pepper veinal mottle virus on chili in India. *Mysore Journal of Agricultural Science, 13*(4), 445–448.

Prasada, R. R. D. V. J., Iizuka, N., Ragunathan, V., & Joshi, N. C., (1980). Occurrence of tomato spotted wilt virus on tomato in Andhra Pradesh. *Indian Phytopathology, 33*(3), 436–439.

Pratap, D., Kashikar, A. R., & Mukherjee, S. K., (2011). Molecular characterization and infectivity of a tomato leaf curl New Delhi virus variant associated with newly emerging yellow mosaic disease of eggplant in India. *Virology Journal, 8*, 305.

Raj, S. K., & Singh, B. P., (1996). Association of geminivirus infection with yellow green mosaic disease of *Cucumis sativus*: Diagnosis by nucleic acid probes. *Indian Journal of Experimental Biology, 34*, 603–605.

Raj, S. K., Khan, M. S., Singh, R., Kumari, N., & Praksh, D., (2005). Occurrence of yellow mosaic geminiviral disease on bitter gourd (*Momordica charantia*) and its impact on phytochemical contents. *International Journal of Food Science and Nutrition, 56*, 185–192.

Ravi, K. S., Kitkaru, A. S., & Winter, S., (2006). Iris yellow spot virus in onions: A new tospovirus record form India. *Plant Pathology, 55*, 288.

Reddy, D. V. R., Ratna, A. S., Sudarshana, M. R., & Kiran, K. I., (1992). Serological relationships and purification of bud necrosis virus, a tospovirus occurring in peanut (*Arachis hypogaea* L.) in India. *Annals of Applied Biology, 120*, 279–286.

Roossinck, M. J., (2011). The good viruses: Viral mutualistic symbioses. *Nature Reviews Microbiology, 9*, 99–108. doi: 10.1038/nrmicro2491. PMID 21200397.

Saha, T., Ansar, M., Nitya, C., & Ray, S. N., (2016). Temporal dynamics of sucking pest and field response of promising insecticidal molecules in okra. *Journal of Applied and Natural Science, 8*(1), 392–397.

Sharma, A., & Kulshrestha, S., (2014). First report of *Amaranthus* sp. as a natural host of capsicum chlorosis virus in India. *Virus Disease, 25*(3), 412–413.

Sherwood, J. L., German, T. L., Moyer, J. W., Ullman, D. E., & Whitefield, A. E., (2000). Tospoviruses. In: Maloyand, O. C., & Murray, T. D., (eds.), *Encyclopedia of Plant Pathology* (pp. 1034–1040). John Wiley & Sons, New York.

Shih, S. L., Tsai, W. S., Green, S. K., & Singh, D., (2006). First report of tomato leaf curl Joydebpur virus infecting chili in India. *New Disease Report, 14*, 17.

Shih, S. L., Tsai, W. S., Green, S. K., & Singh, D., (2007). *First Report of Tomato Leaf Curl Joydebpur Virus Infecting Chili in India*. doi: org/10.1111/j.1365–3059.2007.01540.x.

Shukla, D. D., Ford, R. E., Tosic, M., Jilka, J., & Ward, C. W., (1989). Possible members of the potyvirus group transmitted or whiteflies share epitopes with aphid-transmitted definitive members of the group. *Archives of Virology, 105*, 143–151.

Shukla, D. D., Ward, C. W., & Brunt, A. A., (1994). *The Potyviridae*. Cambridge: CAB International. doi: org/10.1017/S0021859600074736.

Singh, R. B., Srivastava, K. K., Khurana, S. M. P., & Pandey, S. K., (1997). Assessment of yield losses due to potato stems necrosis disease. *Indian Journal of Virology, 3*, 135–137.

Singh, S. J., & Krishnareddy, M., (1996). Watermelon bud necrosis: A new tospovirus disease. *Acta Horticulturae, 431*, 68–77.

Smith, H. A., & Giurcanu, M. C., (2014). New insecticides for management of tomato yellow leaf curl, a virus vectored by the silver leaf whitefly, *Bemisia tabaci*. *Journal of Insect Science, 14*(1), 183.

Sreenivasulu, M., & Gopal, D., (2010). Development of recombinant coat protein antibody based IC-RT-PCR and comparison of its sensitivity with other immunoassays for the detection of papaya ringspot virus isolates from India. *Journal of Plant Pathology, 26*, 25–31.

Srinivasulu, M., Sarovar, B., Anthony, J. A. M., & Sai, G. D. V. R., (2010). Association of a potyvirus with mosaic disease of gherkin (*Cucumis anguria* L.) in India. *Indian Journal of Microbiology*, *50*, 221–224.

Tiwari, N., Singh, V. B., Sharma, P. K., & Malathi, V. G., (2013). Tomato leaf curl Joydebpur virus: A monopartite begomovirus causing severe leaf curl in tomato in West Bengal. *Archives of Virology, 158*, 1–10.

Todd, J. M., Ponniah, S., & Subramanyam, C. P., (1975). First record of tomato spotted wilt virus from the Nilgiris in India. *Madras Agriculture Journal*, *62*(3), 162–163.

Tsompana, M., & Moyer, J. W., (2008). Tospoviruses. In: Mahy, B. W. J., & Van Regenmortel, M. H. V., (eds.), *Encyclopedia of Virology* (3rd edn., Vol. 5, pp. 157–162). Elsevier Ltd., Oxford, UK.

Varma, A., & Malathi, V. G., (2003). Emerging geminivirus problems: A serious threat to crop production. *Annals of Applied Biology, 142*, 145–164.

Venkataravanappa, V., Lakshminarayana, R. C. N., Swarnalatha, P., et al., (2014). *Phytoparasitica, 42*, 109. doi: org/10.1007/s12600–013–0345–6.

Verma, R., (2004). First report of *Zucchini yellow mosaic virus* in Bottlegourd (*Lagenaria siceraria*) in India. *Plant Disease, 88*(4), 426, 3–426.

Ward, C. W., & Shukla, D. D., (1991). Taxonomy of potyviruses: Current problems and some solutions. *Intervirology, 32*, 269–296.

Whitfield, A. E., Ullman, D. E., & German, T. L., (2005). Tospovirus-thrips interactions. *Annual Review of Phytopathology*, *43*, 459–489.

Zitter, T. A., & Murphy, J. F., (2009). Cucumber mosaic. *The Plant Health Instructor*. doi: 10.1094/PHI-I-2009–0518–01.

CHAPTER 22

Advancement of Agrochemical Research and Regulation

PRITAM GANGULY

Department of Soil Science and Agricultural Chemistry, Bihar Agricultural University, Sabour, Bhagalpur, Bihar, India, E-mail: pritam0410@gmail.com

ABSTRACT

Agrochemicals play crucial role in agriculture to increase productivity and quality of produce as well. Pesticides, one of the key farm input, has increased production drastically by protecting crops from dangerous pests and diseases. Another important input is plant growth regulator which is used to control different growth behavior of crop plants. But, continuous use of agrochemicals is detrimental to natural environmental processes and may impose health risk to the humans. Researches have been advanced to bring agrochemicals with new mode of action to ascertain safety to the non-target organisms and also to reduce environmental load. Simultaneously, several countries, including India, have been coming up with more restricted regulation related to pesticide application.

22.1 INTRODUCTION

Agrochemicals are very important inputs for improved crop production technology. Agrochemicals are broadly classified as insecticides, fungicides, herbicides, rodenticides, and nematicides. It does also include synthetic fertilizers, manures, and plant growth hormones. Agrochemicals can benefit crop production if applied at the right time, through the right method and with the right dose. The balanced use of agrochemicals shall not cause any significant damage to the environment. Rather improves livelihood. Nature of soils, types of crop plant and weather condition is the crucial factors for pesticide and fertilizer applications.

Pesticides can kill, manage, eradicate, attract, repel or control pests those who can incur loss at any time of cultivation, storage, transportation and food processing. Pesticides are the substances which not only protect the crops from pest incidence but can enhance production both in terms of quantity and quality. These are subdivided based on the target organisms (e.g., insecticides, acaricides, fungicides, herbicides, molluscicides, rodenticides, etc.), chemical composition (e.g., organic, inorganic, synthetic, biopesticides, etc.), as well as physical properties (e.g., liquid, solid, gaseous, etc.).

22.2 GENERATION OF PESTICIDES

1. **First Generation:** These pesticides were generally composed of inorganic and organic compounds and primarily used before 1940. Minerals like arsenic, mercury, and lead were mainly used as inorganic compounds. But their use created a serious issue as they were not easily metabolized or degraded in the environment. As a result, these pesticides began to persist and being accumulated in the soil which caused the land to be unfertile.

   Organic pesticidal compounds in this generation were produced from plants, and through their use, the plants developed resistance to certain insects. These organic pesticidal compounds are called as botanicals. Pesticides such as pyrethrin come from chrysanthemum flowers. Rotenone is isolated from jewel vine and nicotine is found in tobacco. Botanicals are easily degradable in the environment as compared to inorganic compound and do not persist. However, these botanicals are highly toxic to non-target organisms like fish and bees.

   Synthetic botanicals were also produced as first-generation pesticides. They are created chemically to mimic the structure of plant produced botanicals. Pyrethoid, almost identical to the botanical pyrethrin, was an important synthetic botanical because it was less persistent in the environment.
2. **Second Generation:** These pesticides differ from earlier pesticides as these are produced synthetically. These are started to be applied in the field in the 1940s after DDT was invented in 1939 by Paul Müller. The most commonly used groups of this generation pesticides are Chlorinated hydrocarbons (OCs), Organophosphates (OPs), and

Carbamates. Insecticides are prevalently found in these. Each group acts differently to produce any effect in the target organism. Majority of these insecticides are persistent while some compounds are easily degradable and thus, cannot be accumulated in nature. The degree of toxicity of these substances to mammals depends upon the properties of the substances.

3. **Third Generation:** It includes insect pheromones such as:
   - **Insect Growth Regulators (IGR):** It causes disruption in normal processes of hormonal activity of insects, which in turn inhibiting the growth and development of the target insect.
   - **Chitin Synthesis Inhibitors (CSI):** CSI targets exoskeleton of insects.
   - **Juvenile Hormones (JH):** These compounds mimic hormones and as a result insect stays in juvenile phase only.
   - **Bt:** In this case, toxins are secreted followed by ingestion.
4. **Fourth Generation:** It includes plants' derived chemicals and other bio-pesticides. A resurgent interest in plant-derived compounds to manage insect pests is developed due to the fact that these pesticide products have less negative impacts in nature and human health than the highly effective synthetic insecticides mostly have. Azadirachtin, precocene, phytoecdysone, etc. belong to this generation.

In India, the second generation insecticides are generally used indiscriminately by the farmers as these are quite cheap in nature and to some extent effective also. But, nowadays, the application of new generation molecules is also increasing due to excellent efficacy against the target pest.

22.3 HISTORY OF AGROCHEMICAL RESEARCH

Prior to 1940, inorganic compounds, for example, sodium chlorate, and arsenic-based compounds, and botanical compounds like pyrethrin were usually applied in fields to manage pest attack. Moreover, some pesticides were produced as by-products of natural gas or similar industrial processes. Consequently, compounds like nitrophenols, chlorophenols, naphthalene, and petroleum oils were commonly applied in the farms for having both fungicidal and insecticidal properties. Along with these, sodium arsenate and ammonium sulfate were primarily used as weedicide. High dose rates, non-selectivity, and phytotoxicity were observed as major drawbacks for range of these compounds (A History of Crop Protection and Pest Control in Our Society, Crop Life Canada, 2002). With the simultaneous discovery of DDT molecule (and its analogs), lindane, endrine, chlordane, dieldrin, aldrin, endosulfan, parathion, captan, 2,4-D, use of chemical pesticides got accelerated. These products had shown broad-spectrum activity and were effective at comparatively lower dose and cost as well (The History of Pesticides, 2008; Delaplane, 2000). DDT was most preferred among these compounds as it helped to control insect-borne diseases, like malaria, typhus, dengue, yellow fever, etc., and was apparently considered to be less toxic to mammals. For this achievement, Dr. Paul Muller, inventor of DDT's insecticidal property, was awarded with Nobel Prize in the arena of Medicine in 1949. However, in 1946, as per reports, house flies got resistant to DDT and, due to its widespread use, the compound was found to be harmful on non-target organisms (The History of Pesticides, 2008; Delaplane, 2000).

The negative impact of pesticidal compounds in non-target organisms was still not investigated and remained unexplored up to 1950s. There was no documented evidence of pesticide injury being recorded as the food price got decreased with their "normal" use (Wessels Living History Farm, Farming in the 1950s and 60s). Some cases were reported where misuse of the chemicals caused harm. But, still, the new pesticides were considered safer, than arsenic compounds which caused death of humans in the 1920s and 1930s (Wessels Living History Farm, Farming in the 1930s). After a long period, problems started arising through the improper usage of pesticides and in 1962 these were mentioned by Rachel Carson in her book Silent Spring (Silent Spring, 2002). Potential health risk caused by the excess use of existing pesticides were investigated which was resulted in synthesis of safer and more eco-friendly compounds.

Pesticide related research work continued thereafter and in 1970s and 1980s, major important developments were happened like discovery of world's largest selling herbicide, glyphosate, the low dose rate sulfonylurea, imidazolinone (imi), dinitroanilines aryloxyphenoxypropionate (fop) and cyclohexanediones (dim) groups. As insecticide is concerned, researchers synthesized 3rd generation of pyrethroids, introduced avermectins, benzoylureas, and Bt as a spray application. In the case of fungicide, this era also witnessed the discovery of the triazole, morpholine, imidazole, pyrimidine and dicarboxamide groups. The special feature of these compounds was mostly single mode of action, which led to increase selectivity. Hence resistance was built up with their continuous use and scientists introduced new management strategies to fight against these adverse effects (http://agrochemicals.iupac.org/index.php?option=com_sobi2&sobi2Task=sobi2Details&catid=3&sobi2Id=31).

In 1990s, new molecules of existing groups have been invented to bring higher selectivity, more eco-friendliness and lower toxicity to the non-target organisms. Moreover, new groups of compounds have been identified like chloronicotinyl, spinosyn, fiprole, and diacylhydrazine insecticides, strobilurin and azolone fungicides and triazolopyrimidine, triketone and isoxazole herbicides. The dose rate also got reduced from kilograms per hectare to grams per hectare (http://agrochemicals.iupac.org/index.php?option=com_sobi2&sobi2Task=sobi2Details&catid=3&sobi2Id=31).

New insecticide (New Insecticide Modes of Action: Whence Selectivity?) and fungicide (Morton and Staub, 2008) chemistry-based compounds have helped in better resistance management and enhanced selectivity. In this period, some old molecules had come out with new eco-friendly safer formulations so as their shelf-life got extended (A History of Crop Protection and Pest Control in Our Society, 2002). Integrated pest management (IPM) approach, which adopts all available pest control measures to manage or control the pest populations and reduce the dependency on the use of synthetic pesticides, has been encouraged. This IPM strategy was not only found eco-friendly, economic but was the factor to reduce pesticide use (OECD Series on Pesticides, 1999).

Recently, the major advancements have been achieved by introducing genetically modified crops with a self-defense mechanism or showing resistance to non-selective broad-spectrum herbicide products or pests. These strategies were found successful in crops like soybeans, corn, canola and cotton where varieties of corn and cotton resistant to corn borer and bollworm respectively (A History of Crop Protection and Pest Control in Our Society, 2002). This technique has also become popular along with the adoption of IPM. These changes have opened up the new arena of pest control and played the role as game changer in agrochemicals research (http://agrochemicals.iupac.org/index.php?option=com_sobi2&sobi2Task=sobi2Details&catid=3&sobi2Id=31).

Plant growth regulators (PGRs) are the substances which can encourage or restrict plant growth when added in small quantity. They have high potential to enhance the global food supply. Since 1940s, PGRs have been successfully used to alter crop biology by modifying their usual behavioral responses. This phenomenon led to bring in positive effects on growth and development of crops passing through different stages. Use of PGRs in crops controls plant metabolism, flowering, fruit set, aging, and their strength to sustain in adverse conditions.

PGRs were applied in ancient times when people in the Middle East used to apply olive oil on plants for better growth. The basic reason behind this application is ethylene which is produced due to the decomposition of oil by heat. By the time 1930s, ethylene was found to promote flowering in crops like pineapple and after that use of acetylene in Hawaii was observed to initiate flowering. During 1940s, another compound called auxins, which found to be reported to produce the same result. Naphthalene acetic acid (NAA), similar product of auxin, was successively used as growth regulator in pineapple. Some commonly used PGRs are: abscisic acid, cycloheximide, alachlor, atrazine, barban, benzyladenine, chlormequat, etc. PGRs have been reported for significant uses in agriculture mostly in germination of seeds, fruit production, prevention of lodging in cereals and in tissue culture.

The primitive use for PGRs was to stimulate and/or encourage the rooting of plant cutting. Indole-3-acetic acid is most probably the best and most commonly used chemical for this stimulation. A substantial research work has been carried out on initiation of rooting, and subsequently being published more than thousands in number in reputed journals and books.

Maleic hydrazide has been applied recognizably for sprouting suppression of potatoes and onions. This molecule had been widely used as an inhibitor of grassy growth on golf courses, in residential lawns, in amusement parks and in roadways. Then comes the extensive use of chlormequat/ cycocel (CCC), that retarded the plant height in cereals especially wheat without altering quantitative as well as qualitative yield. This dwarfingness in wheat enables plant to prevent lodging after rainfall and heavy wind in even the case where nitrogen is applied at higher dose.

Another important PGR, the ethephon, (2-chloroethylphosponic acid) an ethylene-producing compound, is used to induce the latex yield in rubber trees. This compound has the potency to enhance the duration of latex flow between tapings so that the production of dry rubber at commercial scale can be increased up to 100%. Application of ethephon preserves the tree life by preventing random cutting of bark.

Recently, ripening of sugar cane is an important subject of research for commercial developments of PGR. Several chemical compounds have been produced promising results to elevate the amount of sucrose in cane at harvest. Etephon is applied at commercial level as a ripener on sugar cane in Zimbabwe and South Africa. The ripeners are mostly used @ 5–25% which depends upon the cane variety, nature of soil and weather condition to increase sugar yield is.

Gibberellic acid (GA) has been found as the potent growth regulator to enhance fruit set in citrus. Aqueous solution of GA was applied in entire citrus trees at full bloom increased the production of seedless fruit of five self-incompatible citrus cultivars. But, however, they were more subjected to dropping after preliminary fruit set compared to seedy fruits developed by cross-pollination. Abscission is a process of dropping of a plant leaf, flower, fruit or stem itself from the plant. This process ensures the effective crop growth for better production (http://www.dawn.com/news/125197/plant-growth-regulators).

22.4 RECENT ADVANCEMENT IN PESTICIDE CHEMISTRY

RNAi-based insecticides (interference RNA) are the latest area of discussion as it is coming out as an interesting subject of research for next-generation pesticide. A protein is produced in a living cell when a specific gene got activated in that cell. The reverse process is absolutely the same, i.e., protein synthesis in that cell got interrupted by turning off that particular gene. Switching off that particular gene is made possible by sending a small fragment of RNA which suppresses the gene's expression into a protein. This process is known as RNA interference, or RNAi. The importance of RNAi is undoubtedly huge. All that scientists need to know the DNA sequence of a gene so that they can develop the specific RNA molecule and send it into the living cell to inhibit protein synthesis by the gene in the cell— eventually switching the gene off. This path-breaking idea has opened up a new scope of research in pest control. Now, researchers are planning to inculcate this process in insect pests to prevent their damage to the crops. Seed companies are playing a major role to harvest this concept and take benefit out of it. They have come out with genetically modified corn plant which can control western corn rootworm, the most harmful pest in the world, by producing small RNAs that silences a gene in that insect. Efforts are already being directed to create RNAi sprays for managing field, households as well as storage pests that will turn off pest genes (The Next Generation of Pesticides, 2013).

22.5 PESTICIDES IN ENVIRONMENT

Fate of plant protection compounds in- a crop eco-system depends upon nature of physical, chemical and biological processes which ultimately decides persistence of the compound. Very little amount of applied pesticides reach target areas, but the rest bigger portion is exposed to several processes like drift, surface runoff, volatilization, microbial degradation as well as by plants and animals, adsorption in soil, bioaccumulation and biomagnifications and photodecomposition (Wilkinson, 1976). Sunlight plays a major role in photodecomposition that may result to various photo-metabolites which are chemically and toxicologically different from the parent compounds. Pesticide molecule adsorbs solar energy and got excited to make or break chemical bonds. The process of photolysis of pesticides is being sensitized by different photo-sensitizers found in environment such as riboflavin, humic acid, etc. They absorb photons and then donate it to acceptor causing decomposition of pesticides. Photochemistry of OPs such as phosphorothioate and phosphorodithiotate compounds was studied vastly. Although not highly susceptible to photodegradation by UV light, Malathion is degraded by UV light to different photo-metabolites such as malathion diacid, o,o-dimethyl phosphorothioic acid, malaoxon, o,o-dimethyl phosphorodithioic acid, and phosphoric acid (Zabik, 1985; Chukudebe, Othman, and Fukuto, 1989). Among these, malaoxon is more toxic compared to its parent compound. Non-target organisms may be unintendedly exposed to insecticides at sub-lethal level but that can damage significantly their population (Haynes, 1988). Sub-lethal dosage of pesticide may show long term chronic effect and that could be expressed at advanced growth stages of the insect (Kariappa and Narasimhanna, 1978; Troitskaya and Chichigina, 1980). But, it may happen that the particular pest can develop resistance within it against the pesticide when exposed to sub-lethal dosages for longer period of time. Troitskaya and Chichigina (Bora, 1998) have found that bacterial and chemical insecticides can pose serious toxicity threat to *Bombyx mori* in silk-producing areas when used in combination.

22.6 PESTICIDE REGULATION

Pesticide residues in environment are now area of serious concern as the application of these xenobiotics is almost inevitable. Pesticide residue is the leftover amount of certain pesticide after application. Although, these plant protection compounds are intended to act as target-specific, but practically, pesticide residues are not confined in their target site. These are generally found in various components of environment such as soil, air, water bodies like pond, lake, river etc., and more importantly getting accumulated in food chain even after good agriculture practice (GAP) is followed. Being toxic in nature, pesticides can pose toxicity threat to the non-target organism including human being if not properly handled. Moreover, the parent compound undergoes several transformations in the surrounding environment and yields highly toxic metabolites.

Thus, pesticide regulation is highly important so as to avoid toxicity risk to the non-target organism associated with its use. Pesticides must be handled with proper care at each step starting from manufacturing, to transport, storage, application and till their disposal in the surrounding environment. Several countries have formed their own regulation standard and follow it strictly. In the US, the Environmental Protection Agency (EPA) regulates pesticides and related compounds under the acts, i.e., the Federal Insecticide, Fungicide, and Rodenticide Act (FIFRA) and the Food Quality Protection Act (FQPA). The main moto of EPA in regulating pesticides is to make sure that these toxic compounds do not create adverse effects in human beings or the environment. European Commission is held responsible for the evaluation of every active substance for safety prior to introduction in the market of the Europe Union (EU).

In India, the first major instance of pesticide poisoning was reported in 1958. The Government of India formed a Commission of Enquiry to recommend remedial action against the casualties that occurred in Kerala and Chennai (Tamil Nadu) by pesticide poisoning contaminated in imported wheat. A committee empanelled with experts of the Indian Council of Agricultural Research, chaired by Prof. M.S. Thacker was appointed in 1964–67 to suggest the way forward related to pesticide legislation. Based on their reports, Insecticides Act was formed and passed in 1968 by Government of India to regulate insecticide (i.e., pesticide) at every stage of import, manufacture, transport, sale, distribution, storage and application of insecticide so as to protect non-target organisms including human beings from risks associated with it. In 1970, the Ministry of Agriculture took the responsibility from the Ministry of Health and family Planning to enforce the Act. The Department of Agriculture of this Ministry had taken immediate steps to frame the Rules related to the Act and formed Central Insecticides Board (CIB) and registration committee (RC). The different states of the country appointed all functionaries as mentioned in the Act. After all such formalities completed, the Insecticides Act were brought into force with effect from 1st August, 1971. As per the Act, pesticide registration has to be dealt at the Central level (by CIBRC) for its manufacture and/or import and for sales and marketing, that would be carried out with at respective State level. With the implementation of the Insecticides Act, pesticides are, at regular basis, thoroughly examined before getting registration and made available in the market after meeting all the criteria as laid down in the Act. After registration, an user guideline is being issued for the pesticides to be applied against agricultural pests, pests of house-hold and storage and the pests of public health that cause human diseases and nuisance in order to minimize the possible health hazards (Central Insecticides Board and Registration Committee).

After several amendments, now the pesticides regulations are governed under the following Acts/Rules:

1. The Insecticides Act 1968 and Rules 1971.
2. Prevention of Food Adulteration Act 1954.
3. The Environment (Protection) Act 1986.
4. The Factories Act 1948.
5. Bureau of Indian Standards Act.
6. Air (Prevention and Control of Pollution) Act 1981.
7. Water (Prevention and Control of Pollution) Act 1974.
8. Hazardous Waste (Management and Handling) Rules 1989.

Role of Central Insecticides Board (CIB)

1. As and when necessary, guides the Central Government on the manufacturing of insecticides under the Industries (Development and Regulation) Act, 1951 (65 of 1951).
2. Classifies and specifies the application of insecticide on the basis of its toxicity as well as suitability for aerial application.
3. Suggests tolerance limits for residues of insecticide in different crops and determines minimum intervals, i.e., Pre-Harvest Interval (PHI) between final insecticide application and harvesting in respect of different commodities.
4. Determines the shelf-life period of insecticide.
5. Suggests on colorization, including coloring matter which may be thoroughly mixed with insecticide's active ingredients, specifically those having high toxicity.
6. Performs any other duties which may be found necessary for strict following of the Act or the Rules (Central Insecticides Board and Registration Committee).

22.7 BAU INITIATIVES ON AGROCHEMICAL RESEARCH [21]

Bihar Agricultural University (BAU), Sabour has very strong dedicated team involved in both fundamental and applied research of agrochemicals. Different experiments have been conducted to determine the bioefficacy of new generation molecules to mitigate the crop loss incurred by major pests.

Experiment	Salient Finding
Herbicides	
Weed management in aerobic rice (direct seeded)	The treatment consists weed free condition in rice plots, have been observed with the highest grain yield of 52.6 q/ha. Solitary spray of pendimethalin @ 1.0 kg a.i./ha as pre-emergence (PE) produced lowest grain yield (35.6 q/ha) among herbicidal treatments. Pendimethalin applied @ 1.0 kg a.i./ha as PE followed by (fb) bispyribac-sodium @ 25 g a.i./ha as post-emergence (PoE) fb one hand weeding found to be the second best yielder (52.2 q/ha) next to weed free condition. Pendimethalin application fb bispyribac sodium with one manual weeding showed lowest weed count, weed dry matter content and higher weed control efficiency (WCE) fb application of (penoxulam + cyhalofop) combination as PoE. Application of pendimethalin @ 1.0 kg a.i./ha as PE fb bispyribac-sodium @ 25 g a.i./ha as PoE fb one manual weeding showed maximum gross income and net return fb pendimethalin fb (pyrazosulfuron + bispyribac sodium) combination. Maximum benefit: cost (B:C) ratio was found as 2.61 with pendimethalin fb (pyrazosulfuron + bispyribac sodium) combination and (pinoxulam + cyhalofop) combination. However, treatment composed of pendimethalin was realized an effective strategy for weed management in any recommendation because it inhibits germination of various weeds.
Integrated weed management in jute	As per the results found in the experiment, fiber yield was enhanced in all treatments as compared to unweeded check. The treatment comprised with PoE application of quizalofop ethyl @ 60 g a.i./ha at 15 DAE plus one manual weeding at 15–20 days after application of herbicide was found best. Treatment with two manual weeding showed highest values in fiber yield but the cultivation cost is high. The B:C ratio was highest in treatment comprised with two manual weeding, fb quizalofop ethyl application.
Effect of chemical weed management on growth, yield and weed dynamics in transplanted rice	Azimsulfuron application @ 12.5 g a.i./ha with bispyribac sodium @ 15 g a.i./ha as PoE was at par with other weed control treatments. Weed control efficiency was highest (89.6%) in case of azimsulfuron @ 12.5 g a.i./ha plus bispyribac sodium @ 15 g a.i./ha as PoE. The maximum gross return was found in weed free plots but B:C ratio was highest in treatment comprised of azimsulfuron @ 12.5 g a.i./ha plus bispyribac sodium @ 15 g a.i./ha as PoE. Based on the findings of present study, the combination of azimsulfuron @ 12.5 g a.i./ha and bispyribac sodium @ 15 g a.i./ha as PoE need further evaluation through on farm trials in the ensuing Kharif season.
Evaluation of comparative bio-efficacy of herbicides in Kharif maize	The weed-free condition produces the highest grain yield (5.22 t/ha) and weedy condition showed the lowest grain yield (3.3 t/ha). Highest WCE (78.6%) was recorded with (glyphosate 40% SL + 2,4-D ethyl ester) applied @ (900 + 300) ml as pre-planting (PP) fb topramezone @ 40 ml as PoE. Among the different herbicides, (glyphosate 40% SL + 2,4-D ethyl ester) combination applied @ (900 + 300) ml as Pre-planting (PP) fb topramezone 40 ml PoE showed highest grain yield (5.05 t/ha) of maize in Kharif season.

(Continued)

Experiment	Salient Finding
Effect of chemical weed management on growth, yield and weed dynamics in lentil	Weed-free situation and spraying of (imazamox + imazethapyr) combination @ 30 g/ha 30 DAS showed highest and lowest seed yield of lentil, respectively. Among the different chemical control options, both spraying of pendimethalin alone and with quizalofop-ethyl were found equally effective in managing the weeds in lentil. Pendimethalin application fb quizalofop ethyl @ 750 g and 50 g/ha as PE and PoE respectively showed significant low weed index with high seed yield of lentil. Application of (imazamox + imezathpyr) combination caused maximum phytotoxicity among different weed control treatments and crop suffered due to poor growth and lower crop biomass. However, maximum B:C ratio of 3.83 was found in treatment comprised of pendimethalin @ 750 g/ha as PE.
Bio-efficacy of imazethapyr alone and its combination with pendimethalin for weed control in irrigated linseed	The maximum WCE (96.6%) was observed for isoproturon @ 1 kg a.i./ha fb manual weeding twice (96%), (pendimethalin 30EC + imazethapyr 2EC) @ 1 kg a.i./ha (95.4%) and (pendimethalin 30EC + imazethapyr 2EC) @ 0.75 kg a.i./ha (94.7%), applied as PE. The maximum grain yield (1225 kg/ha) was observed in manual weeding twice fb PoE spraying of isoproturon @ 1 kg a.i./ha (1106 kg/ha), imazethapyr @ 60 g.a.i./ha (1074 kg/ha) and PE spraying of pendimethalin @ 1 kg a.i./ha. But the highest NMR and B:C ratio was observed for PoE application of isoproturon @ 1 kg a.i./ha fb imazethapyr @ 60 g a.i/.ha and pendimethalin @ 1 kg a.i./ha.
	Among the herbicides, (pendimethalin + imazethapyr) @ 1 kg a.i./ha and 0.75 kg a.i./ha applied as PE effectively controlled broad spectrum of weeds but were found to be highly phytotoxic to linseed crop, thereby reducing linseed plant population drastically.
Comparative bio-efficacy of post-emergence herbicides against weed complex in utera linseed	Comparing all the treatments, highest WCE was found in manual weeding twice (92.9%) fb isoproturon @ 1 kg a.i./ha (89.2%) and imazethapyr @ 60 g a.i./ha as PoE (87.7%), whereas the maximum grain yield (855 kg/ha) was observed for manual weeding twice which was found at par with isoproturon @ 1 kg a.i./ha (820 kg/ha), imazethapyr @ 50 g a.i./ha (783 kg/ha) and imazethapyr @ 60 g a.i./ha (780 kg/ha).
Integrated weed management in aerobic rice	Combination of seven herbicides (pendimethalin, bispyribac sodium, 2,4-D Na salt, (chlorimuron + metsulfuron methyl), butachlor) along with straw mulching, mechanical weeding, need-based manual weeding (NBMW) and unwedded control were judged in aerobic rice at BRC Islampur Farm. Spraying of pendimethalin @ 1.00 kg a.i./ha (3-4DAS) plus bispyribac sodium @ 35 g a.i./ha (15–20 DAS) resulted at par to NBMW and effectively reduced weed leading to gain maximum grain yield.

(Continued)

Experiment	Salient Finding
Bio-efficacy of Platform-385 on wheat	"Platform-385" sprayed at high dose of 6.0 l/ha as PE or as early PoE resulted in getting statistically at par yield with the manual weeded plots, still it was observed to be on the lower side for producing stunting effect on crop vigor without having any visible phyto-toxicity symptoms. Stunting effect was observed to be more prominent with early PoE application.
Effect of pre and post emergence herbicides for control of smell melon (ghurmi) in summer green gram for koshi region of Bihar	Pendimethalin (PE) @ 1.5 l a.i./ha + imazethapyr (PoE) @ 60 g a.i./ha. produced considerably higher grain yield, WCE and weed index (WI) comparing all the treatments except pendimethalin (PE) @ 1.5 liters a.i./ha plus imazethapyr (POE) 40 g a. i./ha which was showing at par result with it.
Minimizing weeds in onion	The highest production (250 q/ha) and the maximum B:C ratio (3.01) were achieved by spraying of glyphosate @ 1 kg a.i./ha at 15 days before transplanting plus PE application of pendimethalin @ 1 kg a.i./ha at 3 DAT plus PoE application of oxyfluorfen 250 @ g a.i./ha at 20 DAT compared to the manually weed-free treatment which yielded (254.00 q/ha) and B:C ratio of 2.25.
Fungicides	
Evaluate the efficacy of fungicides against false smut disease of rice	The experiment showed that the least infection of panicle (5.48%) and spikelet (2.51%) have been observed in the plots spraying with kresoxim-methyl @ 0.1% fb the plots spraying with propiconazole @ 0.1% to the extent of 10.43% and 2.96% respectively. Performance of copper oxychloride @ 0.03% efficiently good in decreasing the panicle and spikelet infection in comparison to the check plot. Similarly, the highest yield (49.60 q/ha) was observed with kresoxim-methyl @ 0.1% fb propiconazole @ 0.1% (46.00 q/ha), which was again followed by copper oxychloride @ 0.03% (43.60 q/ha) whereas the uncontrolled check plot yielded 40.40 q/ha.
Evaluation of efficacy of antibiotics, chemicals and botanicals causing bacterial leaf blight (BLB) of rice	The in-vitro experiments revealed that the combinations of antibiotics and chemicals effectively inhibit the growth of bacterium comparatively than sole application of antibiotics or botanicals. Maximum inhibition zone (15 mm) was found in the treatment, i.e., streptomycin (200 ppm) pluscopper hydroxide 77% WP (0.25%) fb 13.38 mm observed in treatment with streptomycin 200 ppm. However, 13 mm of inhibition zone was observed in treatment with combination of streptocycline (100 ppm) plus copper oxychloride (0.25%). No inhibition zone observed in treatments with botanicals (1 to 5% concentrations).

(Continued)

Experiment	Salient Finding
	The lowest BLB severity (60%) along with max. mean yield (36.12 q/ha) and 1000 grain wt. (28.33 g) was obtained in treatment streptocycline (300 ppm) plus carbendazim (0.15%) fb 61.48% BLB severity along with yield of 35.14 q/ha and 1000 grain wt. of 27.39 g noticed in Streptocycline (300 ppm) plus copper hydroxide (0.25%) in field.
Management of stem rot of jute under integrated crop management system	Lowest stem rot and root rot incidence were recorded with late sown crop coupled with fertilizer doses of 80:40:40 and protection modules, i.e., seed treatment plus butachlor along with spraying of carbendazim @ 0.1% plus spraying of endosulphan @ 0.15% at 15 days gap. Interaction of $D_2F_2P_2$ gives maximum yield, i.e., 23.25 q/ha. D_2 (date of sowing) = 30th March, F_2 (Fertilizer dose) = 80:40:40 (NPK) and P_2 (Protection Module) = seed treatment plus butachlor along with spraying of carbendazim @ 0.1% plus spraying of endosulphan @ 0.15% at 15 days gap. In the trial 'evaluation of new fungicide molecule against *Macrophomina phaseolina*,' seed treated with carbendazim @ 2 g/kg, tebuconazole @ 1.5 ml/kg and with (azoxystrobin plus difenoconazole) @ 1.0 ml/kg of seed and their respective spraying at 45 days effectively controlled the disease. In the experiment disease free seed production, seed sowing in the mid July was found suitable for jute seed production. Spraying of fungicide (carbendazim) at pod setting stage has been shown more beneficial in enhancing seed yield compared to fungicide application at pod maturation stage and untreated plot.
Chemical management of two diseases phytopthora leaf rot and anthracnose leaf spot of betelvine	Recommendation for Phytophthora leaf rot control: Spraying of fungicide group (metalaxyl 8% + mancozeb 64%) WP (0.2%) at 15 day of interval (two spray) or application of bordeaux mixture (0.5%) at 15 day of interval (two spray) Recommendation for Anthracnose leaf spot control: Spraying of fungicide group (carbendazim 12% + mancozeb 63%) WP (0.4%) at 15 days of interval (two sprays) or application of tebuconazole 25 EC (0.2%) at 15 days of interval (two sprays).
Development of a module for integrated disease Management of fungal and bacterial diseases of betelvine	To control fungal diseases such as foot-rot, leaf rot, sclerotium wilt and bacterial diseases like bacterial leaf spot in betelvine, soil treatment with bordeaux mixture (1%) and planting material with bordeaux mixture (1%) + streptocycline sulphate (1000 ppm) has been shown most effective as compared with control.

(Continued)

Experiment	Salient Finding
Efficacy of bio-control agents and chemical treatments on minimizing disease incidence in strawberry	The treatment combined of soil drench and foliar spraying of (metalaxyl + mancozeb) had shown highest initial plant establishment. Minimum disease incidence of root wilt infestation (17.5%) was also observed in this treatment. Minimum disease severity (10%) was observed with soil and foliar spraying of *Trichoderma*. The maximum yield recorded as 205.35 g/plant was in integrated system, i.e., root dip of carbendazim + soil application of *Trichoderma* + foliar application of (metalaxyl + mancozeb) + mulch (black polythene).
Management of foliar blight of wheat though chemicals	Three foliar applications of dithane M 45 @ 0.25% at boot leaf or at disease initiation on flag leaf fb second and third spraying at 10 days gap each was found superior among the treatments.
AICRP (STF)	• 0.1% of carbendazim spray three times at interval gap of 10 days effectively controlled anthracnose disease of mango. • 0.2% of mancozeb sprayed twice at interval gap of 10 days at panicle emergence effectively controlled alternaria blossom blight of mango. • Cost effective management of post-harvest anthracnose of mango by pre and post-harvest treatments indicated that two pre-harvest sprays of carbendazim along with hot water treatment (52°C for 10 min) supplemented with carbendazim (0.05%) was best treatment for control of post-harvest anthracnose in mango. But alone hot water treatment was most effective with high B:C ratio.
Management of Blossom Blight in Mango	In this experiment, it was found that spraying twice with mancozeb (0.2%), or chlorothalonil (0.2%) or (Carbendazim + mancozeb) (0.2%) at a gap of 10 days at flowering reduced the occurrence of blossom blight in mango.
Management of wilt complex in lentil	Maximum control was found in treating the seed with carbendazim @ 1 g/kg seed plus *Trichoderma viride* @ 5 g/kg seed with highest yield.
Impact of nanoparticles on different fungal pathosystems	Chitosan-based nanoparticle tested against different sclerotial fungi (*Sclerotium rolfsii* and *Macrophomina phaseolina*), and a post-harvest pathogen (*Aspergillus niger*). Silver-based nanoparticles were shown promising result to control various fungi under a laboratory assay. In a pot experiment, silver-based nanoparticle treated seedlings were freed from infection by *Sclerotum rolfsii.*

(Continued)

Experiment	Salient Finding
Management of blight diseases of tomato	Spraying of (metalaxyl + mancozeb) @ 0.2% shown better in decreasing the late blight severity (15.29%) followed by (dimethomorph + mancozeb) @ (0.1 + 0.2%) (19.32%).
Management of guava anthracnose	Three sprays of (carbendazim + mancozeb) @ 0.2% at a gap of 15 days starting at bud initiation stage were most affective in managing anthracnose in guava.
Insecticides	
Management of stem borer in maize	Application of deltamethrin 20EC was found most effective in reducing no. of eggs (1.92/10 plants) and less no. of damaged leaves with highest yield (56.94 q/ha).
Status and management of important pests of lentil, chick pea and pigeon pea with special reference to pod borers	Minimum percent pod damage (4.01%) was recorded in the treatment spraying with profenophos which was found at par with the treatment spraying with spinosad (4.98%). Profenophos effectively reduced the population of leaf webber. Population of spiders was observed highest in the untreated control plot which was found at par with treatment spraying with pongamia oil, neem oil and (NSKE + cow urine). Population of coccinelids were observed highest in the untreated control was and found at par with treatment spraying with pongamia oil, neem oil, emamectin benzoate, HaNPV and (NSKE + Cow urine). The yield was recorded maximum in the treatment profenophos (22.11 q/ha).
Integrated management of yellow vein mosaic diseases of okra	Module consisting seed treatment with imidacloprid @ 0.5 ml/l fb spraying of imidacloprid @ 0.5 ml/l as well as neem oil @ 2.0 ml 4–5 times until fruit formation at 10 days gap showed significant results in reducing diseases and vector population and maximize yield.
Evaluation of new insecticide molecules against sucking pests of okra	The experiment was set up to judge new promising compounds against sucking pests of okra under field conditions. Thiacloprid @ 0.5 ml/l and thiomethoxam @ 0.35 gm/l most effective controlled whitefly and leafhoppers. The least percent fruit damage was observed in thiacloprid @ 0.5 ml/l leading to achieve maximum good fruit yield (95.11 q/ha). In another study, different treatments evaluated under field condition of which flonicamid 50 WG @ 0.4 ml/l and flonicamid 50 WG @ 0.3 ml/l most effectively reduced whitefly and leafhopper population. It was fb flupyrifurone 200 SL @ 2.5 ml/l and flupyrifurone 200 SL @ 2.0 ml/l. The maximum yield was achieved in the treatment, i.e., flonicamid 50 WG @ 0.4 ml/l (98 q/ha) which was found at par with flonicamid 50 WG @ 0.3 ml/l (95 q/ha). However, all the newer molecules were found safer to coccinellids.

(Continued)

Experiment	Salient Finding
Management of fruit fly through traps in mango and guava	Wooden block (5 × 5 × 1 cm) that hanged in plastic bottle and soaked in solution ratio 6:4:1 (alcohal: methyl eugenol: DDVP) @ 10 traps/ha (replacement of wooden block at 2 months interval) was found superior as it had shown highest fruit fly catch/trap/week and lowest fruit damage.
Management module against mango hoppers	The module comprising first spray of thiamethoxam 25 WG @ 1 g/31 water at panicle emergence, second spray of profenofos 50EC @ 2 ml/l water at pea size fruit stage and third need based spray of carbaryl 50 WP @ 3 g/l water resulted in highest (309.67 kg/tree) fruit yield and reduced hopper population.
Development of pest management modules for insect pest complex in brinjal	Bio-rational module composed of seedling root dip in imidacloprid 200 SL @ 1 ml/l for 3 hours before trans-planting, then first foliar spray of thiomethoxam 75 WG @ 0.5 g/l at 40 DAT, after that second foliar spray of spinosad 45 SC @ 0.5 ml/l at 60 DAT, fb third foliar application of indoxacarb 14.5 SC @ 0.5 ml/l at 75 DAT and fourth foliar application of emmamectin benzoate 25 WG @ 0.4 g/l at 90 DAT had shown to be best module to decrease the shoot and fruit damage.
Validation of vermiwash against viral disease of Tomato	Application of 10% vermiwash alone significantly reduced (45.45%) the Tomato leaf curl virus (ToLCV) and all the treatment of vermiwash (2%, 3%, 5% and 10%) with imidacloprid (0.03%) significantly managed the ToLCV. 10% vermiwash with imidacloprid (0.03%) significantly reduced the ToLCV incidence by 81.81%, whereas use of imidacloprid (0.03%) reduced incidence by 54.54%.
Management of Brown plant hopper in rice	The treatment containing spraying of imidacloprid 17.8 SL effectively reduced BPH population which was found at par with acephate 75 SP. Maximum yield was obtained (60.63 q/ha) in the treatment containing spraying of imidacloprid 17.8 SL followed by acephate 75 SP (56.29 q/ha).
Bio-efficacy of cassava-based bio-pesticides on insect pest complex of brinjal	Whitefly and leafhopper population was lowest in the plot applied with Quinolphos 25EC @ 3 ml/l which was fb dimethoate 30 EC @ 2 ml/l.
Management of litchi fruit borer	Flubendiamide, spinosad and novaluron were found statistically at par as compared with other insecticides. Flubendiamide fb spinosad and novaluron effectively reduced pest infestation 6.4% resulted in higher yield 84.3 kg/tree than other insecticides.

(Continued)

Experiment	Salient Finding
Studies on Insect and Gastropod Pest complex and their management in Makhana Ecosystem	The experiment recorded that the highest yield (35.84 and 35.37 q/ha), avoidable loss (23.72 and 22.70%) and B:C ratio (6.27 and 5.78:1) were achieved when seed treatment and root dip treatment have been done with imidacloprid 70 WS or thiomethoxam 25 WG @ 5 gm/kg along with foliar spray of NSKE @ 5%.
Field evaluation of different insecticide use strategies as resistance management and control tactics for shoot and fruit borer in brinjal	The results in terms of yield (no. basis and wt. basis) and shoot damage indicated that treatment comprised rotational strategy (rynaxipyr 20 SC @ 0.4 ml/l followed by emmamectin benzoate 5 SG @ 0.5 g/l, spinosad 45 SC @ 0.5 ml/l, chlorpyriphos 20 EC @ 2 ml/l, cypermethrin 25 EC @ 0.5 ml/l) significantly reduced shoot and fruit damage (6.37% and 9.40%) and showed maximum (350 q/ha) yield. As per findings of mean population of coccinellids, it was clear that all the newer molecules were safer to the predator and they were found statistically at par with untreated check.
All India Co-ordinated Research Project (AICRP) on Vegetable Crops (Entomology Component)	The results revealed that all the modules were found significantly superior over untreated control. Among the modules evaluated under field condition, the best module comprised erection of yellow sticky traps (1–2 traps @ 50–100 m^2) + foliar spray of imidacloprid 200 SL @ 0.5 ml/l at 20 and 30 DAT + spray (mixture of malathion 50 EC @ 2 ml/l and 10 g jaggery or gur in 1 l at flowering) + spray of rynaxypyr 20 SC @ 0.5 ml/ l at 15 days interval at the initiation of flowering + spray fenzaquin 10 EC @ 0.25 ml/l at the appearance of mite was found as best treatment to reduce whitefly and jassid population. Minimum percent of fruit damage by *Helicoverpa* was found in this treatment. The incidence of leaf curl virus disease in these treated plots was found significantly less as compared with other plots. Highest marketable yield was also achieved in this module (301 q/ha).
Plant Growth Regulator (PGR)	
Mitigation of high temperature stress in late sown wheat through exogenous application of synthetic compounds	Foliar spray of KNO_3 @ 0.5% both at booting and anthesis stage significantly enhanced the grain yield of late sown wheat (42.99 q/ha) as compared with no foliar spray (34.74 q/ha) and was found statistically at par with the treatments such as foliar spraying of KNO_3 @ 1% was done only at anthesis stage and the foliar application of $CaCl_2$ @ 0.1% was done at both booting and anthesis stage. The yield was found to be increased around 20–23% on average as a result of foliar application of synthetic compounds (KNO_3 and $CaCl_2$). The treatments significantly increased higher yield and caused significant reduction in electrolytic leaf leakage and proline content as well in flag leaf, both at anthesis and grain filling stage. higher level of chlorophyll content was also being maintained. Thus, these treatments significantly improved high temperature stress characteristics of late sown wheat. K salts

(Continued)

Experiment	Salient Finding
	having Nitrate (NO_3–) counterpart / Ca salts are having direct beneficial effect under field condition to enhance yield (20–23%) of late sown wheat· Anthesis stage is the most responsive stage for spray· Spray of these inorganic salts at lower concentration improves high temperature stress tolerance of late sown wheat.
Effect of pre-harvest treatments on postharvest life of strawberry fruits	Comparing all the different treatments of salicylic acid and calcium chloride, pre-harvest application of salicylic acid (1 mM) effectively reduced (9.85%) and weight loss (14.27%) of fruit compared with control, 15 days after storage at 2°C. Fruits sprayed with salicylic acid (1 mM) maintained highest anthocyanins (27.17 mg/100 g), total phenolics (2.074 µg GAE/g) and total antioxidant capacity (20.73 µmol TE/g) than control and other treatments. TSS content was not different among the treated and control fruits.
Management of lodging and yield maximization using nutrient expert (SPL-2)	The recommended wheat variety, HD2967, for timely sown irrigated conditions of this zone was evaluated for lodging and yield maximization using nutrient expert and two applications of chlormequat @ 0.2% of commercial product at first node stage (45DAS) and at flag leaf stage (80DAS) along with two combined sprayings of chlormequat @ 0.2% and tebuconazole at node and flag leaf stage, with recommended dose of fertilizer and NE targeted yield of 6 t/ha and 7 t/ha. Maximum mean grain yield (47.6 q/ha) was found in the plots where NE targeted yield 7 t/ha (140 kg N/ha: 68 kg P_2O_5/ha: 101 kg K_2O/ha) and two spray of chlormequat chloride @ 0.2% and tebuconazole that applied at first node stage (45DAS) and at flag leaf stage (80DAS) respectively. This was observed statistically at par with mean grain yield (46.3 q/ha) achieved from the plots where NE targeted yield 6 t/ha (120 kg N/ha: 63 kg P_2O_5: 82 kg K_2O/ha) and two spray of chlormequat chloride @ 0.2% and tebuconazole that applied at first node stage (45DAS) and at flag leaf stage (80DAS) respectively. Both these treatments were observed significantly superior than other treatments in mean grain yield. Plants of chlormequat applied plots were found comparatively shorter in height than all other treatments.
Pruning for rejuvenation of overcrowded orchards in mango	In this experiment, maximum fruit yield (8.31 t/ha) was observed in treatment heading back up to the crowded branchlet and center opening along with spraying of paclobutrazol (3.5 ml/sq.m) during off season of fruiting. However, the maximum average fruit weight (307.00 g) was observed in treatment of heading back up to secondary branchlet without application of paclobutrazol.
Evaluation of PGR and promising chemicals for early flowering in litchi	The results revealed that different PGRs had profound effect on the early flowering and fruiting in litchi. Foliar spray of ethrel 400 ppm resulted in 5 days advancement of flowering (anthesis). The treatment K_2HPO_4 (1%) + KNO_3 (1%) gave maximum fruits/panicle (21.67) with highest fruit weight (21.54 g), yield (93.33 kg/tree) and TSS (21.54°B).

(Continued)

Experiment	Salient Finding
Light annual pruning and chemical treatment for improving fruit yield and quality of mango	• Significant effect has not been observed in newly emerged shoot length but significantly highest girth diameter (5.26 cm) was recorded under the treatment consists of 25% pruning + 3 g a.i. paclobutazol + 3% KNO_3. • Date of flower bud initiation was varied from 05.02.2016 to 10.02.2016 and fruit setting varied from 09.03.2016 to 11.03.2016. • Significantly highest yield (713.33 fruits/plant) was observed under the treatment consists of 25% pruning + 3 g a.i. paclobutazol + 3% KNO_3 followed by the treatment consists of under 25% pruning along with paclobutrazol (615 fruits/plant) and the treatment, i.e., 25% pruning along with 3% KNO_3 (506.33 fruit/plant).
Enhancement of fruit set and reduction in fruit drop through foliar application of calcium, boron and sorbitol in mango cv. langra	The minimum fruit drop, and higher fruit yield per plant (117.29 kg) was found maximum in treatment calcium nitrate (0.06%) + boric Acid (0.02%) closely followed by calcium nitrate (0.06%). This technology is beneficial for minimizing the fruit drop and inducing the fruit set as well as enhancing the fruit yield.

Several generic and advanced herbicides, insecticides, fungicides, plant growth regulator, etc. are being tested regularly in different crops to improve production and also quality of harvest. Some of the experiments are listed below:

KEYWORDS

- ***Bacillus thuringiensis***
- **bacterial leaf blight**
- **Bihar Agricultural University**
- **Central Insecticides Board**
- **chitin synthesis inhibitors**
- **chlormequat/cycocel**

REFERENCES

A History of Crop Protection and Pest Control in Our Society, Crop Life Canada, (2002). http://www.croplife.ca/english/pdf/Analyzing2003/T1History.pdf (Accessed on 25 November 2019).

Agenda Notes. Research Council Meeting, Bihar Agricultural University, Sabour, Bhagalpur.

Bora, D. S., (1998). Effect of environmental stress with special reference to photoperiod and insecticide on muga worms. *Antheraea assama* Westwood. *PhD Thesis*. Dibrugarh University, Dibrugarh, Assam, India.

Central Insecticides Board and Registration Committee. http://cibrc.nic.in/about_us.htm (Accessed on 25 November 2019).

Chukudebe, R. B., Othman, M., & Fukuto, T. R., (1989). Formation of trialkyl phosphorothioate esters from organophosphorus insecticides after exposure to either ultraviolet light or sunlight. *J. Agric. Food. Chem., 37*, 539–545.

Delaplane, K. S., (2000). *Pesticide Usage in the United States: History, Benefits, Risks, and Trends*. Bulletin 1121, Cooperative Extension Service, The University of Georgia College of Agricultural and Environmental Sciences. http://pubs.caes.uga.edu/caespubs/pubs/PDF/B1121.pdf (Accessed on 25 November 2019).

Haynes, K. F., (1988). Sublethal effects of neurotoxic insecticides on insect behavior. *Annu Rev Entomol., 33*, 149–168.

http://agrochemicals.iupac.org/index.php?option=com_sobi2&sobi2Task=sobi2Details&catid=3&sobi2Id=31 (Accessed on 25 November 2019).

http://www.dawn.com/news/125197/plant-growth-regulators (Accessed on 25 November 2019).

Kariappa, B. K., & Narasimhanna, M. N., (1978). Effect of insecticides in controlling the mulberry thrips and their effect on rearing silkworm, *Bombyx mori*. *Indian J. Seric., 17*, 7–14.

Morton, V., & Staub, T., (2008). *A Short History of Fungicides*. APSnet, http://www.apsnet.org/online/feature/fungi/ (Accessed on 25 November 2019).

New Insecticide Modes of Action: Whence Selectivity? *J. Coats, Iowa State University*. Ames, Iowa, USA http://www.slideworld.org/viewslides.aspx/New-Insecticide-Modes-of-Action-Whence-Selectivity-ppt-42841 (Accessed on 25 November 2019).

OECD Series on Pesticides, (1999). Number 8, Report of the OECD/FAO Workshop on Integrated Pest Management and Pesticide Risk Reduction. http://www.olis.oecd.org/olis/1999doc.nsf/LinkTo/NT00000FBE/$FILE/04E94320.PDF (Accessed on 25 November 2019).

Silent Spring, (2002). *40th Anniversary Edition*. Rachel Carson, Houghton Mifflin Harcourt. ISBN: 0618249060,9780618249060. http://books.google.co.uk/books?hl=en&id=HeR1l0V0r54C&dq=silent + spring&printsec=frontcover&source=web&ots=1r4bWmlR2G&sig=RFBfJr0UBxYcFAS7Y6YdVWkSwwQ&sa=X&oi=book_result&resnum=6&ct=result#PPP1,M1 (Accessed on 25 November 2019).

The History of Pesticides, Organic Pesticides, September 19th, (2008). http://blog.ecosmart.com/index.php/2008/09/19/the-history-of-pesticides/ (Accessed on 25 November 2019).

The Next Generation of Pesticides, (2013). Jonathan Lundgren, https://igrow.org/agronomy/corn/the-next-generation-of-pesticides/ (Accessed on 25 November 2019).

Troitskaya, E. N., & Chichigina, I. P., (1980). The effect of combined insecticidal preparations on silkworm larvae. *Uzbekshii Biologicheskii Zhurnal. 3*, 50–53.

Wessels Living History Farm. York, Nebraska, Farming in the 1930s. http://www.livinghistoryfarm.org/farminginthe30s/pests_04.html (Accessed on 25 November 2019).

Wessels Living History Farm. York, Nebraska, Farming in the 1950s & 1960s. http://www.livinghistoryfarm.org/farminginthe50s/pests_08.html (Accessed on 25 November 2019).

Wilkinson, C. F., (1976). *Insecticide Biochemistry and Physiology*. Plenum Press. New York.

Zabik, M. J., (1985). Photochemistry of pesticides. In: Gilbert, L. I., & Kerkut, G. A., (eds.), *Comprehensive Insect Physiology, Biochemistry and Pharmacology*. Pergamon Press, Oxford. 12776801.

PART IV

Product Development and Extension Education

CHAPTER 23

Institutional Outreach Through Innovative Approaches

ADITYA SINHA and R. K. SOHANE

Department of Extension Education, Bihar Agricultural University, Sabour, Bhagalpur, Bihar, India,
E-mail: inc.aditya@gmail.com (Aditya)

ABSTRACT

The new age innovative approaches including information and communication technologies (ICTs) have the potential to provide varied solutions to the farming community like improving their capacity building towards high-value crops along with proper access to credit and markets, promoting better collaboration between the functioning agencies and so on. India is on a track to embark upon a "Cyber revolution" with effective extension advisory services (EAS) ICTs. Of late, the government is investing heavily on the Digital India program which helps leverage the potential of connectivity to enhance the services delivery mechanism in the country. Along with it, it can be an enabler in bridging the digital divide between urban and rural India along with an effective skill development program on a sustained basis for self-employment of the masses. It also has an ambitious initiative to include everyone digitally by 2019. The present paper deals with the innovative approaches of Bihar Agricultural University (BAU), Sabour in reaching the farmers with ease in the quickest possible time. The review is based on advancements after the year 2000. It discusses the role of Kisan Chaupal, Video conferencing, community radio services, farmers' helpline and related avenues in generating interest and promoting better extension among the farmers. The impact of these innovative experiments offered through ICTs is much more what could be attributed due to traditional means along with lowering of expenses incurred.

23.1 INTRODUCTION

The use of ICTs in agriculture is no more a new area of study particularly in the case of developing countries. Digital developments such as the Internet and mobile phone has capability to empower communities in rural areas along with harnessing and capturing local farmer knowledge and using participatory approaches to provide extension and advisory services. Technology per se does not solve economic problems, but availability of ICT will have a significant impact on rural development in developing countries (Makuleke project, 2009). Beyond connectivity ICT offers security and mobility to owners, and requires that users have basic literacy. It has been reported that the new ways of managing knowledge have emerged across developing countries beyond the traditional farmer-extension systems (Ferrández-Villena, M., and Ruiz-Canales, A., 2017). There are changing demands of farmer's in the current time and complex decisions need to be made to effectively manage their farming and businesses. The range of services for farmers is continuing to grow so as to better meet their needs and help them be more productive and remunerative. It is often debated that the future of extension services is in establishing a central community-focused organization with efficient e-based services of the current times that can help manage these services for farmers. The essence of the current e-extension services lies in shifting to maintain as a personal secretary for the farmers. Agriculture remains the principal activity for sustainability in rural Asia and Africa. The work on providing access to web-linked information on crop prices, quality of agricultural products, various methods of growing seeds and livestock, quality of soil, effective fertilization and up to date weather information via Internet and mobile phones is on the rise. Several mobile applications are developed regularly which are providing localized services to the farmers. The use of participatory approaches can empower groups of farmers collectively, thereby leaving the decision-making in the hands of the farmers helping contribute to the better understanding of farmers' needs. In order to leverage the benefits of ICT enabled extension services, BAU, Sabour has rolled out e-initiatives in a comprehensive manner for the benefit to the farmers. The current paper deals with the application of mobile-based services, community radio and video-conferencing for effective EAS for the farmers.

23.2 NEED FOR ICTS IN AGRICULTURAL EXTENSION AND RURAL DEVELOPMENT

Different empirical studies over the years have shown that knowledge cannot easily reach farmers through traditional extension systems and development projects. This has called for the ways to implement new systems along with various potentialities to figure out the scalable methodologies. Developing countries have implemented several models to bridge the divide among the urban and rural and also to mitigate the challenges faced because of low extension worker to farmer ratio. The use of ICTs for small holder and family farmers in India is no longer constrained by access or ownership to basic ICT tools, such as the mobile phone, at the household level the way it was ten years ago. Studies further indicate that the success or sustainability of ICT based models is associated with a variety of factors such as good management, minimum level of infrastructure, strong local demand, new and relevant content development, availability of innovative and locally relevant services, and external linkages and networking (Benjamin, 2001; Ali, 2012; Etta and Parvyn-Wamahiu, 2003; Herold, 2010; Manda, 2002; Roman and Colle, 2003). There is increasing evidence that infrastructure such as telecommunications network and ICTs help to reduce poverty and provide opportunities to people in developing countries (Torero and Chowdhury, 2005). It is imperative to make information readily and cheaply available which can enhance production, increase bargaining power and incomes, and ultimately, lead to poverty reduction and economic development. In addition, making information readily and cheaply available can enhance production by enabling farmers to plant and harvest at appropriate times (weather information), improve the functioning of credit markets (by facilitating social capital as through increased communications farmers develop and gain trust and reputations), facilitate access to more efficient technologies (through finances obtained through increased credit ratings/worthiness), and in the long run, transform production processes through more rapid and diffuse spread of technological innovations amongst a broader range of interacting agents-consumers (with different tastes and preferences leading to product differentiation or new product development), sellers (of inputs products and services), traders, and processors (Eggleston et al., 2002). The benefits to an organization after managing all extension services for farmers through e-based services can be understood as:

- Constant two-way dialogue between scientists and farmers and the ability to negotiate and adapt services to better meet local farmer's needs.
- Greater ability to aggregate farmers and meet market requirements as per specific crop of the location.
- Regular feedback and increased information about farmers that allows for better matching with available services for enhanced extension services.

23.3 METHODOLOGY

This inventory is limited to documenting innovative farmer information services enabled through innovative means including ICTs. It is focused on projects/services that provide agricultural training and information to farmers directly, in BAU, Sabour; rather than documenting services that facilitate the exchange of information among researchers and policymakers. We have provided an overview of various innovative approaches through which the University is reaching the farmers effectively.

23.4 KISAN CHAUPAL

Kisan Chaupal is an innovative extension program launched in the year 2012 by BAU, Sabour to connect the farmers with the scientists directly at their own place. This program was implemented by all *Krishi Vigyan Kendras* (KVK) and colleges under the jurisdiction of the university. The modus operandi of the program is that a group of scientists belonging to different branches of agriculture and veterinary sciences visit the pre-decided villages where the farmers already gather on a common meeting ground for getting their queries answered by the scientists instantly. The villages are identified with the participation of farmers' representatives and the scientists of the KVK. The program is conducted each week, particularly on Saturday. Over the past years, this program has also focused on creating awareness among farmers regarding various central and state-sponsored schemes in agriculture and allied sectors. Over 3 lakh farmers have been benefitted through this program over a period of three years.

23.4.1 OBJECTIVES OF KISAN CHOUPAL

The specific objective of the Kisan Chaupal is to strengthen the linkages between scientists and farmers with instant problem-solving at their door-steps along with a collection of feedback for the development of research priorities of the university. It also helps in necessary convergence with different agencies working for extension work at the grass root level.

23.4.2 NEW INITIATIVES/INNOVATIONS IN KISANCHAUPAL

The scientists have recognized several new innovations and have incorporated it in the "Kisan Chaupal" to make it more lively and interesting for the farmers so as to fulfill the needs and requirement of maximum farmers in its jurisdiction. Some of the new initiatives are discussed below:

1. **Krishak Sandhya:** It is an innovative attempt made in this initiative to teach the basics of agriculture through entertainment. It was generally observed that farmers were mainly interested in their works in fields and other family work during day time. Hence, they were reluctant to undergo training and avail of service of scientists, who are available only during office hours. The concept of starting a program named as "Krishak Sandhya" (An evening with farmers) was felt in which farmers could be enlightened via entertainment. Folk artists are explained techniques of modern and remunerative farming and are asked to prepare folk songs. The farmers are educated by the scientists in the middle of the program. The main benefit of the program was that it enabled better rapport building with the farmers as they became familiar with the scientists through

"Krishak Sandhya" and started to consider them as their friends. Local women also participated actively. The farmers' enthusiasm is a positive sign which was recognized through this initiative.

2. **Mahila Chaupal:** It was experienced in the "Kisan Chaupal" initiative that women farmers were normally hesitant to ask questions on agricultural practices in a gathering dominated by men. The scientists of the university planned to implement a separate forum exclusively for the women farmers in which the participation of women scientists will be ascertained. The "Mahila Chaupal" has proved to be a popular initiative that has created lots of success stories in such a small span of time in the form of women agri-entrepreneur with expertise on mushroom cultivation, vegetable cultivation among a few.

23.5 TECHNOLOGY WEEK

With the aim to fulfill the fourth mandate of KVK to act as Knowledge and Resource Center, technology week has been started in all the KVKs of BAU. Technology Week has provided a platform to bring a number of stakeholders like farmer, extension workers, input dealers, scientists and other stakeholders under a single umbrella. It is celebrated on Public-Private Partnership model (PPP) to make farmers aware about advance scientific know-how. The following steps are adopted in the celebration of technology week.

- Planning and pre-inception meeting with all the stakeholders.
- Mobilization/publicity of technology week before the celebration.
- Display of different scientific know-how through suitable exhibition materials viz.; posters, objects, charts, models, live demonstration, etc.
- Interaction between scientists, farmers and other stakeholders.
- Brainstorming of farmers for solving their problems.
- Media management: media backstopping, sensitization of media about KVK work.

In addition to it, for further enrichment and boosting up of knowledge, the KVKs have organized film shows for the farmers during the technology week. Regularly *kisangoshthi* were organized in different topics like crop production, horticulture, women empowerment, animal husbandry and agri-entrepreneurship development.

The unique feature of technology week is the focus on the best-fit approach rather than the best technology approach, convergence of different stakeholders and PPP mode. It is indeed a great learning experience for farmers who get relevant scientific understanding and scientist as well through farmers' feed-back. It has also helped in formulating the research priorities of the varsity.

23.6 MOBILE MESSAGING SERVICES

23.6.1 MOBILE BASED AGRO-ADVISORY SERVICES THROUGH "KISAAN SMS PORTAL"

The University is utilizing the mKisan SMS portal launched by the Department of Agriculture and Cooperation and Farmers Welfare, Ministry of Agriculture and Farmers Welfare, Government of India in the year 2013 to send regular advisories to the farmers of various districts through its KVKs. Also, the scientists and the officers of the university are authorized to send crop relevant messages to the farmers in both Hindi and English. This particular platform is benefiting more than 80,000 farmers of the region through the university.

23.7 KISAN HELPLINE

The Kisan Helpline services were started in the university in the year 2012 as an information wing of the Plant Health Clinic. A group of scientists from different disciplines are assigned duties to answer to the queries of the farmers over phone. Also, the university has extended this service on a dedicated number through WhatsApp messaging services so that farmers can also make use of crop images and other characteristics to explain the problem in a better way.

The discipline wise calls received from the Kisan Call Center are mentioned above. Table 23.1 shows that maximum calls were received related to queries on agronomical aspects including weed and meteorology.

The Plant Health clinic is also responsible for testing of soil samples. It has tested a total of 1280 soil samples in the financial year 2015–'16 (Table 23.2).

23.8 AGROMET ADVISORY SERVICES (AAS)

Agromet advisory services (AAS) is provided by the agromet advisory unit, Department of Agronomy, BAU, Sabour. The weather forecast is provided twice in a week on Tuesday and Friday. Parameters forecasted are rainfall, maximum temperature, minimum temperature, cloud cover, relative humidity (morning), relative humidity (evening), wind velocity and wind direction. The forecast is valid for a period of five days. Along with it, advisory services for farmers provide information on management of farming systems according to the weather conditions and application of fertilizer, irrigation, cultural practices, spraying of pesticides, fungicides, harvesting of crops, storage and other farm-related works. Aspects related to fisheries, animal husbandry, poultry, etc. are also advised according to the prevalent weather conditions.

TABLE 23.1 Calls Received at Kisan Call Center from Various Disciplines (April 2015–March 2016)

Sl. No.	Discipline	No. of Calls
1.	Crop Protection	395
2.	Fruits	400
3.	Vegetable	309
4.	Agronomy (Crop, weeds and Meteorology)	785
5.	Animal Husbandry	286
6.	Plant Breeding	103
7.	Bee keeping	27
8.	Soil Science	103
9.	Social Science (Extension, Economics, Statistics)	280
Total		**2688**

TABLE 23.2 Soil Testing Work (April 2015–March 2016)

Sl. No.	Sample	No.
1.	Soil samples tested (Farmers, organizations and others)	439
2.	Soil samples tested for World Soil Day, 2015	841
Total		**1280**

23.8.1 AREA COVERED

Weather forecasting and advisory services are provided for agro-ecological regions III A and III B of Bihar state.

- **Zone III A:** Six districts are covered viz. Bhagalpur, Munger, Banka, Jamui, Sheikhpura, Lakhisarai.

- **Zone III B:** Eleven districts are covered viz. Patna, Nawada, Nalanda, Buxar, Bhabhua, Aurangabad, Gaya, Jehanabad, Rohtas, Arwal, Bhojpur.

The advisories are communicated to the farmers mainly through local newspapers. Along with it, the agro advisory bulletin is sent to the KVK's, NGO's, ATMA, the office of Annadata program for farmers on ETV, the office of District agricultural officer, All India Radio, through e-mail. It is also uploaded on the web site of the university and Indian Meteorological Department (IMD) for wider circulation.

23.9 VIDEO CONFERENCING AND TVU SYSTEM IN RENDERING EXTENSION SERVICES

Videoconferencing can act as a suitable delivery method for responding to the wants and needs of the remote farmers. It is an effective instrument to curtail costs associated with traditional training programs offered by the universities and departments. It has already been experimental successfully on students. It has been argued that educators are increasingly facilitating connections between their clients and educational resources located anywhere in the world through videoconferencing (Valsamidis et al., 2011; Nudell et al., 2005). The university is regularly reaching out with farmers located in different KVKs of the districts through video conferencing.

By using TVU systems which are lightweight, portable video transmission systems capable of transmitting information from the farmers' field, now it is possible to link farmers' fields in real-time with the scientist sitting at headquarters for providing advisory to farmers (Table 23.3).

It can be seen from Table 23.3 that a total of 18 KVKs of the University at various locations are connected through videoconferencing from the university headquarters at Sabour. The trainings on various aspects based on needs of the farmers are organized regularly with the active collaboration of the KVKs. The participation of farmers is on the rise with this innovative method to save time, money, etc. for the want of expert advice. The domain covered under advisory through video conferencing is weed management, vaccination in poultry, orchard management, goatery, nutrient management, mushroom farming, etc. A total of 8645 farmers were directly benefitted through video conferencing through 344 specialized trainings in the year 2015–16.

TABLE 23.3 Trainings Organized Through Video Conferencing (April–March 2016)

Month	No. of KVK Connected	No. of Trainings Through Video Conferencing	No. of Farmers Benefited
April 2015	10	35	875
May 2015	10	35	875
June 2015	10	23	575
July 2015	10	24	600
August 2015	10	30	710
September 2015	10	35	820
October 2015	10	30	700
December 2015	17	40	1000
January 2015	17	22	540
February 2015	17	38	1025
March 2015	18	32	925
Total		**344**	**8645**

The benefits attributed to the video-conferencing facility provided by the university are briefly described below:

- **Increased Productivity:** The use of videoconferencing is the next driver for productivity because it is quite easier for the scientists to keep in closer contact with the farmers. This closeness has led to new ideas on how to speed up the development of new products and services in the agricultural sector.
- **Improved Communication and Reinforces Relationships:** During a videoconference, one can see the facial expressions and body language of conference participants. These are both important aspects of communication that are lost with a basic telephone call.

The video conferencing facility enables better interaction and the possibility to be reflected on the screen connects better with the advices suggested.

- **Reduced Travel Expenses:** In today's scenario cutting down the travel expenses and time of the scientists is of utmost importance. Video-conferencing provides a suitable medium for university scientists to stay connected with the farmers without having to visit them personally.
- **Improved Effectiveness:** The quote "A picture is worth a thousand words" can be very well substantiated by Videoconferencing. A live video call is much more effective than a phone call in many situations. The benefit attributed to the video-conferencing by the KVKs has improved the effectiveness of not only the farmers but also the scientists associated with the university.
- **Time Saved:** The video conferencing facility has enabled saving of time of all the people engaged in the process.

The video conferencing facility has not only facilitated better interaction between the scientists and the farmers but also helped the policymakers have a better know-how of the farmers.

23.10 DIGITAL STORYTELLING

Extension professionals are utilizing various video-sharing platforms like YouTube, Vimeo, etc. to provide expert information to the farming community through the generation of farmer participatory videos. The scientists of the university are providing handy information on several agricultural avenues to the farmers in this way. It is also updated regularly to meet the taste of the viewers from all over the world (Table 23.4).

The web analytics of the YouTube channels reveals that Poultry Farming has the maximum viewership share (1,84,405 views) comprising 34% of the total views of all the videos uploaded followed by Goat farming on the second place (1,02,274 views, 19%) and Quail farming on third place (90,292 views, 17%).

The 19th Livestock census 2012 also reveals the rise of poultry farming in the country with 12.13% increase from the year 2007 to 2012. Though the Goat population has declined by 3.82% over the previous census and the total Goat in the country is 135.17 million numbers in 2012 (Ministry of Agriculture, 2012). There is a rising interest among the farmers in the rural areas in India for poultry and goat farming which can be very much depicted from the statistics in Table 23.1.

TABLE 23.4 Web Analytics of Farmer Friendly Videos Shared on YouTube

Topic	Watch Time (Minutes)	Views	Views (%)
Poultry farming	982,654	184,405	34%
Goat farming	446,832	102,274	19%
Quail farming	350,771	90,292	17%
Honeybee rearing	123,070	26,956	4.9%
Scientific cultivation of pointed gourd	103,747	18,706	3.4%
Tomato ketchup preparation method	91,083	19,809	3.6%
Azolla farming	77,848	17,395	3.2%
Integrated Farming System	71,813	14,086	2.6%
Strawberry cultivation in Bihar	69,291	15,670	2.9%
Insect and Pest control in Paddy	52,025	11,853	2.2%

Source: YouTube web analytics (bausabour channel), September 2016.

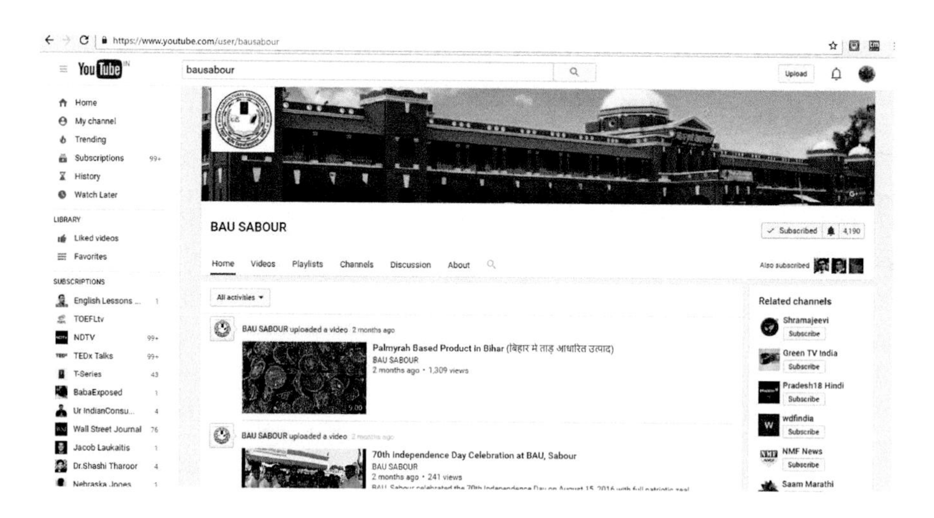

Table 23.5 shows the demographics of viewership on YouTube. Maximum views are largely from the youth in the age group of 25–34 years. The viewership base is mostly from India (64%) followed by Saudi Arabia (9.9%), United Arab Emirates (5%), Pakistan (2.5%) and Kuwait (2.1%).

The web analytics of farmer-friendly videos produced by the university along with the visits from different countries show the widening interest of the youth towards farming through social media channels.

TABLE 23.5 Demographics of Viewership on YouTube

Viewer Age (in Years)	Watch Time (in %)	Male (in %)	Female (in %)
13–17	1.5%	83%	17%
18–24	17%	91%	8.5%
25–34	49%	94%	5.5%
35–44	20%	94%	6.1%
45–54	6.7%	91%	8.9%
55–64	3.2%	91%	9.1%
65 and above	1.7%	91%	9.4%

Source: YouTube web analytics (bausabour channel), September 2016.

23.11 KISAN GYAN RATH (MULTI UTILITY MOBILE VEHICLE FOR ENHANCING FARMERS' KNOWLEDGE)

An approach to mobile technology dissemination facility (a bus vehicle equipped with multimedia content on farm science and technology and means of displaying knowledge and information) initiated by BAU is *Kishan Gyan Rath* which was launched in 2014. It is an important mobile medium for providing expert guidance, consultancy, and method illustration to the farmers on their doorsteps through audio-visual means. The *Kishan Gyan Rath* in addition to the technology dissemination facility also has a fully equipped unit for onsite soil analysis, interpretation, and issuing of soil health cards to the farmers. The benefit of such service is that farmers need not visit the soil testing laboratories located far from their place; instead, the soil testing laboratories reach farmers on their doorsteps.

In addition to this, efforts are also made to bring about change in the knowledge and attitude of farmers through the *Nukkad Natak* (Folk drama). This *Rath* displays technical videos developed by the varsity. Apart from this, soil samples are also collected from the farmers and analyzed reports are given to the farmers in the form of soil health cards at the shortest time.

23.12 PROMOTING INTERACTIVITY THROUGH COMMUNITY RADIO

Community radio is thought of as a platform where the everyday experiences of ordinary people can be shared. Chapman et al. (2003) reported that the growth of rural radio stations reflects both the improvements in information technologies and the shifting of development paradigm towards a more participatory style of information and knowledge transfer. It is an important means to strengthen participatory communication for both economic and social development. The Community Radio of the University is operational at KVK, Barh, Patna, Bihar since May, 2011 on FM 91.2 Mhz bandwidth. The present study was conducted in the year 2014 with sample from eight blocks namely Barh, Pandarak, Athmalgola, Belchi, Bakhtiarpur, Khusrupur, Ghoswari, and Mokama in which services of Community Radio Station, Barh located at Krishi Vigyan Kendra, Barh, Patna is on air. It is broadcasting locally relevant programs for the farming community on a daily basis for a total duration of three hours. It covers a total of 503 villages covering a population of 1,34,0293 individuals. The topics for women, children and farmers are discussed in minute detail. The station is well accessible to the local community and is staffed with local community members for enhanced and sustainable benefits.

23.12.1 FARMERS' REACTIONS TO THE RADIO PROGRAMMES

The reaction of farmers to the content of various radio programs were studied in detail. The results are mentioned in Table 23.6.

TABLE 23.6 Target Audience Understanding of Radio Programmes

Responses	Frequency	Percentage (%)
Understood well	58	77.33%
No response	2	2.66%
Partly understood	12	16.00%
Had little understanding	3	4.00%

The understanding of radio content by the audience was studied which revealed that a maximum (77.33%) audience understood the content very well whether it is in local dialect *Maghi* or Hindi. It was followed by those who understood party (16%), had little understanding of the content (4%) and those who did not respond (2.66%) (Table 23.7).

TABLE 23.7 Perceived Main Messages of the Programme

Main Message	Frequency (n=213*)	Ranking
Importance of health and proper nutrition/sanitation	21	5
Importance of adopting new technologies in agriculture like mushroom cultivation, vermicomposting and honeybee rearing	33	2
Dairying and animal husbandry practices	29	4 =
Organic farming	16	6
Government schemes	29	4 =
Orchard cultivation	32	3
Agroforestry and plantation of trees on the bunds	15	7
Plant protection measures and alert on diseases/pest	38	1

*Respondents were permitted more than one response.

The main message/central theme perceived from the program which was ranked on the first is plant protection measures and alert on diseases/ pests infesting the crops. It was followed by the practical aspect concerning the importance of adopting new technologies in agriculture like mushroom cultivation, vermicomposting and honeybee rearing among a few on the second slot. Horticultural aspect like the cultivation of orchard was on the third rank. Information on various government schemes for the social sector and practices like dairying and animal husbandry occupied the same fourth rank. The importance of health and proper nutrition/ sanitation was on the fifth rank followed by newer methods like organic farming on the sixth. Various methods on agroforestry and plantation of the tree on the bunds were ranked on the last position.

23.13 DISSEMINATION OF VIDEOS THROUGH LOW-COST SD CARDS

It is often said that "Seeing is believing." The importance of visualization is very important in the digital era where videos are getting popular in every domain of learning. To exploit the power of effective videos for the farming community, a collaborative research project by the National Institute of Agricultural Extension Management (MANAGE), Hyderabad and BAU, Sabour is initiated in twenty-five districts of Bihar state under the jurisdiction of BAU, Sabour. The project entitled "Digital Videos for farmers by using mobile phones with low-cost SD cards" is providing 100

Secure Digital (SD) cards each to 100 different progressive farmers having smartphones in 25 districts of Bihar through the KVK. The low-cost SD cards have 20 popular videos in agriculture and allied disciplines (Table 23.1) collected from different sources which would be helpful for the farmers in understanding and initiating different agricultural enterprises in the region. The research project aims to understand the agricultural video-viewing behavior of the farmer along with impact evaluation of the digital videos on farm productivity. The farmer-friendly videos are developed in 3GP file format which is easily compatible/ can be played even on a basic smartphone. The project will further help in bridging the digital divide arising due to low-connectivity in rural areas in which farmers face problems in accessing uninterrupted videos from the internet and paying for the high data charges. The SD card of is of 8GB size in which the videos occupy only 2GB of space. Awareness campaign is run at the 25 KVKs where farmers are made aware on the importance of digital videos before distributing SD cards to them. They are also provided with a flyer containing details on the videos available on the card along with other relevant information on the videos (Table 23.8).

TABLE 23.8 List of Agriculture-Relevant Videos Present in the SD Card

Sl. No.	Title	Source	Duration (Minutes)
1.	Azolla farming practices	Bihar Agricultural University, Sabour	12.58
2.	Goat farming	Bihar Agricultural University, Sabour	13.49
3.	Quail (*Bater*) farming	Bihar Agricultural University, Sabour	15.57
4.	Scientific cultivation of Chickpea	Bihar Agricultural University, Sabour	14.05
5.	Pest and diseases of Paddy	Bihar Agricultural University, Sabour	19.30
6.	Scientific cultivation of Paddy	Bihar Agricultural University, Sabour	34.55
7.	Preparation of Kitchen garden	*Jeevika* and Digital Green	06.11
8.	Successful cultivation of Litchi	Bihar Agricultural University, Sabour	18.47
9.	Honeybee farming	Bihar Agricultural University, Sabour	13.02
10.	Importance of Soil testing	Bihar Agricultural University, Sabour	07.51
11.	Poultry management	Bihar Agricultural University, Sabour	21.48
12.	Nursery bed preparation of Onion	*Jeevika* and Digital Green	08.18
13.	Oyster Mushroom farming	Bihar Agricultural University, Sabour	10.31
14.	Farming of Pointed gourd	Bihar Agricultural University, Sabour	27.04
15.	How to develop Poly house	Bihar Agricultural University, Sabour	14.20
16.	Scientific cultivation of Rabi maize	Bihar Agricultural University, Sabour	12.40

TABLE 23.8 *(Continued)*

Sl. No.	Title	Source	Duration (Minutes)
17.	Integrated Farming System	Bihar Agricultural University, Sabour	20.42
18.	Farming of Strawberry	Bihar Agricultural University, Sabour	14.45
19.	How to prepare Tomato Ketchup	Bihar Agricultural University, Sabour	20.48
20.	Wheat cultivation through Zero Tillage method	Bihar Agricultural University, Sabour	10.16

The program has generated immense interest among the farmers. Most of them have been benefitted greatly by watching the agriculturally relevant videos at any time without worrying on data speed/charges.

23.14 CONCLUSION

It is important to realize that farmers and agricultural laborers should not be treated as mere beneficiaries of generic information. The agricultural sector requires a well-organized learning community in the form of farmers' associations, cooperatives, women's groups, etc. for better results out of various advisories provided. Innovative farmer information systems are a blended learning process in which face-to-face interaction, learning by doing, learning through evaluation and experience, participatory research, brainstorming exercises, etc. convert the generic information into location-specific, need-based knowledge and then empower its members through horizontal transfer of knowledge. It should enhance the self- directed learning among the rural community. There will never be a 'one fit for all' system. ICT application for an extension in the university has a long way to go. While the list above is not an exhaustive list of tools, it shows how the university is harnessing the potential of ICTs to further build and connect communities. To make the most of ICT tools, new age Extension professionals should consider a variety of outreach methods and choose those that will provide the widest outreach for the time they have available to produce quality content.

KEYWORDS

- **agromet advisory services**
- **Bihar Agricultural University**
- **extension advisory services**
- **information and communication technologies**
- ***Krishi Vigyan Kendras***
- **public-private partnership**

REFERENCES

Ali, J., (2012). Factors affecting the adoption of information and communication technologies (ICTs) for farming decisions. *Journal of Agricultural and Food Information, 13*(1), 78–96.

Benjamin, P., (2001). Does 'telecenter' mean the center is far away? Telecenter development in South Africa." *The Southern African Journal of Information and Communication, 1*(1). Retrieved from: http://link.wits.ac.za/journal/j-01-pb.htm (Accessed on 25 November 2019).

Chapman, R., Blench, R., Kranjac-Berisavljevic' G., & Zakariah, A. B. T., (2003). "Rural radio in agricultural extension: The example of vernacular radio programs on soil and water conservation in northern Ghana." *Agricultural Research and Extension Network, Network Paper No. 127.* ISBN: 0 85003 640 2.

Eggleston, K., Jensen, R., & Zeckhauser, R., (2002). *"Information and Communication Technologies, Markets and Economic Development, Working Paper."* Tufts University, Department of Economics.

Etta, F. E., & Parvyn-Wamahiu, S., (2003). "Telecenters in Uganda." In: Etta, F. E., & Parvyn-Wamahiu, S., (eds.)", *Information and Communication Technologies for Development in Africa: The Experience With Community Telecenters* (Vol. 2, pp. 71–113). Ottawa, ON, Canada: International Development Research Center (IDRC).

Ferrández-Villena, M., & Ruiz-Canales, A., (2017). Advances on ICTs for water management in agriculture. *Agricultural Water Management, 183*, 1–3.

Herold, D. K., (2010). "Imperfect use? ICT provisions and human decisions: An introduction to the Special Issue on ICT adoption and user choices." *The Information Society, 26*(4), 243–246.

Makuleke Project. http://www.faraafrica.org/media/uploads/File/NSF2/RAILS/Innovative_Farmer_Advisory_Systems.pdf (Accessed on 25 November 2019).

Manda, P., (2002). "Information and agricultural development in Tanzania: A critique." *Information Development, 18*(3), 181–189.

Nudell, D., et al. (2005). "Non-traditional extension education using video conference." *Journal of Extension [On-line], 43*(1). Available at: http://www.joe.org/joe/2005february/tt3.php (Accesses on 25 November 2019).

Roman, R., & Colle, R. D., (2003). "Content creation for ICT development projects: Integrating normative approaches and community demand." *Information Technology for Development, 10*(2), 85–94.

Torero, M., & Chowdhury, S., (2005). "*Increasing Access to Infrastructure for Africa's Rural Poor: 2020 Africa Conference Brief 16.*" Washington, DC: International Food Policy Research Institute.

Valsamidis, S., et al., (2011). "*A Framework for e-Learning in Agricultural Education.*" HAICTA.

CHAPTER 24

An Appraisal on Quality Honey Production

RAMANUJ VISHWAKARMA and RANJEET KUMAR

Department of Entomology, Bihar Agricultural University, Sabour, Bhagalpur, Bihar, India

E-mails: entoramanuj@gmail.com (R. Vishwakarma); rkipm06@gmail.com (R. Kumar)

ABSTRACT

Honey is a good source of carbohydrate, and usually contains a rich diversity of minor elements viz., minerals, proteins, vitamins and others. Quality is the major factor of honey that determines domestic market price alongside its export potential in other countries; it must meet the quality standards. Proper management of bee colonies and honey production, extraction, processing and packaging techniques can only assure us to have a good grade honey having with its natural characteristics such as color, flavor, aroma, and enzymes are preserved. Under this chapter, we have tried to accumulate the information about different aspects of quality honey production and its standard measurement which may immensely be helpful to different stakeholders including producers, retailers and consumers.

24.1 INTRODUCTION

It is a proven fact that beekeeping improves the economic condition of the farmers; helps in the holistic development of rural society and restrict the migration of rural youth to urban areas. It is the only endeavor which does not require more capital investment and skilled labor, and does not create any problem to the nature. This enterprise has minimum land and structural requirements (Singh and Singh, 2006). It is a most ideal industry for upliftment of people. In addition to meet our local health and industry requirements, this has vast potential to exports honey, wax, royal jelly, bee pollen, bee venom and propolis to the USA and Europe.

The total honey production of the country was 81,000 metric tonnes during the year 2014–15, out of that 29,578.52 metric tonnes exported in USA, Saudi Arabia, UAE, Libya and Morocco and worth of Rs. 535.07 crore (Anonymous, 2015a, b). Out of total apiary honey production of the country, 63% contributed by West Bengal, Punjab, Uttar Pradesh, and Bihar; however, Bihar produces 7800 metric tonnes during the year 2014–15 (Anonymous, 2015a).

24.2 QUALITY HONEY PRODUCTION

Honey is a natural product that has no substitute. It is a sweet liquid prepared by honey bees from the nectars of plants. In Vedas, it was considered one of the nature's most precious gifts. It had a high place in the society since its great food value, valuable medicinal effect for human beings and for its use in a number of ritual occasions. Its unique taste, aroma has sprung off many cooking ideas and recipes. It also offers incredible antiseptic, antioxidant and cleansing properties for our body and health - from eye conjunctivitis to athlete foot. Conventional beekeeping is a complementary agricultural activity with diverse socio-economic benefits to reduce the risk involved in depending solely on conventional crop and animal production as the only source of income (Olarinde et al., 2008).

The major markets for Indian honey are Germany, the USA, the UK, Japan, France, Italy, and Spain. Indian not used honey in the form of daily food as its per capita per year consumption only about 8.40 grams, while in other countries it is considered as a food for instance, in Germany, per capita honey consumption is 1.80 kg. In the world, on an average, its per capita consumption is 200 grams whereas in Asia, Japan has the highest per capita honey consumption, i.e., about 600 grams. The National Commission on Agriculture had visualized the need for deploying about 150 million Bee colonies for pollinating the agricultural crops in the country. There is so huge potential and opportunities for the growth of beekeeping industry in India. Beekeeping industry has great self-help potential for the rural people, tribals, marginal and small farmers, land-less laborers, etc. The enormous possibilities in beekeeping are given as under:

1. As per the cropped area under the major insects pollinated crops, about 200 million bee colonies are required in the country to enhance the yield levels of these crops at par with the yield levels of developed countries. It will provide jobs to about 215 lakh persons.
2. As compared to honey cost, the beeswax is twice.
3. Other hive products, such as bee pollen, royal jelly, and bee-venom are more costlier as compared to honey and beeswax.
4. Enhancing crop productivity and quality produces through intervention of bee pollination in farmers fields (Vishwakarma and Ghatak, 2014; Vishwakarma and Singh, 2017; Vishwakarma and Chand, 2017).
5. Maintenance of biodiversity by pollination of flowering plants.
6. Apitherapy medicine using bees' products.

The honey production of our country may be increased to many folds by developing and providing the advanced bee management technology. However, production alone will not determine the prosperity of this industry until supported by ready market. Rather, it is the quality honey production which will provide a good market for the produce. However quality is the major factor which will determine our export market as in a number of importing countries, quality is the most important consideration. So our honey to have export potential in particular and for our domestic market in general, it must meet the quality standards. Therefore, it has become imperative to pay due attention to proper management of bee colonies to have a good grade honey so that all its natural characteristics such as color, flavor, aroma, and enzymes are preserved.

A number of faulty practices are being adopted by the beekeepers. Resorting to unrecommended practices hamper the quality of honey and create problems in its export. China's export market earlier was badly affected because of the contamination of its honey. If Indian honey fails to meet the international standards then India may also face a similar situation. During honey flow period the colonies may manage efficiently and the beekeepers should have knowledge of honey flow trends and the status of colonies. The main function of the beekeeper is to keep the colony moral high congestion in the hive must be avoided and surplus house bees are drawn in supers. Therefore, drawn combs should be provided liberally. Many times the queen enters in the super chamber and lay eggs, and honey extraction become difficult as well as to maintain the quality of honey. At least 2–3 weeks before honey extraction a queen excluder should be placed in between brood and super chamber so that queen is confined to brood chamber. The combs which are completely sealed or two-thirds capped may be taken out for honey extraction and again placed to the supers after honey extraction. Such extraction, apparently gives an incentive to the colonies and helps to activate the bees to store more honey with quality. It is strongly recommended to harvest honey only from supers and leaving the stores in the brood chamber for colony development. To maintain quality in honey production beekeepers should have to follow the following points in beekeeping.

- Honey extraction should be always made from the bee colonies which 70–75% cells sealed.
- Honey extracted be strained and kept in stainless steel containers or food-grade plastic canes.
- Honey should never be exposed to air.
- Raw honey should be kept in cool dry places never be stored in places exposed in sunlight.
- Treatment of colonies against disease/pest should be stopped 4–6 weeks prior to the honey flow season.
- No extraction should be made from the diseased colonies.
- Maintain the colony hygiene.
- Maintain the strength of colonies.
- Stimulating diet should be provided to maintain high moral of bees.

The present article deals bee management technology for production of quality honey-based on scientific experience in beekeeping.

Production of honey is the prime objective of beekeeping industry. Its production was negligible 50 years ago from hived bees, i.e., with modern beekeeping with *Apis cerana* in hives. The pattern of honey production was gradually increased over the years. Most of the beekeepers in India are illiterate and untrained. They maintain very small number of bee colonies for honey production. However, the trend is changing and large numbers of commercial beekeepers have come up, especially, with *Apis mellifera.* Bee colonies of *Apis mellifera* are real boost to the honey production in Punjab. Beekeeping with *Apis mellifera* has also gained good popularity in Bihar, West Bengal, Uttar Pradesh, and Jammu and Kashmir. Increasing awareness about the honey among the beekeepers, honey packers and to meet the increasing demands of consumers, new patterns of honey production have developed. Migratory beekeeping is becoming more popular with *A. mellifera* in Himachal Pradesh, Punjab, Jammu and Kashmir, and Bihar. Beekeepers are obtaining 50–70 kg honey yield per colony through this technology. With increasing in area under different surplus honey crops and monocultures, it has become possible for the beekeepers to produce unifloral honeys. Beekeepers are also aware of the advantages of unifloral over mixed or multifloral honeys. High honey yielding, *A. mellifera* is extended too many potential and more suitable areas and honey production scenario is fast changing.

The production can further be increased to many folds by developing and providing the advanced bee management technology. However, production alone will not determine the prosperity of this industry until supported by ready market. Rather, it is the quality honey production which will provide a good market for the produce. When we talk about quality honey production, it starts right from the apiary, selection of right apiary and proper bee management is necessary. Therefore, it has become imperative to pay due attention to proper handling of honey to ultimately have a good grade honey so that all its natural characteristics such as color, flavor, aroma vitamins and enzymes are preserved.

A number of faulty practices are being adopted by the beekeepers. Their resorting to un-recommended practices hampers the quality and creates problems in its export as well as growth of the colonies. China's export market earlier was badly affected because of contamination of its honey. If Indian honey fails to meet the global standards then India may also face a similar situation.

24.3 QUALITY OF HONEY

For maintaining the quality of honey the following practices may be adopted:

- Honey shall be well-ripened, a natural product extracted with the help of centrifugal extractor; it shall be clear and free from objectionable flavor, foreign materials and other contaminations such as mold dirt, bees wax, fragments of plant and bee bodies, etc.
- Color of the honey shall be uniform throughout and may vary from light to dark.
- Honey shall not contain any food additives such as color, vitamins, minerals and saccharine, etc.
- Honey shall not be exposed to higher temperatures at any condition.
- Honey shall also comply with the essential requirement as per AGMARK.
- Adulteration may contain any sweet syrup called honey-like corn, cane, rice, etc. most simple adulterations can be detected if certain characteristics exceed the legal standard, for example:

  - Sucrose content, higher than 5%, if added by cane/beat sugar.
  - High HMF (80 ppm) value, if acid hydrolyzed corn syrup is added.
  - L/D ratio of corn syrup is similar to honey but contains very high HMF > 200 mg/kg.
- However detection of isotopes is regained this can be detected any kind of adulteration in honey.

24.4 STIMULATION FOR NECTAR COLLECTION

Productive efficiency in this industry is therefore dependent of several factors that affect worker bees and their activity. An understanding of these factors is important to stimulate bees for increased nectar collection such as;

- New combs encourage nectar collection and purity in honey.
- Similarly, an increase in comb space encourages nectar collection and facilitates moisture reduction in the process of ripening of honey.
- The genetic makeup of the worker bees, which is obtained from the laying queen, also plays an important role in stimulating nectar collection activity.
- Young queen bee lay more eggs, resulting in increased workforce that augments food gathering.

24.5 CONVERSION OF NECTAR TO HONEY

The nectar is collected from the plant sources and has about 25% sugar and 73% water. In this condition it gets fermented. In order to store nectar for future use, bees remove some water from nectar by an elaborate process that results in its evaporation. In this process, bees add some digestive enzymes secreted by their mouthparts. These enzymes convert sucrose in the nectar into inverted sugars, namely, glucose and fructose. After the conversion of nectar into honey and reduction in water/content, bees seal with honey cells of the comb with wax. This honey in sealed combs is the ripe honey and is ready for extraction.

1. **For quality honey production, the following management practices may be followed: Before Honey Flow:** The essence of good management before honey flow, i.e., during build-up period is to help bee colonies reaching peak strength just before the onset of the honey flow. During this period a beekeeper essentially needs to re-queen/unite weak bee colonies, prevent/control swarming, give supplement feeding and control disease and pests.

2. **Provision of Comb and Comb Foundation Sheet:** Honeybee colonies should be provided with enough space for brood rearing and food storage. Queens are reluctant to lay eggs in the older combs. Due to continuous brood-rearing, the cells are narrowed down and the bees reared in older combs are smaller in size. Honey stored in old dark combs by bees also becomes darker in color. It is strongly recommended that older and darker color combs should be discarded. Beekeepers should provide a comb foundation sheet to save the energy of bees. New raised combs should be ready for placing in the super who will be utilized for depositing quality honey and to promote the bees' efficiency in productive work.
3. **Honey Flow:** It is the period when bee forage is available in plenty and surplus is stored by bees. Honey flow is monitored for its start and end by observing changes in weight of hives. Honey flow is indicated by gradual increase in weight but weight gain ceases as the flow end. For better honey production shortage of space should be avoided. Number of combs to be added in supers depends on strength of the colony. If thin population in the colony only few frames are places at a time because too large space also leads to demoralization of the colony. It is advisable to keep nine frames in ten frame beehive. This allows expansion of top of the combs and this simplifies both uncapping and harvesting. It is strongly recommended to harvest honey only from supers and leaving the stores in brood chamber for the colony development. Only ripe honey having less than 20% moisture should be harvested when two-third of the comb cells are capped.
4. **Inspection of Colony:** The inspection of the colony helps to overcome the problem. However, during one season, the bee colonies should examine at weekly interval. Undesirable interference and mishandling of the bees during inspection may result into absconding of colonies. The bees should, therefore, be disturbed as little as possible. Before approaching to bees for inspection of the honey stores in the combs during honey flow period, he or she should put an overall, a bee veil and gloves also to prevent any contamination.
5. **Production Techniques:** Every beekeeper has to ensure the production of quality honey with purity. The following are the few steps in this direction.

    - It is very important not to use chemicals for control of bee diseases or for cleaning supers of bees. Chemicals easily get into the honey and contaminate it.

- Production apiaries should not be placed in or near crops on which chemicals are applied.
- All the equipment and implement used for extracting or handling honey should be hygienic and if needed sterilized.
- Containers for storing honey should be of materials of food-grade quality.
- Honey should be stored in airtight containers, which are kept in hygienic places.
- At any case, honey should not be extracted from brood chamber's comb.

India has a large potential for the production of honey and other hive products in beekeeping sector. By utilizing all these advantages there will be a unique opportunity for rural development through the promotion and extension of beekeeping. Beekeeping is a good profitable venture requiring a small investment of capital and skilled labors and high yield enterprise in comparison to other poverty reduction activities.

24.6 PATTERNS OF HONEY PRODUCTION

Increasing awareness about the honey among the beekeepers, honey packers and to meet to the increasing needs of consumers, the following new patterns of honey production have developed:

1. Traditional honey hunters or tribes, generally kill the bees by fire, comb is cut and squeezed along with pollen, brood, wax, etc. Such honey is highly contaminated and gets spoiled due to fermentation, however, obtained honey by pressing or squeezing of combs is called "pressed" or "squeezed' honey.
2. From hives, honey frames are removed and are extracted with the help of centrifugal machine. This can be done repeatedly without disturbing the colonies. Honey collected from the combs by centrifugal machine is generally termed as "apiary honey" or "extracted honey."
3. Honeys are classified mainly as per the floral origin (litchi, jamun, mustard honey etc.), seasonal honey (spring, greeshma or kartika honey) or according to the place where honey is produced (Mahabaleshwar honey in Maharashtra, Coorg honey of Karnataka, Kashmir honey of J and K, English honey of United Kingdom).
4. Comb honey may be available in pieces of the original comb, cut and wrapped (cut comb) or pieces of comb in glass jar of liquid honey

(chunk honey). Very small frames filled with honey by the bees or sealed honey in comb is termed as "section comb."

5. Blended honey, certified organic honey, granulated or crystallized honey, creamed honeys are gaining great demand among consumers.

24.7 QUALITY CONTROL OF HONEY

It's an important aspect of mass production of honey and thereafter its sale or consumption. The literal meaning of 'quality' is superiority in kind which involves three components, skill, relativity and reference and 'control' includes notion of checking or verification or comparison or regulation, etc. In fact, the success of quality program lies in mass honey produce by bees from natural flora. The aim of quality control should be to determine whether the honey is still in a condition to consume by the end consumers. Keeping this in view, one should not consider the maximal or optimal quality but somewhat like term 'acceptable quality' (Anonymous, 1995). It is unnecessary and expensive goal to pursue for keeping quality of mass produced honey.

Quality control of honey includes controllable factors that either positively or negatively influence the finished product. Identification of die critical points is essential since the process control relates to the processing of good and sound raw material. If the raw material is of poor quality then even good processing will not give desired quality of finished product. Hence, in food industry, Hazard Analysis Critical Control Point (HACCP) and Total Quality Management (TQM) are recommended under International Standard Organization (ISO) 9000 series. The principles of quality control in honey are considered under the following points:

- Raw material control;
- Process control;
- Finished product inspection.

24.7.1 RAW MATERIAL CONTROL

Quality control does not imply that a poor raw material can be converted into a good finished product. Invariably, once a food product has been through a manufacturing process, little can be done to alter its quality. This means that raw material of the desired standard should be accepted to maintain the quality of the finished product. Careful planning, formulation of the type of sampling and the test applied must reflect in the finished product. In the case of processed honey, the dominant raw material is ripe honey present in the comb so it should be tested in relation to its contribution to product quality. Following are some of the points which need attention to maintain the desired quality of honey:

- Raw honey should be taken from those combs in which there is more than 75% sealed honey.
- During honey flow season, queen excluders should be used to ger superior quality honey in super chambers.
- Extraction from unsealed cells reduces keeping quality of honey which eventually will ferment.
- Old and blackish frames used for honey production, decreases the yield and color of honey.
- Few beekeepers use if excess sugar feeding is given to bees before extraction, it is bound to affect the quality.
- Extracted honey has better quality than squeezed honey because the latter has many contaminations and high moisture due to open comb.
- Unifloral honeys are preferred over multifloral honey because of distinctive taste and flavor.
- In the case of comb honey and chunk honey, fresh combs containing honey should be preferred over honey collected in older combs.
- Drained honey and raw honey should be collected in sterilized conditions. Persons involved in this process should change into sterilized clothing and gloves.
- Testing of raw honey for purity is must before it is sent for processing.

24.7.2 HONEY PROCESSING

Honey processing should be ensured as soon as possible after removal from the hive. The processing is a sticky operation, where time and patience become key points to achieve the best results. For destruction of the sugar tolerant yeast to avoid fermentation and to check granulation in honey processing must be done alongside the reduction in moisture level. It is recommended that the processing of honey at 60°C for 20 min, at 65°C for 10 min and at 70°C for 2.5 min does not affect its quality. It is well known that the honey is a food commodity, so for it must be handled carefully and all processing equipment must be perfectly clean. After processing, honey is bottled in wide-mouth glass bottles and food-grade plastic jars of different sizes.

Process Control:

- Extraction should always be done under hygienic and sterilized conditions to avoid contamination.
- Honey is very sensitive to heat, hence the thermal property of honey needs to be considered very carefully during liquefaction, filtration, processing, and handling of honey.
- Excess heating should be avoided as it increases HMF and reduces enzyme activity. The heat does also affect sensory qualities and reduces the freshness. Heat processing can darken the natural honey color (browning), too.
- No or low heating than recommended time will not kill yeast cells present in raw honey. Such type of honey is liable to ferment early.
- Direct heating should be avoided in bee products to maintain their texture.
- Complete mechanized system of extraction, processing, and packaging (bottling) has good quality.
- Creamed honey is a fine crystallizes honey which is used as spread. Processing should be such to maintain the desired texture.
- These days, non-thermal processing alternative for honey are being tested and ultrasonicated honey is available in which yeast cells are destroyed.
- Some honeys (*Brassica* honey) granulate within a few days of extraction due to high dextrose content. Crystallization can be delayed by liquefying dextrose crystals at 40–45°C and by eliminating insoluble matter by filtration.

24.7.3 FINISHED PRODUCT INSPECTION

Honey is a biologically active substance so it should be regularly tested at different storage periods for fermentation and moisture content. The sugar tolerant yeast cells are present in honey which causes fermentation. It can be seen by the presence of air bubbles or foam of carbon dioxide on the surface. With the presence of partial granulation, increase in moisture content, storage temperature and time, rate of fermentation in honey is increased.

Wide variations in constituents of honey are reflected in the color, flavor, and taste of honey (Nair, 1980). In India, 5000 honeys are available depending upon floral sources, areas and bee species. In general, *Codex Alimentarius* Honey Standard (1993) is applied to honey traders of all over the world;

however, regional or country wise honey standards are also fixed. In India, the Bureau of Indian Standards (1994) or ISI standards are applicable.

- Processed honey should be according to BIS, ISI standards (Table 24.1).
- Through visual inspection of honey, it must be ensured that the honey is free from any foreign matters viz., molds, pieces of beeswax, exuviae bees, etc.
- The honey color is generally varied from light to dark brown.
- Dark-colored honeys contain more minerals than light honeys.
- HMF content should be tested on regular basis. Storage stability depends on this content.
- If the finished product is Crystallized honey then crystallization should be uniform.
- Honey produced in coastal areas (rubber honey of Kerala) contains high moisture due to hygroscopic nature. Such honeys are liable to fermentation so they have shorter shelf life.
- During storage, if the crystallization occurs and it is in layered form, then it means the presence of contaminants. Such honeys are considered of low quality. If the crystallization is complete, honey is considered pure and of high quality.
- Honey gives sour or acidic taste due to increased acidity on long storage or on fermentation.

TABLE 24.1 Essential Composition and Quality Factors for Honey

Characteristics	B.I.S. (IS: 4941: 1994)			Agmark (1984)		
	Special	'A'	Standard	Special	'A"	Standard
Specific Gravity at 27°C	1.37	1.37	1.37	1.40	1.40	1.35
Moisture (%) (max)	20	22	25	20	22	25
Reducing Sugars (%) (min.)	70	65	65	65	65	65
Sucrose (%) (max)	5	5	5	5	5	5.
Ratio L: D	1	1	1	1	0.95	0.95
Ash (%) (max)	0.50	0.50	0.50	0.50	0.50	0.50
Acidity (%) (max) per	0.20	0.20	0.20	0.20	0.20	0.20
Fiche's test	Negative			Negative		
Total count of pollen	50,000	50,000	50,000			
H.M.F. mg/kg (max)	80	80	80	--	-	--
Optical density (max) 660 nm	0.3	0.3	0.3			

24.8 GRADE DESIGNATION MARK

Organic farming should be integrated with the organic honey production. Organic honey can be produced by a countrywide campaign to explore the forest flora for honey production in various geographical zones. Product from such local niches may be labeled as produce of a particular niche. Such an approach will definitely give quality honey and better price for the beekeepers. Exploring the forest areas for beekeeping will help in bio-diversity conservation in long run. Product which is offered in the market is very important in marketing as the product quality packaging and brand are the major factors. Honey packaging is needed to be delicately handled during purification and processing so that the purest honey produced by the bees is not deteriorated. All concerns should note that 'least processing is the best processing.' Honey by definition is a substance made when the nectar and sweet deposits from the plants are gathered and stored in honey-combs by honey bees. The quality honey is possible by following scientific management practices, implementing IPM strategies in the crop ecosystems, combating pests and diseases of honey bees by using safe methods, impro-vising beekeeping technologies and educating beekeepers regarding scien-tific-technical know-how. Honey is needed to be popularized as food and not only as medicine. This approach shall expand the domestic market of the honey and will give a boost to the apiculture industry as a whole thereby paying a good price to the beekeepers. Widespread promotional efforts are required to expand domestic market. More aggressive approach, on the lines of promotional efforts for dairy and poultry products both by public and private sectors should be essential to expand the domestic market. India will need more than 5 lakh tons of honey for domestic consumption if average per capita consumption increases to about 500 grams per year. It is worth mentioning that in the USA and Germany the honey consumption is one and 1.8 kg per capita per year respectively. It will be of great advantage to the beekeepers and honey traders if domestic market expands since higher prices can be fetched in domestic market.

Quality honey can be produced if beekeepers get higher price for their produce. So far the same price is given to the beekeepers whatever quality of honey they produce. Honey is required to be graded and then priced like other commodities. It is worth mentioning that companies get 3–5 times more price in comparison to price given to beekeepers. Unified market concept for procurement and grading of honey is vital to ensure quality honey. About 1.2 million tonnes of honey is produced annually. About 400,000 tonnes is traded internationally. The honey offered for sale to importers and packers contain antibiotic residues and have to be rejected. The industry and consumers want honey free of residues because it is perceived as a pure and natural product. Beekeepers will destroy the market for honey unless they begin to understand the seriousness of the situation. It is absolutely vital that beekeeper understand they have to stop using antibiotics in such a way that residues reach the honey. Beekeepers depend on exporters, importers and packers to sell their honey and exporters, importers and packers depend upon beekeepers to produce the honey. Thus good cooperation between everyone in the supply chain is essential if consumers are going to continue receiving honey. India is richly endowed by nature with an imaging range and variety of flora and fauna. Aiming to help promote an economically viable and environmentally sound ecosystem and to improve the living standards of rural area.

24.9 QUALITY ASPECTS

In today's scenario, trading and consumption of honey in its raw-form is restricted to only tribal communities- and living within the boundaries of the forest areas and such volume could be miniscule. Largest part of Honey enters the markets and consumer's home after a processing comprising of physical measures like heat treatments and filtration. By virtue of such commercialization- the end product is required to meet both regulatory requirements and consumer's expectations. In a nutshell, these requirements may be summarized as under:

24.9.1 REGULATORY REQUIREMENTS

- Purity standards as stipulated by concerned regulatory authority and mandatory upon the processing company/ manufacturers. In India, the erstwhile PFA (which covered Honey) was done away with. As such, the subject of Honey is now entrusted to a wider and autonomous body, the Food Safety and Standards Authority of India (FSSAI).
- In addition, the processed honey is bound by other regulations like weights and measurements act at par with other marketed food products.
- When the product is meant for exports, the regulatory requirements of importing country become mandatory upon the export house. In those cases, the local regulatory authorities give adequate flexibility to the exporter and assist them in avoiding any conflicts.

24.9.2 CONSUMER'S EXPECTATIONS

- The consumer's expectations by and large do not deviate from the regulatory standards. However, they may have additional expectations in terms of sensory factors, appeal of packaging, product information and accessibility to claim support. Sometimes, these expectations may be drawn basis their knowledge on global trends.

The basic requirements of purity in honey have been time tested across the globe. These parameters include but not limited to; Specific Gravity, Moisture content, Sucrose content, Total reducing sugars, FG ratio, Ash Content, HMF, etc. Optimal/acceptable values for these parameters are set by different countries in consideration to the local factors.

KEYWORDS

- ***Apis cerana***
- ***Apis mellifera***
- **Food Safety and Standards Authority of India**
- **hazard analysis critical control point**
- **International Standard Organization**
- **total quality management**

REFERENCES

Agmark, (1984). *Sr. No. 183, Rules 1979*. Gazzette Part, I. I., sub, sec. 10.

Agmark, (1985). *Third Amendment Published as S. O. 754 Dated 28–2–1984 in the Gazette of India, Part, I. I., Section 3 (ii) Dated 10–4–1985*.

Anonymous, (1995). *Swiss Food Manual, (Schweizerisches Lebensmittelbuch) Chapter 23 A: Honey*. Eidg. Drucksachen und Materialzentralle, Bern.

Anonymous, (2015a). *Indian Horticulture Database 2015*. All India 2014–2015 (Final Estimates), Department of Agriculture and Cooperation. http://nhb.gov.in/MISDailyAreaProduction.aspx?enc (Accessed on 25 November 2019).

Anonymous, (2015b). *Agricultural and Processed Food Products Export Development Authority, Ministry of Commerce & Industry, GOI*. http://apeda.gov.in/apedawebsite/SubHead_Products/Natural_Honey.htm (Accessed on 25 November 2019).

BIS, (1994). *Extracted Honey-Specifications* (pp. 1–10). Bureau of Indian standards (IS 4941: 1994).

Codex Alimentarius, (1993). *Standard for Honey*. Ref. Nr. CL 1993/14-SH FAO and WHO, Rome.

Codex Alimentarius, (1998). *Draft Revised for Honey at Step 6 of the Codex Procedure*. CX 5/10.2, CL 1998/12-S.

Conte, L. C., Miorini, M., Giorio, A., Bertacco, G., & Zironi, R., (1998). Evaluation of some fixed components for unifloral honey characterization. *J. Agric. Food Chem., 46*, 1844–1849.

EU Daft, (1996). *Proposal for a Directive of the European Council Relating to Honey*, EU document 96/0114, 1996.

Mishra, R. C., (1993). Handling, processing and storage of honey. In: Goyal, N. P., & Sharma, R. K., (eds.), *I Nat. Conf. Beekeeping, Chandigarh, NHB* (pp. 88–97).

Nair, K. S., (1980). Physico-chemical characteristics of rubber honey in India. In: *Proc. II Int. Conf. Apic. Trop. Climates* (pp. 676–684). New Delhi.

Nanda, V., Sarkar, B. C., Sharma, H. K., & Bawa, A. S., (2003). Physico-chemical properties and estimation of mineral content in honey produced from different plants in Northern India. *J. Food Comp. Anal., 16*, 613–619.

Olarinde, L. O., Ajao, O. A., & Okunola, S. O., (2008). Determinants of technical efficiency in beekeeping farms in Oyo State, Nigeria. A stochastic production frontier function. *Research Journal of Agricultural and Biological Sciences*, *4*(1), 65–69.

Singh, D., & Singh, D. P., (2006). *A Handbook of Beekeeping* (p. 287). Agrobios (India).

Vishwakarma, R., & Chand, P., (2017). Foraging activity of insect pollinators and their impact on yield of rapeseed-mustard. *Bioinfolet*, *14*(3), 222–227.

Vishwakarma, R., & Ghatak, S. S., (2014). Impact of foraging activity of pollinators including honeybees on seed yield of sunflower. *Indian Journal of Entomology*, *76*(2), 136–141.

Vishwakarma, R., & Singh, R., (2017). Foraging behavior of insect visitors and their effect on yield of mango var. Amrapali. *Indian Journal of Entomology*, *79*(1), 72–75.

CHAPTER 25

Empowering Rural Youth Through Agripreneurship

RAM DATT

Department of Extension Education, Bihar Agricultural University, Sabour, Bhagalpur, Bihar, India

ABSTRACT

Every era has its own challenges. In order to deal with the situation, each challenge demands specific responses. Presently, Indian Agriculture is facing a major challenge of 'how to make agriculture more lucrative or profitable as a profession and an equally respected employment option in our money-driven society. Agripreneurship plays a crucial role in transforming subsistence agriculture into a commercial agriculture it also helps in creating jobs at the grassroots level. Ultimately this will act as a catalyst for empowering the rural community and overall growth of the economy. Fundamentally this chapter deals about concepts of entrepreneurship and agripreneurship, innovative initiatives of the varsity and success stories vis-à-vis agripreneurship promoted by the varsity.

25.1 BACKGROUND

The term *entrepreneur* originates from the French word *entreprendre* meaning "undertaker, adventurer and projector" (Gündoğdu, 2012; Wadhera and Koreth, 2012; Landströn, 2005). Hence the term 'entrepreneur' did not find any prominence in the history of economic thought (Gopakumar, 1995). Entrepreneurship appeared in the economic thought after writing of Richard Cantillon (1680–1734) -an Irish economist and whose work *"Essai Sur la Nature du Commerce en General"* published posthumously in 1755. Richard Cantillon is recognized as the father of entrepreneurship. The fundamental assumption was, an entrepreneur buys inputs at fixed price, makes value addition and sells them at uncertain price. An individual who takes advantage of these unrealized profits is known as entrepreneur (Landströn, 2005). Similarly, the work of an 18th-century French writer, Bernard F. deBelidor defined entrepreneurship as buying labor and material at uncertain prices and selling the resultant output at contracted prices (Gopakumar, 1995; Hoselitz, 1960). Further, John Stuart Mill (1848) advocated that risk is a prime ingredient of entrepreneurial activities, whereas J.B. Say (1816) stated that entrepreneur is an economic agent.

Modern use of entrepreneurship is credited to Schumpeter (1934) proposed dynamic theory of entrepreneurship and who considered entrepreneur as the catalyst that disrupts the stationary circular flow of the economy and thereby initiates and sustains the process of economic development. Similarly, Peter F. Drucker (1985) emphasized innovation as an instrument for entrepreneurship development. McClelland's (1961) emphasized that entrepreneur is having a high need for achievement—preference for challenge, acceptance of personal responsibility for outcomes, innovativeness—characterized successful initiators of new businesses (Shaver and Scott, 1991).

There is a long history of entrepreneurship research but agripreneurship research came somewhat late in the literature in the year 1980s (Wortman, 1990). Slowly, the worth of entrepreneurial research and/or activities was recognized on a massive scale (Kahan, 2012; Vaillant and Lafuente, 2007; Fuller-Love et al., 2006; Kulawczuk, 1998).

25.2 WHAT IS ENTREPRENEURSHIP AND AGRIPRENEURSHIP?

There is a plethora of definitions of entrepreneurship available in the literature. Pioneer contributor Richard Cantillon (1755) defined entrepreneur as a person who takes risk and he inherently believes in providing one's own economic well-being. Cole (1949) stated that entrepreneur is person who conceive, initiate and maintain a social institution for a long period of time which produces economic goods. McClelland (1961) defined entrepreneur as a person with a strong need for achievement and preference for moderate risk. Stevenson (1983, 1985, and 1990) discussed fundamentally three key components of entrepreneurship viz; (1) pursuit; (2) opportunity, and (3) beyond the resources control. Drucker (1985) stated that 'the entrepreneur always searches for change, responds to it, and exploits it as an opportunity.' He identified seven sources of innovations namely-the unexpected; the incongruity; innovation based on process need; changes in industry structure or market structure; changes in perception, mood and meaning and new knowledge, both scientific and nonscientific. Mamat and Raya (1990) emphasize that an entrepreneur undertakes a venture, organizes it, raises capital to finance it and assumes all or major portion of the risk. Pareek and Rao (1978) describe entrepreneurship as a creative and innovative response to the environment.

Agripreneurship means entrepreneurship in agriculture. Entrepreneurs are innovative, take calculated risks, having strategic vision and change themselves as per the need of the existing market and/or entrepreneurial environment (GFRAS, 2016). Most of the time farmers are creative but unable to incubate their ideas due to lack of venture capital, unable to access services, market and other necessary inputs (Wongtschowski et al., 2013). There are many other internal and external factors, which greatly affect the agripreneurship. Entrepreneurial mindset is a prime factor for small land-holding farmers to ensure their food, nutritional and financial security under changing climatic and market conditions. When, farmers see their farm as a profitable venture and take calculated risk then they efficiently use available innovations (Kahan, 2012). Stephen C. Mukembo (2016) defined agripreneurship as "the application of entrepreneurial principles to identify, develop, and manage potential agricultural enterprises optimally and sustainably for profit and improved livelihoods."

With the help of aforementioned key factors we can summarize agripreneurship is a process in which an agripreneurs accept changes and risk, develop strategic vision and arrange necessary capital and other resources to convert farming as a profitable venture.

25.3 DETERMINANTS OF AGRIPRENEURSHIP

There are many motivational or personal factors which greatly influence entrepreneurial process. It is also quite obvious that entrepreneurial activity not solely depends on motivational factors; but external factors also play a key role (availability of primary goods, venture capital, communication channels, market situation, government policy, etc.) in influencing it.

25.3.1 PERSONAL FACTORS

- **Need for Achievement:** The need for achievement is one of the social motives which have been studied since long (McClelland,

1961; McClelland and Winter, 1969). McClelland stated that persons who are having high need for achievement are involved in managing overall enterprises, have a high level of accountability, strives for maximum outcomes, required self efficacy, perseverance, calculated risk taking ability and clear strategic vision on performance (Shane et al., 2003). Further, McClelland (1961) (mentioned in Wadhera and Koreth, 2012) stated that achievement motivation is not merely achieving success rather achievement motivation is characterized by a need to:

1. Take moderate risks in setting goals to improve results;
2. Measure the results achieved;
3. Seek and use 'feedback' about interim performance and about the environment; and
4. Try new things in order to achieve better results.

Collins et al. (2000) reported that achievement motivation is an effective variable for differentiating successful and unsuccessful entrepreneurs. Many studies thus highlight the importance of a need for achievement as a characteristic of entrepreneurs and an influence on entrepreneurial success (Barba-Sánchez and Atienza-Sahuquillo, 2012).

- **Willingness to Take Risk:** Economic theory suggested that risk taking ability is an important attribute of an entrepreneur. McClelland (1961) argued that individuals with high need for achievement prefer to take calculated risk. Similarly, Atkinson (1957) stated that entrepreneurs, who have high need for achievement, take medium level of risk which helps them to maximize their profits. Whereas Liles (1974) also claimed that entrepreneurs always take risk in terms of finance, emotional, career and family relations. Moreover, several studies claimed entrepreneur always takes risk (Venkataraman, 1997).
- **Locus of Control:** Entrepreneur is an individual with a high degree of 'internal locus of control.' If an individual perceives that event and/or his activities in his control this belief is termed as internal control. Whereas, external locus control is when an individual do not entirely rely on his action or believe on luck/fate (Rotter, 1966). As McClelland (1961) mentioned earlier, persons who are having high need for achievement prefer a situation in which they feel that they have direct control over the outcomes or in which they feel that they can directly see how their efforts are going to affect the outcome(s) of a certain activity.

- **Entrepreneurial Self-Efficacy (ESE):** Bandura (1994) defined self-efficacy as people's beliefs about their capabilities to produce desired levels of performance that exercise influence over events that affect their lives. Further, Bandura (1994) mentioned four major processes related to self-efficacy which includes cognitive, motivational, and affective and selection processes, whereas, Chen et al. (1998) defined ESE as to the strength of a person's belief that (s) he is capable of successfully performing the various responsibilities related to entrepreneurship. Further, he pointed out five factors namely marketing, innovation, management, risk-taking, and financial control.

  Self-efficacy reflects task-specific self-confidence of an individual. A person having high level of self-efficacy towards any task depicts his tendency to exert more efforts for a greater length of time, persist through setbacks, set and accept higher goals, and develop better plans and strategies for the task.
- **Flexibility:** Most of the successful entrepreneurs measure the pros and cons of a decision and tend to change if the situation demands. They never feel reluctant to revise their decisions. They are the persons with open mind without rigidity. Especially, in the case of farming (where climate change happens to be one of the biggest concerns, agripreneurs have to be ready to shape their decisions accordingly), which highly depends on climate with high rate of change. Agripreneurs always adopt and/or act on these changes/variability and find viable solutions. This is one of the important attribute of an entrepreneur (Kahan, 2012).
- **Innovativeness:** Land, labor, capital and entrepreneurship are considered as four fundamental factors of production process. Among these factors entrepreneurship is added as fourth factor in the production process. Sometime, entrepreneurship factor of production is a debatable issue that how this factor contributes towards production process. Successful entrepreneurs are innovators. They constantly put their efforts in introducing new products, new method of production, opening new markets and recognizing the enterprise (Schumpeter, 1949). To move towards 21st century agriculture, we must view agriculture and allied field as a business instead of doing it in subsistence mode. For this transformation, agripreneurs need to make new combination of their resources and/or production factors. Similarly, Schumpeter (1949) clearly mentioned that entrepreneur is someone who disrupts the circular flow of economy and make new

combinations (equal to innovation). Drucker (1985) stated that innovation is an important tool of an entrepreneur. Innovativeness is one of the core attributes of an entrepreneur.

- **Perseverance:** It is an important ingredient of entrepreneurship development process. Entrepreneur is possessed with the characteristic of sticking to the job he decides to undertake. Once committed to a specific goal and course of action, entrepreneurs become absorbed in it. They personally solve the problems that come across their way while setting up the project. They also work sincerely until the whole project is successfully implemented. Entrepreneur always carries out relentless efforts with same level of energy and motivation as well as never give up to the situation.
- **Strategic Vision:** Unlike routine tasks for which structures and processes provide a frame of reference, creation of enterprises deal with unmet needs and gaps that are relevant in the present and the future around which both competitive forces and uncertainty operate. Mitton (1989) suggests that entrepreneurs have a knack for looking at the usual and finding the unusual. This would lead to what is commonly referred to as strategic vision. Strategic vision in this context is to set a future-oriented goal, based on environmental analysis, for determining the content of enterprise action.

In addition to the afore-mentioned characteristics of entrepreneur, researchers have identified some other characteristics that are decision maker, independent, arbitrageur, tolerance for ambiguity, coordinator, positive self-concept, analytical mind, creative, interpersonal communication skill ability, etc. Kahan (2012) identified nine competencies for an agripreneur including initiative, taking risks, ambition, creative thinking, focused problem-solving, flexibility and adaptability, interpersonal abilities, readiness to learn and networking.

25.3.2 SOCIAL AND CULTURAL FACTORS

Some scholars have stressed the importance of socio-cultural milieu in entrepreneurship development (Cochran, 1949; Cole, 1949; Lamb, 1952; Williamson, 1966). Jenks (1949) and Cochran (1949) suggested that the socio-cultural history accounts for the performance of entrepreneurial functions by a considerable number of individuals. Several writers have used a comparative framework to highlight the ways in which different societies, with differing interests, attitudes, systems of stratification and the like operate to produce different kinds of business and different patterns of entrepreneurial behavior.

25.4 FRAMEWORK OF AGRIPRENEURSHIP

There is no denying the fact that entrepreneurial activities originate from the individuals. Hence understanding of their personality traits, skills and background are crucial for the development of a comprehensive theory. Thus if agripreneurship is the individual response to a situation, that is the environment around him, and creation of an organization is essential for carrying through that response, the agripreneur, environment and the organization must be regarded as crucial elements of any framework relating to agripreneurship (Figure 25.1) (Kanungo, 1997).

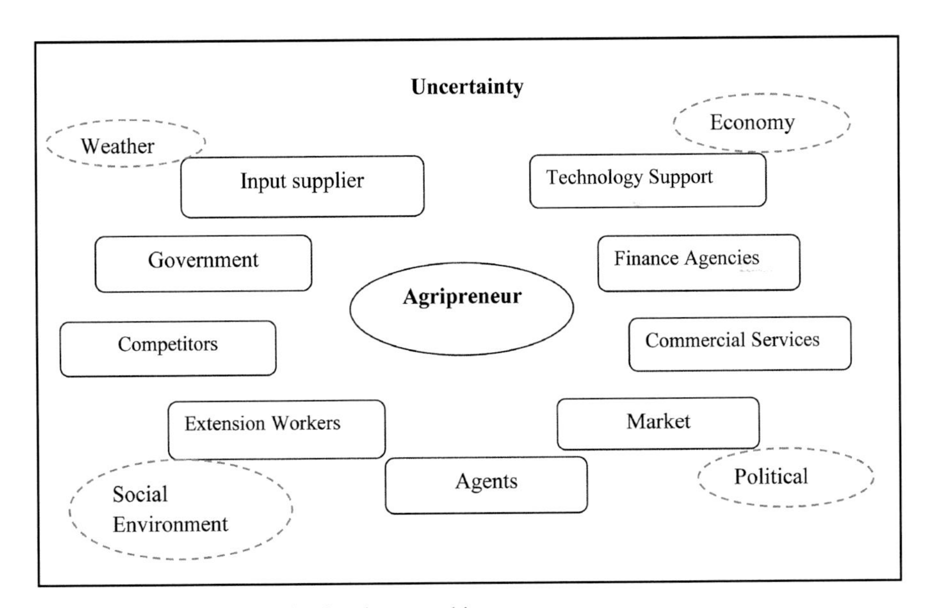

FIGURE 25.1 Framework of agripreneurship.

Agripreneurs are independent in nature but they do not work alone. They operate in a complex and dynamic environment. They are part of a larger collection of people including other farmers, suppliers, traders, transporters and processors; each one of them has a separate role to play in the value chain. Similarly, Kanungo (1997) mentioned that rural entrepreneurship development process involves individual attributes, micro environmental factors such as family, social and cultural values. These factors are often adequate in themselves for enterprise creation where institutional requirements are well developed as in developed economies. At the same time, macro and contingency factors affect at both creation and development stages.

25.5 AGRIPRENEURSHIP DEVELOPMENT PROCESS

Agripreneurship development is a dynamic process. One phase is interconnected with another phase of agripreneurship development process.

1. **Identification of Potential Agripreneurs:** In this project, we approached two villages through key contact person and remaining three villages of research area were selected sans key contact person. Those two villages approached by key leaders, we have been more successful in terms of generating people participation, trust building and commencing of entrepreneurship development process. A number of psychological variables were reported by researchers, achievement motivation is one of the prime variables to become a successful entrepreneur. Under this study, we selected only those respondents who were having high need for achievement.
2. **Opportunities Identifications:** Under this study for exploring opportunities, we developed individual and village resource maps, which helped us in identification of opportunities.
3. **Selection of Enterprises:** Respondents themselves conducted local market survey to understand the scope of different potential enterprises. At the same time, they also critically analyzed their own resources and availability of resources in their villages. Then, agripreneurs selected agri-venture themselves and we only facilitated the process.
4. **Entrepreneurship Development Programme (EDP)/Entrepreneurial Motivation Training (EMT):** Under EDP, we conducted training in three phases viz; pre-training, training and post-training. In pre-training phase firstly grouping of respondents on the basis of enterprises was done and then skill gaps of respondents were analyzed. Before we started the training, EMT through different videos and FGD were conducted. Then, we conducted different training sessions according to their enterprises.

5. **Venture Capital:** Mostly the selected respondents were from poor families lacking the resources, so at the initial stage they needed seed capital. Respondents who were linked with Amrit Hatcheries Pvt. Ltd. started their businesses but the people who did not receive venture capital; they could not start entrepreneurial activities.
6. **Venture Startup:** This is a very crucial phase for agri-enterprise development. The selected agripreneurs were provided with regular hand-holding activities. Majority of successful agripreneurs got assured markets through Amrit Hatcheries Pvt. Ltd.
7. **Growth Phase:** Those agripreneurs reached in sustainable phase in their enterprise development. In this phase, agripreneurs started their own marketing channels and/or marketing management (Figure 25.2).

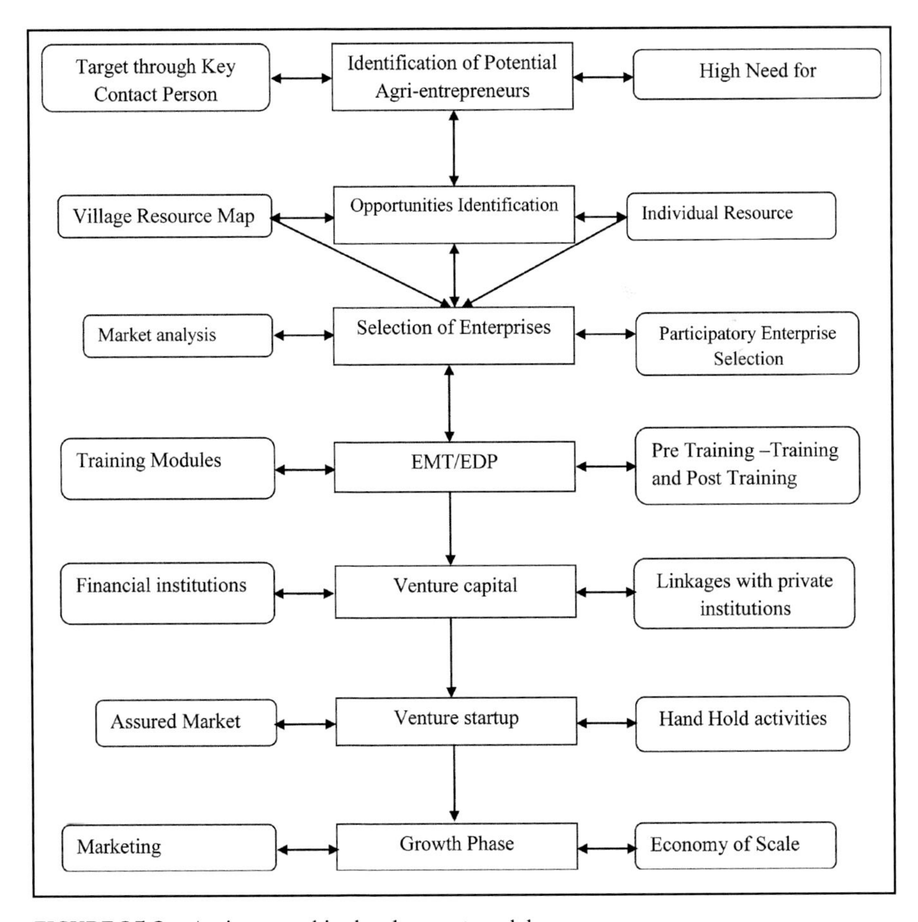

FIGURE 25.2 Agripreneurship development model.

25.5.1 ENTREPRENEURIAL OPPORTUNITIES IN AGRICULTURE

Agriculture plays a pivotal role in India's economy. Majority of 54.6% of the population is engaged in agriculture and allied activities (census, 2011). Majority of Indian farmers are marginal and small, while more than 90% farmers of Bihar are having less than one hectare of land. Given the critical role of agriculture in the local economy and the centrality of small and marginal farmers in agriculture, encouraging entrepreneurship in agriculture with focus on small and marginal farmers would help the sector cope with the rapidly changing global environment (Philroy et al., 2014). Bihar, with a population of 104.1 million as per the Census of India, 2011, is the third most populated state in country. It comprises 9% of India's population and covers 3% of its landmass, making it the most densely populated state. Migrants have predominantly moved from rural to urban destinations and from agricultural to non-agricultural occupations. More than 90% of households to be deleted reported migrant members in urban destinations for long durations from the state (Rodgers et al., 2013 mentioned in Datta, 2016). Consequently, a large number of defacto women are living in the villages of Bihar State and they are actively involved in farming activities and decision making. At the same time, it is a challenge to transform rural women into agripreneurs especially that are having less land or no land.

There are many lucrative enterprises existing at farm level such as integrated framing system, high value horticulture, food processing, fruit and vegetable processing, livestock and dairying, fisheries poultry, participatory seed production, mushroom lab/production, bee keeping, vermicompost, etc.; as service providers such as input dealers, local service providers supplementary ventures such as soil testing, cold-storage services, transportation, trading and microcredit agencies run by farmer groups; and processing and marketing ventures, etc., (Philroy et al., 2014). There are some promising agri-enterprises discussed in subsections.

25.5.1.1 INTEGRATED FARMING SYSTEMS (IFS)

Integrated Farming System (IFS) is an emerging sustainable and profitable approach of farming and there is appreciable interest raised in IFS among small and marginal farmers. Integration of enterprises like crops, fruits, vegetables, poultry, goat-rearing, piggery, duck farming and dairying is advantageous as these are mutually beneficial to each other, since crop residues can be used as animal and poultry feed, while animal manure can be utilized to enhance soil fertility that enhances agricultural productivity (Reddy, 2016). This complementary combination helps in reducing risk, enhancing soil fertility, improves nutrient cycle, intensifies soil use and increases the overall income of the farm. The "win-win" outcomes can be achieved by small and marginal farmers' in-terms of nutritional security and farm income. Bihar Agricultural University (BAU), Sabour has developed a number of IFS models at its constituent colleges and KVKs as a demonstration unit.

Sri Hari Mohan Jha is a farmer of Khutha Baijnathpur village of Bhargama block Araria. He started farming in the year 1986 in 15 acres of his own land. In the year 2006, he got training on different components of IFS model from KVK Araria. Presently, from IFS model he is earning over INR 30,000/annum from mango orchard, INR 20,000/annum from fishery, INR 40,000/annum from livestock. He also earns 20,000/annum from 10 pits of vermicompost. He is also cultivating potato, ginger and turmeric and earning INR 25,000/annum. Altogether from IFS model he is earning INR 2.5–3.0 lakhs/year.

25.5.1.2 HIGH VALUE CROPS

High value crops refer to non-staple agricultural crops such as vegetables, fruits, flowers, ornamentals and medicinal and spices (Temu and Temu, 2005 and CGIAR, 2005). High value agricultural crops are those which are having higher return per unit of land in comparison to traditional crops. High value crops could be potential opportunities for emerging entrepreneur in the field of agriculture. Bihar state is also having a lot of potential for fruits, vegetables, medicinal plants and flower cultivation.

Dilip Kumar Singh is an agripreneur from Mehaddiganj near Sasaram in Rohtas district. He has ownership of 2 acres of land. He proved his persistency and innovativeness in his farm. Rohatas district which is known as bowl of rice and rice-wheat cropping pattern is prominent across the district. In the year of 2004, he came in the contact with Krishi Vigyan Kendra, Rohtas, Bikramganj and participated in training programs like scientific cultivation of vegetables including okra, brinjal, tomato, cauliflower, broccoli, cabbage, potato, onion, chili, bottle gourd, bitter gourd, capsicum, etc. There was some tracts of barren land in the adjoining villages. He started taking those lands on lease in a very nominal rent for few years of contract. At the same time, he used to persuade the landowners that after the cultivation of vegetables the land will gradually convert into a cultivable land. Presently, he is producing different offseason and on season vegetables both organic and non-organic based on 100-acres of leased land and earning around Rs. 7–10 lakhs in a year. Apart from that, he also provides employment to 10–15 thousand laborers each year in his own vegetable farming. He is also using a number of modern farm implements. In addition to farming he is guiding and motivating farming community for vegetable production. Indian Council of Agricultural Research, New Delhi recognized his noble work, and awarded him with prestigious Jagjeevan Abhinav Kisan Award in the year of 2012. His successful journey clearly reflects his hard work, motivation, innovativeness, and risk taking ability.

25.5.1.3 FOOD PROCESSING

About 15–50% of agricultural produce get damaged due to post-harvest losses in the state. Food processing related enterprises not only help in improving the socio-economic condition of farming community, but also help in reducing wastage of agricultural products. There is huge potential of food processing in the states like Bihar. Over the last two decades after globalization of Indian economy, demands of processed food have been increasing day-by-day. As a result, the future of food processing industry is increasing tremendously.

Mrs. Radha Devi is from village Gore Gaon, Danapur, Patna. Coming from a very poor family; she had to struggle for even life- saving necessities. Situation improved only when she came into the contact with KVK Barh, Patna and received training on food processing. She started preparing value-added products of rice, wheat, maize, chickpea, soybean, and barley flour. Initially, she was producing meager quantity of value-added products due to which it was difficult for her to access the market. Then she started contacting other women and developed a SHG. Now, she has improved her soci-economic condition and is providing employment to many poor women.

25.5.1.4 FRUIT AND VEGETABLE PROCESSING

India is the second-largest producer of fruit and vegetables in the world. But commercial processing of these commodities is only 2% of the total production. Despite the fact that our country is the second-largest producer of fruit and vegetables, its share in the world's horticultural products is less than 1% (Sidhu, 2005). Another side, demand of processed horticultural products has been increasing tremendously in the last two decades. So, there is a huge market of processed horticultural products. In Bihar there is a high range of variability in terms of soil, water and weather conditions to produce a large quantity of mango, litchi, banana, and vegetables. The state produces 18.56MT accounting for 7.72% of horticultural production of the country. It is encouraging to note that Bihar is the leading state in litchi and fourth-largest producer of mango and guava. The state is also fifth largest producer of pineapple, seventh largest in banana, third largest producer of okra and potato and fourth largest in onion production. There are immense opportunities for agripreneurs in the area of processing and value addition of horticultural products.

Mrs. Manorama Devi is from Sabour block Bhagalpur. She is founder of Srijan Mahila. After, death of his husband in the year of 1991, she started Srijan with only five women. Initially, she adopted Bapu's Charkha as means of earning and started selling product door-to-door. In the year of 1994, Manorama Devi joined hands with a few women of a self-help group. For starting the venture, they needed venture capital and she tried for many financial institutions because of mortgage problem many of them denied to lend the loan. Finally, the Primary Agriculture Co-operative Society, Rajandipur, gave her a loan of Rs 10,000. Presently, Srijan has taken a shape of successful women's co-operative and producing a number products like silk cloth, silk saree, silk suit, Bindi, different types of spices, incense sticks, etc. Even the Ministry of Textile, Govt. of India procures silk sarees and bedsheets from this society. In addition to these products, this cooperative is making a number of processed fruit and vegetable products like pickles, jam, jelly, morabba, etc. Presently, she is providing direct employment to more than 600 women.

25.5.1.5 COMMERCIAL DAIRY FARMING AND VALUE ADDED MILK PRODUCTS

India has a very vast resource of livestock, which plays a significant role in the national economy making it the largest milk producer in the world accounting 18.5 of total world milk production (Economic Survey, 2015–16). Animal husbandry is the backbone of small and marginal farmers' livelihood especially for states like Bihar where more than 90% of farmers are marginal and small. The state is producing about 6516 thousand MT per year and availability of milk is 175 gram/capita whereas ICMR recommendation is 220 gram/capita. Presently, the demand of value-added products of milk like paneer, ice cream, cheese, butter, bottled milk, yogurt, etc., is increasing continuously. There is a huge potential of this sector especially in states like Bihar which abounds in fertile soil and freshwater.

Er. Indra Kishore Chourasia is from Village Dogacchi, Kasba block of Purnea district of Bihar. His village is 6 km away from Purnea district. After completing his study he started the business of computer supplier. But he realized that he should do something in his village only. He started dairy farming with 15 cows and two buffaloes. Presently, he is procuring 100–110 liters of milk and also having an automatic milking machine. After the positive response of consumers, he started the value addition of milk in forms of different sweets and sugar-free sweets (Aditya et al., 2015).

25.5.1.6 POULTRY FARMING

Poultry is a more profitable and popular venture among the rural entrepreneurs because it gives provides remunerative returns on the amount invested, time and energy spent and labor involved. This venture requires less investment in comparison to dairy farming and also provides quick return within short span of time. The demand of egg and poultry is increasing over the years. India produced about 78.48 billion eggs in 2014–15, and availability of eggs per capita is 58 eggs/year. Bihar state produces 11002 lakhs eggs and per capita availability is 11 eggs/year, whereas recommendation ICMR is 180 eggs/year (Agricultural Road Map-2012–17). This is another promising sector through which agripreneurs can earn a lot.

Mr. Shantanu Kumar is from Pachgachhia, Gopal block of Bhagalpur district. He is a young agripreneur who started his poultry farming with 1000 of birds. He started his poultry farming with Amrit Hatcheries Pvt Ltd and developed linkage with market. Presently, he is having 5000 birds and exploring his own business. He is directly providing employment to 10–15 people and 4.5–5.0 lakhs/annum.

25.5.1.7 FISH FARMING

Fish farming is a gainful venture where there is plenty of water available. Bihar is endowed with vast inland aquatic resources including rivers, ponds, reservoirs, ox-bow-lakes, flood plains (chaur) and tanks. Fish farming is a traditional profession especially in flood-prone areas of Bihar. Still, the state has not been optimally exploited of its water resources having the large untapped potential of fish farming. The annual domestic fish demand is about

5.2 lakh tones where we are producing only 2.88 lakh tones only. There is huge gap between the demand and supply of fish within the state. At the same time there is an enormous demand of fingerlings in the state, but small portion of this demand is met indigenously as few firms and agripreneurs are into this business, rest demand is fulfilled from West Bengal.

Dr. Sanjivanand Thakur is an agripreneur of Balua Deorhi Village of Palasi Block from Araria district of Bihar. He is a doctorate in Zoology. After, completion of his study he chosen his own profession instead of service. He started fish farming in his own 7 acres of low land. He attended a training programme at Govind Ballabh Pant University of Agriculture and Technology, Pantnagar (U.K.) on carp breeding and culture. He started fish seed production through induced breeding by using low cost improvised hapa method. He converted his seven acres of low land into two stocking ponds, one brood stock and three nursery ponds. Then in a tough competition with West Bengal, he linked with market. Presently, he is successfully doing breeding and seed production of catla, rohu and common carp and earns net profit of INR 5.5–6.0 Lakhs/annum (Aditya et al., 2015).

25.5.1.8 MUSHROOM PRODUCTION

Mushroom cultivation is very popular among the landless farming community especially for women farmers. Mushroom is nutritionally very rich food and has high market value. In addition to good quality protein, no cholesterol, high fiber, low sodium, good quantity of vitamins and minerals, the mushrooms also have bioactive compounds like b-glucans, protein polysaccharide complexes that impart with unique medicinal values like anti-cancer and anti-viral properties (Singh et al., 2011). Integrating mushroom cultivation in family farming not only increases income of family but also helps in decreasing the disguised unemployment of the family. Over the last decade, mushroom production in Bihar has gained popularity among the farming community. Oyster and Button mushroom both can be grown in the state. But, oyster is widely grown in comparison to button mushroom because production of button mushroom needs compost and high skilled human resources. The cost of production of button mushroom is also much higher than oyster mushroom. However, market value of button mushroom is much higher and it is also suitable for canning process. Now a day, oyster mushroom production is more popular among the women members of SHGs. Among the districts of Bihar, Nalanda district is the leading in mushroom production.

Mrs. Madhu Patel, Panditpura, Rajgir, Nalanda (Bihar) started her entrepreneurial journey in year 2008. Mrs. Patel has now become a leader of about 1,00,000 rural women by greatly transforming their lives. When Mrs. Patel found Mushroom as highly beneficial venture, she promoted other women of her locality by motivating them on account of the excellent taste of mushroom but as the women of the locality started realizing its huge profits with minimum or negligible inputs the industry took up a quick boom. As the cost of the spawn reached very high due to its transportation from Delhi or Ranchi, she thought of manufacturing it on her own. Succeeding in her motives she is now capable of distributing the spawn on a very large scale. She has influenced the lives of women headed household by providing about 8 to 10 thousand per month of income. For marketing initially, she approached the Brahmin and Mahuri community people who were strictly vegetarians and convinced them about high protein and nutrient content of mushroom. Later as an effective way to spread and popularize the product, she gave Rs. 10.00 as margin per kg of mushroom to the newspaper vendor as a marketing strategy. This proved to be a boon for the mushroom marketing. She has also remarkably improved the status of poor women by distributing the spawn free of cost. She skillfully provides training on how to sterilize the straw of wheat, way of spawning, method of proper hanging, maintenance of optimum temperature and humidity and maintaining the hygiene throughout the cultivation process. Initially started as a small group, which has now transformed into Mushroom Grower Association led by Mrs. Madhu, is successfully providing training to the rural women not only for proper cultivation and its proper sale, but also for its processed products like mushroom pickles, mushroom papad, murabba, etc., which provides for the premium returns of the farmers and thus helps raise their living status significantly. Her mushroom labhas got modern mushroom equipments like laminar airflow, BOD incubator, autoclave, refrigerator, AC, etc. on account of which she is able to produce quality spawn on a large scale. She is also supplying spawn not only to the farmers, but also to the District Agricultural Officers for its wider distribution. Her area of coverage not only includes many districts of Bihar but extends even to many other states like Jharkhand, Uttar Pradesh, etc. Rajgeer, a very well known tourist place serves as a big market for the sale of the mushroom and its processed products. This vital factor of marketing provides a great incentive for its production as the cultivators get remunerative prices and high demand for their produces. This highly proteinaceous food on account of its excellent taste and versatile uses is being greatly demanded not only in the domestic market but also in abroad. Now, Nalanda is being recognized nationally as well as globally for its mushroom. She contributed a lot for mushroom revolution in Nalanda district.

25.5.1.9 BEE KEEPING

Beekeeping is one of the old traditions in our country for collecting the honey. Beekeeping is becoming very famous among peasants because of its demand in national and international market. Beekeeping is not only profitable venture for beekeepers but helps in increasing cross pollination of crops which increases the farm productivity. Besides honey, we get venom, royal jelly, wax, propolis, etc. from bee keeping which are used in medicine, cosmetics and other purposes. These products are in great demand in indigenous and abroad markets. Bihar is the leading state in honey production and litchi honey of the state is very popular in the market.

> *Sri Shashi Kumaris from Surhari, Gaya. He is a successful agripreneur in the field of beekeeping and honey processing. He started beekeeping in the year of 1995 with 10 boxes. He struggled a lot during his initial days of beekeeping but sticked to his goal. Within two years, he increased from 10 Apis mellifera* colonies *boxes to 400 colonies and established "Shiva Honey" with financial support of 1 lakh rupees from Bank of Baroda, Manpur, Gaya under PMRY scheme. In the year of 2004, he started M/S Kunwar Apiary Pvt. Ltd company with project cost 1.24 crore. He established a processing, testing and packaging plant and started selling his produce under the brand name of "Shiva Agro" (Aditya et al., 2015). Presently, he is having linkage with more than 200 beekeeper of Bihar. He is consistently receiving scientific inputs from BAU, Sabour and its KVK Gaya. In addition to his venture, he is inculcating basic skills of beekeeping among the farmers and proving market linkage through his company.* Sashi *Kumar's contribution to honey production and processing in Bihar has earned him awards from the Agricultural Produce Export Development Agency, APEDA, BAU, Sabour, the National Bee Board and the State government.*

25.5.1.10 COMMERCIAL VERMICOMPOST PRODUCTION

Vermicompost is the product of the composting using earthworm. Vermicomposting is the process by which worms are used to convert organic materials (usually wastes) into a humus-like material known as vermicompost. Commercial vermicompost production is one the promising enterprises for the agripreneurs.

> *Sri Santosh Kumar is from village-Shekhwara, block-Bodh Gaya of Gaya district. He started his dairy farming only with two cows in the year of*

2003. Presently he is having 186 milch cows. He is utilizing the cow dung to establish a large vermicompost unit of 3000 mt capacity. He is earning net return from vermicompost 10.0 lakhs/year (Aditya et al., 2015). He received Jagjivan Ram Innovative Farmers Award-2012 by ICAR, New Delhi.

25.5.1.11 COMMERCIAL MAKHANA PRODUCTION

Bihar state is the leading state for *Makhana (Euryale ferox)* production. The state of Bihar alone, accounts for 90% of world's total *Makhana* production. Districts of Bihar where it is popular and cultivated by farmers are Darbhanga, Madhubani, Purnea, Katihar, Sitamarhi, Saharsa, Supaul, Araria, and Kishanganj. Makhana is culturally intertwined with *Mithila.* This is a very profitable venture but the collection and processing of *Makhana* are difficult. A few agripreneurs from Bihar started processing of *Makhana* and earning millions.

Sri Sahid Parwez is from near B.G.P. School, Saharsa. He went to Ambala to see the machine which is especially developed for Makhana popping by M/S Jwala Engineering and Consultancy Services, Ambala. Makhana popping is most tedious and labor intensive task during the makhana processing. He also got basic training about Makhana processing and prototype for popping machine from Central Institute of Post Harvest Engineering and Technology, Ludhiana. The available popping machines in the market were expensive and needed uninterrupted supply of electricity. But, due to erratic electricity supply in his locality, he decided to install 12 HP generators to run the popping machine. He established whole popping unit with the cost of 6.2 lakhs. Now he is earning about INR 42 lakhs annually and sending processed makhana to Varanasi, Lucknow and Indore market (Aditya et al., 2015).

In addition to aforementioned agri-enterprises, many other promising agri-enterprises like nursery production, cultivation of medicinal and aromatic plant, protected cultivation, service provider through farm implement banks, participatory seed production, animal feeds, goat farming, etc. are available as profitable ventures to rural entrepreneurs.

25.6 CONCLUSION

Modernizing agriculture and transforming subsistence farming into a profitable venture, this process could be a strong engine for enhancing the socio-economic condition of the grassroots. This will also help in attracting and retaining rural youth towards agriculture because majority of the farmers believe that agriculture is no longer a lucrative venture. Undoubtedly, there are many constrains like lack of achievement motivation; risk taking ability due to poor socio-economic condition, lack of skill and knowledge, poor linkage with markets, availability of venture capital, policy issues, social stigma faced by the agripreneurs while enterprise development process. The growth of agripreneurship is very much the need of the hour. However, there are plenty of limiting factors which hinder the entrepreneurial process. There is a strong need of mechanism or policy which encourages agripreneurial process.

KEYWORDS

- **agripreneurship**
- **entrepreneurial motivation training**
- **entrepreneurial self-efficacy**
- **entrepreneurship**

REFERENCES

Aditya, Sohane, R. K., & Jaiswal, U. S., (2015). *Agripreneurs of Bihar*. Directorate of extension education, BAU, Sabour.

Agricultural Road Map, (2012–2017). *Department of Agriculture*. Government of Bihar.

Atkinson, J. W., (1957). *Motives in Fantasy, Action, and Society*. Princeton, NJ: Van Nostrand.

Bandura, A., (1997). *Self-Efficacy: The Exercise of Self Control*. New York: Freeman.

Barba-Sánchez, V., & Atienza-Sahuquillo, C., (2012). Entrepreneurial behavior: Impact of motivation factors on decision to create a new venture. *Investigaciones Europeas de Dirección y Economía de la Empresa, 18*(2), 132–138.

Cantillon, R., (1755). Henry, H., (ed.), *Essai sur Ia Nature de Commerce en General*. London: Macmillan, 1931.

CGIAR, (2005). *CGIAR System Research Priorities 2005–2015* (p. 96). CGIAR Science Council Secretariat.

Chen, C. C., Greene, P. G., & Crick, A., (1998). "Does entrepreneurial self-efficacy distinguish entrepreneurs from managers?". *Journal of Business Venturing, 13*(4), 295–316.

Cochran, T. C., (1949). Approaches to entrepreneurial personality. In: *Change and the Entrepreneur* (pp. 97–112). Prepared by the research center in entrepreneurial history. Cambridge, MA: Harvard University Press.

Cole, A. H., (1949). Entrepreneurship and entrepreneurial history. In: *Change and the Entrepreneur* (pp. 85–107). Prepared by the research center in entrepreneurial history. Cambridge, MA: Harvard University Press.

Collins, C., Locke, E., & Hanges, P., (2000). *The Relationship of Need for Achievement to Entrepreneurial Behavior: A Meta-Analysis*. Working paper, University of Maryland, College Park, MD.

Datt, R., (2015). *Agripreneurship Development: An Action Research*. Report submitted to Directorate of Research, BAU, Sabour.

Datta, A., (2016). Migration, remittances and changing sources of income in rural Bihar (1999–2011): Some findings from a longitudinal study. *Economic and Political Weekly, II*(31), 85–93.

Drucker, P. F., (1985). *Innovation and Entrepreneurship*. Heinemann, London.

Economic survey 2015–2016.

Fuller-Love, N., Midmore, P., Thomas, D., & Henley, A., (2006). Entrepreneurship and rural economic development: A scenario analysis approach. *International Journal of Entrepreneurial Behavior and Research, 12*(5), 289–305.

Global forum for rural advisory services. www.g-fras.org (Accessed on 25 November 2019).

Gopakumar, K., (1995). Entrepreneurship in economic thought: A thematic review. *Journal of Entrepreneurship, 4*(1), 1–17.

Gündoğdu, M. C., (2012). Re-thinking entrepreneurship, intrapreneurship, and innovation: A multi-concept perspective. *Procedia - Social and Behavioral Sciences, 41*, 296–303.

Hoselitz, B. F., (1960). The early history of entrepreneurial theory. In Spengler, (ed.), *Essay in Economic Thought* (pp. 234–257). Chicago: Rand McNally & Co.

Jenk, L. K., (1949). Approaches to entrepreneurial personality. In: *Change and the Entrepreneur* (pp. 80–96). Prepared by the research center in entrepreneurial history. Cambridge, MA: Harvard University Press.

Kahan, D., (2012). *Entrepreneurship in Farming: Farm Management Extension Guide*. Rome: Food and Agriculture Organization of the United Nations. Available: http://www.fao.org/docrep/018/i3231e/i3231e.pdf (Accessed on 25 November 2019).

Kanungo, R. N., (1997). *Entrepreneurship and Innovation. Sage* Publications, New Delhi.

Kulawczuk, P., (1998). *The Development of Entrepreneurship in Rural Areas* (pp. 97–106). The transfer of power. Local Government and Public Service Reform Initiative, Budapest, Hungary.

Lamb, R. K., (1952). Entrepreneur and community. In: Miller, W., (ed.), *Men in Business* (pp. 91–119). Cambridge, MA: Harvard University Press.

Landströn, H., (2005). *Pioneers in Entrepreneurship and Small Business Research*. New York: Springer.

Liles, P. R., (1974). *New Business Ventures and the Entrepreneur*. Homewood, IL: Irwin.

Mamat, K., & Raya, R., (1990). The Malaysian entrepreneur: His qualities, attitudes and skills. *Malaysian Management Review, 25*(2), 11–18.

McClelland, D. C., & Winter, D. G., (1969). *Motivating Economic Achievement*. New York: The Free Press.

McClelland, D. C., (1961). *The Achieving Society*. Princeton: Van Nostrand.

Mill, J. S., (1848). *Principles of Political Economy*. London: Longmans, Green and Co.

Mitton, D. G., (1989). The complete entrepreneur. *Entrepreneurship Theory and Practice, 13*(3), 9–20.

Mukembo, S. C., (2016). Available at: www.g-fras.org (Accessed on 25 November 2019).

Pareek, U., & Rao, T. V., (1978). *Developing Entrepreneurship, a Handbook for Policy Makers, Entrepreneurs, Trainers and Development Personnel*. New Delhi: Learning System.

Philroy, J., Karuppanchetty, S. M., Divya, N. G., Bhubesh, K. R., & Selvaraj, A., (2014). *Agribusiness Incubation Transforming Indian Agriculture: A Business Incubation Approach of NARS–BPDs*. ICRISAT, Hyderabad. Retrieved from: http://www.aipicrisat.org/wp-content/uploads/2014/12/Agribusiness%20Incubation%20Transforming%20Indian%20Agriculture.pdf (Accessed on 25 November 2019).

Reddy, P. P., (2016). *Integrated Crop-Livestock Farming Systems*. Springer.

Rodgers, G., Datta, A., Rodgers, J., Mishra, S. K., & Sharma, A. N., (2013). *The Challenge of Inclusive Development in Rural Bihar.* New Delhi: Institute for Human Development and Manak Publications.

Rotter, J. B., (1966). Generalized expectancies for internal versus external control of reinforcement. *Psychological Monographs: General and Applied, 80*, 1–28.

Say, J. B., (1816). A treatise on political economy or the distribution and consumption of wealth, A. M. Kelly, Publishers, New York. In: Ruhul, A. S., (ed.), *Modeling entrepreneurship in, small-scale enterprises, Curtin Business School (CBS), GPO Box U1987, Curtin University of Technology, Perth, WA 6845, Australia, Applied Economics Letters* (Vol. 12, pp. 51–57).

Schumpeter, J. A., (1934). *The Theory of Economic Development*. Cambridge, MA: Harvard University Press.

Schumpeter, J. A., (1949). "Economic theory and entrepreneurial history." In: *Change and the Entrepreneur: Postulates and Patterns for Entrepreneurial History*. Cambridge, Mass.: Harvard University Press.

Shane, S., Locke, E., & Collins, C. J., (2003). Entrepreneurial motivation. *Human Resource Management Review, 13*(2), 257–280.

Shaver, K. G., & Scott, L. R., (1991). Person, process, choice: The psychology of new venture creation. *Entrepreneurship Theory and Practice*, 23–45.

Sidhu, M. S., (2005). Fruit and vegetable processing industry in India: An appraisal of the post-reform period. *Economic and Political Weekly, 40*(28), 3056–3061.

Stevenson, H., & Jarillo, J., (1991). "A new entrepreneurial paradigm." In: Etzioni, A., & Lawrence, P., (eds.), *Socioeconomics: Toward New Synthesis*. New York: M. E. Sharpe, Inc.

Stevenson, H., (1983). *A Perspective on Entrepreneurship*." Harvard Business School Working Paper, 9–384–131.

Stevenson, H., (1985). *"The Heart of Entrepreneurship"* (pp. 85–94). Harvard Business Review.

Temu, A. E., & Temu, A. A., (2005). *High Value Agricultural Products for Smallholder Markets in Sub-Saharan Africa: Trends, Opportunities and Research Priorities*. International Workshop on how can the poor benefits from the growing markets for high value agricultural products? International Center for Tropical Agriculture, Cali, Colombia.

Vaillant, Y., & Lafuente, E., (2007). Do different institutional frameworks condition the influence of local fear of failure and entrepreneurial examples over entrepreneurial activity? *Entrepreneurship and Regional Development, 19*(4), 313–337.

Venkataraman, S., (1997). The distinctive domain of entrepreneurship research: An editor's perspective. In: Katz, J., & Brockhaus, R., (eds.), *Advances in Entrepreneurship, Firm Emergence, and Growth* (Vol. 3, pp. 119–138). Greenwich, CT: JAI Press.

Wadhera, K., & Koreth, G., (2012). *Empowering Rural Women*. Micro Enterprise through Achievement Motivation, Sage Publication, London.

Williamson, H. F., (1966). Business history, economic history. *Journal of Economic History, 26*(4), 407–417.

Wongtschowski, M., Belt, J., Heemskerk, W., & Kahan, D., (2013). *The Business of Agricultural Business Services: Working with Smallholders in Africa.* Amsterdam: Royal Tropical Institute, Rome: Food and Agriculture Organization of the United Nations, Arnhem: Agri-ProFocus. Available: http://www.kit.nl/sed/wpcontent/uploads/publications/2080_the_business_of_agricultural_business_services.pdf (Accessed on 25 November 2019).
Wortman, Jr. M. S., (1990). Rural entrepreneurship research: An integration into the entrepreneurship field. *Agribusiness, 6*(4), 329–344.

Index

A

Abiotic
  factors, 228, 298
  stresses, 10, 16, 23, 32, 54, 70, 133, 229, 297, 300, 303, 433
Abscisic acid, 462
Abutilon indicum, 416
Acari, 382, 385
Acaricides, 458
Acetamiprid, 416
Acetic acid, 127, 128, 462
Acidic soil, 324
Acrididae, 385, 386
Actinomycetes, 272
Adhwara basin, 350
Adulteration, 55, 57, 506
Aerobic
  rice, 15–17, 132, 467, 468
  soil
    conditions, 17
    organisms, 126
  surface layer, 127
Aeroponic
  cultivation, 56
  techniques, 301
Afforestation, 353
Agaricaceae, 205
Agricultural
  crops, 99, 175, 321, 323, 325, 330, 502, 527
  earth worm role, 271
  emissions, 138
  land fertility status, 43
  policymakers, 4
  productivity, 137, 151, 152, 527
  topics, 152
  training, 484
  waste, 274, 275
Agripreneur/agripreneurship, 517–519, 522, 523, 524, 527, 530, 531, 533, 535
Agro meteorological field units (AMFUs), 153, 155, 156, 168, 169, 171, 172
Agro techniques, 105
Agro-advisories, 154
Agrobacterium, 9, 273
Agrochemicals, 131, 189, 191, 194, 288, 368, 457, 461, 466
Agro-climatic
  conditions, 324
  regions, 13, 70, 76, 78, 79
  situations, 21, 26, 27, 103
  zones, 6, 15, 21, 75, 122, 153, 154, 286, 346, 347, 432
Agroecosystem, 56
Agro-industries, 274, 282, 354
Agro-industry waste, 274, 275
Agromet
  advisory, 151, 153, 154, 156–159, 162, 168, 169, 171, 487, 498
    bulletin, 151, 154, 156–159, 168, 169
    methods, 153
    services (AAS), 149, 150, 152, 153, 156, 158, 159, 162, 487, 498
  field unit, 172
Agrometeorological services, 151
Agromyzidae, 409
Agronomic practices, 128, 129, 133
  planting methods, 129
  protected cultivation, 129
Alachlor, 462
Aldrin, 460
Aliyar Nagar
  1 (ALR 1), 69
  2 (ALR 2), 69
Alkathenes, 218
All India Co-ordinated Research Project (AICRP), 16, 25, 32, 48, 50, 70, 71, 73–75, 80, 217, 471, 474
Allele mining, 56
Allergy toxicity, 10
Alleviation, 3, 63, 131, 151

Allicin, 47
Allied crops, 44
Alluvial
plains, 339, 346–349, 351
soil, 142, 144
Amalgamation, 9, 231, 270, 395
Amaryllidaceae, 47
Amblyseius ovalis, 412
Amelioration, 356
Amino acids, 223, 432
Ammonia, 126, 177
Ammonium sulfate, 459
Amphibolite, 341, 352
Amplified fragment length polymorphism (AFLP), 9, 49
Anaerobic
conditions, 126, 213, 276, 281
sub-surface layer, 127
Angiosperm, 203, 205
Annual, 90, 103, 240, 243
chrysanthemum, 101
crops, 212, 222, 240, 241
flowers, 112
leaf production, 79
manorial potential, 278
water requirement, 221
Anthesis, 124, 125, 130, 474, 475
Anthocyanine pigment contents, 265
Anthracite, 323
Anthracnose leaf spot, 470
Anthropogenic activities, 146, 174, 183
Anthurium, 99, 112, 114, 115, 311–313
Anti-inflammatory properties, 41
Antimicrobial properties, 41
Antioxidant properties, 41, 368
Aphid, 53, 107, 110, 162, 383, 384, 387, 390, 395, 421, 431, 433, 438–441, 446, 447, 450
Apical dominance, 263
Apis
cerana, 504, 515
mellifera, 504, 515, 533
Arbuscular mycorrhiza (AM), 204–208
Arbuscules, 205
Archaeosporales, 206
Archaeosporomycetes, 206
Arecaceae, 81
Aroma, 17, 20, 21, 47, 86, 132, 501–503, 505
Aromatic, 39, 42
crops, 44
flower, 90
oil extraction unit, 275
plants, 44, 91, 103
rice, 15, 17, 21, 388
Arsenic compounds, 460
Artificial mating, 8
Ascomycota, 205
Aspergillus niger, 369, 471
Asphondylia capparis, 398
Assam green tall (AGT), 77
Atmospheric nitrogen, 238
Atomic
absorption (AA), 53, 191, 201
force microscopy (AFM), 191, 201
Atrazine, 462
Attar, 103
Aulacophora foveicollis, 422
Auxin, 128, 273, 462
Avena sativa, 370
Avermectins, 460
Azadirachtin, 459
Azimsulfuron, 467
Azolla, 280, 285, 290, 492, 496
Azotobacter, 272, 273, 280, 285, 286, 290
Azoxystrobin, 470

B

Bacillus thuringiensis (Bt), 401, 477
Bacteria, 45, 56, 127, 272, 273, 280, 285
Bacterial leaf blight (BLB), 13, 16, 20, 469, 470, 477
Bactrocera cucurbitae, 396, 421
Bagasse, 274
Barban, 462
Barcoding, 55
Base exchange capacity (BEC), 339, 364
Basidiomycetous, 205
Basin
application, 79
system, 214–216
Bean golden mosaic virus, 433
Beauveria bassiana, 405, 424
Bedding plants, 90
Bee pollen, 502, 503
Beeswax, 503, 512
Begomoviruses, 431, 433–436, 451

Benzoylureas, 460
Benzyladenine, 462
Beta vulgaris, 327
Betasatellite, 434, 435
Betulaceae, 205
Bhagalpur Katarni, 20, 22
Biennial cycle, 259
Bihar
  Agricultural University (BAU), 15, 17, 23, 25, 27, 30, 32, 34, 75, 113, 124–126, 129, 130, 132, 133, 141, 154, 155, 185, 189, 195, 206, 208, 223, 248, 278, 280, 284–286, 311, 312, 329, 367, 369, 387, 436, 446, 466, 477, 481, 482, 484, 486, 487, 493, 495–498, 527, 533
    agronomic approaches, 129
    crop management, 129
    initiative, 206
    nutrient management, 130
  coconut-based cropping system, 80
  crops development, 10
    maize, 28
    rice breeding, 10
    wheat breeding, 21
  geological information, 342
  river basin, 350, 353
  soil, 342
    carbon sequestration, 141
    pollution, 182
Bioagents, 53
Bioavailability, 57, 204, 371
Biochemical
  oxygen demand (BOD), 175, 187, 532
  processes, 222
  reactions, 213
Biodegradability, 131
Biodegradable waste, 275
Biodegrading, 373, 375
  nanoparticles, 372
Bioefficacy, 466
Biofertilizers, 50
Biogas slurry, 275
Bioinformatics tools, 56
Bio-inoculants, 269, 272, 288, 290
Bio-intensive pest management (BIPM), 397
Biological
  nitrogen fixation, 273
  sequestration, 136, 146
  soil remediation, 181
    compost remediation, 182
    fungal remediation, 182
    microbial remediation, 181
    non-remedial options, 182
    phytoremediation, 181
Biomagnifications, 463
Biomass, 50, 75, 130, 138, 139, 193, 282, 322, 324, 327, 370, 468
Bionomics, 417
Biopesticides, 53, 328, 374, 401, 458
Bioprospecting, 56
Bio-regulators treatments, 259
Biosystematics, 54
Biotechnological methods, 45
Biotechnology, 8, 9, 48
Biotic
  factors, 451
  problems, 368
  stresses, 70, 229, 297, 300, 379, 433
Biotypes, 9, 387
Bipartite, 433, 435
Bipolaris sorokiniana, 372
Bispyribac sodium, 467, 468
Bitter gourd leafhopper, 398
Black spot tolerant, 107
Blight, 16, 23, 28, 53, 372, 442, 471, 472
Boletus, 205
Bombyx mori, 464
Borassus flabellifer L., 65, 81
Botanical, 458, 469
  insecticides, 409
  pesticides, 402
Botrytis cinerea, 369
Brassica
  honey, 511
  junceca, 322
Brazilian garlic germplasm, 49
Breeding
  hybrid, 15
  program, 12, 113, 448
Brevennia rehi, 381
Brinjal shoot/fruit borer, 395, 413, 424
Brown
  manuring, 130
  planthopper (BPH), 20, 171, 172, 380, 387, 389, 392, 473

Bulb, 47, 48, 90, 98, 103, 184
characters, 49
diameter, 47
formation, 49
weight, 47
Bulbil, 49
Bulbing period, 47
Bulk density, 140, 321, 323, 325, 326, 331
Burkholderia, 273
Butachlor, 468, 470

C

Calcium hydroxide (CaO), 323, 331
Calcschist, 341, 352
Calendula, 112
Calmodulin-binding protein, 16
Calopogonium muconoides, 74
Calotropis gigantean, 402
Canopy, 214, 215, 222, 223, 239, 241–244, 246, 251, 255, 257, 258, 261, 262, 265, 407, 416
Capsicum
annum, 328
chlorosis virus (CaCV), 444, 451
Captan, 460
Carbamates, 459
Carbendazim, 470–472
Carbohydrate, 258, 501
Carbon, 56, 123, 124, 131, 135–142, 144, 146, 190, 300, 319, 320, 323, 354, 367, 511
assimilation, 204
dioxide, 124, 135, 136, 139, 146, 300, 511
management index, 141
organic farming, 141
retention, 137, 140, 146
sensitive indicators, 141
sequestration, 56, 131, 135–137, 139–142, 144, 146
chemical transformation, 136
engineering techniques, 136
potential, 140
soil carbon sequestration, 136, 137
terrestrial sequestration, 136
sink, 140
storage potential, 138
trapping, 137
Carnation, 93, 99, 101, 307
Carotenoids, 83, 107
Cartap hydrochloride, 419
Caterpillar instars, 417
Cation, 325, 326
exchange capacity, 326
Cecidomyiidae, 381, 385
Cellulose, 144
Cenoccocum, 205
Center for Protected Cultivation Technology (CPCT), 302, 316
Central
cotton growing zone, 406
Insecticides Board (CIB), 465, 466, 477
Institute of Plastic Engineering and Technology (CIPET), 270
Pollution Control Board (CPCB), 270
Centrifugal extractor, 505
Cereal, 3, 6, 9, 16, 36, 122, 129, 163, 184, 195, 198, 272, 336, 356, 379, 380, 386–388, 391, 392, 406, 432, 462
crops, 3, 10, 198, 379, 380, 387, 391
productivity, 122
System Initiatives for South Asia (CSISA), 16
Chemical
fertilizer, 73, 144, 328, 329
growth retardants, 257
modification, 56
pesticides, 368, 397
Chemopreventive properties, 41
Chemotaxonomic categorization, 47
Chilean garlic germplasm, 49
Chiling stress, 128
Chilli thrips, 424
Chilo partellus, 384–387, 392
Chinese garlic germplasm, 49
Chitin synthesis inhibitors (CSI), 459, 477
Chitosan, 131, 189, 199, 200, 373, 375
Chlorantraniliprole, 419
Chlordane, 460
Chlorimuron, 468
Chlorinated hydrocarbons (OCs), 458
Chlormequat, 462, 475, 477
Chlorofluoro carbons, 124
Chlorophenols, 459
Chlorophyll, 130, 223, 474
Chlorotic spots, 415
Chlorpyriphos, 171, 388, 390, 420, 474

Chour land, 364
Chowghat
  green dwarf (CGD), 70
  orange dwarf (COD), 70, 77
Chrysanthemum, 76, 93, 98–101, 104, 105, 112, 113, 116, 307, 311, 314, 409, 458
Chrysomelidae, 381, 382, 422
Cicadellidae, 381
Cicer arientinum, 322
Citric acid, 83, 84
Citrus
  crops, 262
  species, 216
  tree, 223
Climate change, 56, 121, 122, 124–126, 128, 130, 131, 133, 135, 137, 139, 140, 142, 144, 146, 149, 150, 152, 153, 521
  Bihar, 122
  impact, 124
    major cereals, 124
    natural resource management, 125
  scenario, 121, 131
  stress, 128
Clonal
  propagation, 48
  selection, 47, 48, 50
Cnaphalocrosis medinalis, 381
Coarse grain, 13, 19
Coat protein (CP), 436, 437, 440, 451
Coccinellids, 472, 474
Coccinia grandis, 402
Coconut research
  Bihar status, 75, 76
    crop improvement, 76
    crop production, 79
  India, 68
    biotechnology interventions improvement, 71
    crop improvement, 70
    germplasm collection and conservation, 68
    varieties and hybrids, 69
Codex Alimentarius, 511
Coir
  industry, 275
  pith, 275
Cold chain system, 104
Coleoptera, 381–386, 402, 422
Colloidal
  formulation, 373
  suspension, 373
Colonization, 206, 207, 272
Comb foundation sheet, 507
Commercialization, 57, 86, 89, 91
Confiscates anoxic stress, 129
Conglomerate, 352
Conjunctivitis, 502
Conservation agriculture, 130, 136–138
Consolida ambigua, 102
Conventional
  breeding, 54
  crops cultivation, 94
  flood irrigation, 218
  plant breeding, 8, 9
Copper, 176, 178, 286, 341, 352
  nanoparticle treatment, 372
Corms, 98, 101, 103, 108–110
Correlation coefficient, 163, 164
Cortical cells, 205
Costeffective nutrient formulation, 131
Cotesi aplutellae, 418
Cotton
  fields, 414
  hybrid, 5
Cotyledon stage, 423
Crambidae, 381, 384–386, 399
Crop
  breeding, 3, 4
  combination, 239, 251, 324
  debris, 414
  drying method, 112
  duration, 52, 263, 307
  geometry, 261, 262
  growth, 56, 124, 129, 150, 305, 328, 331, 347, 463
  plants, 6, 8, 10, 203, 319, 386, 406, 441, 457
    breeding methods, 6, 8
  production technology, 303
  productivity, 6, 138, 174, 300, 319, 327, 328, 353, 503
  protection strategy, 398
  rotation, 6, 138
  season, 19, 150
  stages, 222
  yield, 222

Cropping
intensity per unit land area, 236
patterns, 360, 379, 386, 391
system, 6, 50, 56, 75, 79–81, 98, 99, 129, 130, 142–144, 195, 231, 239–241, 243, 245, 336, 350, 360, 398, 431, 445, 451
Cross-pollination, 8, 463
Crotolaria
retusa, 74
striata, 74
Cryopreservation technique, 71
Crystallization, 512
Cucumber mosaic virus (CMV), 440–442, 451
Cucumovirus, 451
Cucurbitaceous, 398, 407, 422, 431, 433, 438
crops, 398, 422, 431
Cucurbits, 238, 360, 421, 423, 431, 435, 436, 438, 439, 441
Cultivars, 17, 32, 41, 50, 51, 53, 69, 113, 127, 254, 256, 263, 265, 312, 327, 401, 412, 448, 462
Curculionidae, 382, 383, 386, 402
Curcuma longa, 51
Curcuminoids, 56
Cut flower crop, 101
Cyanobacteria, 326, 327
Cyclic market fluctuations, 55
Cyclohexanediones, 460
Cycloheximide, 462
Cycocel, 257
Cyhalofop, 467
Cypermethrin, 390, 403, 405, 414, 416, 445, 474

D

Dahlia hortensis, 102
Daira land, 364
Dalma lava, 352
Defoliation, 416
Deforestation, 125, 135, 139
Defragmentation, 104
Degree of milling, 20
Dehumidification, 305
Dehydration, 113, 223
Dehydrogenase, 193, 204, 326
Deltamethrin, 390, 416, 472
Demographic changes, 151
Demonstration and seed production farm (DSP Farm), 72
Dengue, 460
Denitrification, 126, 127
Deo-geowoo-gen, 13
Desertification, 139
Deterioration, 137, 179, 180, 369, 397
Diadegma semiclausm, 418
Diamond back moths (DBM), 417, 424
Diara lands, 360
Diarrhea, 41
Dicarboxamide groups, 460
Dicladispa armigera, 381
Dicot wetland plants, 370
Dicotyledonous plants, 433
Dicyandiamide (DCD), 127
Dieldrin, 460
Digital
filming, 152
image processing (DIP), 341
Dinitroanilines aryloxyphenoxypropionate, 460
Diptera, 381, 383–386, 421
Direct seeded rice (DSR), 129, 133
Disc filters, 309
Discomycetes, 205
Disease management, 80, 367–369, 371, 373–375, 415, 445
Dissemination, 360, 433, 446, 493
Distillation methods, 103
District agromet advisory bulletin, 156
Diversification, 54, 100–104
Diversisporales, 206
Dolerite, 341, 352
Domestic
demands, 5
market, 43, 90, 94, 501, 503, 513, 532
Double haploid (DH), 25
Draught stress alleviation, 131
Drip
fertigation, 79
irrigation, 72, 73, 126, 214, 219, 299, 308
system, 216, 217
trials, 73
system, 217
Drippers, 73, 217, 308, 309
Drug modeling, 55

Dry matter, 51, 467
Dwarf
cultivars, 69, 256, 257, 266
indicas, 13
pyramid, 258, 265
Dwarfing rootstock, 258, 263

E

Earias
insulana, 413
vittella, 413
East coast tall (ECT), 69, 70, 77
Eco-friendly compounds, 460
Economic
security, 68
yield, 133, 240, 242
Ectomycorrhizae (ECM), 204, 205, 208
Ectomycorrhizal root, 205
Ectotrophicmycorrhize, 204
Eisenia foetida, 280, 281, 324
Electrical
conductivity, 220, 326
tools, 303
Eleusineindica, 371
Embryo, 71, 82
Emission control devices, 319, 320
Endomycorrhizae, 203–206
Endomycorrhizal fungi, 203
Endophytes, 206
Endosperm, 82, 85, 86
Endosulphan, 460, 470
Endrine, 460
Entrepreneurial
motivation training (EMT), 524, 535
self-efficacy (ESE), 521, 535
Entrepreneurship, 100, 517–519, 521–524, 526, 535
development program, 524
Environmental
oxidation, 138
Protection Agency (EPA), 189, 465
requirement, 281
Epidemics, 55, 417, 444, 446
Epidermal leaf tissues, 418
Epidermis, 418
Epigeic worms, 281
Epilachna
beetle, 402, 405
dodecastigma, 402
Equilibrium, 131, 192, 194, 195
Ericales, 205
Erosion
map, 357
prone area, 233
Erwinia, 273
Essential oils, 51, 53, 89, 90, 103
Ethephon, 462
Ethylene, 461, 462
Eudrilus eugeniae, 280, 281
Eulophidae, 405, 410
Europe Union (EU), 465
Evaporation, 73, 194, 215–218, 230, 325, 371, 506
Extended x-ray absorption fine structure (EXAFS), 191
Extension advisory services (EAS), 481, 482, 498

F

Federal Insecticide, Fungicide, and Rodenticide Act (FIFRA), 465
Fermentation, 84, 139, 508, 510–512
Fertigation, 72–74, 79, 217, 218, 222–224, 303, 308, 309, 316
levels, 79
studies, 222
system, 309
unit, 303
Fertile clones, 47, 48
Fertilization, 47, 143, 261, 482
Fertilizer, 125, 126, 130, 141, 174, 180, 192–194, 196, 204, 217, 219, 227, 246, 247, 271, 273, 286, 288, 309, 319, 321, 322, 325, 330, 371, 389, 414, 416, 457
application, 246
compounds, 212
consumption, 5
control order (FCO), 285
Fiberglass-reinforced polyester, 304
Flonicamid, 472
Flood
irrigation methods, 214
plains, 350, 530
prone, 350
resistant, 12
system, 215

Florets, 107–111
Floricultural
  commodities, 90, 93, 94, 104
  crops, 89, 90
  division, 104
  products, 91–93, 102, 105
  sector, 98, 99
    diversification potentiality, 100–103
    profitability, 98
    scope, 98
  units, 101
Floriculture, 89–94, 98–100, 102–106, 116
  Bihar scenario, 94
  business constraint and challenges, 104
    improvement strategies, 104
  crops, 90, 92
  industry, 89, 90, 104
  research work, 105
  sector, 90–92, 94, 99, 103
    global scenario, 92
Floriculturist, 103
Flower crops, 93, 98–100, 105, 106, 112, 116, 243, 299, 300, 311
  cultivation, 98, 100, 298, 300
  improved varieties, 106
    chrysanthemum, 110
    dry flower technology, 112
    gladiolus, 108
    marigold, 107
    rose, 106
Flubendiamide, 402, 403, 409, 419, 473
Flupyrifurone, 472
Fly ash, 319–331
  application, 324–328, 331
  classification, 323
  utilization, 322
Foliage, 89–91, 93, 94, 98, 100, 220, 408, 409, 415, 418, 422, 439
Foliar
  application, 193, 424, 471, 473, 474, 476
  disease, 34
  spray, 410, 473, 474
  systems, 308
Food
  grade plastic canes, 504
  Quality Protection Act (FQPA), 465
  Safety and Standards Authority of India (FSSAI), 514, 515
Forest degradation, 125
Forestry crops, 321, 325, 362
French marigold, 108
Fruit
  crop based cropping system, 239, 241–243, 251
    benefit, 240
    components, 242
    laying procedure, 243
  fly, 395, 398, 424, 473
  orchard, 230, 237–239, 260
  plantation, 230, 232
Fulvic acid, 144, 146
Fungal
  pathogens, 16
  pathosystems, 372, 375, 471
  remediation, 181, 187
Fungi, 45, 182, 195, 203–206, 272, 369, 372, 373, 405, 406, 423, 424, 471
Fungicidal wound dressing, 260
Fungicides, 162, 175, 217, 227, 230, 367, 374, 457, 458, 460, 461, 469, 470, 477, 488
Furrow system, 215
Fusarium oxysporum, 372

G

Gall midge, 380, 387, 398
Gandak basin, 350
Gangetic plains, 25, 351
Garlic
  breeding, 47
  genetics, 48
  germplasm, 49
Garlicky flavor, 47
Gasteromycetes, 205
Gazania flowers, 110
Geminiviridae, 433
Geminivirus, 431, 433, 435–437, 446
Genetic
  diversity, 4, 46, 47, 49, 82, 90, 261, 435
  engineering, 9, 10
  fingerprinting, 57
  improvement, 3, 4, 36
  makeup, 8, 506
  stock, 12
  transformation, 8, 9, 10
  variability, 46, 48, 84

Genomics, 3, 36, 56
Genotypes, 6, 9, 25, 45, 47, 48, 51, 52, 54, 68, 113, 128
Geographic information systems (GIS), 335, 351, 354
Geomorphologic situations, 356
Geo-morphology, 357
Geophagus worms, 281
Geraniums, 90
Gerbera, 93, 101, 112, 307, 312
Germination, 72, 82, 128, 199–201, 324, 325, 370, 372, 373, 462, 467
Germplasm, 9, 12, 15, 16, 23, 45, 49, 55, 57, 68, 69, 71, 76, 83, 87, 131, 301
Ghujia weevil, 383
Gibberellic acid (GA), 128, 133, 462
Gibberellins, 273
Gigasporales, 206
Ginger
  candy, 55
  shreds, 55
Gladiolus, 101, 105
Glasshouse, 303
Glimpses, 372
Global
  emissions, 122, 123, 138
  floriculture, 92
  gene bank, 55
  GHG emissions, 123
  market, 44, 93, 103
  pests, 398
  scenario, 92, 302
  warming, 121, 123, 124, 131, 133, 135, 138, 139, 323, 379, 386, 391
    potential (GWP), 124, 129, 133
Glomalin, 208
Glomerales, 206
Glomeromycetes, 206
Glomeromycota, 206
Glomus
  coronatum, 206
  mosseae, 206, 207
Glume surface glabrous, 24
Glycine, 223
Glyphosate, 17, 371, 460, 467, 469
  herbicide, 17
Glyricidia, 74
Gneiss, 348, 352
Gomphrena globosa, 102
Gondwana rocks, 352
Good agriculture practice (GAP), 464
Gora, 12
Gramin Krishi Mausam Sewa (GKMS), 150, 152, 153, 155, 161, 162
  economic impact, 162
    case study, 162
Granite, 339, 341, 352
Grass
  alleys, 231
  sod, 234
Grassy alleys, 233
Grazing land, 142, 340, 344
Green
  foliage, 24, 26
  growth, 231
  manure crop, 74
  manuring, 74, 409
  revolution, 5, 22, 271, 288, 335, 360, 371, 380, 391
  vegetation, 231, 233
Greenhouse (GHG), 91, 99, 122, 124, 125, 127, 129, 133, 138–140, 144, 189, 192, 295, 296, 299–302, 304, 306–308, 316
  cultivation, 299, 301, 304
  damage causes, 306
  emissions, 123–126, 133, 140, 144, 151
  gases, 122, 133, 138
  technology, 307
Gross domestic product (GDP), 3, 5, 10, 91
Groundnut bud necrosis virus, 432, 443, 444, 450
Groundwater exploitation, 336, 353, 355
Growing ornamental plants, 310
Growth
  phase, 525
  regulators, 52, 84, 461
  retardants, 257
Guatemalan garlic germplasm, 49
Guava trees, 262
Gulkand, 103
Gundhi bug, 160, 380, 381
Gypsum, 204, 220, 272, 321, 322

H

Hadda beetle, 405
Harmful ultraviolet, 124, 133

Hatching, 400, 405, 411, 421
Havoc infestation, 410
Hazard analysis critical control point (HACCP), 509, 515
Heat
conductivity, 368
stress, 124, 133
Hedging, 257, 258, 262
Helichrysum bracteatum, 102
Helicoverpa armigera, 383, 385, 396–398, 407
Hemicelluloses, 144
Hemiptera, 381–386, 415, 420
Henosepilachna vigitioctopunctata, 402
Herbicide, 10, 53, 175, 219, 232–234, 371, 457, 458, 460, 461, 467–469, 477
glyphosate, 371
resistance rice, 16
tolerance, 10
Heterosis, 46
Heterozygous population, 108
Hibiscus ficulensus, 416
High density
orcharding, 261
planting (HDP), 56, 57, 253–266
Hi-tech horticulture, 311
Hoeing, 171, 230
Holistic growth, 42
Holotrichia consanguinea, 384–386
Homogeneity, 164, 446
Honeycombs, 513
Horblende schist, 352
Hornblende
gneiss, 352
schist, 341, 352
Horticultural
crops, 42, 102, 212, 218, 221, 224, 295–297, 306, 307, 354, 398, 437
interventions, 56
operations, 253, 255
Horticulture, 42, 56, 90, 91, 98, 212, 224, 253, 265, 296–298, 311, 486, 526
production, 224
sector, 42, 296
Host
cells, 205
plant, 206, 399, 406, 407, 409, 433
tissue, 205
Hot stress, 229
Human resource development, 189, 201
Humic acid, 144, 146, 464
Humid tropics, 73
Humidity, 65, 66, 149, 213, 220, 247, 300, 303–305, 310, 420, 488, 532
Hybrid, 12, 16, 29–34, 46, 70, 78, 79, 102, 109, 128, 132, 133, 243, 295, 297, 298, 397, 415
rice, 15
vigor, 29
Hybridization, 8, 12, 13, 22, 50
Hydraulic conductivity, 323
Hydrocyclone, 309
Hydrogen producers, 127
Hydrolysis, 371
Hydrolytic process, 213
Hydro-meteorological natural disasters, 122
Hydroponics, 56, 301
Hygroscopic nature, 512
Hymenomycetes, 205
Hymenoptera, 405, 410

I

Ideotypes, 55
Imazamox, 468
Imidacloprid, 389, 410, 411, 445, 447, 472–474
Imidazolinone, 460
In vitro germination, 71
Inactive larva, 412
Incompatible rootstock, 258
Incubation period, 199, 399, 413
Indian
Institute of Spices Research (IISR), 45, 46, 50, 52
scenario, 93
spices, 52
Indigenous flora, 89
Indigofera parviflora, 74
Indo-Gangetic plains (IGP), 122, 335, 342, 343, 401
Indole-3-acetic acid, 462
Indoor plants, 89, 90
Inductively coupled plasma (ICP), 191, 201
Industrial solid waste, 320
Information and communication technologies (ICTs), 152, 481–484, 497, 498

Inorganic
compounds, 458, 459
fertilization, 139
fertilizers, 74
inputs, 328
salts, 475
soil, 322
Insect growth regulators (IGR), 459
Insecticides, 160, 162, 175, 177, 217, 300, 325, 388, 389, 395, 401, 403, 408, 411, 415–417, 421, 422, 424, 432, 445, 447, 448, 451, 457–461, 463–466, 472–474, 477
In-situ rainwater harvesting, 224
Integrated
cropping system models, 79, 80
farming system (IFS), 526, 527
pest management (IPM), 54, 56, 80, 186, 187, 387, 401, 421, 424, 461, 513
Integration, 8, 105, 329, 368
Intelligent packing materials, 131
Inter SSRs (ISSRs), 9
Intercropping, 75, 236, 237, 240, 244, 247, 248, 408, 411, 414
Intercrops, 53, 75, 76, 227–231, 236–239, 242, 243, 245, 251
International Standard Organization (ISO), 44, 509, 515
Iris yellow spot virus (IYSV), 444, 451
Irrigation, 17, 25, 43, 53, 57, 72, 73, 75, 79, 80, 104, 107, 108, 110, 112, 126, 129, 132, 150, 151, 159, 160, 171, 174, 194, 211–220, 222, 224, 230, 232, 233, 243, 246, 247, 258, 283, 285, 303, 308, 321, 325, 338, 345, 347, 353, 355–358, 416, 488
channels, 216
frequency, 220
management, 57, 211
potential, 218, 345, 353
studies, 222
system, 214, 218, 308
time, 220
water, 211, 216–220, 230
Isozymes, 49

J

Juvenile
hormones (JH), 459
period, 237, 239
phase, 82, 84, 231, 242, 459
stage, 236

K

Katarni rice, 20, 21
Kinnow plants, 263
Krishi Vigyan Kendras (KVK), 57, 207, 484, 486, 489, 490, 494, 496, 498, 527, 528, 533

L

Labor-intensive operations, 260
Labyrinth, 308
Laccaria, 205
Lakshadweep ordinary (LCT), 69, 70
Land use planning (LUP), 335, 336, 338, 342, 350, 351, 357, 364
Landfilling, 320
Landraces, 4, 6, 21, 448, 451
Landscape industry, 89
Landscaping, 90, 103, 105, 110
Larva, 400, 407, 413
Larval-pupal endoparasitoid, 405
Larvicidal toxicity, 405
Late
maturing, 21, 391
sown irrigated (LS-IR), 132
Lavendula augustifolia, 102
Leaf
lamina, 423, 435
petioles, 87
spot, 106, 259, 470
Legumes, 227, 237, 239, 243, 451
Leguminous
crops, 234, 235, 237, 238, 243
hosts, 433, 437
Lentil collar rot disease, 374
Lepidoptera, 381, 383–386, 399, 407, 409, 413, 417, 419
Leptocorisa acuta, 381
Leucinodes orbonalis, 396, 399, 402, 404
Liberalization, 89, 91
Lichens, 90
Lignite quality, 319
Limonium sinuatum, 102
Lindane, 460
Lipid, 369, 443
Liriomyza trifolii, 397, 409

Litchi, 161, 162, 206, 207, 208, 237, 243, 257, 265, 342, 361, 362, 496
Low
cost structure, 304
density orcharding, 261
dose rate sulfonylurea, 460
tunnel polyhouse, 303
Lycopersicon esculentum, 399

M

Macromode, 54
Macronutrients, 204, 320, 321, 354
Macrophomina phaseolina, 470, 471
Magnesium, 184, 357
Mahananda basins, 350
Maize, 6, 7, 10, 28–34, 132, 133, 141, 160–162, 172, 185, 198, 204, 206–208, 221, 239, 278, 287, 322, 325, 327, 336, 342, 345, 355, 357, 360–362, 379, 380, 383, 384, 387, 390, 422, 446, 447, 467, 472, 496, 528
composite maize, 29
growers, 30
hybrid, 32
maize, 29
rabi maize, 30
Malaoxon, 464
Malaria, 460
Malathion, 107, 391, 405, 414, 422, 424, 464, 474
Malayan yellow dwarf (MYD), 70, 77
Maleic hydrazide, 462
Mancozeb, 52, 470–472
Marble schist, 341, 352
Marigold, 95–102, 104, 105, 107, 112, 113, 408, 410
Marker-assisted selection (MAS), 9, 16, 23
Maximum permissible limit (MPL), 176, 187
Meadow orcharding, 259, 266
establishment, 260
Medicinal
crops, 42
plants, 75, 98, 527
Medium
cost structures, 305
high density, 254
Melanin synthesis, 372
Melons, 406, 409
Mesocarp, 86
Mesophilic process, 271
Metabolic
compounds, 223
reactions, 213
Metabolism, 126, 263, 461
Metallic
foils, 278
nanoparticles, 372
Meteorological
department (IMD), 150, 151, 154, 156, 168, 169, 171, 489
seasons, 163, 167
Methane, 125–127, 135, 139
Methanogens, 127
Mica, 192, 341, 352
Microbes, 127, 181, 276
Microbial
count, 204
degradation, 140, 144, 371, 463
enzymes, 368
inhibitors, 127
population, 137, 204, 278, 285
Micro-ecosystem, 310
Microencapsulation, 56
Micro-irrigation system, 263
Micronutrients, 230, 246, 247, 276, 319–321, 325, 326, 354
Microorganisms, 180, 203, 229, 272, 273, 280, 368, 408
Microrhizomes, 52
Microsatellites, 9
Microtubes, 218
Mildly fragrant blooms, 106
Millennium, 152
Millets, 221, 336, 345, 357, 380
crops, 28
Milling, 19, 20
Mimic hormones, 459
Mimosa pudica, 213
Mission for integrated development of horticulture (MIDH), 42, 54
Mite population, 412
Mitigation, 56, 122, 126, 128, 133, 138, 139, 183
Moisture
adequacy index, 335
sensitive stage, 222

stress, 131, 217, 220, 282, 359
Molasses, 274, 422
Molecular
biology, 8, 55
breeding, 8
characteristics, 47
farming, 55
markers, 9, 55
profiles, 8, 55
profiling, 56
Molluscicides, 458
Monocot wetland plants, 370
Mono-cropping, 360
Monocrotophos, 80, 389, 390, 413
Monocultures, 504
Monopartite, 433, 434, 438
begomoviruses, 433
Morpholine, 460
Morphological
character, 47
diversity, 49
traits, 47
Morpho-phenological features, 239
Mosaics, 433
Mosses, 90
Mottling, 435, 438, 442, 444
Mulching, 128–130, 218, 224, 230–232, 234, 235, 245, 247–250, 468, 471, 284
Multinutreint nanoclay polymer composite formulation (MNCPC), 198
Multiplication rate, 281, 285
Mungbean, 324, 327
Municipal
solid waste (MSW), 269, 270, 282
waste, 270, 275
Muscidae, 383–386
Mutation, 10, 17, 48, 51, 175
breeding, 10, 48
Mycelium, 372
Mycorrhiza, 204, 205, 208
arbuscular mycorrhiza (AM), 205
ectomycorrhizae (ECM), 205
Mycorrhizal
colonization, 204
fungi, 203, 206
hyphae, 204
plants, 204
root, 206
Mycotoxin, 55
Mythimna separata, 381, 383–385

N

Nano-emulsion, 371
Nanobiosensors, 131
Nanochemicals, 374
Nanochitosan, 367, 373, 374
Nanoclay, 192, 193, 198, 374, 375
application, 374
polymer composites (NCPCs), 192, 193, 195, 198
Nanofertilizers, 131
Nanoformulation, 131, 196, 197, 199, 200, 371, 373, 375
Nanomission, 369
Nanoparticles, 131, 189, 190, 192, 367–375, 471
Nanopolymeric
coating materials, 131
hydrogel, 131
Nanoscale
characterization techniques, 191
synthesis, 190
Nanoscience, 190, 191, 367, 369
Nanosensors, 131
Nanosilver, 131
Nanotechnology, 55, 131, 189–191, 195, 201, 368–371
agricultural applications, 192
nanoclay polymer composites (NCPC), 192
application, 191, 370
BAU initiatives, 195
achievements, 195
education, 195
research projects, 195
Nanotitanium dioxide, 131
Nanotoxicology, 131
Naphthalene, 459
acetic acid (NAA), 50, 52, 462
National
Centre for Medium Range Weather Forecast (NCMRWVF), 149, 150, 153, 156, 159, 163, 164
Horticultural
Mission (NHM), 42, 54, 214, 296, 307, 411, 413

Research and Development Foundation (NHRDF), 48
Nanotechnology Initiative (NNI), 190
Thermal Power Corporation (NTPC), 329, 331
Natural
dyes, 103, 104, 116
enemy fauna, 408
plowman, 271
resource management, 125
climate resilient varieties, 128
fertilizer application method, 126
hormones application, 128
microbes use, 127
nitrification inhibitors use, 126
seed inoculation, 127
site specific nutrient management techniques, 125
suitable varieties selection, 127
tillage management, 126
water management, 126
Nausea, 41, 180
Necrotic
lesions, 50
region, 421
Need-based manual weeding (NBMW), 468
Neem seed kernel seed extract (NSKE), 410, 472, 474
Negative correlation, 47, 51
Nematicides, 457
Nematodes, 55, 288, 398
Neochrysocharis farmosa, 410
Nephotettix nigropictus, 381
Nilaparvata lugens, 381, 387
Nitrobacter, 272
Nitrogen
metabolism, 258
cycle, 126
rich cover crop, 234
Nitrogenous fertilizer, 130, 223
Nitrophenols, 459
Nitrous oxide
emission, 127
flux, 127
gases, 127
Noctuidae, 381, 383–386, 407, 413, 419
Non-edible products, 82, 86
Non-soluble manures, 217
Non-structural protein (NSs), 443
Non-target organisms, 175, 457, 458, 460, 464, 465
North-east alluvial plains, 355
Novel
agrochemicals formulations, 131
fertilizer formulations, 192
fungicide, 373
nano Fe fertilizer, 196
nano P fertilizers, 196
nanoformulations, 199
nanopesticides, 131
superabsorbent hydrogel, 195, 196
Nucleic acids, 212, 223
Nucleotide, 9, 441
Nutraceuticals, 41, 46
Nutrient
compounds, 371
index, 354
Nutritional
security, 269, 270, 432, 527
standards, 432
Nymphal stages, 415
Nymphs, 411, 412
Nymphula depunctalis, 381

O

Odontotermes obesus, 381, 383
Okra, 128, 239, 322, 398, 406, 414, 415, 432, 445–447, 472, 527, 529
stem fly, 398
Oleanders, 278
Oligonychus oryzae, 382
Open
field cultivation, 105, 297
pollinated population, 108, 109
reading frames (ORFs), 433, 451
Optimum
growth, 171, 230, 231
high density, 254
plant density, 264
Orchard floor management, 228, 265
benefits, 229
factors affecting, 229
irrigation water, 230
pest management, 230
soil stabilization, 229
weed population, 229

objectives, 228
system, 231
clean cultivation, 233
grass alleys, 233
inter crop cultivation, 236
mulching, 234
solid grass cover, 233
weed control, 232
Organic
acids, 271, 273
amendments, 136, 138, 141–144
carbon, 130, 141, 143, 241, 285, 325, 326
compounds, 458
farming, 54, 56, 74, 141, 142, 174, 186, 271, 272, 495, 513
fertilizer, 86, 273
impurities, 309
inputs, 328
manure, 74, 140, 144, 218
matter, 50, 74, 130, 137, 138, 144, 146, 175, 179, 186, 193, 227, 232, 233, 281, 325, 353
mulches, 218
pesticidal compounds, 458
sources, 74, 137, 269, 288
substances, 321, 324
Organization for Economic Cooperation and Development (OECD), 320, 331, 461
Organophosphates (OPs), 458, 464
Ornamental, 90, 94
crops, 90, 406
gardening, 103
industry, 90
plants, 46, 90–93, 106, 288
Orthoptera, 385, 386
Orthotospovirus, 451
Oviposition, 399, 400, 405, 412, 418

P

Paclobutrazol, 257
Paddy transplanter, 126
Palmyra
national level research status, 83
crop improvement, 83
crop production, 84
palm, 65, 81, 82, 84, 88
products, 83
Para-chlorophenoxy acetic acid (PCPA), 128, 133
Parasitoid, 390, 401, 410, 418
Parathion, 107, 460
Parthenocarpic
cucumber, 299, 307, 311, 312
gynoecious lines, 312
Partial root drying (PRD), 211, 218, 224
Pathogen, 3, 9, 16, 45, 55, 56, 276, 288, 303, 307, 367, 369, 421, 471
Pathogenic fungi, 199
Pearl millet, 127, 379, 380, 386, 390
Pediobius foveolatus, 405
Pegmatite, 341, 352
Pendimethalin (PE), 53, 467–469
Pepper veinal mottle virus (PVMV), 438, 440
Percolation, 217, 219
Pest
free crop, 416
management, 53, 186, 187, 211, 247, 251, 265, 387, 388, 395, 397, 398, 401, 404, 421, 424, 461, 473
resistance, 10
Pesticidal compounds, 460
Pesticides, 53, 55, 56, 151, 174, 175, 177, 179, 181, 186, 192, 219, 227, 230, 322, 327, 328, 367, 368, 371, 372, 374, 386, 397, 398, 401, 405, 409, 457–461, 463–466, 473, 488
Pest-resistant varieties, 408
Petals, 106–109
Petroleum oils, 459
Pharmaceutical
industries, 47, 108
properties, 46
Pharmacological activities, 55
Phenacoccus solenopsis, 406
Phenolics, 475
Phosphatase, 203, 204, 326
dehydrogenase, 204
enzyme, 203
Phosphogypsum, 272, 280, 285
Phosphoric acid, 195, 464
Phosphorodithioic acid, 464
Phosphorodithiotate compounds, 464
Phosphorothioic acid, 464
Phosphorus (P), 83, 126, 203, 204, 206–208, 273, 325, 326, 336, 356

inorganic, 203
nutrition, 206
organic, 203
solubilizing
bacteria, 273
potential, 273
Photodecomposition, 463, 464
Photoemission, 368
Photo-metabolites, 464
Photo-period sensitive, 17
Photo-sensitizers, 464
Photosynthates, 223, 308
Photosynthesis, 212, 213, 223, 303
Phyllite, 341, 352
Physicochemical
characteristics, 320, 324, 328, 331, 371
constituent, 83
properties, 57, 325, 331
Physiographic levels, 354
Physiological
activities, 212, 222, 223
maturity, 124, 257, 263
parameters
fertigation effect, 223
plant status, 223
Phytochemicals, 56
Phytopathogen, 367, 369
Phytopathogenic fungi, 369, 372, 373, 375
Phytopathosystem, 367–369, 374
Phytophagus, 281
Phytophthora, 54, 55, 470
Phytotoxicity, 459, 468
Pisciculture, 360
Pisolithus, 205
Pisum sativum, 322, 443
Pit
filling, 244
marking, 244
preparation, 244
Plant
breeding, 3, 6, 8–10, 56
canopy, 262
cells, 223
density, 253, 254, 263
geometry, 255
growth regulator (PGRs), 50, 230, 457, 461, 462, 475, 477
life water importance, 213
metabolic activities, 212
nutrient, 214, 270, 271, 288, 290, 300, 319, 321, 323, 325, 329, 331
nutrition, 308
population density, 255
Plantation, 63, 65, 66, 72, 84, 105, 212, 214, 227–240, 242, 243, 246–248, 255, 261, 265, 275, 296, 330, 356, 361, 495
crop, 63, 65, 69, 212, 214, 296
Planthopper, 380, 381
Planting materials, 45, 54, 56, 98, 105, 131, 189, 409, 438, 446
Plastic cladding materials, 310
Policymakers, 484, 491
Pollination, 50, 303, 503, 533
Polyacrylamide/clay (PAM/clay), 194
Polycarbonate, 304, 310
Polycondensation, 131
Polyethylene, 248, 304, 305, 310
Polyhouse, 91, 112, 113, 116, 129, 298–307, 309–311, 313, 315, 316
cooling, 310
heating, 309
Polymerase chain reaction (PCR), 8, 49, 57, 435–437
Polymorphism, 49
Polyphagous, 387, 405, 415, 422
Polysaccharides, 144
Polythene sheets, 80, 302
Polyvinyl
alcohol (PVA), 195
chloride films, 304
Popularization, 56, 57, 105
Population dynamics, 231, 232, 280, 388
Post
emergence (PoE), 467–469
flood period, 360
harvest technology, 84, 356
neera collection and storage, 84
Kharif cultivation, 5
rabi cultivation, 5
vedic period, 40
Potentiality, 34, 94, 100, 174, 227, 233, 234, 239, 263, 320, 322, 323, 345, 373, 432
Potpourri, 103, 116
Potyvirus, 431, 433, 437–440, 449, 451
Poultry manure, 276
Powdery mildew, 53, 106, 108, 162

Pre-emergence (PE), 467–469
Pre-harvest interval (PHI), 466
Pre-planting (PP), 467
Press drying, 112
Pressurized irrigation methods, 214
Production technology refinements, 71
  coconut organic farming, 74
    bio-fertilizer recommendation, 74
    organic recycling, 74
  inter cropping, 75
    coconut-based cropping system, 75
    multiple cropping systems, 75
  inter-cultural operation weed management, 74
  planting material, 72
    mother palm selection, 72
  seed nut collection and planting, 71
  water management, 72
    drip irrigation and fertigation, 72
Propolis, 502, 533
Protected
  cultivation, 56, 99, 295, 297–300, 302, 304, 312, 534
    advantages, 300
    status, 301
  horticulture
    components, 302
    plant containers, 307
  structures
    cladding material, 310
    design and orientation, 303
    irrigation and nutrition systems, 308
    site selection, 306
Pruning, 74, 107, 230, 246, 248, 257, 258, 260–262, 265, 266, 303, 401, 475, 476
Pseudococcidae, 381, 406
Pseudomonas putida, 127
Pseudostem, 52
Pteridophytes, 205
Pumpkin beetle, 423, 424
Pupae, 400, 405, 407, 408, 413, 415, 420, 422
Pupal
  period, 399, 400, 412, 418
  stage, 399, 400, 411, 414, 419
Pupate, 400, 407
Pupation, 400, 407, 417, 419, 421
Pyralidae, 381, 385, 386
Pyrazosulfuron, 467
Pyrethrin, 458, 459
Pyrethroids, 416, 460
Pyrilla perpusilla, 385
Pyrimidine, 460
Pyrole rings, 223

Q

Quality
  protein maize (QPM), 29, 31
  up-gradation, 43
Quantitative verification analysis, 163–165
  cloud cover, 166
  maximum/minimum temperatures, 167
  rainfall, 166
  wind
    direction, 167
    speed, 167
Quartzite, 341, 352
Quinalphos, 52, 160

R

Radiation filter, 300
Rajendra hybrid makka, 30, 31
Random amplified polymorphic DNAs (RAPDs), 9
Rashtriya Krishi Vikas Yojana (RKVY), 214, 312
Ratoon crop, 414
Raw materials, 63, 90
Recommended dose (RD), 74, 198, 269, 273, 282, 288, 328, 329, 331, 475
Recurrent blooming, 106
Red spider mite, 398
Redox potential, 127
Registration committee (RC), 465
Rejuvenation, 80
Relative
  humidity, 47, 84, 149, 297, 300, 310, 488
  water content (RWC), 220, 223, 224
Remote sensing and GIS, 339, 351, 364
Rhizobium, 272, 286
Rhizomes, 43, 51, 52
Rhizosphere, 56, 131, 196, 198, 204, 241
Riboflavin, 83, 464
Ribonucleoproteins (RNPs), 443
Rice
  ecosystems, 14
  research priorities, 15
  residues, 274
  tungro virus, 13

Ricinus communis, 405
Ridge guard, 435, 436
Ring system, 216
Ripeners, 462
Rodenticides, 175, 457, 458
Root
  colonization, 206
  mean square error (RMSE), 164, 163, 165, 172
  restriction, 258
Rootstocks, 128, 232, 253, 254, 256–263, 265, 266
Root-zone irrigation, 218

S

Sabour
  Ardhjal, 17, 18, 132
  deep, 19, 132
  hybrid maize, 32–35
  Nirjal, 25, 26, 132
  Samriddhi, 26, 27, 132
  Shreshtha (BRW 934), 27
  Surbhit, 17, 18, 132
Saffron, 41, 108
Saishin, 417
Salicylic acid, 475
Saline
  alkali soils, 322
  soils, 354
Salinity hazards, 353
Sand stone, 341, 352
Saturation, 140
Scanning electron microscopy (SEM), 191
Scirpophaga incertulas, 381, 392
Scleroderma, 205
Sclerotia, 199, 201, 372–374
Sclerotial germination, 200, 373
Sclerotium rolfsii, 199, 200, 372, 374, 471
Scotinophara coarctata, 382
Secure digital (SD), 495, 496
Seed
  fertilizer, 130
  rate, 54, 108
Seedling, 72, 80–82, 88, 103, 127, 160, 193, 194, 235, 287, 295, 298–300, 307, 372, 388, 395, 411, 418, 423, 445, 446, 471
Semi-dwarf
  varieties, 23, 24, 132
  wheat, 23
Semi-erect flag leaf, 27
Senile orchard, 80
Sensitivity index, 141, 144
Sesbania cannabina, 327
Shaktiman
  -2, 29, 30, 31
  -3, 29, 30, 31
  -4, 29, 30, 31
Sheath blight, 20
Silicon dioxide, 131
Silver nanoparticles, 368, 369, 372
Single nucleotide polymorphism (SNPs), 9
Skeletonization, 405
Slow-release phosphate fertilizer (SAPSRPF), 192, 195
Sodic soil, 273
Sodium arsenate, 459
Sogatella furcifera, 381
Soil
  aeration, 72, 230
  amendments, 319, 321, 322
  association map, 354
  carbon, 136–140, 142, 144, 146
    sequestration, 137, 139, 144
    stock, 137
  characteristics, 176, 180, 323, 326, 328, 337, 347
  erosion, 174, 179, 186, 217, 228, 230–232, 239, 273, 353
  fertility, 138, 141, 146, 174, 186, 204, 228, 229, 234, 237, 242, 260, 262, 263, 270, 319, 321, 328, 527
  health, 135, 137–139, 141, 174, 179, 185, 238, 271, 288, 290, 320, 321, 324, 326, 331, 353, 354, 493
    indicator, 354
  management system, 353
  metal contents, 177
  micro flora, 130
  microbes, 126
  moisture, 126, 220–222, 224, 229, 233, 246–248, 263, 321, 359, 371
    tension, 220
  organic carbon (SOC), 140–143, 146
  pH, 325, 326, 341
  phosphorus, 204
  plant
    atmosphere continuum (SPAC), 175, 187
    system, 273

pollution effects, 180
porosity, 326
quality, 137
salinity, 175, 353
signs, 220, 224
temperature, 128, 130, 233, 248
tensiometers, 220
wet/dry pollutant deposition, 185
Solanaceous, 407, 431–433, 443
Solanum
anomalum, 399
incanum, 128
indicum, 399
melongena, 399, 443
nigrum, 399, 416
pimpinellifolium, 128
tuberosum, 399, 443
Solar
cookers, 57
dryers, 57
energy, 264, 464
harvesting, 264
panel driven devices, 309
radiation, 253, 255, 264, 265, 300, 351
Sole mineral fertilizer, 143, 144
Solid vegetation covers, 231
Solubilization, 204
Somaclonal variations, 48
Sooty mold, 406, 415
Sorghum, 127, 287, 324, 379, 380, 384, 387, 390, 411, 414
Sowing dates, 130, 133
Spice, 39, 41–45, 55
crops, 41, 42, 45, 56
black pepper, 50
chili, 46
garlic, 47, 49
ginger, 51
seed spices, 53
turmeric, 51
extracts, 57
farming, 54
history, 39
import, 45
importance, 41
research, 45, 54
challenges, 55
opportunities, 54
Spodoptera
litura, 398, 419
mauritia, 381
Spoilage bacteria, 56
Spongy haustorium, 82
Standardized precipitation index (SPI), 169, 171, 172
State Agricultural University (SAU), 42, 105, 149, 369
Steady flow rate, 308
Steam distillation units, 103
Stenodiplosis sorghicola, 385, 387
Streptocycline, 469, 470
Stress Tolerant Rice for Asia and South Africa (STRASA), 16
Substrate, 272, 273, 275, 278, 280, 282–285, 289
Subtropical
climate, 6, 48, 53, 137, 341
clones, 49
Subtropics zones, 63
Succinic acid, 223
Sugarcane trash, 274
Sulfonylurea, 460
Sulfur nutrition, 272
Sulfuric acids, 141
Super absorbent hydrogel, 194
Superfine slender grain, 17
Superior
cultivars, 45
grade, 56
grain, 13
Surface
catalytic activity, 368
evaporation losses, 230
Synthetic
compounds, 130, 474
dyes, 104

T

Taiwanese garlic germplasm, 49
Tal
areas, 360
land, 331, 364
Tanymecus indicus, 383
Tarsonemidae, 382, 412
Taxonomic
origins, 49
status, 49

Tephrosia purpurea, 74
Terrain, 6, 347, 350, 351
Terrestrial carbon sequestration, 136
Thatching roofs, 86
Thelephora, 205
Thermal
  plants, 320
  power plants, 319, 331
  regime, 140
Thermophillic stage, 277
Thesis period, 125
Thiamin, 83
Thinning, 230, 257, 258, 263
Thripidae, 381, 383, 384, 410
Thysanoptera, 381, 383, 384, 410
Tiptur tall (TPT), 69, 77
Titanium, 341, 352
Tobacco, 33, 336, 355, 395, 416, 443, 458
Tomato
  leaf curl virus (TLCV), 446, 449
  spotted wilt virus (TSWV), 442, 443
Topography, 215, 218, 219, 337, 339, 342, 343, 352
Topping, 257, 262
Tospoviruses, 433, 443, 444
Total
  geographical area (TGA), 339
  quality management (TQM), 509, 515
Toxic
  chemicals, 327
  elements, 175, 319, 322, 326, 327
  metabolites, 464
Toxicity, 176, 181, 297, 336, 356, 459, 461, 464–466, 469
Traditional
  aromatic varieties, 14
  breeding programs, 10
  irrigation practices, 224
  planting system, 254
Transcriptomics, 56
Transfer of technology (TOT), 57
Transmission electron microscopy (TEM), 191
Transpiration
  loss, 214
  rate, 212, 223
Transplanting, 110, 111, 125, 126, 130, 160, 287, 308, 314, 388, 408, 418, 469, 473
Tree
  thinning, 257
  vigor, 233
Trellis, 87, 233
Trichoderma, 471
Trichogramma
  chilonis, 390, 401, 409
  evanescens, 401
Trickle irrigation system, 216
Trifolium alexandrinum, 370
Triticum vulgare, 322, 327
Tropical
  flowers, 89
  fruit, 266
    crops, 255, 259
  plants, 39, 301
Tuberculation, 407
Tuberose, 98, 100, 101, 103, 105
Tubers, 90, 103, 184, 296, 336, 355, 432, 438, 446
Turgidity, 213
Turgor pressure, 213
Turmeric rhizomes, 52

U

Ultra-high density, 254, 259
  guava orcharding, 262
Ultraviolet (UV), 124, 133, 300, 302, 305, 310, 316, 464
Unfertile lands, 213
Urban solid waste, 275
Urbanization, 91, 151, 183, 269, 298
Urease, 204, 326

V

Valorizations, 45
Vegetable
  crops, 212, 215, 238, 287, 299, 395–398, 406, 407, 409, 417, 424, 432, 433, 448
  pathosystem, 435, 444, 446, 451
  waste, 274, 275
Vegetative
  growth, 77, 78, 218, 222, 223, 232, 260, 264
  parameters, 224
    fertigation effect, 222
Veinlets, 405
Ventilation, 289, 303–306, 310

Venture
capital, 525
startup, 525
Vermicasts, 281
Vermicompost, 50, 125, 269–276, 278, 280–288, 290, 324, 328, 330, 526, 527, 533, 534
application, 285, 287
epilogue, 288
pests and parasites, 288
influence, 273
nutrient content, 285
production, 280, 282, 533
used materials, 278
Vermicomposting, 271–273, 275, 278, 280, 282–285, 288, 290, 495, 533
importance, 272, 273
materials, 274
methods, 282
heaps above the ground production, 283
in-situ method, 285
open method, 284
pit production, 283
tanks above the ground production, 283
techniques, 282
Vermiculture, 271, 274
Vertical farming, 56
Viruses, 45, 431–433, 435, 437, 438, 441–444, 446, 448, 451

W

Water
absorbency (WA), 131, 192, 194, 195
harvesting, 224
holding capacity, 242, 273, 321, 325, 326, 331, 339
influencing factor, 213
cultivation, 214
fertility level, 213
light, 214
soil moisture and humidity, 213
management, 211, 225, 247
percolation, 126
scarcity, 222, 230
technology center for eastern region (WTCER), 218
use efficiency (WUE), 43, 212, 216, 218, 249
Waterlogging, 16, 20, 33, 129, 160, 353
Watermelon bud necrosis virus (WBNV), 444, 445
Wavelengths, 351
Weather forecast, 149–151, 153, 156, 162–164, 171, 487
analysis, 163
qualitative verification analysis, 163
analysis verification, 163
rainfall forecast verification, 163
verification, 163
Weed, 17, 53, 74, 129, 227–230, 232–235, 243, 247–249, 276, 298, 303, 354, 371, 416, 467–469, 487, 489
control
efficiency (WCE), 467–469
process, 232
index (WI), 468, 469
population, 227, 229, 232–234
West coast tall (WCT), 69, 70, 73
Wheat
genotypes, 17, 25
research, 22, 23, 25, 30, 32

X

Xenobiotics, 464
X-ray absorption near edge structure (XANES), 191

Y

Yeast hydrolysate, 422
Yellow
fever, 460
mosaic virus, 450
vein mosaic (YVM), 415, 416, 445, 447, 450, 472
Yield
fertigation effect, 224
potential, 13, 17, 19, 21, 23, 26, 27, 29, 34, 66, 132, 238, 300
ranging, 22, 50
Young orchards, 237

Z

Zero
energy
chamber, 304

polyhouse, 303
tillage, 129, 140
Zincated nanoclay polymer composites (ZNCPC), 193, 196
Zingiber officinale, 51
Zucchini yellow mosaic virus (ZYMV), 438–440